Urban Air Pollution – European Aspects

ENVIRONMENTAL POLLUTION

VOLUME 1

Urban Air Pollution – European Aspects

edited by

Jes Fenger, Ole Hertel and Finn Palmgren

National Environmental Research Institute,
Roskilde, Denmark

KLUWER ACADEMIC PUBLISHERS
DORDRECHT / BOSTON / LONDON

Library of Congress Cataloging-in-Publication Data

Urban air pollution : European aspects / edited by Jes Fenger, Ole Hertel, and Finn Palmgren.
p. cm. -- (Environmental pollution)
Includes index.
ISBN 0-7923-5502-4 (hardcover : alk. paper)
1. Air--Pollution--Europe. 2. Urbanization--Environmental aspects--Europe. 3. Air--Pollution--Government policy--Europe. I. Fenger, J. (Jes) II. Hertel, Ole. III. Palmgren, Finn. IV. Series: Environmental pollution (Dordrecht, Netherlands)
TD883.7.E85U73 1999
363.739'2'091732--dc21 98-49062

ISBN 0-7923-5502-4

Published by Kluwer Academic Publishers,
P.O. Box 17, 3300 AA Dordrecht, The Netherlands.

Sold and distributed in North, Central and South America
by Kluwer Academic Publishers,
101 Philip Drive, Norwell, MA 02061, U.S.A.

In all other countries, sold and distributed
by Kluwer Academic Publishers,
P.O. Box 322, 3300 AH Dordrecht, The Netherlands.

Printed on acid-free paper

Printed in the Netherlands.

TABLE OF CONTENTS

PREFACE

The present book treats the various aspects of urban air pollution with the emphasis on European cities. It is aimed at university students at graduate level, but it should also be useful for technical experts in international, governmental and municipal institutions, in private consulting companies, and in non-governmental organisations.

The individual chapters are written by European experts in the various fields, only edited to give a comprehensive presentation. The viewpoints expressed by the authors are their personal ones and they do not necessarily represent the official opinion of their respective institutions. In several cases the authors received material and other information from colleagues. This is here gratefully acknowledged.

The project was initiated within the COST CITAIR - Programme "Science and Research for Better Air in European Cities". COST is a framework for scientific and technical co-operation, allowing the co-ordination of national research on a European level. Many of the authors were recruited via this programme, and the workshops and collaboration, which the programme involved, provided much useful information and collaboration.

The general structure and level of the book were planned at a Workshop in Copenhagen in February 1997. This workshop was financed by the Danish National Science Research Council (Ministry of Research and Information Technology) and the Danish Transport Council (Ministry of Transport).

The preparation of a camera-ready manuscript from a pile of contributions requires a considerable effort in layout and redrawing of figures in order to produce a homogeneous product. The editors thank the Tuborg Foundation for providing the necessary financial means.

Last, but not least, we thank our own institution, The National Environmental Research Institute (Ministry of Environment and Energy), for providing excellent working conditions - including the invaluable collaboration with the Graphic Design Centre (Beatrix Rauch, Britta Munter and Jennifer Aas) and the secretariat at the Department for Atmospheric Environment (Pernille Carlsson, Helle Rosvang Fomsgaard and Bodil Chemnitz).

Roskilde, Denmark, September 1998

Jes Fenger　　　Ole Hertel　　　Finn Palmgren

PREFACE

I

STATEMENT OF THE PROBLEM

Human impact on the environment is determined by three main factors: The material standard of living, the applied technology and the number of people.

Since 1950 the world population has more than doubled and - as a measure of standard of living - the global number of cars has increased by a factor of 10. The resulting increases in emissions of air pollutants have led to large scale phenomena as acidification, depletion of the ozone layer and threats of global change.

Still however, the most direct impacts of air pollution are felt in cities. The fraction of people living in urban areas has increased by a factor of 4. This development is expected to continue in the coming years, and the United Nations estimate that in year 2000 about 47% of the world population will live in urban areas.

The situation is different in developing and industrialised countries and depends upon climate, location, infrastructure and many other factors. The present book focus on air pollution in European cities, which mainly have a temperate climate and pollution levels, which are increasingly dominated by traffic emissions.

The background and structure of the book, including practical information on references, units and symbols, which are common for all chapters, are given in Chapter 1. Next follows a review (Chapter 2) of the historical development of urban air pollution with special reference to Europe. Finally the scene is set with a description (Chapter 3) of the general characteristics of Europe and its cities.

Chapter 1

ABOUT THE BOOK

JES FENGER, OLE HERTEL and FINN PALMGREN
National Environmental Research Institute
Department of Atmospheric Environment
P.O.Box 358, DK-4000 Roskilde, Denmark

1.1 Delimitation of the subject

Air pollution is enacted on various geographical and temporal scales, ranging from strictly "here and now" problems to global phenomena, which over the next centuries can change the conditions for man and nature on the entire globe. The present book focus on problems at *urban scale*, and regional phenomena are only treated to the extent that they influence urban conditions. Global phenomena are not considered, and e.g. traffic is thus only treated as a source of "classical" pollutants; its important role as a source of greenhouse gases - especially carbon dioxide - is not discussed.

The book further concentrates on *European cities*. In spite of differences in climate and infrastructure, they have generally a common background and - via the European Commission - a related, environmental legislation. It has therefore been natural to carry out the writing within the framework of the COST- CITAIR programme "Science and Research for Better Air in European Cities".

The emphasis is laid on the *technical and scientific aspects* i.e. sources, atmospheric pollution and impacts. *Legislation and policy* are discussed in terms of limit values, and a short chapter on *city planning* and *economic measures* as useful tools is included. Socio-economic aspects and e.g. a treatment of the possibilities of influencing human behaviour via changing attitudes have been found outside the scope of the book.

One of the results of the UN conference on environment and development in Rio de Janeiro in 1992 was an action plan for the attainment of a sustainable global development - the so called Agenda-21. As a consequence many cities and administrative units have embarked on local programs. They are not explicitly treated, although of course the planned savings in energy and resources will have beneficial effects on urban air quality.

1.2 The structure of the book

After the general statement of the problem (Chapters 1-3) the structure of the book follows by and large the chain of events in air pollution, with the chapters grouped according to the main themes, each with a short common introduction:

1.2.1 SOURCES OF POLLUTION

A series of sources emit various compounds. The main contributions are due to the use of fossil fuels. Emissions from agricultural activities do not play a direct role in urban air pollution, but may have an indirect influence via i.a. large scale transformations and transport. Stationary sources (Chapter 4) comprise industries, power plants, space heating systems of various sizes etc. Mobile sources (Chapter 5) are dominated by different types of cars. Other mobile sources such as trains, aeroplanes, and ships are normally of minor importance for urban air pollution. In all cases there are various technical and regulatory ways to reduce the emissions.

Emissions can be measured individually from single sources, but often emission inventories (Chapter 6), based on empirical "emission factors" and consumption of fuels or raw materials are carried out.

1.2.2 BASIC ATMOSPHERIC PHENOMENA

In the atmosphere the pollutants are advected, dispersed and finally deposited on the soil or other surfaces (Chapter 7). During these processes the pollutants may undergo chemical transformations, and new "secondary" pollutants are formed (Chapter 8). Particles (Chapter 9) are formed by condensation, coagulation and chemical processes or are emitted directly from the sources. The particle size is important for their transport and deposition. The size further determine their impact on visibility (Chapter 21).

1.2.3 AIR POLLUTION MODELLING

Mathematical modelling of dispersion and transformation is a useful tool both for scientific description of pollution phenomena and in planning and decision making. Although the basic processes are the same, the details of the models depend upon the geographical and temporal scale. Only dispersion models relevant for urban air pollution are treated i.e. models at regional scale (Chapter 10), urban scale (Chapter 11), and street scale (Chapter 12). In addition stochastic models, which are used for predicting pollution episodes are described (Chapter 13).

Specific phenomena i.a. dispersion around complex building configurations may be highly difficult to explore in field experiments. In such cases physical models in wind tunnels (Chapter 14) may serve as an important tool. They may also yield useful information in development of mathematical models.

1.2.4 AIR QUALITY MEASUREMENTS

Many different chemical and physical measurement methods have been applied for measuring and monitoring of air pollutants. The main categories of measurement methods are described (Chapter 15). However, within the scope of this book it has not been possible to describe the methods in details. For further details we refer to the special literature.

Monitoring networks have been established in nearly all large European cities. The main principles are described. They include network design, application of air quality models, maintenance, quality control and quality assurance, data presentation and treatment and reporting (Chapters 16 and 17).

1.2.5 IMPACTS OF AIR POLLUTION

Cities are in the nature of things concentrations of people, materials and activities. This leads not only to elevated levels of pollution, but also to extensive exposure. Of most importance are the impact on human health and well-being (Chapter 18), which may be either immediate (respiratory diseases) or delayed (cancer). But also the material damage (Chapter 19) on both economical/technical structures and cultural heritage may be considerable.

Less noticed are impacts on urban ecosystems (Chapter 20), which however, can both be seen as an environmental strain in itself and as an indication of unacceptable pollution levels. In European cities reduction of visibility (Chapter 21) are not any more as dramatic as it was seen in e.g. the London "smog" up to about 1950. But especially in the southern part of Europe "photochemical smog" may result in significant effects.

1.2.6 POLICIES AND LEGISLATION

Air pollution can be reduced not only by technical means, but also via planning and control of the urban structure and the resulting traffic. The various tools are shortly described in Chapter 22. A comprehensive description of the complex pollution and its development can be given in the form of indicators (Chapter 23). The actual regulation of air quality is carried out by means of limit values, some of which have EU-standing The basic principles for establishment of the new EU limit values are described in Chapter 24.

1.2.7 OVERVIEW AND CONCLUSIONS

To demonstrate the complex interplay between climate, infrastructure and emissions in the formation of urban air pollution An overview of the air pollution situation in a series of major European cities is given in Chapter 25.

The emission pattern, the industry and transport structure and the climate vary significantly between the different European cities, and the dispersion and the chemical processes governing the air quality are different in the different regions of Europe. These differences between European regions are demonstrated. Consequently, the air quality is very different from North to South and from East to West. It has been outside the scope of this book to give a full description of the air quality in European cities, which has been given in the so-called Dobris reports prepared by the European Topic Centre for Air Quality (ATC-AQ). However, an overview of the present situation is given, i.a. based on data from ATC-AQ. In the final, concluding Chapter 26 the results of Chapter 25 are summarised and put in perspective.

1.3 Literature, references and index

Each chapter has its own *list of references* comprising typically 20 entries arranged according to the "Harvard system" i.e. author(s) and year of publication. Not all technical details are documented, but they can in most cases be traced via references to review-papers or textbooks.

All *figures, tables and formulas* are numbered within each chapter with the chapter number as prefix.

An *index* at the end of the book is common for all chapters and has about 1000 entries.

1.4 Units and symbols

Generally the SI-system of units with standard prefixes is used. (*International Standard ISO 1000 Third edition 1992-11-01: SI units and recommendations for the use of their multiples and of certain other units).* But there are a number of exceptions. Partly because the book to a large extent is based on published material, and changes in units may be misleading. Partly because parameters of different orders of magnitude are involved; therefore it may be reasonable to use units, which directly give impression of the magnitude. The number of digits in quantitative measures does not always indicate the accuracy; this applies especially to emission inventories, where the possible errors may be 10% or more.

Emissions are normally given in weight, and they can be related to various parameters: fuel consumption, activity, time, area etc. *Concentrations* are given as "weight/volume" (typically $\mu g/m^3$ or mg/m^3) and "volume/volume" (typically ppbv or ppmv - normally just written ppb or ppm). Note that the two types of units can only be related for a specific temperature and pressure. At 25°C and 760 mm Hg the most important relations are:

Nitrogen dioxide (NO_2)	1 ppb	= 1.882 $\mu g/m^3$
Nitrogen monoxide (NO)	1 --	= 1.227 --
Ozone (O_3)	1 --	= 1.963 --
Sulphur dioxide (SO_2)	1 --	= 2.620 --
Carbon monoxide (CO)	1 --	= 1.146 --

The *symbols* used in the book are currently defined. The editors have attempted a consistent use, but the large number of individual contributions may have introduced some differences from chapter to chapter. On some figures, for the sake of brevity, countries are indicated with the licence plate code.

Chapter 2

HISTORY OF URBAN AIR POLLUTION

PETER BRIMBLECOMBE
School of Environmental Sciences, University of East Anglia
Norwich NR4 7TJ, United Kingdom

Sometimes the history of air pollution seems to offer little more than anecdote to enliven discussions. It requires more relevance than this if it is to warrant a full chapter in a textbook. Here I try to show the features that characterise air pollution at various periods. An understanding of these characteristics can aid the way we view the sources of air pollutants and our response to them. Although I try to follow a chronological pattern, I will not stress this as it is my intention to focus on significant features, not chronology.

This book takes a regional focus, so we need to consider that Europe has a particularly long and well-documented history of air pollution. Although there are hints of early occurrences of air pollution outside Europe, our writers and scientists have provided us with a written account of their insights that spans more that two thousand years.

2.1 Air pollution in antiquity

It is easy to imagine that crowded interiors in the distant past were filled with smoke from fires for heating and cooking. Evidence that they were polluted comes from remnants of blackened walls and paleopathological examination of human remains. Mummified lung tissue frequently shows anthracosis as evidence of people who lived in smoky dwellings, but mummies are not typical of Europe's humid temperate climate. There is skeletal indication of air pollution effects on humans. The roughening of the bone that forms the base of the maxillary sinus is an indication of sinusitis in past populations. Studies suggest that a remarkable prevalence of sinusitis in some periods might be the product of exposure to high concentrations of indoor smoke particles (Brimblecombe 1987a). Reconstructed dwellings from third century Sweden (Figure 2.1, next page) suggest that high indoor concentrations of pollutants are likely. Such conditions may have been common in dwelling places in Northern Europe of the first millennium.

Figure 2.1 Swedish reconstruction of an "Iron age" house. At certain wind directions the smoke from the open fireplace cannot escape (Edgren, Herschend 1982).

In classical antiquity, environmental problems such as deforestation, overgrazing and urbanisation were very familiar and widely described (Hughes 1994). Vitruvius (De Architectura) argued that town planning should consider the relevance of winds to urban health. Warfare in the ancient world often placed cities under siege. Death under these situations was as often from disease and starvation, as it was from enemy action. In the 8th Century BC Hermopolis was driven to surrender because of the unbearable stench.

2.1.1 AIR POLLUTION IN ROME

Ancient Rome may have had a population over one million. This, even in a warm climate meant an enormous consumption of wood, so temples were blackened by soot and the health of sensitive individuals deteriorated. Martial wrote that a sun tan gained in the country would soon be lost on return to the city.

Physicians appreciated the role that the air could affect health as this had long been incorporated into the writings that studied, most particularly the Hippocratic Corpus' Air Water and Places. Urban air pollution in Rome was directly implicated in health by Nero's tutor, Lucius Annaeus Seneca (Epistulae Morales CIV). He wrote that his doctor had ordered him to leave Rome, and that no sooner had he escaped its oppressive atmosphere and awful culinary stenches, than he found his health on the mend.

2.1.2 EARLY MEASURES

A most characteristic approach to pollution control in early societies is the notion of removing the offending source beyond the city boundaries. This method of control is common throughout the classical Mediterranean (Brimblecombe 1987b); Aristotle, in Athenaion Politeia, describes the duties of the astynomoi (controllers of the town) who, among other things, were to ensure that dung was removed two kilometres beyond the city walls. Some administrators insisted that offensive industries are removed from the populated parts of towns and be placed downwind. The principles of zoning, isolating sources from dwelling places and not allowing operations, such as tanning, to be conducted up-wind of towns, seem to be well developed in Hebrew thought. Similar practices are found in medieval times as in the towns of Beverley and York England where industries such as tilemaking, brickmaking were placed outside the walls and downwind (Brimblecombe, Bowler 1992). Even today we can find this logic in the placement of an industrial zone around cities.

Frontinus, Rome's Commissioner of Water in the late 1st Century was convinced that the air pollution (gravioris caeli) had decreased and the infamis aer that offended previous generations had gone (De Aquis Urbis Romae II.88). The claimed improvement seems to have come from better sanitation that arose with a good supply of water.

There are early references to legal attempts to compensate for air pollution. Roman law had no equivalent to the modern concept of nuisance, but was sensitive to wanton interference with ancient rights that gave, for instance, access to water or right of way. Some of these rights (iura praediorum urbanorum) concerned urban problems and legal decisions against the production of smoke are known (Brimblecombe 1987b), e.g. "Aristo... did not think smoke could legitimately allowed to penetrate from a cheese factory into building higher up the road... no more than you can throw water or anything else from a building higher up onto those lower down".

We can compare this to a case in medieval London (Brimblecombe 1987a) "Thomas Yonge and Alice his wife complain ... the chimney is lower by 12ft than it should be ... and the stench of the smoke from the sea-coal used in the forge, penetrates their hall and chambers, so that whereas formerly they could let the premises for 10 marks a year, they are now worth only 40 s". The need for high chimneys had been known since classical times. Strabo (3.2.8 C146) tells us that the metallurgists of Spain built "their silver-smelting furnaces with high chimneys, so that the gas from the ore may be carried into the air because it is heavy and deadly".

2.2 Pre-industrial cities

Although combustion became the dominant issue in framing modern air pollution problems, odour was a major concern in cities of the past. In many ways it is an important contemporary issue on a local level as many complaints to civic environmental officers relate to odour.

In pre-industrial cities, the stench from rotting organic matter was an ever-present problem. Sir Kenelme Digby said it was so bad in Paris that silverware rapidly tarnished due to airborne sulphides (Brimblecombe 1987a). The culinary odours that so troubled Seneca in Imperial Rome remained fifteen centuries later. Police regulations in Papal Rome discouraged cooking some foods at home, so tripe was cooked in enormous cauldrons set up in front of the church of San Marcello and cabbages boiled near the Piazza Colonna. Coffee was roasted at the foot of the column of Marcus Aurelius. This solution was not typical in 16th Century England, where civic authorities had to handle continual objections to the smells from beef boilers, soap makers (Brimblecombe, Bowler 1992). In modern times the production of chocolate, beer or coffee brings frequent complain, but the occurrences are as often rural as urban.

2.2.1 CHANGES IN FUEL

It can be argued that changes in fuel use are responsible for changes in air pollution. Novel fuels bring with them novel difficulties in their use. On a detailed level there are deficiencies in such a view, but as a broad generalisation there are many examples where transition in fuel use have led to sharp perceived changes in air quality. The new fuel can change the quality of the smoke and this is more noticeable than a gradual increase in smokiness that occurs with the general growth of population of towns.

A radical shift in fuel use was a move from wood to coal. This occurred in medieval England after the loss of convenient wood supplies near cities. An early transition from wood to coal depends on the geological availability of the fossil fuel. The unfamiliar smell of coal smoke led to early fears about the health risk, through the belief that disease was carried in miasmata (malodorous airs). Coal smoke made London's citizens aware of problems of air pollution in the late 13th Century. The fuel was used mostly in the production of lime for mortar. Despite pressures on artisans to return to wood, its high price compared with coal meant that control measures failed even where harsh penalties were imposed (Brimblecombe 1987a).

It was not until the 16th Century, with the widespread construction of chimneys, that the fossil fuel began to be used domestically. Initially it was used by poorer people who could not afford the more desirable wood. However, fuel shortages and changing attitudes were such that it was increasingly adopted (Brimblecombe 1987a). The transition to coal was virtually complete in London by the early 17th Century and all the affects of the resultant air pollution well described in John Evelyn's Fumifugium of 1661. This is not typical of European cities where wood was abandoned much later. In Paris, for instance, coal remained unpopular. In 1714 the fuel was prohibited and later that century English coal tended to be used by the poorer people. However there was a general outcry against the fossil fuel that it was claimed: vitiated the air, soiled linen set to dry, caused chest infections and impaired the delicacy of the female complexion. Attitude towards coal remained negative despite a favourable ruling by the Academies of Sciences and Medicine and visitors to Normandy in the 19th Century noted adverse reactions to those who wished to use coal domestically.

2.2.2 IMPACTS

In medieval times it was primarily concern about health that drove the administrative responses towards air pollution. There were also economic losses entailed (the case of Thomas and Alice Yonge noted above). Damage to interior furnishings, vegetation, washing set to dry and enhanced corrosion of building materials were all frequent complaints by the beginning of the 17th Century in London. Religious authorities in particular were often forced to repair the damage caused by smoke. Sometimes the church administrators had access to the legal tools to enable them to fine the producers of smoke. This was especially true where the ecclesiastical authorities had jurisdiction over areas beyond the confines of their buildings.

It was not only large public buildings that suffered; superficial soiling by coal smoke caused economic loss to ordinary householders. In London of the 18th Century the rate of darkening to the paintwork of some houses was so rapid that repainting needed to be frequent (Brimblecombe 1987a). In Paris somewhat later there were complaints that dyes faded rapidly in the polluted air (Witz 1885).

An indication of dramatic increases in the smoke pollution in London is shown by comments on damage to clothing. This occurred not only when clothing was set out to dry, but merely when walking through the streets. Small businesses were set up to refurbish clothes that had been "smoked" and white coloured clothing was avoided. The umbrella came as a black accessory, because it was required to ward off soot-laden rain (Brimblecombe 1987a).

2.2.3 TRANSITION FROM TRADE TO INDUSTRY

Air pollution is frequently related to industrialisation, but pre-industrial cities could be smelly, smoky places, so it is not entirely clear what distinguishes the industrial and pre-industrial situation. The smoke and smell of early trades have sometimes been termed "pollution artisanale", while pollution that came with the large scale industrialisation of the 19th century can be classified as "pollution industrielle". Pfister and Brimblecombe (1990) talked about the difficulty of this distinction based on scale and qualitative changes.

It may be that this desire to ascribe differences to the pre-industrial pollution, can be better satisfied by considering the approaches adopted when controlling the pollution. In the earlier period pollution control took place through specific local regulation, not through the national control we now see as dominant. A need to limit pollution from small workshops and cottage industries typifies the situation up to the 18th Century, in cities such as Paris that had established a large body of local regulations (Fournel 1813). This approach addressed local concerns and resulted in specific proclamations and edicts to restrict emissions from identifiable industrial sources.

It must be remembered that civic government at the beginning of the 19th Century was loaded with older structures that hampered administration of the growing metropolises. In Manchester of 1801 the medieval Court Leet prosecuted eleven factories for failing to consume their own smoke, and this same near-obsolete body continued its jurisdiction over nuisance in the city for another three decades (Bowler, Brimblecombe 1998). Technological control was also limited and showed only the slowest of advances. A smokeless stove invented by Dalesme and exhibited at St. Germans Fair in Paris in the 1680's was still considered novel enough to be described seventy years later (Brimblecombe 1987a). Chimney designs were explored scientifically by Count Rumford in 1796, but one feels more from the view of architecture than effectiveness.

2.3 Industrialisation and sanitary reform

One of the earliest general laws covering environmental pollution was enacted on 15 October 1810, by Emperor Napoleon. This decree covered the permission required by any establishment that emitted offensive odours. This imperial law would have been relevant to much of Europe and by 1845 more than three hundred types of industry in France were classified as "etablissements insalubres, incommodes ou dangereux".

We might argue it is this legislative development that most characterises pollution industrielle. Additionally, the Imperial German Civil Code Paragraph 906 allowed property owners to object to smoke, although only when it caused material damage.

2.3.1 THE EMERGENCE OF STEAM ENGINES

Despite developing national and supra-national laws, protest remained local. People did not perceive a wider interest in their environment, which meant that local factories, particularly those in populated parts of towns, were the target of most protest. In France, 1820 almost a hundred citizens complained of chemical works at Menil-Montant, and in Berlin, 1829 there were similar protests from seventy citizens. In London of the late 1700's many people reacted strongly to the awful clouds of smoke released by the new steam driven pumps that lined the Thames which seemed, "in the rare intervals that they were working, more determined to suffocate the inhabitants of London than to supply them with water". The general public disliked the "new-fangled" steam engines that were noisy, dirty and dangerous (Brimblecombe 1987a).

The steam engine was an important new change to the urban life. Its development led to the possibility of smoke pollution on a much larger scale than before. It also drew workers into towns of ever increasing population density. Factories could grow very large as power was now available to run these gigantic establishments.

Civic administrators rapidly realised that they needed to respond to smoke pollution. The Manchester and Salford Police Act of 1792, gave an increasing range of powers to control nuisance, which emphasised chimney construction "to burn the Smoak arising" in a report circulating five hundred copies among local industries. The concerns in France ran parallel to those in Britain. An ordinance of 1815 specified that steam engines would fall within the scope of the Napoleonic Decree of 1810 (Payen 1985). Smoke abatement became a continuing theme; the French ordinance of 11 November 1854 required steam engines in Paris to consume their own smoke, but the regulation was largely ineffective. A national ordinance of 19 January 1865 applied to smoke abatement from all furnaces, but it also failed. This may have been due to resistance of manufacturers, but probably from the lack of workable "smoke consuming apparatus". In England, many clauses concerning smoke were to be embedded in parliamentary legislation of the 19th Century, where smoke abatement appeared as clauses forming part of Acts whose principle concern were issues such as Public Health, civic administration &tc.

An interest in large single sources, such as steam engines, drew attention to the problems of industrial chimney heights. Chimney heights had been seen as an issue even in medieval times, but it wasn't until the 19th Century, that heights began to be a characteristic of laws. In Belgium, a Royal Decree 29 January 1865 prescribed the height for some industries e.g., boiler factories 15 m, glass factories 30-40 m. Often the heights appear to have been set arbitrarily. In some areas, such as Saxony, where there was considerable conflict between agricultural and industrial interests, the construction of very high chimney stacks became the preferred solution in the late 19th Century (Schramm 1990).

2.3.2 STOVES, FURNACES AND CHIMNEYS

The steam engine forced engineers to examine the way in which smoke was generated. Benjamin Franklin, who was interested in the design of stoves and in fuel economy in the 1760's, almost inevitably became involved in the debate over smoke from the earliest Boulton and Watt steam engines. Franklin emphasised the need to burn all the smoke and gave two reasons: the first because the smoke which escapes represents unburned and therefore wasted fuel, and the second that the smoke is likely to form an insulating crust on the lower surface of the boiler. These early notions of "burning your own smoke" embody a philosophy that was to remain a central theme of smoke abatement until the present century. Smoke prevention was seen as good practice and sound economy, rather than as a means of preserving the quality of the air.

The widespread adoption of the steam engine meant that the design and construction of furnaces, which had once taxed the skills of the best engineers of an age, was left almost totally in the hands of the brickmaker. It is little wonder that the increasing prevalence of furnaces, boilers and steam engines in towns meant that they became hideously smoky.

In parallel the skilled fireman and stokers of the 18th Century were replaced by less skilled and underpaid labourers through the 19th Century. In Prussia, courses for stokers were introduced in the first years of the 20th Century and instructors from the Boiler Supervision Societies went around the country spreading good practice. Hamburg's "Verein für Feuerungsbetrieb und Rauchbekämpfung" employed skilled instructors and wrote valuable reports. British advocates of cleaner air saw Hamburg's interest in smoke abatement as a mark of considerable leadership.

2.3.3 SMOKE ABATEMENT IN THE 19th CENTURY

By the first half of the 19th Century most people thought smoke an undesirable aspect of urban life. It was sometimes seen as a "necessary evil", but the idea of smoke abatement began to be broadly welcomed. There were few technological developments, so most smoke abatement advocates pinned their hopes on legislative change.

This was accompanied by a growing interest in the health of towns as the locus of population shifted to cities. Medical topographies were common and these often noted increased mortality in urban areas. It is argued that the statutory regulation of pollution during the last century was part of a general movement to improve the sanitary conditions of urban life, perhaps typified by the movements that followed the reformer Edwin Chadwick.

Although some air pollution control in the early 19th Century arose as a direct response to the problems of industrial emissions, much developed along side sanitary reforms that typified last half of the 19th Century. The increasing discrepancy between the rate of mortality in rural and urban areas often induced urban administrators to take an interest in improving the health of their towns. The new sanitary laws, with their smoke abatement clauses were intended to go beyond being the simple local rules that had typified earlier attempts to improve the urban environment (Diedericks, Juergens 1990).

There were administrative changes also. A general relaxation of the Crown in the 18th Century had strengthened local government in Britain. In the early 19th Century this became formalised with legislation such as the Municipal Corporation Reform Act (1835). Nevertheless central government pressures need to be maintained if municipal governments are to implement smoke abatement clauses in various Acts relating to public health (Brimblecombe, Bowler 1990). On the continent air pollution tended to be supervised by local police authorities on a case by case basis or licensed (Stolberg 1994), as in France's Napoleonic Decree 1810 or the national ordinance of 1865.

As we have seen the passage of many laws concerning smoke control did not necessarily abate smoke. Governments placed so much emphasis on industrial progress that industry occupied a privileged position (Melosi 1980). It is claimed that environmental laws rarely developed where they impeded industrial development. However, industrial enterprise and environmental legislation could arise together. In Manchester, for example, a rapidly industrialising city, there was a Steam Engine Smoke Committee of the council by 1822 (Bowler, Brimblecombe 1998).

Figure 2.2 November fog in London 1872.

The emerging environmental laws of the 19th Century lacked power. Even where there was enthusiasm, there were administrative and technical barriers to the abolition of smoke. Some British law of the late 19th Century tried to set up administrative mechanisms for its implementation (e.g. the Health of Towns Act of 1853 and the Public Health Acts of 1875 required local governments to appoint an Inspector of Nuisances). Although the administrative procedures were well defined and enthusiastically followed, the lack of appropriate smoke control technology seemed to prevent both the administrators and industrialists from achieving a substantial improvement in air quality (Brimblecombe, Bowler 1990). The genuineness of their concern is in line with the wider scale interest in sanitary reform of the times; the health of towns was regarded as having a high administrative priority (Brimblecombe, Bowler 1998).

Responses of manufacturers, in England, when warned of producing too much smoke varied: blaming (1) the use of poor quality coal, (2) bad stoking or firing of the furnace or (3) insufficient ventilation of the furnace. The remedy for this was most commonly seen to be: (1) change of fuel from coal to coke, (2) warning or sacking the fireman or (3) installation of a patent smoke apparatus. While the first two may well have been effective in reducing the smoke, the third remedy was no panacea as smoke consuming apparatus although readily available at this time, was not necessarily effective unless used with great care. This was an enduring difficulty everywhere.

The development of railways also brought new problems of smoke pollution. The laws that related to their development were often quite different to those applicable to steam engines in factories. Such engines could obviously cross national boundaries.

The steam engine had allowed protest to focus, but it also caused the early environmentalists to take a naive approach to atmospheric pollution. Their activities became directed at very specific problems relating to a single source not general air pollution issues. Beyond this, the environment was still not a unified concept and issues remained firmly anthropocentric, most typically related to the physical health of the urban population.

2.4 Modern development

Air pollution, in terms of practical interest in the 19th Century, concentrated on smoke, and there was relatively little regulatory concern over gases except for hydrogen chloride dealt with under the UK Alkali Act of 1863. This preoccupation with smoke was no doubt a result of its public visibility and the lack of any scientifically based monitoring activity. While scientists were increasingly employed by governments in the 19th Century, and the role of experts was increasingly apparent, the invisibility of many agents affecting public health remained a difficulty when justifying actions to sceptics. This was well displayed in Ibsen's play "An Enemy of the People" (1882). The emerging smoke abatement laws seem the result of much forethought and observation, but rarely saw monitoring as a significant part of improving the environment, so quantitative regard was first focused on emission limits.

The opening decades of the 20th Century saw smoke abatement legislation present across much of Europe, but experience with it had been unpromising. This didn't stop the passage of new legislation, but made proponents aware of the need for more sophistication and prescriptive limits. In France an Act of 1932 dealt with the suppression of industrial smoke, specifying that the density must not exceed Ringlemann No.1 for 5% of the working time, waste gas must not exceed: 1% CO, 2% SO_2 1.5 g/m^3 dust. A Ordinance "le loi Morizet" was adopted in 1939, but its flat prohibition of smoke may have rendered it as unsuccessful as the 19th Century regulation. In the UK the Public Health Act of 1936 contained statutory provisions to control smoke, grit and dust from industry.

On the local level their attention had to shift among the many smaller sources. Issues arose, became important for a decade or so and then vanished. It was not so much that a distinct solution was found, rather the problem was superseded.

Sometimes new technology or plant rendered a troublesome and emotive issue, which angered the public and annoyed local administrators, quite superfluous. Improvements came, but with imperceptible slowness.

The lack of formal monitoring of the changes taking place in the atmosphere meant that the effectiveness of legislation remained almost impossible to assess. The most widespread early observations were simple descriptions of smoke from chimneys. Occasionally these were timed and used Ringelmann charts, but the frequency was a biased product of public and political interest rather than a usefully distributed sampling strategy. They were not particularly useful when trying to gauge the improvements.

2.4.1 THE BEGINNING OF MONITORING

Some amateur enthusiasts had made measurements of air pollutants last century. Ducrois and Smith undertook rainfall analyses that showed rain could be acid, but also contaminated and even to the point in the UK that the acids could be neutralised by grit. In the 1880's Russell in London and Ladureau in Lille analysed SO_2 in the air. Such observations continued in a sporadic and inaccurate manner well into the 20th Century. The measurements suggest concentrations of smoke and sulphur dioxide that are far too high to be believable (Brimblecombe 1987a).

It is only the measurement of deposited solids and solutes in industrial towns that are reliable from an early date. However, deposit measurements tend to focus on the coarse local fraction of the emission. Some soot deposit measurements were made in the 1880's in England, 1926 in France, 1932 in Germany and 1944 in Denmark (Brimblecombe 1982; Andreasen, Gravesen 1949). The first UK national air deposit gauge network was established early in the 20th Century with apparatus that was used for more than 50 years.

Through the 1920's and 1930's government agencies were designing of a range of air pollution monitoring equipment: the jet dust counter, lead candle (SO_2 deposit), filter paper black smoke monitors and the bubblers. This pioneering work was influential in the development of instruments for the earliest networks in the US (Ives et al. 1936) and Europe (Andreasen, Gravesen 1949). Thus, from the 1930's there were limited continuous measurements of SO_2 and smoke concentrations. The number of sites increased substantially in the years that followed the London smog of 1952.

Data from the early networks suggest a decline in the deposited matter in large cities, such as London for much of the century (Figure 2.3). The concentrations of classical primary pollutants, such as sulphur dioxide and smoke tend to have declined also, although the length of this decline varies for individual cities across Europe. Nevertheless these are not unexpected given urban changes that have pushed major industries away from city centres, where traffic has increasingly dominated as a source of air pollution. The history of air pollution in other cities around the world, often shows similarities with the patterns that have been seen in European cities. At present high concentrations of smoke and sulphur dioxide from stationery sources are found in industrialising regions, where regulatory policy has yet to play an active role.

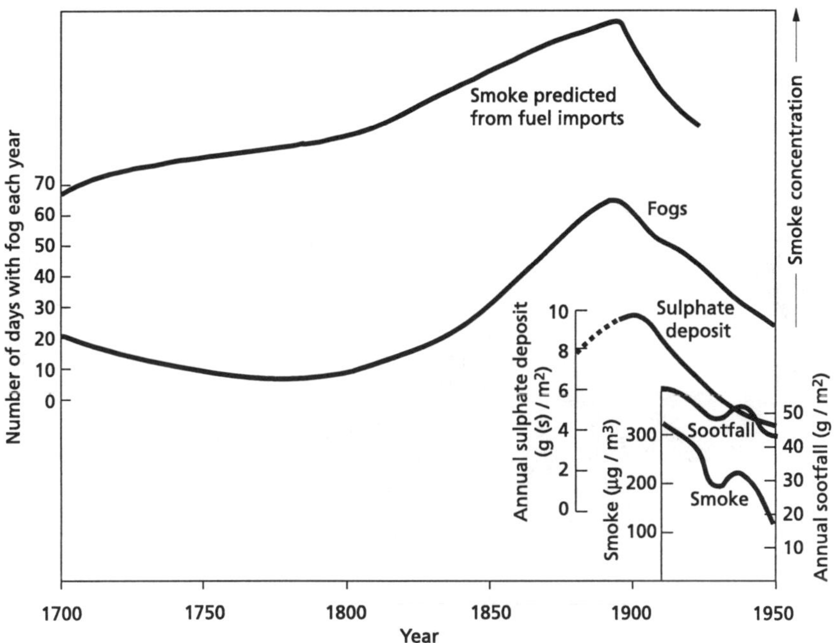

Figure 2.3 Air pollution in London since the seventeenth century, comparing predicted values with fogs and later measurements (Brimblecombe 1987a).

2.4.2 SMOG DISASTERS

It was clear from medical statistics of the late 19th Century that fogs enhanced the death rates in large cities and this was increasingly known across Europe. However, a few major air pollution episodes in the 20th Century have had the strongest influence on developing air pollution policy. The first was probably the disaster in the Meuse Valley in Belgium between the 1st and 5th of December 1930 (Firket 1931). This occurred when air stagnated in the densely populated and highly industrialised valley. Although the valley wall was little over a hundred metres high pollutants became trapped. People became ill and sixty died, while many cattle had to be slaughtered. The causative agents have never properly been identified, although sulphur dioxide or sulphuric acid were obvious candidates. Some thought that fluorine from as many as fifteen factories in the valley may have been an important factor.

Of comparable significance was the London smog of 1952. This winter event caused as many as four thousand excess deaths, which led to public outcry and a strong political reaction. A subsequent enquiry resulted in Clean Air Act of 1956 (Brimblecombe 1987a). The smog and its resultant legislation are often viewed by many as a turning point in dealing with the problems of urban pollution. However the effectiveness of the Act has often been questioned. Some have felt (e.g. Brimblecombe, 1987a) that the Act simply reinforced shifts to less smoky fuels that were well underway, while others were critical of its inability to cover anything but smoke emissions. However, regardless of these arguments the years that followed have seen a reduction in domestic coal use (as it was replaced by gas and electricity) and the growing importance of emissions from liquid fuels used in the private automobile fleet. Reductions in traditional pollutants (SO_2 and smoke) masked an increase in new components in urban air.

Figure 2.4 Although air pollution is a nuisance, it has also been seen as a symbol of growth and prosperity (advertisement by Gerda Wegener).

In the early part of the century while much attention was paid to the combustion of smoke there were some significant advances in botanical studies of air pollution in UK and Germany. There was little interest in the medical study of human health effects before the tragedy in the Meuse Valley. French work began to appear at this time and the dangers of air pollution were discussed in the Chamber of Deputies. The 1920s and 1930s also saw a shift away from vegetation to health interests in Germany, while British workers concentrated on instrument design. There was a growing interest in carbon monoxide from automobiles and an awareness in French workers such as Milliard and Pardoe studies the emission of aldehydes. The period also saw important developments in the understanding of pollutant dispersion from stacks (Halliday 1961).

2.4.3 EMERGENCE OF PHOTOCHEMICAL POLLUTION

In the years that followed the Second World War, the US Donora incident sparked considerable increase in activity, this was reinforced by the London smog of 1952. However the most radical advances were made following Haagen-Smit's proposal that Los Angeles smog was similar to the products of the ozonolysis of gasoline vapours. This idea was confirmed by long path length IR (Stephens 1987), and Leighton's review (1961) unravelled the photochemistry of Los Angeles smog. The pivotal role of OH in atmospheric chemistry soon emerged (Levy 1971).

Unique conditions that favour photochemical smog formation in Los Angeles, may have prevented its early detection elsewhere, although it was obvious that the precursors would be commonly present in modern cities (Leighton 1961). In the late 1960's measurements began to be made in Europe which soon identified photochemical products. High ozone concentrations took on almost continental dimensions in the summers of the 1970's (e.g. Cox et al. 1975; Guicherit, Van Dop 1977). The heat wave of 1976 produced an impressive episode across southern England that particularly captured the public imagination. An increasing frequency of ozone episodes started a debate about its role in forest damage along with that of acid rain. By the 1980's the ozone problems found on a synoptic scale in north-western Europe, were seen at a more urban scale in Mediterranean cities such as Athens.

The transition from a situation, where air pollution is dominated by primary pollutants for one where photochemical reactions are important, has had to be incorporated into policy. The older legislation dominated by emission control have become inadequate. Control of secondary pollutants and their long range transport has raised multi-national protocols and agreements. The legislative responses have broadened and as the millennium approaches we will see the notion of air quality management firmly within European Directives (e.g. 96/62/EC).

2.5 References

Primary sources are often not referenced because these can be such obscure older works, so I have tried to refer the reader to more general reviews or accessible materials (where original sources will usually be cited). Nevertheless some readable primary material is given because of its wide general interest. Classical references are generally cited in the text because they are so widely accessible it would be restrictive to refer to specific editions or translations.

Andreasen, A.H.M., Gravesen, P. (1949) Preliminary report of investigations of atmospheric pollution, *Trans. Danish Academy of Technical Sciences* **no. 4,** 1-72.

Brimblecombe, P. (1982) Trends in the deposition of sulphate and total solids in London, *Science of the Total Environment* **22**, 97-103.

Brimblecombe, P. (1987a) *The Big Smoke*, Methuen, London.

Brimblecombe, P. (1987b) The antiquity of smokeless zones, *Atmospheric Environment*, **21**, 2485.

Brimblecombe, P., Bowler, C. (1990) Air pollution history, York 1850-1900, in: Brimblecombe, P., Pfister, C. (editors), *The Silent Countdown*, pp. 182-195, Springer Verlag.

Brimblecombe, P., Bowler, C. (1992) The history of air pollution in York, England, *J. Air Waste Management Association*, **42**, 1562-1566.

Bowler, C., Brimblecombe, P. (1998) Control of air pollution in Manchester prior to the Public Health Act 1875 (submitted *Environment and History*).

Cox, R.A., Eggleton, A.E.J., Derwent, R.G., Lovelock, J.E., Pack, D. (1975) Long range transport of photochemical ozone in North-western Europe, *Nature*, **255**, 118-121.

Diedericks, H., Juergens, C. (1990) The environment in The Netherlands in the 19th Century, in: Brimblecombe, P., Pfister, C. (editors), *The Silent Countdown*, pp. 167-181, Springer Verlag.

Edgren, B., Herschend, F. (1982) Eketorpför Fjärde gången, *Forskning och Fremsteg* **no. 5**, 13-19.

Firket, M. (1931) Sur les causes des accidents survenus dans la valee de la meuse lors des brouillards de Decembre, 1930, *Bulletin de L'academie Royale de Medicine de Belgique*, **11**, 126.

Fournel, F. (1813) Traite du Voisinage, Tome II, Paris.

Guicherit, R., Van Dop, H. (1977) Photochemical production of ozone in Western Europe (1971-1975) and its relation to meteorology, *Atmospheric Environment*, **11**, 145-156.

Halliday, E.C. (1961) *Historical review, in Air Pollution*, pp. 9-37, WHO Geneva.

Hughes, J.D. (1994) *Pan's Travail - Environmental Problems of Ancient Greeks and Romans*, John Hopkins University Press, Baltimore.

Ives, J.E., Britten, R.H., Armstrong, D.W., Gill, W.A., Goldman, F.H. (1936) Atmospheric Pollution of American Cities for the Years 1931 to 1936, *US Treasury Department, Public Health Bulletin No. 224*.

Leighton, P.A. (1961) *Photochemistry of Air Pollution*. Academic Press, NY.

Levy, H. (1971) Normal atmosphere: large radical and formaldehyde concentrations predicted, *Science*, **173**, 141-143.

Melosi, M.V. (1980) Environmental crisis in the city, in: Melosi, M.V. (editor), *Pollution and Reform in American Cities,* pp. 1870-1930, University of Texas Press, Austen.

Payen, J. (1985) *Technologie de l'Eenergie Vapeur en France dans la Premiere Moitie du XIXe Siecle - La Machine a Vapeur Fixe*, Comite des Travaus Historiques et Scientifiques, Paris.

Pfister, C., Brimblecombe, P. (1990) Introduction, in Brimblecombe, P. and Pfister, C. (editors), *The Silent Countdown*, p.p. 1-6, Springer Verlag.

Schramm, E. (1990) Experts in the smelt-smoke debate, in: Brimblecombe, P., Pfister, C. (editors), *The Silent Countdown*, Springer Verlag (in press).

Stephens, E.R. (1987) Smog studies of the 1950's, *EOS*, **68**, 89-93.

Stolberg, M. (1994) *Ein Recht auf saubere Luft? Umweltkonflikte am Beginn des Industriezeitalers*, Fischer, Erlangen.

Witz, M. (1885) Sur la presence de l'acide sulfureux dans l'atmosphere des villes, *C. R. Acad. Sci.*, **100**, 1385-89.

Chapter 3

EUROPE AND ITS CITIES

KNUT E. GRØNSKEI
Norwegian Institute of Air Research
P.O.Box 100, N-2007 Kjeller, Norway

3.1 Regions and mountains

Europe is separated from the continent of Asia by the Ural mountains, the Caspian sea, Caucasus and the Black Sea. To the south, to the west and to the north Europe has coastlines to the Mediterranean sea, the Atlantic and the Polar sea, respectively. The continent extends from 37°N to 61°N.

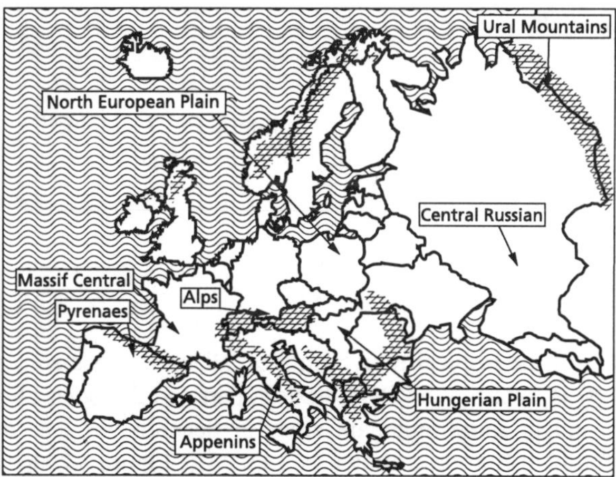

Figure 3.1 Mountain ranges and regions in Europe.

The continent is divided in parts by several mountain ranges like the Pyrenees and Apennines, the Alps, the Carpathians and the mountainous areas in Scandinavia, Scotland and Balkan. The flat lowland in-between the mountain ranges and along the coastline to the west and to the north includes the North European Plain, the Po valley and the Hungarian Plain. The coastal area to the west of the European mainland consists of flat lowland. The variety of physical landforms, the change in latitude and coastal versus inland areas influence the urban environments and the dispersion of air pollution in the regions.

3.2 Climate and main wind regimes

Meteorological conditions include a number of parameters such as wind, turbulence, wind fluctuations, temperature, humidity, cloudiness, sunshine and rainfall. The first three are particularly important for transport and dispersion of air pollution. The practical implications of climatic parameters include emissions as well as dispersion. In this chapter, the description is used for classification of European cities.

Köppen's classification scheme (Ahvens 1993; Køppen 1936), account for important parameters regarding emissions, transport and reactions in the atmosphere. The scheme includes the following main zones characterised by temperature and moisture:

A : Tropical rainy climates
B : Dry climates
C : Temperate rainy climates
Cw : Warm, with dry winter
Cf : Warm, moist in all seasons
Cs : Warm, with dry seasons
D : Cold snow-forest climates
Df : Snow forest, moist in all seasons
Dw : Snow forest, dry winter
E : Polar (snow) climates

A better description of yearly variation in temperature is obtained by adding a third letter to the zone classification with the following meaning:

a : Hot summer
b : Warm summer
c : Cool, short summer
d : Very cold winter

With the exception of the tropical rainy climates, all these types are found in Europe. The dry climate (B) is found in the south-eastern part and the polar climate is found in the northern part and in some mountain areas in Scandinavia. The definition of the different classes found in Europe is shown in Table 3.1. Figure 3.2 shows the extension of the main climatological regions in Europe.

Table 3.1 Climatological regions in Europe. The symbols are defined in the text.

1. letter	2. letter	3. letter	
Cfa/Cfb/Cfc			***Marine west-coast climates***
C			Coldest month > -3°C but < 18°C, at least one month has an average temperature above 10°C
	f		moist, adequate precipitation in all months and no dry season
		a	hot summer, warmest month > 22°C
		b	warm summer, warmest month < 22°C
		c	less than four months over 10°C
Csa/Csb			***Mediterranean climates***
C			Coldest month > -3°C but < 18°C, at least one month has an average temperature above 10°C
	s		dry season in summer
		a	hot summer, warmest month > 22°C
		b	warm summer, warmest month < 22°C
Dfb (Dfc)			***Humid continental climate (continental sub-arctic climate)***
			Coldest month < -3°C, warmest month has at average temperature above 10°C
	f		moist, adequate precipitation in all months and no dry season
		b	warm summer, warmest month < 22°C
		c	less than four months over 10°C
E			***Polar climate.***
			The average temperature for the warmest month is lower than 10°C and greater than 0°C

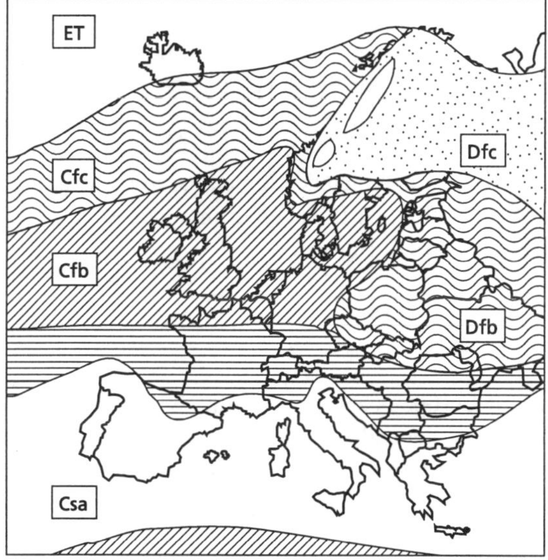

Figure 3.2 The extension of climatological regions in Europe. See Table 3.1 for legends.

The Mediterranean region is characterized by warm and dry summers and wet winters. During summer time this region is influenced by the subtropical high-pressure area favouring sinking air, stagnation and low level inversion. These conditions favour accumulation of local pollution emissions. During winter time the area is more often influenced by the cyclones and large scale wind systems.

Further, the major wind system is modified by local differences as a result of the distribution of land and sea, i.e. over the *Iberian peninsula* a local low-pressure area develops over land as a result of the solar heating during the day (the Iberian Thermal Low). The resulting wind system, a so-called sea breeze, flows from the sea to land. During the night the wind may reverse as a result of the cooling of the land surfaces.

As the large scale wind is characterised by stagnation, a closed regional circulation may develop in different parts of the Mediterranean area during the summer. This is important when the effects of local emissions are considered. The diffusion and transport of atmospheric pollutants in the Mediterranean regions may be significantly different from those observed in Central and Northern Europe.

Western Europe is influenced by moist air from the sea transported by the general wind system coming from west. This type of climate also favours long range transport of air pollution.

In *Central and Eastern Europe* the continental type of climate dominates, with cold winter and warm summer. High pressure situations are often found in these areas both in summer and in winter and are characterised by high frequency of air stagnation with accumulation of local air pollution emissions.

In *Northern Europe* the low temperature in particular during winter require energy for domestic heating. At some distance from the coastlines the small amount of incoming solar radiation during winter favour the development of persistent inversions and poor dispersion conditions.

In addition to meteorological parameters Köppen based the classification system on the characteristic type of vegetation. Different environmental effects of air pollution may be expected in the different zones of Europe.

3.3 Population and resources

The total population of Europe is about 750 mill. people of which more than 500 mill. live in urban areas. They depend on energy and raw material for work and living conditions. The locations of larger European cities are shown in Figure 3.3. The figure also gives an impression of the population density in Europe.

The industry in Europe is based on a high degree of import of raw material. Several of the most industrialised areas are located close to sources of energy and centralised areas regarding transport and communications. Import of raw material and export of products are important activities in many urban areas. Harbour activities by the coast and means of inland transportation cause emission of pollution in addition to emissions from the industrial activities.

Domestic activities including home heating cause emissions that vary with the energy requirements and availability. Energy for home heating increase with decreasing temperature from the warm Mediterranean climate to the cold polar climate.

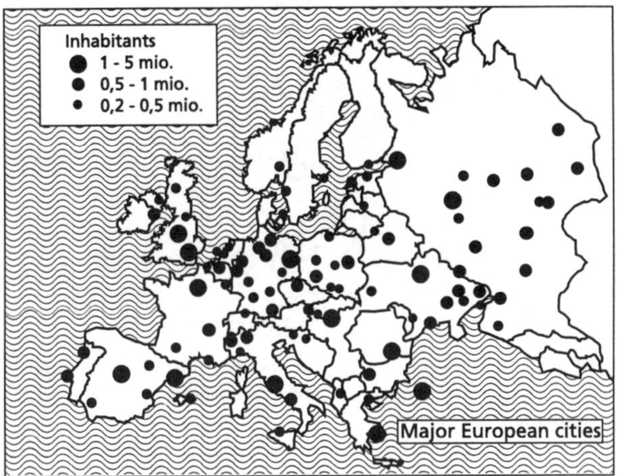

Figure 3.3 Location of large cities in Europe.

Fossil fuel and minerals. The large resources of coal in England, Central and Eastern Europe have been the primary source of energy during the industrial revolution. In the last decades resources of oil and gas have been found in the North Sea and in Russia, and pipelines have been constructed for transport to centralized areas. In the urban areas of Western Europe the consumption of solid fuel in small units have been replaced by consumption of gas and oil. In Eastern Europe the consumption of solid fuel in private houses and in industries cause elevated concentration of SO_2 and particulates in urban areas.

Wind and hydroelectric power. The amount of precipitation and the variation in topography determine the availability of hydroelectric power. Similarly the wind climate will in the future determine the possibility to apply wind production of energy. Hydroelectric power is used primarily close to mountainous regions, and coal is used close to the existing deposits. Pipelines transporting oil and gas are also important for the availability of energy.

Industrial activities in urban areas depend on localisation, resources of minerals and energy in the regions. In this way many factors influence emissions. Ventilation and dilution of pollution varies throughout Europe as a consequence of differences in urban climate, local topography and interaction between land and sea. Figure 3.4 (next page) shows industrial areas based on energy from oil and coal and zones characterised by hydro electric power production.

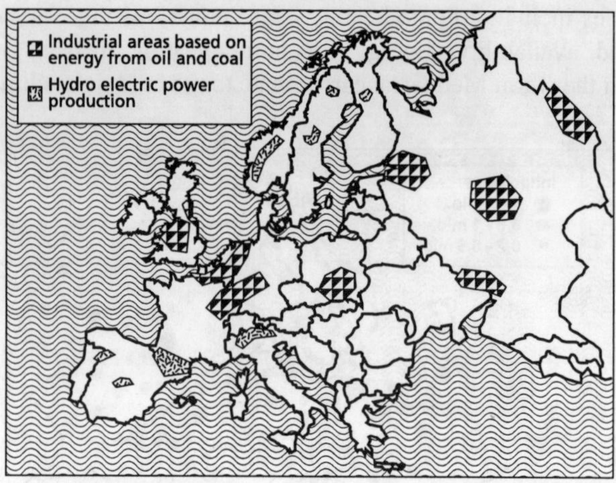

Figure 3.4 In central Europe industries based on energy from oil and coal are found in several zones. In Southern and Northern Europe availability of hydroelectric power becomes more important.

3.4 Regions climates and natural resources of energy and minerals

3.4.1 THE MEDITERRANEAN REGION

The southern part of Europe is characterised by hot and dry summers. Major rainfalls occur during the winter. These areas are influenced by the subtropical high-pressure systems during summer time. Subsidence and light large scale winds suppress long range transport. Local and regional scale transport is influenced by local wind systems, i.e.

- land/sea breeze systems (the Iberian Thermal low, and several local scale systems)
- mountain/valley winds and drainage winds (the Mistral which blows southward from the Rhone Valley in France)
- urban scale circulation (e.g. in the Athens area).

Some urban areas are characterised by high emissions, in particular from industry and traffic. Local wind systems may lead to accumulation of pollution by recirculation. Urban and local scale emission result in regional pollution problems, particularly of secondary pollution. Most of the soil in Southern and Eastern Europe has a high capacity to absorb N- and S-deposition. The emissions are high in a few larger cities and in some industrialised areas. This may cause exceedances of critical loads and adverse environmental effects occur.

In *Eastern Europe* (Balkan), in Italy and on the Iberian Peninsula mountainous areas and complex topography influence the transport and dilution of local emissions. Large cities are located at the coast, i.a. Athens and Thessaloniki in Greece, Rome, Naples and Palermo in Italy, Marseilles in France, Barcelona and Valencia in Spain, Lisbon and Porto in Portugal. The characteristic conditions for Summer smog episodes are characterised by high temperature, high solar radiation intensity (low cloud cover) and low capacity of dispersion. Under these conditions secondary pollution components develop as a result of emissions of nitrogen oxides and VOC (volatile organic compounds).

Winter smog episodes are characterised by low temperatures and poor dispersion. High emission of primary components, in particular particles and SO_2, causes exceedances of pollution guidelines due to local-urban scale emissions. Some cities in the Mediterranean area are also exposed to *winter smog* conditions in addition to cities in the northern and eastern part of Europe.

3.4.2 THE CENTRAL AND WESTERN PART OF EUROPE

In these areas the dispersion conditions are relatively good, both during winter and summer time. Moderate temperatures reduce emissions from many urban activities. The most industrialised regions are localised here and maximum emission intensities occur in some conurbations. Exceedances of critical loads in particular for deposition of nitrogen compounds are found in these areas. The large scale episodes with pollution with ozone and other photochemical oxidants occur in particular in this region. Emissions of nitrogen oxides and volatile organic compounds primarily from urban areas are the main cause of these problems.

The Atlantic region (Western Europe) is characterised by moist and mild winds coming from the Atlantic and the main wind direction is from west to east. This type of climate also favours long range transport of air pollution. The consumption of solid fuel have been replaced by oil and gas. The chemical and petrochemical industries are located mainly in the coastal areas.

The Central European region is characterised by cold winters and moderately warm summers. Solid fuels is still used for heating in small units. Some industrial areas have been located where the primary resources of coal and minerals exist.

3.4.3 THE CENTRAL AND EASTERN PART OF EUROPE

These areas have a continental subarctic climate with low temperature in winter, requiring increased use of energy. In urban areas the pollution situation may be aggravated by an increased number of inversions due to topographical influence. Emissions of SO_2 and particulates are larger than in Western Europe, and winter smog situations occur with high frequency in some urban areas of this region. Considering secondary pollution components and exceedances of critical loads of sulphur deposition occur close to many urban areas in this region.

3.4.4 THE POLAR REGION

This area is characterised by very cold winters and few people live there. The eastern part is characterised by cold winters and warm, dry summers. Figure 3.3 (page 25) shows that only few large European cities are located in the Polar region. However, small towns in this region may experience air pollution problems as a result of emissions from domestic heating, car traffic and industry, in particular in areas where solid fuel is used in small (heating) units for domestic heating. The problems are aggravated by very poor dispersion conditions during the winter.

3.4.5 TYPICAL DIFFERENCES

The difference in temperature between polar zones and the Mediterranean zones in the need for domestic heating; naturally the requirements for energy in the polar regions are larger than in the southern parts of Europe.

Different types of vegetation develop in the different regions as result of the climatic conditions, and the plants have different tolerance to air pollution. The ecosystems in the polar regions are more susceptible to pollution stress than further south.

The ability of the ground to absorb pollution is also a function of the chemical composition of the ground, i.e. for sulphur and nitrogen the tolerance for deposition is high in the southern and eastern part of Europe, but much smaller in the northern and the central parts. Since the major part of the large scale pollution comes from urban areas, the influence of urban emissions varies according to large scale pollution problems. The emissions are larger in the central parts, and the exceedances of critical loads are consequently more frequent in these areas.

The urban ecosystems including emissions, dispersion, deposition and adverse effects are a complex function of the location and urban activity in Europe. Some characteristics may be summarised for various regions. They must be accounted for when sustainable development of urban areas are considered.

3.5 Local organisation of urban areas

Modern life is unquestionably urban life, but it is different in the different regions in Europe. The cities may be grouped according to location, climate and natural resources. Alternatively they can be grouped according to human activities in urban areas. Activities that are important for urban stress and pollution include:

- energy supply
- industrial production and emission
- transportation, i.e. car traffic, harbour and air ports.

In addition to the activities causing emissions, the urban structure and land use may limit or aggravate environmental problems. The urban shape, structure and land use (city morphology) are important variables for environmental effects. Patterns of urban activities, their spatial arrangement and the consumption of natural resources. City morphology includes segregated and integrated land use.

Segregated land use is often applied to improve the environmental conditions in the residential areas by concentrating polluting activities in industrial zones and main traffic lanes.

Integrated land use is useful for minimising transport demands and consequently for reducing total urban emissions. Open spaces and parks are used to improve environmental conditions, in particular in the living areas.

Examples are the Green Belt in London and buffer zones between polluting activities and in many residential urban areas. Considering low level emissions, the intensity in the neighbourhood is of primary importance. As a rule of the thumb the pollution contribution from sources within a distance of one kilometre is more important for seasonal average concentrations than emissions in the rest of the urban area.

3.5.1 LOCATION OF CITY CENTRE

Most cities have one and only one centre. The functions of the city centre include the administrative, the commercial and the entertainment activities. This generates a lot of traffic to these centres from the residential and the industrial areas.

When the city is located close to the sea, the centre often remains close to the harbour (Figure 3.5). This structure may be efficient from many point of views. However, the emission intensity concentrates towards the centre. The emission intensity in the densely populated areas may be reduced when the industrial areas are well located regarding local topography and main wind directions.

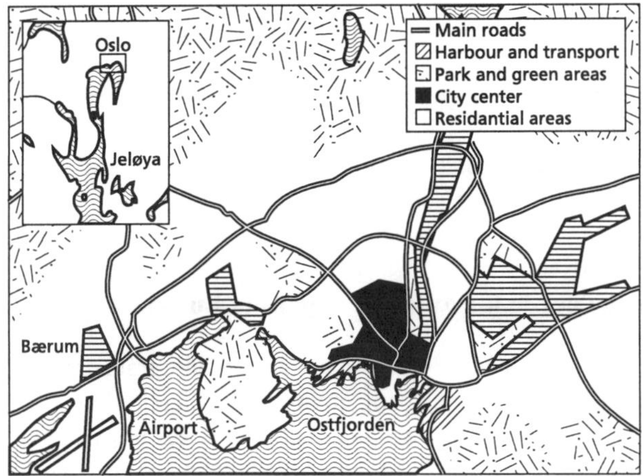

Figure 3.5 The map show examples of the location of centralised heating plants, industrial areas, airport, harbour and road system in an urban area (Oslo, Norway). Green areas are located in-between the activities producing pollution in order to improve environment.

Climate also modifies the structure of urban areas. In Southern Europe the houses are normally built closely together, whereas in Northern Europe, there is more space between the buildings.

3.5.2 EXAMPLES OF URBAN MORPHOLOGY

Figure 3.6 (page 31) shows two examples of urban morphology. The urbanised area in Prague is scattered in a wider area than in Sevilla. The emission of air pollution in parks and open areas is very small and consequently the air quality is improved close to these areas. The local climate and the dispersion conditions are also influenced by the city morphology.

In Sevilla an urban heat island is expected to be located above the city centre. In Prague a less pronounced urban heat island or several heat islands may be observed depending on the heat balance close to the ground.

When buffer zones are located close to the main roads system, the local scale influence of air pollution emissions is reduced considerably. The local scale influence on emissions from chimneys depend on frequency of wind directions and dispersion conditions. This effect can be reduced considerably by increasing the chimney height.

Other types of cities are:

- *London*. One central zone and a city zone separated from the outer zone by a green belt. Industrial zones are located in the outer zone.
- *Athens*. One city centre located about 5 km from the harbour. The city is densely populated with only few green areas. The industrial areas are located outside the city centre.
- *Warsaw*. One city centre including the commercial area. The centre is surrounded by residential areas. The industrial zones are located outside the centre.
- *Helsinki*. One city centre close to the harbour surrounded by green areas and residential areas. The industries are located in several zones. Main streets have been constructed with buffer zones.
- *Lille (France) and St. Petersburg (Russia)*. Several city centres and a prolonged commercial area.
- *Oslo*. One city centre located close to the harbour. Several industrial zones and green areas are located in the surroundings of the city.

3.6 Urban morphology in the different regions of Europe

In the different climate zones the local meteorological parameters may influence the choice of pattern of buildings, improving environmental conditions.

In Southern Europe *solar radiation* is a factor of stress in particular during summer time and shadow is considered to improve urban environment. Cities in the Mediterranean area are characterised by narrow streets and a densely populated areas in the city (built-up areas). In Northern Europe on the other hand solar radiation is considered to be a positive factor of the environment and the urban pattern is developed to maximise sun exposure on buildings.

The *wind conditions* may also influence the urban structure. When strong winds prevail by the coast, the built-up areas are located in the lower parts of the terrain to get shield. High wind velocity in mountainous terrain results in complex air streams around hill sides and recirculations in valleys. These air streams often cause impact of high pollution concentrations from stack emissions.

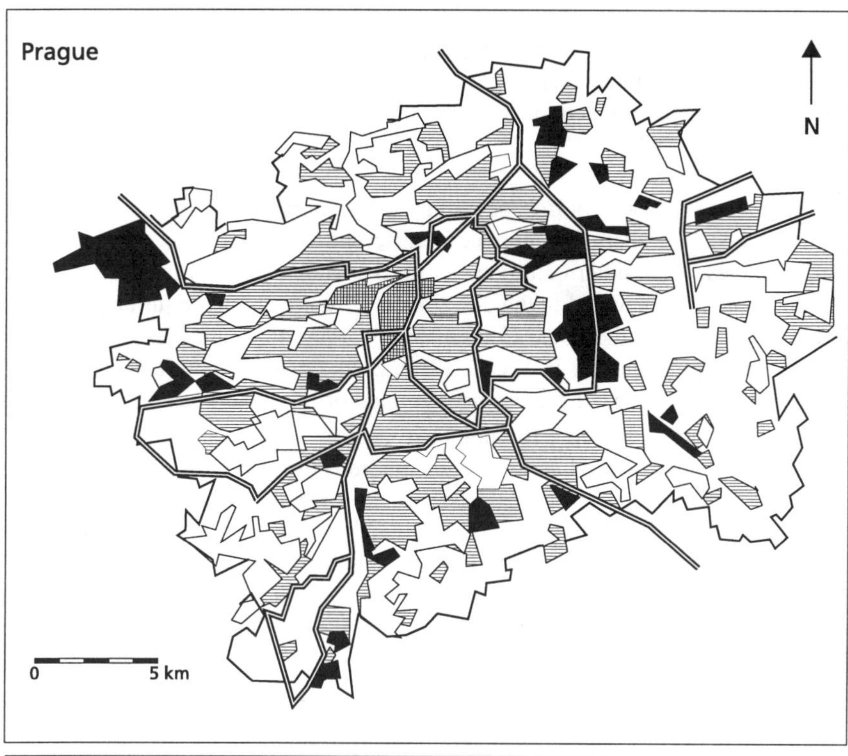

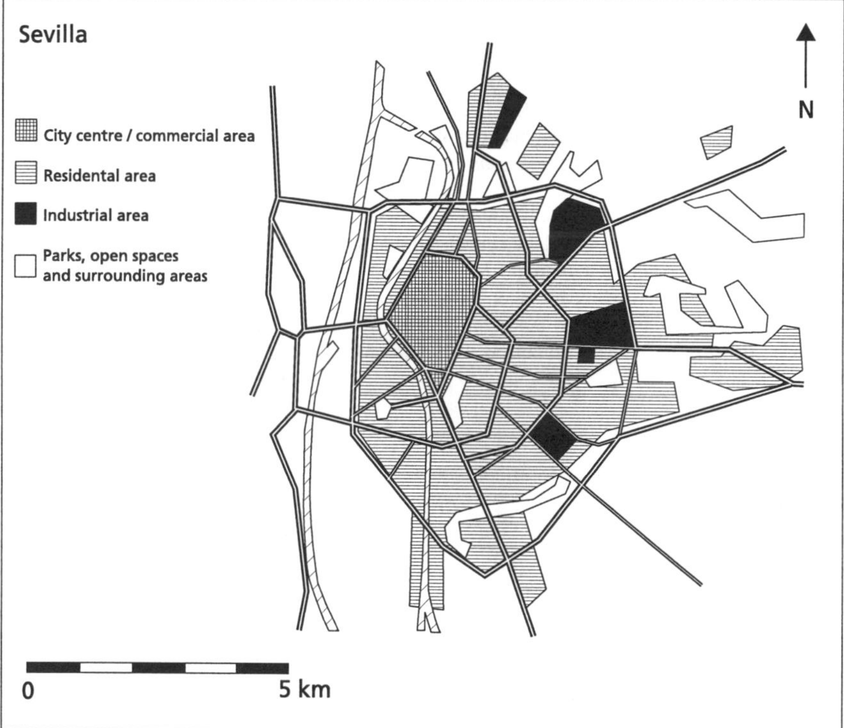

Figure 3.6 Two examples of city morphology that influence the spatial structure of emission and in particular pollution conditions in the living areas.

In areas influenced by high pressure systems and stagnating air masses (i.e. Dfc and Dfb, Figure 3.2, page 23), the lower parts of the terrain may be influenced by the stagnant air, which is characterised by poor ventilation of local emissions. Under these conditions location of residential areas at the hill sides improve the environmental conditions for the population.

Low temperatures in Northern Europe require more energy for *domestic heating* than in Southern Europe. The energy is supplied in different ways in Europe, depending on the natural resources. In Southern Europe the need for air conditioning to reduce indoor temperatures must be included in the description of future energy requirements.

In many cities in Northern and Central Europe domestic heating has been one of the main sources of air pollution. In Western Europe the use of heavy oil and bituminous coal in small systems caused high concentrations of SO_2 and particulates. During the last decades the urban systems for domestic heating have been centralised, cleaning equipment have been installed and reduced emissions through tall stacks minimise local air pollution from home heating. In Eastern and Central Europe domestic heating in cities still rely on the use of solid fuel in small units.

Reducing urban scale emissions will also improve the large scale air quality. While the pollution impact from space heating has decreased in many cities with the technological development, the car traffic has increased, emitting other types of pollution. The same trend is expected to appear in Eastern Europe.

3.7 References

Ahrens, C.D. (1993) *Essentials of Meteorology. An Invitation to the Atmosphere*, West Publishing Company. St. Paul.

EEA (1995) *Europe's Environment. The Dobris assessment*. Stanners, D., Bordeau, P. (editors), Luxembourg, Office for Occicial Publications of the European Communities.

Köppen, W. (1936) *Das geographische System der Klimate*. vol. 1, part C. Handbuch der Klimatology, Berlin (Sec. 2).

Oke T.R. (1978) *Boundary Layer Climates*, London. Methuen.

RIVM (1995) Calculation and mapping of critical thresholds in Europe: in: Posch, M., De Smet, P.A.M., Hettelingh, J.-P., Downing, R.J. (editors), Status report 1995, Bilthoven, *RIVM Report 259101004.*

Sluyter, R.J.C.I. (1995) Air Quality in major European cities. Part 1: Scientific background document to Europe's environment. Part 2: City report forms, Bilthoven, *RIVM Report 722401004 and 722401009.*

WHO (1995) Concerns for Europe's Tomorrow. WHO European Center for Environment and Health, Weiss, Stuttgart.

URBAN SOURCES OF AIR POLLUTION

Air pollution arises from a series human activities. To a large extent the various sources emit the same compounds, only in different proportions. The table below indicate the relative importance of *urban sources* for the main *urban pollutants*. It should be stressed that the importance of a source as an *emitter* of a specific compound does not necessarily mean that is has a corresponding impact on the *air quality*. This also depends on the dispersion as described in Chapters 7-14. Thus traffic in a street canyon has a relatively much larger impact than e.g. a power plant with high stacks.

Main emission sources for the different components.

Source category		Compounds						
		SO_2	NO_2	CO	TSP	Organic	Pb	Heavy metals[1]
Power generation (fossil fuel)		◎	○	○				○/◎
Space heating	- coal	◎	○	◎	◎	◎/○		○/◎
	- oil	◎	○					
	- wood				◎	◎/○		
Traffic	- gasoline		◎	●		◎	●	
	- diesel	○	◎		◎	◎		
Solvents						○		
Industry		○		○	○	○	○	◎/●

○ Between 5 and 25% of total emissions in commercial cities
◎ Between 25 and 50% of total emissions in commercial cities
● More than 50% of total emissions in commercial cities
[1] With the exception of lead (Pb)

Stationary urban sources are treated in Chapter 4 and comprise public power and district heating plants (often combined in co-generation systems), industrial plants and processes, waste incineration plants and minor plants for individual domestic heating. The emphasis is on larger plants and the technical possibilities of reducing the emissions by application of clean technology and flue gas rinsing.

Mobile sources, which play an increasing role for urban air quality, are treated in Chapter 5 with the emphasis on automobiles i.e. passenger cars, buses and lorries. Other mobile sources are only described shortly.

In Europe most local trains - and especially subways - are electrical and do thus not count as local sources. Intercity trains may be diesel powered, but their impact on air quality is modest, and normally localised around railway lines, stations and shunting areas. Ships, in the nature of things, only play a role in harbour cities, but may here (e.g. in the Mediterranean area) be an important source, since they often use low quality fuels.

The impact on urban air quality from air traffic is generally modest; partly because airports are normally located in the outskirts of the cities and in flat areas with good possibilities of dispersion, partly because the main part of the emission takes place somewhat above ground. In most cases the significant impact from airports arises from the generation of additional road traffic.

Emission inventories (Chapter 6) are carried out with different purposes, where the degree of spatial and temporal resolution depend upon the application of the data. For compounds with so long lifetimes that they can have global effects e.g. carbon dioxide, which is the most important greenhouse gas, or CFC's, which attack the ozone layer, only total emissions averaged over longer times are necessary.

In studies of urban air pollution however, the emissions are often used as input data to dispersion modelling (Chapters 10-12). Here the time resolution required may be 1 hour or less and the spatial resolution the size of a single street.

Chapter 4

STATIONARY SOURCES

JAN E. JOHNSSON
Department of Applied Chemistry
Technical University of Denmark, DK-2800 Lyngby, Denmark

This chapter describes the formation and release of air pollutants from different stationary sources. Different methods to reduce emission of air pollutants from stationary sources to the atmosphere will be treated: First reduction of the formation of air pollutants using the principles of clean technology and life cycle assessment then the methods for control of particulate and gaseous emissions from stationary sources.

4.1 Sources

In European emission surveys the sources are described in different source categories (Corinair '94 1997). Each main source category is divided into different subsectors. Examples on subsectors relevant for emission in cities are given below in each stationary main source category:

- Combustion in energy and transformation industries: Public power, district heating plants.
- Non-industrial combustion: Commercial and institutional plants, residential plants.
- Combustion in manufacturing industries: Boilers, gas turbines, stationary engines.
- Production processes: Petroleum industries, iron and steel industries, chemical industries, food, drink and other industries.
- Extraction and distribution of fossils fuels/geothermal energy: Gasoline distribution.
- Solvent and other product use: Paint application, degreasing, dry cleaning and electronics.
- Waste treatment and disposal: Waste incineration, cremation.

4.2 Fuels and materials

The major sources of air pollutants in cities are transportation, energy conversion and industrial production processes. This means that combustion of solid, liquid and gaseous fuels is very important for the emission of air pollutants. All fuels contain carbon and hydrogen, and there may be a content of oxygen, sulphur, nitrogen, water and mineral substances also, dependent on the fuel type. The mineral parts of the fuel form ash, but the remaining constituents are converted to gaseous components. An overall reaction scheme for complete combustion of a fuel, on an ash and water free basis, can be written as follows:

$$C_aH_bO_cS_dN_e + (a + b/4 + d + xe/2 - c/2)\, O_2 \rightarrow$$

$$a\, CO_2 + b/2\, H_2O + d\, SO_2 + xe\, NO + (1-x)e/2\, N_2 \qquad (4.1)$$

The content of carbon, hydrogen and sulphur is oxidized to CO_2, H_2O and SO_2, and if ash is present, it may capture a small part of the SO_2. In most stationary sources an almost complete combustion to CO_2 and H_2O is usually obtained, but smaller amounts of CO, CH_4 and NMVOC (Non Methane Volatile Organic Compounds) may be emitted, (Flagan, Seinfeld 1988; Turns 1996; Warnatz et al. 1996). In the internal combustion engine used in mobile sources a larger part of the carbon content leaves the combustion chamber as CO, CH_4 and NMVOC.

Only a fraction, x, of the nitrogen content in the fuel is converted to NO, and the major part forms N_2. The degree of conversion to NO depends on the design and operation of the combustor and can be lowered by low-NO_X burners as described later (Figure 4.12). At high combustion temperatures NO may be formed from O_2 and N_2 in the combustion air (thermal NO_X), even if the fuel is without a nitrogen content. Some trace metals like Hg, Cd and Pb may be emitted as aerosols or partly vaporised dependent on the mineral content in the fuel and the flue gas cleaning techniques. The conclusion is, that the emissions from combustion depends on the fuel and plant type (Flagan, Seinfeld 1988). To illustrate the influence of fuel type, the emission factors for different fuels and air pollutants are given in Table 4.1.

Table 4.1 Emission factors (in g/GJ for public power combustion plants of 50 to 300 MW thermal input in Denmark (CORINAIR '94 1997, CORINAIR DK 1995 1997).

Fuel type	SO_2	NO_X	NMVOC	CH_4	CO
Steam coal	695	283	1.5	1	10
Wood	25	130	48	32	160
Municipal waste	76	150	9	6	10
Straw	18	131	35	24	115
Residual oil	325	323	3	3	15
Gas oil	31	100	2	2	12
Natural gas	0.3	240	3	3	20

A large variation in the emission factor for SO_2 is seen, as the sulphur content of coal is of the order of one wt%, but is negligible in natural gas. The main part of the sulphur is oxidized to SO_2 and emitted with the flue gas, because there is no flue gas desulphurization on combustion plants below the size of 300 MW thermal input. On the other hand, the emission factor for NO_X varies only by a factor of about three. In this case the fuel type is less important, but the type of combustion plant can be determining. It is seen that the emission factor for residual oil and natural gas is almost the same, even if the hydrocarbons in natural gas have no nitrogen content, while residual oil may contain one to two wt% nitrogen. The reason is the very high flame temperature in natural gas combustion and the formation of thermal NO_X. However, the formation of NO_X can be lowered by the use of low-NO_X burners.

As the emission factors depend on fuel type, type and size of combustor and flue gas cleaning techniques, it is difficult to estimate general emission factors, even for the same fuels or the same type of combustors, and even more difficult with emission factors for production processes. In the Corinair emission inventories described in Chapter 6, local emission factors for each country have been applied.

4.3 Clean technology

The ideal way to avoid the emission of pollutants is a complete recycling of all materials in the complex ecosystem encompassing the whole world, but so far our state of technology does not allow us to do so. Further, the energy consumption for a 100% recycling of material may be very high, giving rise to high emissions from energy conversion processes, so as a consequence the total emissions may increase. However, it is possible to come a long way in reducing the emission of pollutants to the environment from production and use and disposal of products by proper planning of the production processes. Figure 4.1a shows a production process, where all steps from the processing of raw material to the disposal of the used product will give rise to emissions to the environment.

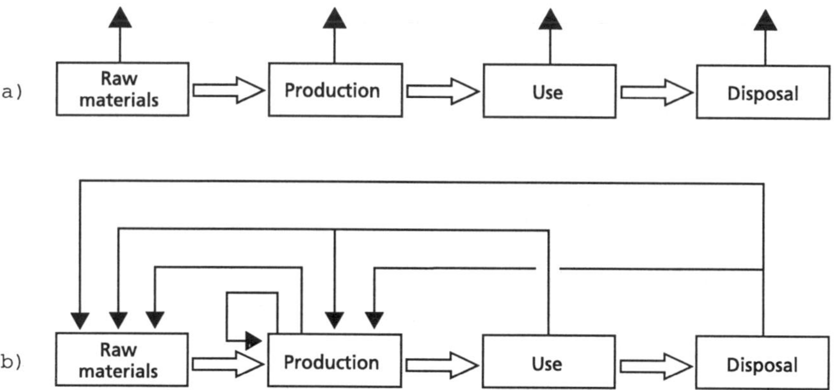

Figure 4.1 a) Traditional production process with emissions to the atmosphere.
b) Clean technology, recirculation of waste materials.

Traditionally focus has been on the production process, which in many cases is a major source of emissions. The use of new or different processes or technologies with lower emissions has been termed "clean technology". It is possible to develop technologies with very low emissions in many different ways by:

– substitution of raw materials.
– substitution of products.
– choice and optimisation of production processes.
– use, re-use and recycling of by-products.

The prevention of the formation or the reduction of the emission of pollutants will often be more efficient than gas cleaning or dispersion in the atmosphere, and the use of clean technology should always be considered, when a new production or other activity with the potential for emissions to the environment is planned. In existing processes it may be very costly and in many cases the technology for zero emission has not yet been developed - therefore gas cleaning and dispersion in the atmosphere will still be important methods to improve the air quality.

So far the use of clean technology has been focused on the production process, whereas emissions from the production of raw materials, use of the produced items and final disposal was not considered. However, there is the potential for emission of pollutants from all steps in the total life cycle of the product from the processing of raw materials to the final disposal, and the ultimate way to reduce emissions would be to recirculate all waste and close the total process completely or to use all waste and by-products in other production processes, as illustrated in Figure 4.1b (previous page). The use of an alternative raw material to reduce emissions from the production process might increase the emission from the processing of raw materials; thus the reduction of the emissions from the whole process is a much more difficult task than to introduce clean technology. The way in which the environmental impact of the complete process is addressed in a formal matter is by a so-called lifecycle assessment (LCA), a family of methods for looking at materials, services, products, processes and technologies over their entire life. This is described in section 4.4.

To illustrate the concept of clean technology a few examples from the field of energy use and conversion will be given. From the emission inventories in Chapter 6 it appears that transportation and space heating are major sources of urban air pollution. Transport is thus the main reason for the high NO_x concentrations measured in most larger cities. Individual space heating has always been a major contributor to the SO_2-concentration in cities at geographical locations where heating is necessary.

In the first half of this century the major energy resource for heating purposes was solid fuels, especially coal in larger installations and so-called smokeless fuels like coke in individual houses. Air pollution with dust, smoke and odorous compounds was a major problem in cities as described in Chapter 2. From the beginning of the sixties solid fuels were gradually substituted with different oil products: Residual oil was used in larger installations and gas oil in smaller individual installations for space heating or industrial purposes.

In one respect this was a clean technology, because the emissions of particulate matter decreased dramatically. However, with respect to emission of SO_2 it was not, since the crude oil came mainly from the Middle East, where the sulphur content in the oil is high. As a consequence the ambient air concentration of SO_2 did not fall, but rather increase, because at the same time the energy consumption in Europe was increasing. However, after the energy crisis in 1972-73 the energy consumption in many European countries levelled off, and simultaneously the interest for different clean technologies started. Over the last 25 years many of these clean technologies have been implemented to a different degree in the many European cities, depending on local circumstances and needs. The description of the clean technology options for heating will be given under the headings mentioned above starting with the choice of raw material.

4.3.1 SUBSTITUTION OF RAW MATERIAL

During the combustion of liquid or gaseous fossil fuels their sulphur content is converted into SO_2 and emitted with the flue gas. One way of reducing the emission of SO_2 in Europe has therefore been to use oil and gas with a low sulphur content. Over the years oils with a relatively high sulphur content have been substituted with oils with a lower sulphur content. This change has been forced by legislation.

There are two ways of obtaining oil with a low sulphur content. One way is a more effective desulphurisation at the oil refinery. In modern refineries the sulphur compounds removed are converted to sulphur, which is then sold and used as raw material for production of sulphuric acid. The other way is to use a crude oil with a lower sulphur content. Fortunately the North Sea oil is relatively low in sulphur, and therefore the last option has been possible in many European countries the last 15 years.

Substitution of oil with natural gas reduces the emission of SO_2 from the heating system to zero. The natural gas may have a content of H_2S when it comes from the gas field, but the H_2S is removed at a gas treatment plant, converted to sulphur, and used in production of sulphuric acid. The oil may of course be substituted with other energy sources like straw and wood, but these means of heating are more common in the countryside. In Table 4.1 it was shown that the emission factors for natural gas, straw and wood are very low, and the substitution of coal and oil with natural gas in cities, has been an important reason for the improved air quality with respect to SO_2.

4.3.2 SUBSTITUTION OF PRODUCTS

The fuel used for space heating can partly be substituted by better insulation of buildings. In Europe the interest in insulation increased dramatically after the energy crisis in 1972-73. Old houses were insulated, and standards for insulation of new houses were enforced during the next years. The purpose was to reduce energy consumption and in many countries to reduce the dependence on oil import. There was a very short pay back time on insulation of existing houses because of the high energy prices, and as a result insulation was carried out to a great extent and the energy consumption for heating was substantially reduced. As a side effect the emissions of SO_2, NO_X and other air pollutants were reduced.

4.3.3 CHOICE AND OPTIMIZATION OF PRODUCTION PROCESSES

In many cities district heating is used for space heating. Hot water is produced in a central unit and distributed to the individual houses. The use of district heating usually will improve the air quality locally, because the emission is moved from many small and low chimneys to one large and high chimney, which may be located outside the city. But the benefit is even larger, if the hot water is produced by cogeneration or combined heat and power production.

The most important method of electricity production in Europe is coal combustion, even if some countries get all or some of their electricity from hydropower, nuclear power or gas and oil combustion. Figure 4.2 shows a modern coal fired power plant located at the sea or at a river. Coal is transported to the plant, crushed to a fine powder and blown into the combustion chamber with air. The heat of combustion is transferred to water flowing in tubes along the walls of the combustion chamber. The water evaporates and steam at high pressure and high temperature is formed. The steam expands in a steam turbine making it rotate and the generator produces electricity. After expansion the steam is condensed with cooling water from the sea or from a cooling tower and the water is pumped back to the combustion chamber. In a typical conventional power plant about 50% of the energy content of the fuel is lost with the cooling water, about 40% is converted to electricity and the rest is lost as heat, mostly with the flue gas.

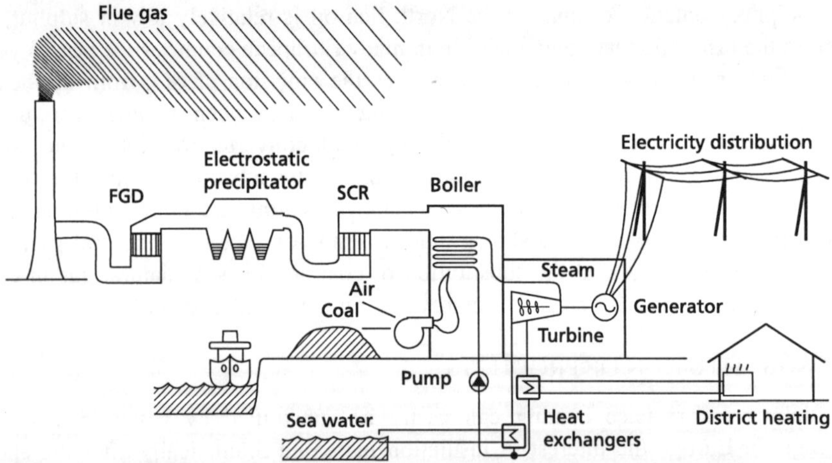

Figure 4.2 Schematic presentation of a modern power plant for combined heat and power production.

If the steam is taken from the turbine before it is completely expanded, its temperature is high enough to heat water for district heating. This is what is called co-generation or combined heat and power production. The Carnot cycle efficiency of the process decreases a little and only about 35% of the energy content of the fuel will be converted to electricity. But 55% will be converted to district heating, and nothing will be lost with cooling water. As a result there is a large saving in fuel consumption compared to production of electricity and hot water in separate plants.

As shown in Figure 4.2 the flue gas from the combustion chamber is cleaned for NO_x, fly ash and SO_2 before going to the chimney. These flue gas cleaning processes will be described in the last section of this chapter. District heating with combined heat and power production results in a major improvement of the air quality. The fuel consumption is much lower than for individual heating or conventional district heating, the flue gas is cleaned before entering the atmosphere, and the chimney is very high. Combined heat and power production in other types of plants is also possible, and many municipal waste combustors and gas fired plants with combined heat and power production are being built.

The use of clean technology options described above and the general trend towards building higher stacks have been the major reasons for the very substantial improvement in the air quality with regard to SO_2, which has been observed in many European cities over the last 25 years. Details about the air quality are given in Chapter 25.

4.4 Life cycle assessment (LCA)

LCA may be defined as follows: "The life-cycle assessment is an objective process to evaluate the environmental burdens associated with a product, process, or activity by identifying and quantifying energy and materials usage and environmental releases, to assess the impact of those energy and materials uses and releases on the environment, and to evaluate and implement opportunities to effect environmental improvements. The assessment includes the entire life cycle of the product, process or activity, encompassing extracting and processing raw materials; manufacturing, transportation, and distribution; use, re-use, maintenance; recycling; and final disposal." (Graedel, Allenby 1995). Using LCA it is possible to ensure that a proposed clean technology will not increase emissions from another step in the life cycle and eventually increase the total emissions. It is very useful for assessment of the environmental impact when comparing different technologies, production processes and products (Wenzel et al. 1997).

Some perspective on LCA is provided by an example comparing the use of fuel to provide power for gasoline and electric automobiles. Providing gasoline involves crude oil production, transporting crude oil, refining the oil, transporting and delivering gasoline and use of the gasoline in the vehicle. For an electricity powered automobile, the energy comes mostly from fossil fuel combustion at large stationary power generation facilities, but in some countries electricity is produced mainly from hydropower and/or nuclear power. In this way LCA for vehicles is including emissions from stationary sources.

The LCA inventory analysis considers the emissions from each type of operation and compares the total emission from the whole process with criteria relevant for the specific emissions. It is important to note how much this comparison rests on the way the scope of the analysis is defined. Although electric vehicles are sometimes regarded to be without emission, this is not true from an LCA point of view. Their operation requires electricity from pollution generating power plants. Thus, an important consequence of the comparison is that most emissions attributable to the gasoline powered vehicle occur where it is being operated, often in the cities, while most emissions attributable to the electric vehicle occur at power plants which may be placed away from urban areas. Hence for pollutants with a short lifetime the population exposure is greater from the gasoline powered vehicles. Moreover control technologies are more efficient for a few large point sources than for a large number of small sources.

The scope of the analysis in this case determines the conclusion: If emissions from the actual vehicle is all that is considered, the electric vehicle obviously is superior. If emission of greenhouse gases are considered, a different conclusion might be the result, although recent estimates indicate that the CO_2 emission will be lowered slightly, even if the electricity is produced on coal fired power stations (Horstman, Jørgensen 1997). Another complication is that the state of flue gas cleaning on the power stations is different from country to country and some countries use mostly hydropower, and therefore the analysis is only valid for the specific local conditions it is made for.

This example was actually only part of a LCA. A comprehensive LCA would consider - in addition the emissions and environmental impacts of the vehicles themselves - their materials, their manufacture and their eventual recycling. And also the energy consumption and the consumption of resources goes into the LCA.

To perform LCA is not easy, and the result will depend on many factors. In recent years a large effort has been put into the development of standard methods, and now consensus about many issues seems to be emerging (ISO 1997). LCA is an excellent method in product development and choices between different new clean technologies or products taking the total environmental impact into account.

4.5 Control of primary particulates

Even if clean technology may be the best way to improve air quality in the long run, control of air pollution by gas cleaning still is a very important method to reduce emissions to the atmosphere. The trend is that emission limits will become more stringent, and therefore gas cleaning may become even more important than today. The following description of control of air pollution from stationary sources will fall in two parts, the first one dealing with particulate removal and the second one treating gaseous compounds. At last some examples of important gas cleaning methods in power production will be discussed. In the text the term gas cleaning efficiency will be used. The definition of this term is straightforward:

$$\eta = \text{efficiency} = \frac{\text{mass of pollutant removed by the cleaning device}}{\text{mass of pollutant entering the cleaning device}} 100\ \% \qquad (4.2)$$

As it will be discussed in Chapter 9, a large part of the fine particulates in the atmosphere are secondary particles. Nonetheless, the control of primary particles is a major part of air pollution control. Although primary particles are generally larger than secondary particles, many primary particles are small enough to be respirable and are thus of health concern. The most important types of control devices are gravity settlers, cyclones, scrubbers, bag filters and electrostatic precipitators (Wark, Warner 1981; Nevers 1995).

The *terminal gravitational settling velocity* is an important characteristic of particles suspended in gas or liquid. At steady state, i.e. zero acceleration, three forces act on the suspended particle: the gravitational force, the drag force and the buoyancy. From a force balance the terminal velocity can be calculated. In Stokes' flow regime, i.e. for particles in air in the size range 5 to 50 μm, the terminal velocity can be written as given in Equation (4.3):

$$V_t = g d_p^2 (\rho_p - \rho_g)/(18\mu) \quad (4.3)$$

g: gravity, d_p: particle size, ρ_p: particle density, ρ_g: gas density, μ: gas viscosity. The equation can be used as an approximation in the particle range from 1 to 100 μm (Nevers 1995). This size range is very important for control of particulates, because the majority of the particles to be removed will be in that interval. Table 4.2 shows the terminal velocity of spherical particles in air at ambient conditions.

Table 4.2 Approximate terminal velocities of spherical particles in air at ambient conditions (298 K and 101.3 kPa). The specific density of the particles is 2000 kg/m^3.

Particle size, μm	0.01	0.1	1.0	10	100	1000
Terminal velocity, m/s	1.3E-7	1.8E-6	6E-5	6E-3	0.5	7

4.5.1 GRAVITY SETTLERS AND CYCLONES

A *gravity settler* is simply a long chamber through which the contaminated gas passes slowly, allowing time for the particles to settle by gravity. It is an old unsophisticated device, but it is simple to construct, requires little maintenance, and has some use in industries treating very dirty gases, e.g. some smelters and metallurgical processes. In Stokes' flow regime the terminal velocity depends on the square of the particle diameter, and the terminal velocity is low for small particles. As a consequence fine particles will pass the gravity settler and only large particles with large settling velocity will be captured. The collection efficiency is low for particles with sizes below 40 to 50 μm, and respirable dust is not removed.

A *cyclone* is a centrifugal separator, where the centrifugal force is used to move the particles, and because the centrifugal force is more powerful than the gravity force, the particle velocity is much larger than the terminal velocity, and smaller particles can be removed from the contaminated gas. A cyclone consists of a vertical cylindrical body and a conical bottom. The gas enters through a rectangular inlet arranged tangentially to the cylindrical part of the cyclone, so that the entering gas flows around the circumference of the cylindrical body, not radially inwards (Figure 4.3, next page).

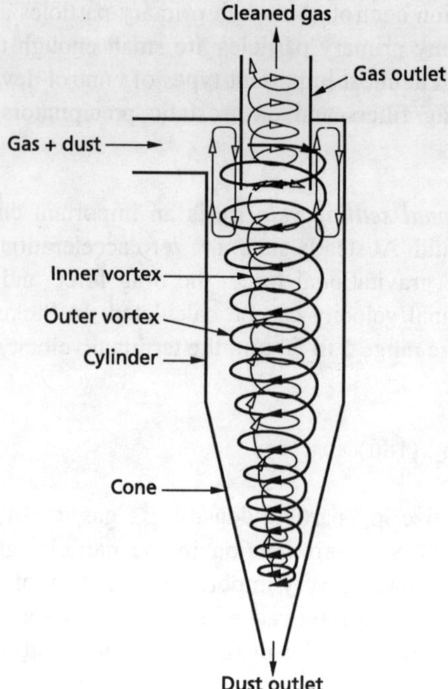

Figure 4.3 Cyclone; arrows indicate the gas flow pattern. Adapted from Wark and Warner (1981).

The gas spirals around the outer part of the cylindrical body with a downward component, then turns and spirals upward, leaving through the central outlet at the top of the cyclone. During the outer spiral of the gas the particles are driven to the wall by centrifugal forces. They collect, attach to each other, and form larger agglomerates that slide down the wall and collect in the dust hopper in the bottom of the conical part. The particles are removed from the cyclone through a rotary valve to avoid any gas leakage in the bottom. Cyclones are useful for gas cleaning, when the particles are relatively large. They are low cost and cheap to maintain, but not efficient for fine particles. As a rule of thumb, they have a good efficiency for particle sizes above 10 μm. A better efficiency for small particles is obtained with many small cyclones in parallel, a so-called multitube cyclone. In this case the centrifugal force is larger because the cyclone diameter is smaller, and the travelling distance for the particles to the wall is short.

4.5.2 SCRUBBERS

In a *wet collector or scrubber* a liquid, usually water, is used to capture particulates. The principle of wet scrubbing is:

- Formation of liquid droplets with a size of 100 to 1000 μm.
- Contact between gas and liquid and transfer of particulates from the gas phase to the liquid phase.
- Separation of droplets from the gas phase.

Fine particulates, both liquid and solid, ranging from 0.1 to 100 μm, can be effectively removed from a gas stream by scrubbers. The actual mechanism of particulate removal may be impaction, direct interception or diffusion as illustrated in Figure 4.4 (Wark, Warner 1981). Particles are carried along at approximately the same velocity as the gas. Owing to its extreme lightness the gas moves in streamlines around any object in its path. However, particles with a much larger mass resist changes in motion. The larger the particle, the less the tendency to change direction.

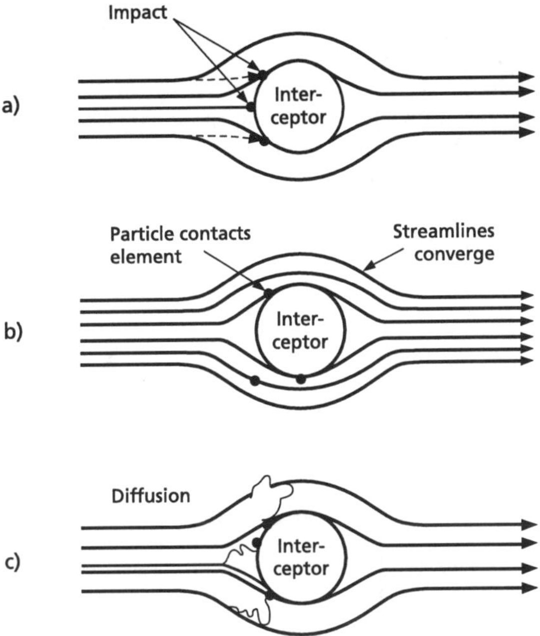

Figure 4.4 Mechanisms of particulate removal. a) Inertial impaction; b) Interception; c) Diffusion. Adapted from Wark and Warner (1981).

Inertial impaction (Figure 4.4a) is associated with the relatively larger particles which travel on collision course with the drops. Inertia keeps them on the path, even though the gas and the smaller particles tend to diverge and pass around the drop or another interceptor. In direct interception (Figure 4.4b), some of the smaller particles, even though they tend to follow the streamlines, may contact the interceptor at the point of closest approach. This occurs because the streamlines tend to converge as the gas passes around the element, and the particle size is greater then the distance between the streamline and the element. Finally, in collection by diffusion (Figure 4.4c) very small particles (usually less than 1 μm) impinge upon the drop as a result of random molecular (Brownian) motion or diffusion.

Impaction is the most important removal mechanism in scrubbers. Consequently a large relative velocity between the gas and the droplets is favourable and particulates with a large terminal velocity are easy to separate. The size and concentration of droplets is important for the removal efficiency. Droplet sizes between 100 and 1000 μm are optimal for particulate collection. If the droplets are smaller than 100 μm, they tend to follow the gas and the relative velocity between particulate and droplets approaches zero. If the droplets are larger than about 1000 μm their concentration will be small, and the chance of collision between a particle and a droplet becomes minor. When the particulates have been captured by the droplets, the problem has changed from removing fine particulates from a gas stream to separation of much larger droplets, an easier task. Many different arrangements of basic equipment are available, but only spray scrubbers will be described in the text.

The *spray tower* or *spray scrubber* is one of the simplest devices for wet collection of particulate. It is an empty vessel, where liquid droplets are produced by suitable nozzles located across the gas flow passage. The polluted gas flows countercurrently, cocurrently or in cross-flow to the liquid and the particles collide with the droplets which will in turn settle by gravity to the bottom of the scrubber. The most common construction is a circular or rectangular tower with at least two levels of spray nozzles over the cross section (Figure 4.5).

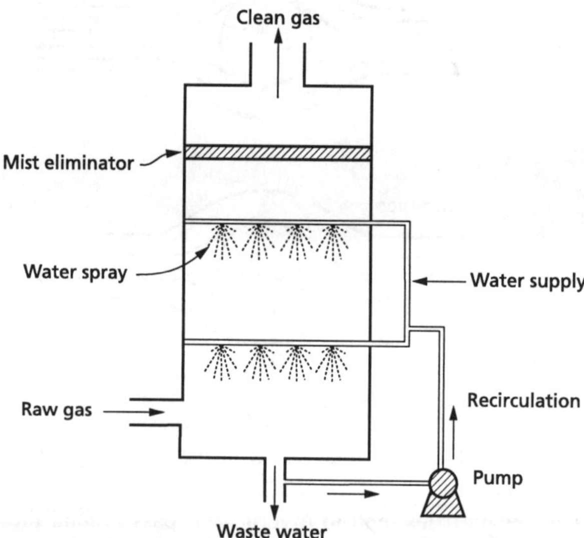

Figure 4.5 Schematic presentation of a spray of scrubber for particle removal. The height can vary between 2 and 50 m.

The gas enters close to the bottom, and the gas velocity is low enough to allow the droplets to settle. For a 500 μm droplet the terminal velocity is about 2 m/s and the gas velocity is usually below 2 to 3 m/s. To avoid carry-over of small droplets there is a mist eliminator in the top of the scrubber. The liquid is recirculated from the bottom of the scrubber to the spray nozzles to reduce water consumption and the amount of waste water. However, the gas becomes saturated upon contact with the liquid and fresh water must be added because of this evaporation. It is also necessary to separate the solids from the liquid, and there may be waste water if water soluble compounds are present in the gas.

The collection efficiency of a spray scrubber is good for particle sizes above 10 µm but other types of wet scrubbers have better efficiencies. The pressure drop of a spray scrubber is low and the more efficient scrubbers have a higher pressure drop and higher operating costs (Table 4.3, page 50). The scrubber is a cheap and simple device and it has widespread use in industry. One important advantage compared to other methods is, that gaseous components may be removed at the same time.

4.5.3 BAG FILTERS

Filtration is one of the oldest and most widely used methods of separating particulates from a gas. A filter generally is any porous structure composed of granular or fibrous material or a membrane which tends to retain the particulate as the carrier gas passes through the voids or the fine holes of the filter material. It is exactly the same principle used in a conventional household vacuum cleaner. The filter material can be of any material compatible with the gas and particulate, and especially the temperature of the gas is important for the choice of filter material. Textile, glass, polymers, metals and ceramics may be used for filter material, depending on process conditions. The cheapest materials are cotton and wool, but they are not applicable at temperatures above 80 to 90°C, and they are not resistant to acids and bases. Many polymers can stand 200°C, but at higher temperatures metals or ceramics may be necessary (Theodore, Buonicore 1994).

Fabric filters are usually formed into cylindrical tubes and hung in rows to provide large surface areas for gas passage. The filter bags may be 100 to 350 mm in diameter and 10 to 15 m long, but also very small filters are used frequently (Figure 4.6).

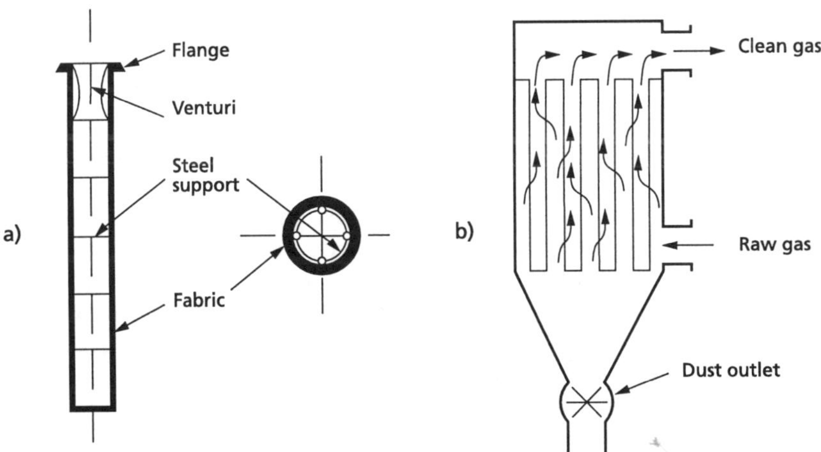

Figure 4.6 Bag filter.
a) A single bag with steel support. b) Chamber with several parallel rows of bags.

The mechanisms for separation of the particles from the gas using a fibrous filter material are impaction, interception and diffusion as illustrated in Figure 4.4 (page 45), but electrostatic effects may play a role also. The gas flow is perpendicular to the fibres and Figure 4.4 shows the cross section of a single fibre. As filtration goes on, a dust cake builds up on the surface of the filter and improves the filtration efficiency. However, the pressure drop increases, and at some point the filter cake must be removed. This must be done on-line and automatically; the most common practice for this operation is illustrated in Figure 4.6 (previous page). The filter bags are supported by a metal frame and the gas to be cleaned passes from the outside of the bags to the clean inner side. The filter cake builds up on the outside and is removed by a short pulse of compressed air from the top of each filter bag. The filter material expands a little during pulsing and the dust cake breaks off, falls to the bottom and is removed automatically. One bag or one row of bags may be cleaned at a time depending on filter size and dust loading.

Fabric filters have a very good efficiency also for small particles. More than 99% of 0.5 μm particles are removed, and even particles as small as 0.01 μm may be removed to a great extent. This is very important from an air pollution point of view, because respirable dust is removed efficiently. Fabric filters are versatile and can be used for many different purposes, ranging from very small to very large volumes of gas, and they have widespread use in industrial production processes.

4.5.4 ELECTROSTATIC PRECIPITATORS

Particle collection by electrostatic precipitation is based on movement of electrically charged particles in an electrical field. The electric force in an electrostatic precipitator is very strong compared to the centrifugal force of a cyclone and the collection efficiency for small particles is much higher. An electrostatic precipitator consists of a large number of emission and collecting electrodes. Figure 4.7 shows schematically the wires used for emission electrodes and the plates used for collection electrodes.

The plates are grounded, and the wires are charged with 20-70 kV below ground potential. The gas flows horizontally between the plates, and the gas molecules are ionized by the corona from the emission electrodes. The next step is that the ions charge the particles, and the negatively charged particles will be moved by the electric force towards the grounded collecting electrodes. A dust layer builds up on the collecting electrodes and adhesive, cohesive and electrical forces prevent reentrainment of the particles in the gas stream. One property of the dust layer which is extremely important in precipitator operation is the dust electrical resistivity. When the dust resistivity is too low there is a rapid movement of charge from the deposited dust to the collector plate. Thus insufficient electrostatic forces remains on the collected dust particles to hold them together. Reentrainment back into the gas stream frequently results and the collection efficiency is reduced.

On the other end of the scale, a high resistivity dust layer acts as an insulator and a part of the voltage drop occurs over the dust layer and only part of the total corona power is available to ionize and drive the charged particles to the collection electrode. As a rule of thumb the resistivity should be within the limits of 10^4 to 10^{10} ohm·cm. The

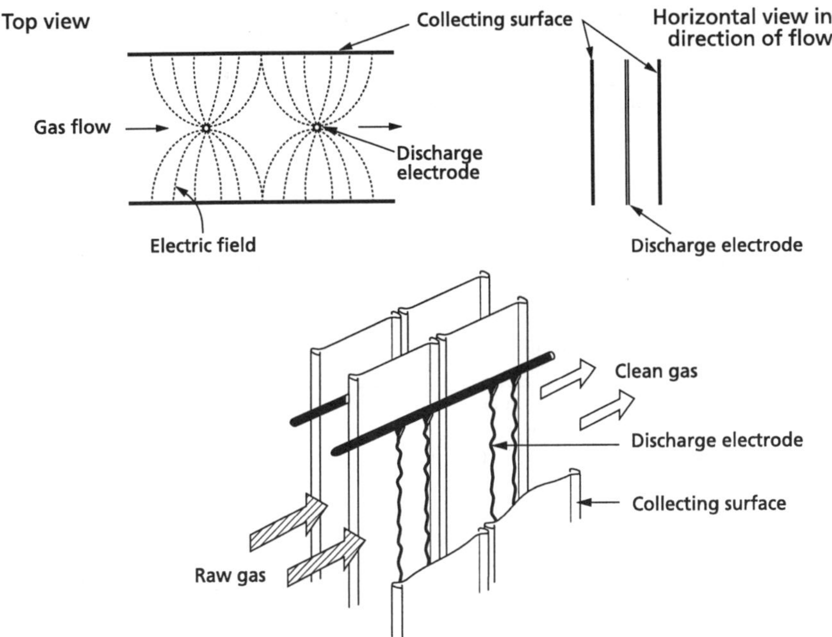

Figure 4.7 The principles of electrostatic precipitation. Adapted partly from Wark and Warner (1981).

dust layer is removed regularly with a rapping system placed near the lower end of the plates.

Electrostatic precipitators are used mostly at power plants, waste incinerators, cement factories and other large scale processes, because the investment is high. A typical size of a precipitator at a power plant could be height·width·length = 16·30·20 m^3. The collection efficiency is very good, for new plants it is usually more than 99.5%, and also respirable dust with a size below 10 μm is removed effectively.

Table 4.3 (next page) gives a comparison of the different methods of particulate air pollution control.

4.6 Control of gases, vapours and odorous compounds

Many different methods are used for removing gaseous air pollutants, and the choice of method depends very much on the type of compound to be removed. The main processes are: oxidation, absorption, adsorption and biological methods (Johnsson 1991). Some of the important properties of the pollutants are: are they combustible, are they water soluble, are they acid or basic, are they present at high or low concentration and last but not least, are they valuable so that regeneration for re-use in the production process is a must?

Table 4.3 Comparison of methods to control particulate air pollutants.

	Gravity settlers	Cyclones	Scrubbers	Bag filters	Electrostatic precipitators
Particle sizes with a high removal efficiency	> 50-100 μm	> 5-10 μm	> 2-5 μm (> 0.3 μm for venturi scrubbers)	> 0.1 μm	> 0.3 μm
Temperature limit	Depends on the material of construction (> 1000°C is possible)	Depends on the material of construction (> 1000°C is possible)	Water consumption increases with temperature due to evaporation	200-250°C (higher with fibres of metals or ceramic material)	400-500°C
Influence of water content	Condensation must be avoided	Condensation must be avoided	No influence	Condensation must be avoided	The efficiency depends on the water content
Pressure drop, Pa	5-10	1000-2000	500-2000 (Venturi scrubbers ~10000)	1000-1500	< 250

4.6.1 OXIDATION

Combustible air pollutants like hydrocarbons, solvents and odorous substances may be oxidized to H_2O and CO_2, as illustrated in Equation 4.1 (page 36). If they contain S, Cl, P or N secondary air pollutants like SO_2, HCl and NO_X may be formed, and thermal NO_X may be formed from O_2 and N_2 in the combustion air, if the flame temperature is above 1500°C. If the concentration of combustibles is above the lower explosion limit, the gas will burn if ignited, and in this case we have a fuel and not an air pollution problem. For safety reasons, the concentration must be below the lower explosion limit, if an oxidation process is used for control of air pollutants, and usually the concentration is well below this limit. As a consequence it is necessary to heat the gas to the temperature necessary for oxidation to take place.

The oxidation can be either thermal at a temperature of 850 to 1000°C or catalytic at a temperature of 250 to 400°C. The catalytic material can be noble metals like Pt or Pd, or it can be oxides of the transition metals like CuO, Cr_2O_3, or NiO. In both cases a heat exchanger is used to increase the temperature of the gas to save fuel. Figure 4.8 shows a catalytic process with exhaust fan, heat exchanger, start-up burner fired with gas or oil and catalyst layer.

The inlet gas temperature is typically 100-120°C, and 99% or more of the fly ash (dust) has been removed in the electrostatic precipitator upstream the FGD plant. The flue gas is cooled to 70-80°C in a heat exchanger counter currently with the cleaned flue gas and enters in the bottom (countercurrent) or in the top (cocurrent) of the absorption tower. The tower may be with or without packing. In the process in Figure 4.11 (page 56) we have a countercurrent absorption tower without packing, a spray tower. The temperature of the gas drops to about 50°C due to evaporation of water from the scrubbing liquid, and simultaneously a number of important steps in the absorption process takes place:

- SO_2 from the gas phase is absorbed in the scrubbing liquid.
- HSO_3^- is oxidized to SO_4^{--}.
- $CaCO_3$ is dissolved.
- Crystals of $CaSO_4,2H_2O$ (gypsum) are formed.

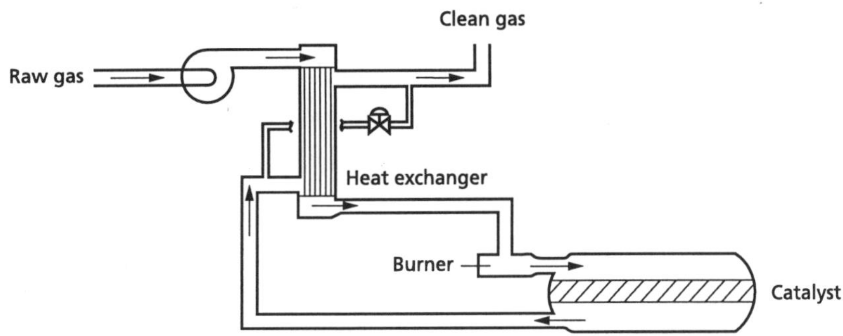

Figure 4.8 Catalytic oxidation process with heat exchanger for raw gas heating.

If the concentration of air pollutants is high enough to give an adequate adiabatic temperature rise, no additional fuel is needed, but if the concentration is low, fuel must be burned to compensate heat losses, or regenerative heat exchange must be used. Oxidation is an attractive process because a very high efficiency can be obtained, typically above 99%. The process is not sensitive to changes in gas flow or pollutant concentration, and the performance is not dependent on the type of pollutant, if the temperature and the residence time is high enough. Heat recovery may be possible, but regeneration of valuable compounds is out of the question and secondary pollutants may be formed as mentioned above. Thermal oxidation or incineration has been used in the chemical industry for many years, and now catalytic oxidation has a wide range of applications.

4.6.2 ABSORPTION

The source control of gaseous pollutants by absorption or wet scrubbing involves bringing the gas into contact with the scrubbing liquid and subsequently separating the cleaned gas from the contaminated liquid. The method has widespread use because it is relatively cheap and very robust. The most common scrubbing liquids are water or water solutions. The method is especially well suited for very soluble gases like NH_3, HCl and acetone, but it is also useful for slightly soluble gases, which can react with chemicals in the scrubbing liquid. A good solubility or a fast reaction in the liquid is important to reduce the amount of scrubbing liquid and thereby reducing the amount of waste water, and to get a fast absorption and thereby reducing the size of the absorption tower.

The most common chemicals are acids, bases and oxidants. An important example is the removal of SO_2 from flue gas by scrubbing with a solution or slurry of $CaCO_3$ in water (Takeshita, Soud 1993). Oxidizing compounds like ozone and NaOCl are used for removal of odorous pollutants in the food and fish meal industry. There are many types of scrubbers and they show good removal efficiency for gases and also for particles as described above.

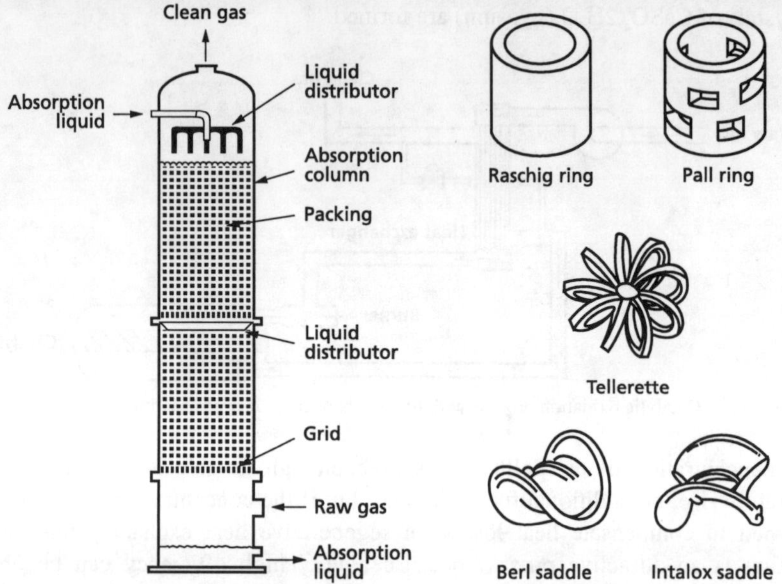

Figure 4.9 To the left: Counterflow absorption tower with packing. To the right: Different types of packing. Adapted from Wark and Warner (1981).

The scrubber may be without a packing as shown in Figure 4.5 (page 46) or with a packing as shown in Figure 4.9. The purpose of the packing is to give a large interfacial area between the gas and the liquid and to ensure an even distribution of liquid over the cross section, to improve the cleaning efficiency. However, the installation cost and the pressure drop is higher than for a spray scrubber without packing, and there is a risk of contamination with dust at high dust loading.

4.6.3 ADSORPTION

The principle of an adsorption process is to utilise the ability of solid surfaces to remove one or more gases selectively from a mixture of many gaseous compounds. Adsorption is especially well suited to control air pollution in the cases: 1) the pollutant is not combustible or forms a secondary pollutant upon combustion, 2) the pollutant is valuable and must be regenerated, or 3) the pollution concentration is very low. There are many types of adsorbents: Activated carbon, zeolites, polymers and silica gel, but in air pollution control activated carbon is the most common. It is a porous material with an extremely high specific surface in small pores. The gas passes through a layer of activated carbon particles, typically with a diameter of 2-5 mm. In an industrial plant the activated carbon is placed in a vessel as seen in Figure 4.10a.

The adsorption can go on for several hours, but after some time the activated carbon becomes saturated and breakthrough of the pollutants will occur. Now, an important feature of the adsorption process is its reversibility. Adsorption is most effective at high concentration and low temperature. The relationship between the concentration in the gas phase and the solid phase at a given temperature is named the adsorption isotherm.

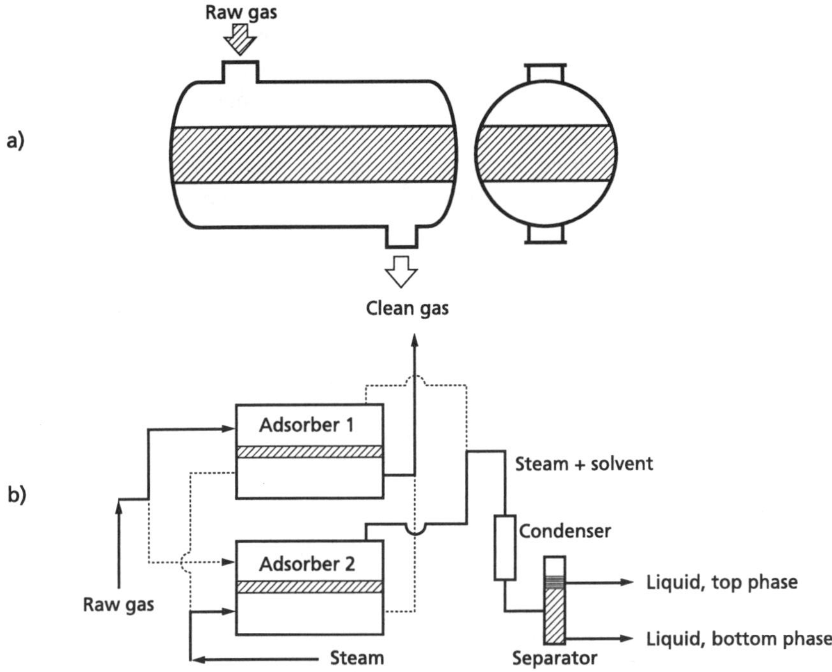

Figure 4.10 Adsorption process. a) Vessel with fixed bed of adsorbent. b) Parallel adsorbers with regeneration equipment. Adapted partly from Wark and Warner (1981).

In general the concentration in the solid phase increases with molecular weight, increasing concentration of the pollutant and decreasing temperature. The adsorption is most effective for nonpolar (hydrophobic) organic compounds like hydrocarbons, and less effective for polar (hydrophilic) compounds like organic acids and lower alcohols.

In practice adsorption is carried out at room temperature and when the activated carbon becomes saturated, desorption or regeneration is done with steam at 100 to 150°C. In Figure 4.10b two parallel adsorbers are shown, one is being regenerated while the other is used as adsorber. The mixture of organic vapours and steam is cooled, condensation occurs, and if the organic pollutant is not soluble in water, the two phases are easy to separate and the organic phase may be reused directly. If the organic compound is water soluble a separation process like distillation is necessary. As a consequence adsorption by activated carbon is a very efficient process for nonpolar compounds, because they are removed very effectively and the are easy to separate from the condensed steam for direct reuse in the process.

4.6.4 BIOLOGICAL METHODS

The principle of biological air pollution control is the same as for biological waste water treatment, i.e. the air pollutants are removed by biological decomposition by micro-organisms. The primary products are CO_2 and H_2O but there is also a production of biomass. The method is especially well suited to solve problems with odorous compounds at relatively low concentration.

There are two different principles of biological air pollution control: filters and scrubbers. In biological filters the micro-organisms are fixed to a porous organic material. The gas flows upward through a layer e.g. peat moss mixed with twigs and small branches to give stability and porosity. The odorous substances are adsorbed to the organic material and used as feedstock by the micro-organisms. The micro-organism adapt to the pollutants after some time, but in some cases special micro-organisms have been grown in the laboratory to ensure the filter performance. The biological filter has a high efficiency, a low pressure drop and there are no expenses to chemicals, but it takes a lot of space and the filter material may need a change after a few years.

The biological scrubbers are of the same type as the scrubbers described earlier, but in this case micro-organisms are suspended in the scrubbing liquid or fixed to the packing in the scrubber. The analogy to biological waste water treatment is obvious, but in many cases N and P has to be added to the scrubbing liquid as nutrients for the micro-organisms.

Table 4.4 gives a comparison of the different methods for control of gaseous air pollutants.

4.7 Flue gas cleaning

One way of lowering the impact from heat and power production on the air pollution in cities has been to centralize the production in large power stations with very tall stacks, as described in Section 4.3. But already in the seventies it was found that a local air pollution problem was solved at the expense of a regional air pollution problem, the acid rain. As described in Section 4.4, it is important to look at the whole life cycle, to make a life cycle assessment of the whole process, and take all environmental impacts into account. The most important local and regional air pollutants from heat and power production are SO_2, NO_X and dust, and as seen in the emission inventories in Chapter 6, heat and power production are among the major sources of emission for these components.

There is a long tradition of controlling dust emissions from power stations with electrostatic precipitators, to avoid local problems, and at the same time regional air pollution problems due to dust have been reduced. Since the beginning of the eighties flue gas desulphurisation and NO_X abatement has been introduced on modern European power plants to improve the local air quality and to reduce transboundary air pollutant transport. Figure 4.2 (page 40) showed a sketch of a modern power plant with flue gas cleaning. The combined heat and power production was explained in Section 4.3, and now the principles of the flue gas cleaning processes to remove SO_2 and NO_X will be described (Soud 1995).

Table 4.4 Comparison of methods to control gaseous air pollutants.

	Oxidation		Absorption		Biological methods	
	Thermal	Catalytic	(Scrubbers)	Adsorption	Scrubbers	Filters
Components removed with high efficiency	VOC and other combustibles (e.g. hydrocarbons, solvents, odorous substances)	VOC and other combustibles (e.g. hydrocarbons, solvents, odorous substances)	Water soluble compounds (e.g. SO_2, HCl, NH_3, organic acids, mercaptans, amines	VOC, hydrocarbons, nonpolar solvents and odorous substances	Water soluble and biodegradable compounds, odorous substances	Biodegradable compounds, odorous substances
Dust content	A low dust load is tolerable	A low dust load is tolerable	No harm	Dust must be removed before adsorption	No harm	Dust must be removed before the filter
Water content	Increases the cost of operation	Increases the cost of operation	No harm	A maximum of 70% relative humidity is tolerable	No harm	Beneficial
High temperature	Beneficial	Beneficial	May lower the efficiency	Cooling is necessary	Cooling is necessary	Cooling is necessary
Advantages	Potential for heat recovery	Potential for heat recovery		Possibilities for reuse of adsorbed compounds		
Disadvantages	High fuel costs in some cases	Risk of catalyst poisoning in special cases	Waste water and/or sludge		Waste water and/or sludge	Takes up a large area

4.7.1 FLUE GAS DESULPHURISATION (FGD)

There are many commercial FGD processes, but wet scrubbers have a share of 80% of the total FGD capacity world-wide. Other processes are spray dry scrubbers (semi-dry systems) sorbent injection processes, regenerable processes and combined SO_2/NO_X removal processes (Takeshita, Soud 1993). A short description of the wet scrubber FGD process will be given. The wet scrubber or absorption tower for removal of gaseous pollutants has been shown earlier, but there are some special features of FGD scrubbers. Sulphur dioxide has a low solubility in water and a reaction with a basic reagent is necessary to obtain a high degree of desulphurisation without the need of an absorption tower of enormous size and an excessive water consumption and waste water production.

The most common reagent is limestone with a high content of $CaCO_3$, often above 95wt%, because it is cheap and because it is possible to produce saleable gypsum as a by-product. The overall chemical reaction can be written as:

$$SO_2(g) + O_2(g) + 2CaCO_3(s) \rightarrow CaSO_4{,}2H_2O(s)(gypsum) + CO_2(g) \qquad (4.4)$$

In practice there are many reactions with influence on the performance, and the optimal process conditions are difficult to assess. The objectives of the process are to:
- obtain a high degree of desulphurisation,
- have a low consumption of $CaCO_3$ and
- produce saleable gypsum.

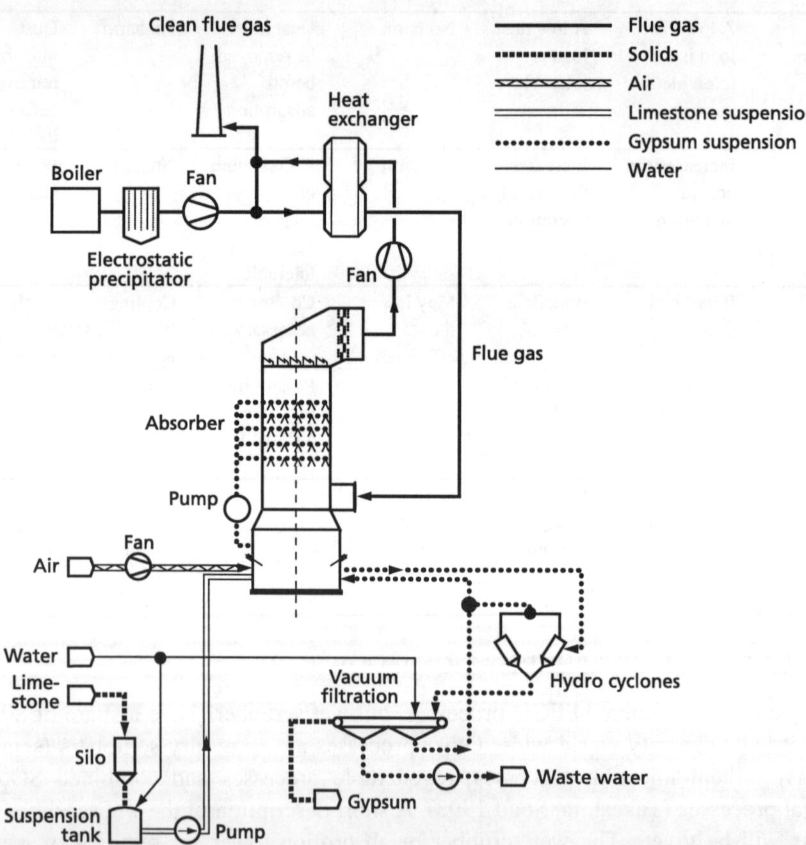

Figure 4.11 A process flow sheet for a limestone based FGD plant on a coal fired boiler. Adapted from ELKRAFT (1990).

The inlet gas temperature is typically 100-120°C, and 99% or more of the fly ash (dust) has been removed in the electrostatic precipitator upstream the FGD plant. The flue gas is cooled to 70-80°C in a heat exchanger counter currently with the cleaned flue gas and enters in the bottom (countercurrent) or in the top (cocurrent) of the absorption tower. The tower may be with or without packing. In the process in Figure 4.11 we have a countercurrent absorption tower without packing, a spray tower. The temperature of the gas drops to about 50°C due to evaporation of water from the scrubbing liquid, and simultaneously a number of important steps in the absorption process takes place:

- SO_2 from the gas phase is absorbed in the scrubbing liquid.
- HSO_3^- is oxidized to SO_4^{--}.
- $CaCO_3$ is dissolved.
- Crystals of $CaSO_4,2H_2O$ (gypsum) are formed.

Limestone is added to the bottom of the absorption tower and is partly dissolved in the scrubbing liquid, which is pumped to the top of scrubber and distributed over the cross section using spray nozzles. The liquid recirculates over the tower and part of the limestone dissolves during the contact between gas and liquid in the tower. Air is blown to the bottom of the absorber to ensure complete oxidation of sulphite to sulphate. A side stream is taken out for removal of gypsum using hydro cyclones and vacuum filters. All the sulphur removed from the flue gas leaves the process as gypsum of high quality. The gypsum is washed with water, which is partly recirculated to the process and partly taken out for waste water treatment where heavy metals - mainly from the small amount of fly ash - are removed. The cleaned flue gas is heated to 70 to 80°C by heat exchange with the incoming flue gas to ensure a proper plume rise and good dispersion of the remaining air pollutants in the atmosphere. The efficiency of flue gas desulphurisation with scrubbers is typically about 90% removal of SO_2, but in newer plants 95% removal efficiency can be obtained. As a side effect there is a good efficiency for capturing the particles remaining after the electrostatic precipitator and an almost complete removal of HCl and HF. Flue gas desulphurisation and simultaneous production of gypsum is a gas cleaning process, but it is in a way also a clean technology, because the gypsum is sold and used as a raw material in the production of cement and wallboard. It is thus an example on the use of by-products.

4.7.2 NO_X REMOVAL FROM FLUE GAS

The formation of NO_X in combustion processes is complicated and depends on a number of parameters as illustrated in Equation 4.1. Because of this dependence, the formation of NO_X can be lowered by proper design and operation of the power plant and other combustion systems. The important parameters are temperature, O_2-concentration and residence time in the hot zone. By staging the air to the burner, the temperature is lowered and reducing conditions are obtained in the primary combustion zone. Figure 4.12 (next page) shows a low-NO_X-burner.

Secondary and tertiary air or overfire air is then used to ensure complete burn-out of the fuel (Figure 4.13, next page). This method is often called the use of low-NO_X-burners, even if more than burner design and operation is involved. It is in reality a clean technology and not a gas cleaning method, and it is now widely used to lower the formation of NO_x on modern power stations (Soud, Fukusawa 1996).

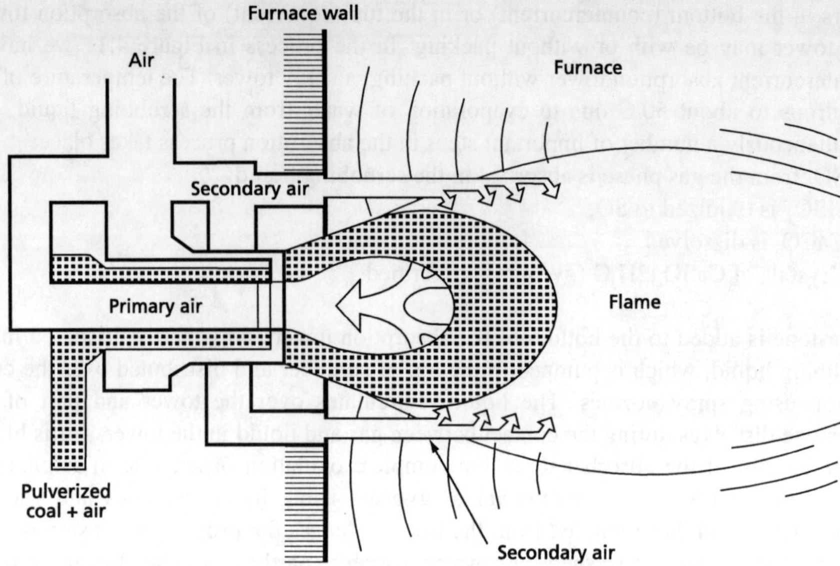

Figure 4.12 Low-NO_x-burner. A mixture of coal dust and enters from the left and the flame is seen to the right. Adapted from Strauss (1992).

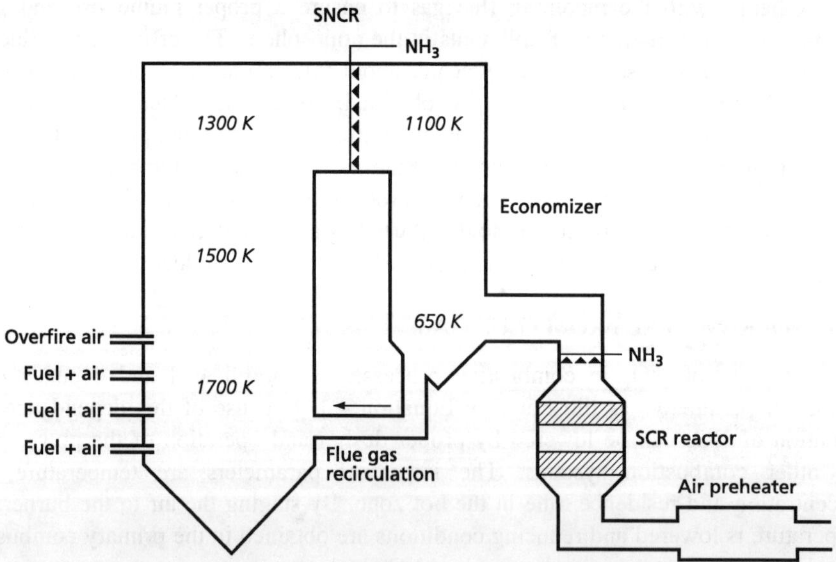

Figure 4.13 Schematic diagram of coal fired boiler showing the position of overfire air, SNCR and SCR.

Unfortunately, it is not possible to eliminate NO_X completely from the flue gas with low-NO_X-burners, and therefore flue gas cleaning is still necessary. More than 95% of the NO_X in flue gas is NO and only a minor part is NO_2. As a consequence NO_X is not removed from the flue gas by conventional flue gas desulphurisation with scrubbers, because NO has a very low solubility in water and is not an acid. The commercial processes for NO_X removal utilises the possible reduction of NO and NO_2 to N_2. The most common reactant for this reduction is NH_3, and the reaction can be carried out at a relatively low temperature (300-400°C) using a catalyst, the SCR-process (Selective Catalytic Reduction) or at a higher temperature (850-1000°C) without a catalyst present, the SNCR-process (Selective Non Catalytic Reduction). A short description of the two processes will be given below (Soud, Fukusawa 1996).

The *principle of the SCR-process* is that NO and NO_2 reacts with NH_3 over a solid catalyst and N_2 and H_2O is formed. The main reactions are:

$$NO + 2/3\ NH_3 \rightarrow 5/6\ N_2 + H_2O \tag{4.5}$$

$$NO_2 + 4/3\ NH_3 \rightarrow 7/6\ N_2 + 2\ H_2O \tag{4.6}$$

with the undesirable side reaction:

$$O_2 + 4/3\ NH_3 \rightarrow 2/3\ N_2 + 2\ H_2O \tag{4.7}$$

The overall reaction is often written as:

$$NO + NH_3 + 1/4\ O_2 \rightarrow N_2 + 3/2\ H_2O \tag{4.8}$$

in good agreement with an observed almost stoichiometric consumption of NH_3. The reaction takes place on the surface of a solid catalyst. The catalyst is porous to give a large specific surface and most of the reaction takes place on the interior surface. The most common catalyst consists of a carrier of TiO_2 with V_2O_5 as the catalytic material. Because the optimum temperature is 300-400°C the catalyst is placed in the flue gas stream before the air preheater where usually the flue gas has this temperature. At this position the flue gas has a high content of fly ash, because it is before the electrostatic precipitator (Figure 4.13), and therefore the catalyst is in blocks with small channels, to ensure that the fly ash is carried through and not deposited in the catalyst (Figure 4.14, next page).

In some cases, especially when retrofitting an existing power plant, the SCR-process is placed after the FGD plant, a tail-end SCR, but then the flue gas has to be heated again to at least 300°C. The catalyst lifetime is an important parameter for the economy of the SCR-process. Industrial experience from a number of European power stations indicate that the average life time is 5 to 6 years, but its difficult to give a precise figure, because the catalyst is replaced gradually over many years of operation.

The *principle of the SNCR-process* is the same as described above in reaction (4.5) and (4.6), but without a catalyst the reaction temperature is higher, 900-1000°C, and another reaction is possible:

$$NH_3 + 5/4\ O_2 \rightarrow NO + 3/2\ H_2O \quad (4.9)$$

The undesirable and detrimental oxidation of NH_3 to NO takes place at temperatures above 950°C and dominates at 1000°C. At temperatures below 900°C the reaction rate in reaction (4.5) becomes low and unreacted NH_3 may be present in the flue gas. The conclusion is, that it is extremely important to inject the NH_3 at a proper place in the boiler, where optimum temperatures of about 950°C prevail. However the temperature changes with fuel type, boiler load and fouling of heat transfer surface, and therefore the NH_3 consumption is higher and the efficiency is lower compared to the SCR-process.

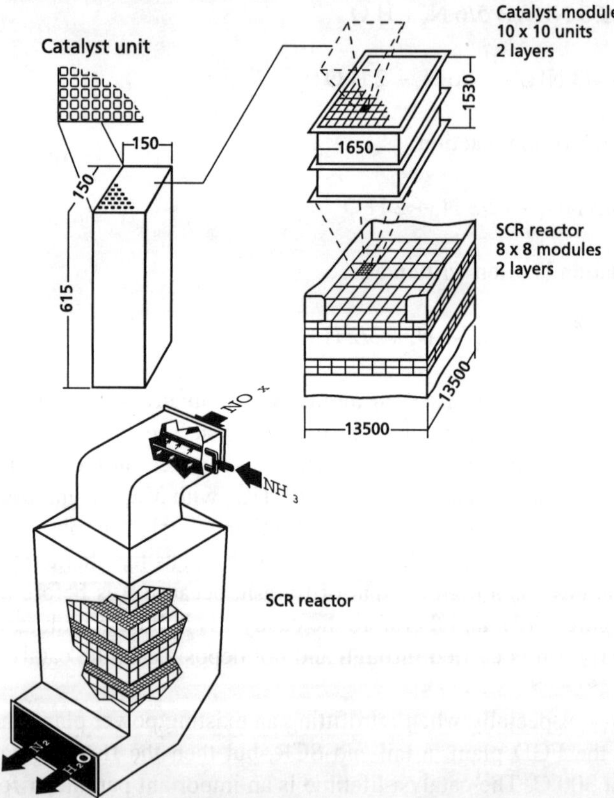

Figure 4.14 SCR catalyst units and modules and SCR reactor, all measures in mm. Adapted from Deutsche Babcock (1983).

4.7.3 FLUE GAS CLEANING SUMMARY

Flue gas cleaning is costly, and Table 4.5 gives a summary of the costs involved. Low-NO_x-burners and fly ash removal by electrostatic precipitators are in the low cost end while SCR and especially desulphurisation are costly processes. On a modern power station flue gas cleaning may add 19 to 38% to the cost of electricity production (Takeshita 1995).

Table 4.5 Air pollution control costs for coal-fired power stations. Adapted from Takeshita (1995).

	Capital costs		Levelised costs	
	ECU/kWe	As a proportion of the capital cost of the plant %	0.001ECU/kWh	As a proportion of the cost of electricity %
Low-NO_x-burners + overfire air	20 - 40	2 - 3	1 - 2	1 - 4
SCR (Selective Catalytic Reduction of NO_x)	70 - 110	5 - 8	4 - 5	6 - 12
Particulate control	40 - 50	3 - 4	2 - 3	4 - 8
Flue gas desulphurisation	130 - 220	10 - 17	5 - 6	8 - 14
Total	260 - 420	20 - 32	12 - 16	19 - 38

It is common practice to calculate the stack height based on emissions without flue gas cleaning, because situations might occur, where the flue gas cleaning systems are not functioning, but the power plant still has to produce heat and electricity because of public demand. This means that during normal operation of a modern power station with efficient flue gas cleaning techniques the impact on the local air pollution level will be very small.

4.8 References

CORINAIR '94 (1997) *Summary Report 1, European Emission Inventory for Air Pollutants*, European Environment Agency, Copenhagen, Denmark.

CORINAIR DK 1995 (1997) *CORINAIR National Annual Data DK 1995*, Risø.

Deutsche Babcock Anlagen AG (1983) *II Symposium Rauchgasreiniging, 29. November* (in German).

ELKRAFT (1990) *Flue gas cleaning on Asnæs power station* (in Danish).

Flagan, R.C., Seinfeld, J.N. (1988) *Fundamentals of Air Pollution Engineering*, Prentice Hall, New Jersey, USA.

Graedel, T.E., Allenby, B.R. (1995) *Industrial Ecology*, Prentice Hall, Englewood Cliffs, New Jersey, USA.

Horstman, J., Jørgensen, K. (1997) *The future for electric vehicles in Denmark*, Orientering fra Miljøstyrelsen, Nr. 1 (in Danish).

ISO (1997) Life Cycle Assessment - Principles and Framework, *ISO 14040*.

Johnsson, J.E. (1991) *Air Pollution Control - Vapours and Odours*, Technical University of Denmark (in Danish).

Nevers, N. de (1995) *Air Pollution Control Engineering*. McGraw-Hill, USA.

Soud, H.N. (1995) Suppliers of FGD and NO_x control systems, *IEA Coal Research, IEACR/83*.

Soud, H.N., Fukusawa, K. (1996) Developments in NO_x Abatement and Control, *IEA Coal Research, IEACR/89*.

Strauss, K. (1992) *Kraftwerkstechnik*, Springer-Verlag, Berlin, (in German).

Takeshita, M., Soud, H.N. (1993) FGD performance and experience on coal-fired plants, *IEA Coal Research, IEACR/58*.

Takeshita, M. (1995) Air pollution control costs for coal-fired power stations, *IEA Coal Research, IEAPER/17*.

Theodore, L., Buonicore (editors) (1994) *Air Pollution Control Equipment*, Springer-Verlag, Berlin, Heidelberg.

Turns, S.R. (1996) *An Introduction to Combustion*, McGraw-Hill Inc., Singapore.

Wark, K., Warner, C.F. (1981) *Air Pollution - Its Origin and Control*, Harper & Row, Publishers, New York.

Warnatz, J., Maas, U., Dibble, R.W. (1996) *Combustion*, Springer-Verlag, Berlin Heidelberg.

Wenzel, H., Hauschild, M., Alting, L. (1997) *Environmental Assessment of Products - Methodology, Tools and Case Studies in Product Development*, Chapman and Hall, London.

Chapter 5

MOBILE SOURCES

ZISSIS SAMARAS
Laboratory of Applied Thermodynamics
Aristotle University, GR-54006 Thessaloniki, Greece

SPENCER C. SORENSEN
Department of Energy Engineering
Technical University of Denmark, DK-2800 Lyngby, Denmark

5.1 Description and micro-inventories of mobile sources in cities

5.1.1 VEHICLE TYPES

Emissions from motor vehicles dominate to a very large extent the emissions over urban areas. From both an emissions and usage standpoint the following major motor vehicle categories have to be distinguished:

Passenger Cars are vehicles used for the carriage of passengers, they comprise not more than eight seats and have a weight not exceeding 2.5 tonnes. Passenger cars can be either petrol (spark ignition engine) or diesel (compression ignition engine) powered. In some countries (e.g. The Netherlands and Italy) a substantial share of LPG powered passenger cars is found. In addition, passenger cars can be used as private cars or as taxis, the latter being almost exclusively diesel powered. During the recent years an important part of private automobiles consists of "company" cars.

Urban Buses are used for the carriage of passengers and have a maximum weight exceeding 5 tonnes. They are almost exclusively diesel powered (However there are also few electric trolley buses and recently in pilot projects natural gas powered buses started to make their appearance), while the most important categories are rigid (around 11 m long), articulated and double deckers. In addition to the urban buses, which support the mass transit inside the cities, the usage of coaches may also have an effect on total emissions. Despite the fact that these vehicles are mostly used for inter-city transport, in some cases they are also used for tourist transport inside the urban areas.

Lorries are used for the carriage of goods and are usually classified according to their maximum permissible weight. From an urban environmental point of view the following categories are of importance:

- Light duty vehicles with a maximum weight up to 3.5 tonnes which can be either gasoline or diesel powered and which are usually equipped with engines and technologies coming from passenger cars.
- Delivery trucks, with gross vehicle weight usually in the range of 3.5 to 7.5 tonnes, exclusively diesel powered.
- Other heavier duty vehicles (above 10 tonnes) which are also diesel powered; however, as a general rule their usage inside urban areas is restricted either to specific time periods or not allowed at all.

Two wheeled vehicles comprise mopeds with a cylinder capacity less than 50 cm^3 and an upper limit in vehicle speed and motorcycles with cylinder capacities above 50 cm^3. Mopeds are usually equipped with two stroke engines, a type of engine which can also be found in the lower range of motorcycles (as a rule up to 250 cm^3). The bigger motorcycles are normally equipped with four stroke engines.

As shown in the following, gasoline passenger cars are primary sources for CO and HC emissions in urban environments, while diesel powered vehicles are major sources for particulate and NO_x emissions. In some - mainly southern cities - two wheelers can make up for a substantial part of HC emissions from road traffic.

In addition to the above, there is an important number of other vehicles which are not covered by the road transport statistics but which may have substantial contribution to total emissions. These mobile sources include railway transport, aircrafts, ships, and speciality vehicles (such as garbage trucks, ambulances etc.), industrial mobile machinery (such as excavators, bulldozers, portable generator sets etc.) as well as home and gardening machinery (such as lawn mowers etc.).

Railway sources generally make a limited contribution to local urban air pollution. This is due to the fact that most urban rail traffic, commuter trains, etc. in Europe is electrically powered. The pollution problems associated with this form for rail traffic, therefore, are those of electrical power generation, and are normally remote from urban centres.

There are some situations where rail traffic can contribute to urban air pollution. These includes diesel powered rail traffic, which can contribute to the overall emissions in an urban area, and on a local basis. In the latter case, this corresponds to locally high emissions along railway lines, or near trains stations and shunting areas. The local emission sources are directly dependent on the mix between diesel and electric power on the railway system involved. The degree of electrification of railway systems varies widely throughout Europe (Jørgensen, Sorensen 1997).

Emissions from trains in motion are highly dependent on the train speed and type of driving (Jørgensen, Sorensen 1997). Energy consumption of urban trains ranges between 200 and 300 kJ/tonne-km, with the highest energy consumption being for operation with many stops. Emissions can be estimated using emission factors for electrical power generation or for diesel engine operation, where the emission factors are in terms of g/kW-h energy produced.

Air traffic is generally not considered to be a major contributor to overall urban air quality problems, since airports are located at the periphery of most cities, and aircraft spend a small amount of time in a urban area, since they rapidly climb, and are routed away from urban areas to avoid noise. Locally, airports can contribute significantly to emissions, as VOC emissions can be high during starting procedures, and also due to smoke and NO_x emissions, which are at their highest during takeoff (Kalivoda, Feller 1995).

Ship emissions vary widely, depending on the city involved. For inland cities without waterways, they are totally insignificant, whereas in seaports they can be very important. A study of some ports in the Mediterranean area has shown that emissions of NO_x and SO_x can be on the order of several thousand tonnes per year (Trozzi et al. 1995). Due to low quality fuels often used for shipping, particulate emissions from marine diesel engine can also be high. This can be a significant load on the local environment. Diesel ships normally have very low emissions of VOC and CO. Inland ports with river traffic experience emissions from this type of traffic. However, magnitude of the emissions from this kind of traffic is not well known. These ships are usually driven by diesel engines, often of an older design.

5.1.2 VEHICULAR DISTRIBUTION

There is a great variation of the car density in the different European countries. Figure 5.1 (next page) presents the density of passenger cars and two wheeled vehicles for 15 EU countries in 1990. Here, and in the following figures, the countries are indicated by the license plate code.

Figure 5.2 (next page) illustrates the passenger car fleet breakdown in gasoline (further split in capacity classes), diesel and Liquefied Petroleum Gas (LPG) vehicles, for each EU 15 Member State. While gasoline passenger cars constitute the major part of the fleets, diesel cars constitute about 15% of the total on average. Especially during the recent years a trend towards dieselisation has occurred, particularly in France, UK and Austria.

There is a significant usage reduction with age, as presented in Figure 5.3 (page 67). The oldest vehicles are, on average, driven half the mileage of the new ones. This is important since older vehicles were built to satisfy much less stringent emissions requirements, and therefore have higher emissions than newer vehicles.

Moreover, there is a relationship between annual mileage and engine size and type. On a European average, cars from 1.4 to 2.0 litre engine capacity are driven about 10 % more than those under 1.4 litre, those over 2.0 litres about 25 % more. Diesel cars are driven about twice as much as the smaller petrol cars. In GR a factor of about 6 is observed between gasoline and diesel passenger cars, but this is due to the fact that until recently diesels were allowed only for taxis.

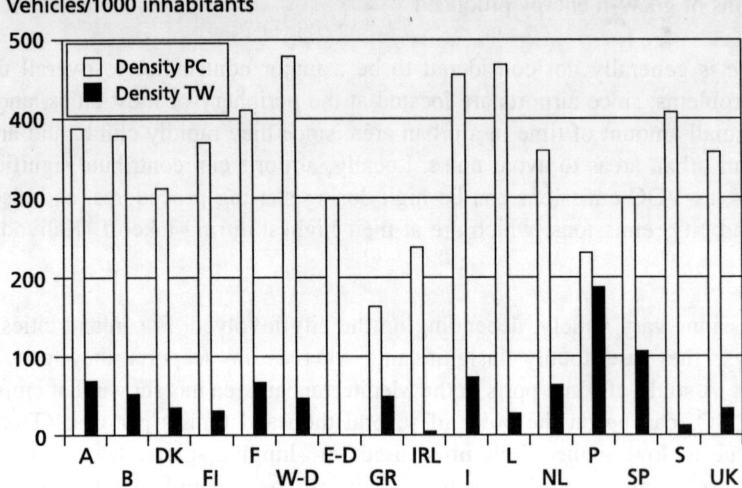

Figure 5.1 Passengers car (PC) and two wheeler (TW) density in the 15 EU countries 1990 (MEET 1997; ECMT 1995).

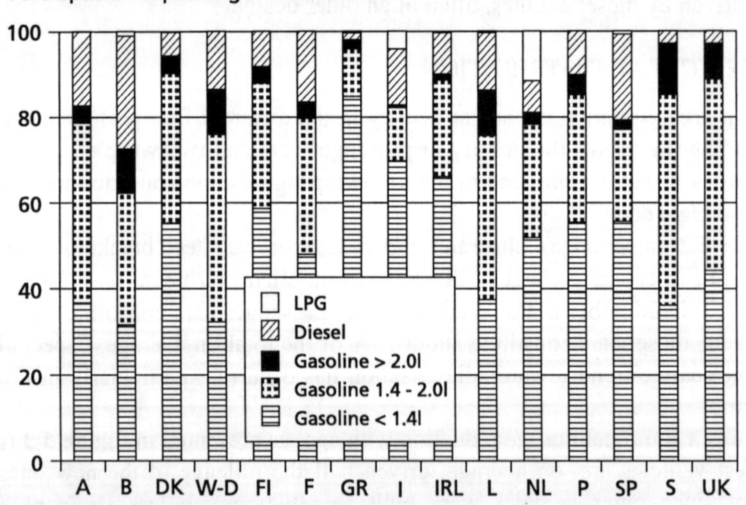

Figure 5.2 Passenger car fleet distribution (1990 data) for EU 15 (MEET 1997).

The different traffic conditions as well as the different driving habits affect also the average urban speed of the passenger cars and the urban share of mileage. This effect is illustrated in Figure 5.4. The urban mean speed ranges from about 20 km/h (i.e. very similar to that of the urban part of the European driving cycle for emissions testing) in E, GR and I up to 40 km/h in DK and L. The mileage share ranges from about 25% in

IRL and P up to 45% in GR, L and UK. Average vehicle speed plays a major role in determining emission factors for the individual vehicles as shown in Section 5.3.

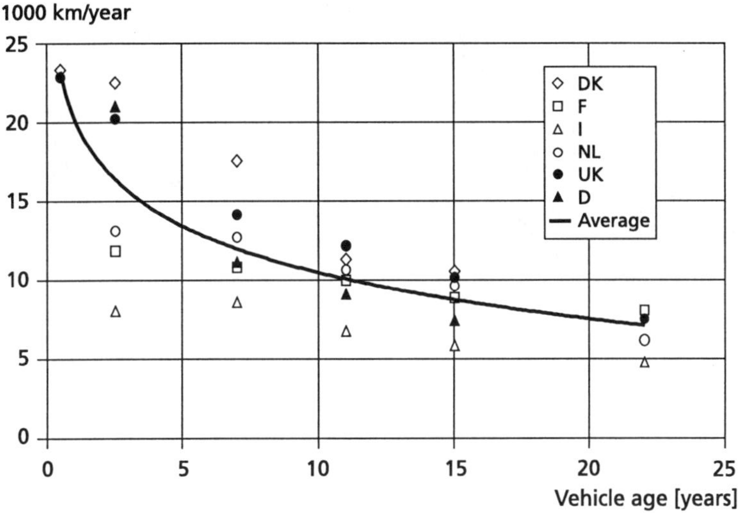

Figure 5.3 Annual mileage as a function of the vehicle age (1990 data) (MEET 1997).

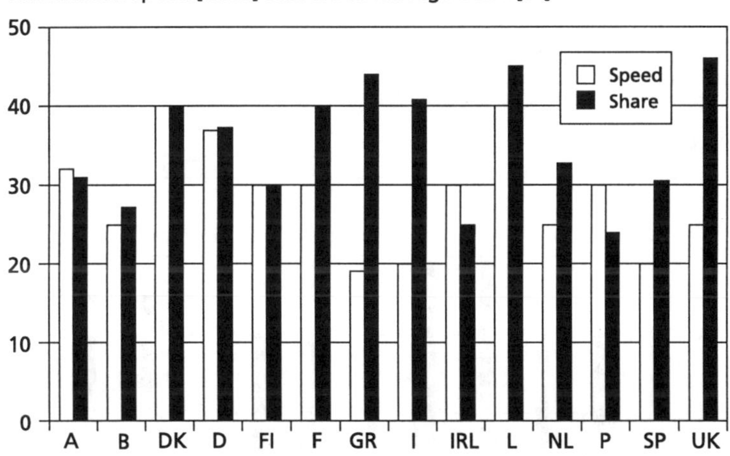

Figure 5.4 Urban mean speed and mileage share of passenger cars (1990 data) (MEET 1997).

Figure 5.5 (next page) shows typical daily traffic load variation profiles in a number of selected cities in Europe. Shown is the fraction of total daily traffic as a function of time. While in most northern cities the "twin peaks" of the traffic load are very clear during a typical work day, in most southern cities there is a rather flat profile encountered during the working hours of the day.

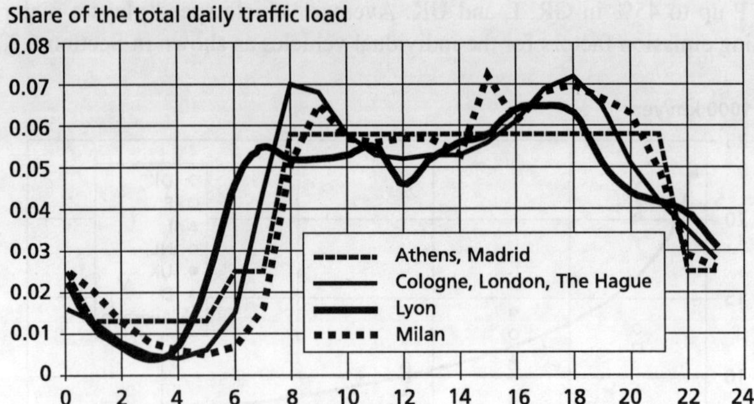

Figure 5.5 Typical hourly profiles of traffic density for selected cities in Europe in 1990 (European Commission 1996).

There is a certain variability of *occupancy rates* of the different passenger vehicle categories as shown in Figure 5.6. On average, in urban areas passenger cars have an occupancy rate of about 30%, the two wheelers about 70%, the urban buses about 50% and the metro about 30%. Since passenger weight is a relatively minor factor in determining the weight of most of the vehicle types in Figure 5.6, there is a substantial potential for emissions reduction on a person-km basis by increasing the load factor for all the types of transport. In practice, this has proven to be exceedingly difficult to accomplish, due to a variety of social and societal factors related to the choice of person transport method.

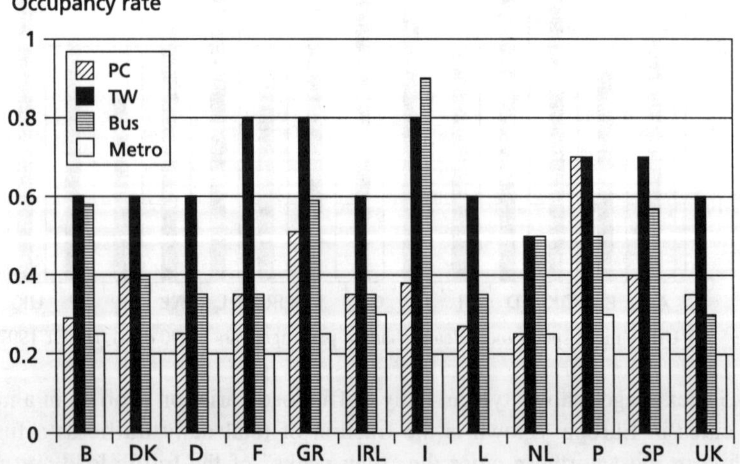

Figure 5.6 Estimated occupancy rates for different passenger transport modes in urban areas 1990 (Samaras, Zierock 1994).

The passenger car dominates the transport of passengers in all European countries, followed by the urban bus, as shown in Figure 5.7. In the EC12 in 1990, an estimated total of about 1,250 billion passenger kilometres is done in urban areas each year. Even in areas of high population density and excellent public transportation, the passenger car remains the preferred mode of transport for about ¾ of urban traffic.

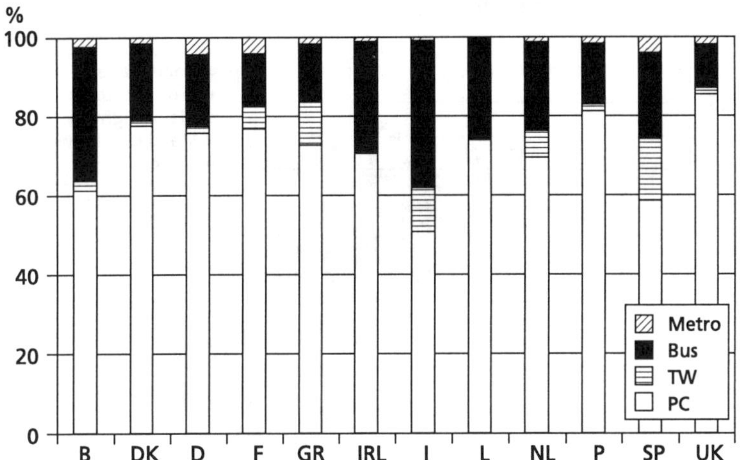

Figure 5.7 Estimated share of different transport modes to total urban passenger kilometres in 1990 for selected European countries (Samaras, Zierock 1994; ECMT 1995).

5.1.3 TRAFFIC CHARACTERISTICS

Emissions from road traffic are affected by the type of driving. In an urban environment, a wide range of driving patterns are experienced, ranging from very low speed "stop and go" traffic in congested urban centres, to motorway conditions on non-congested motorways.

Several studies have indicated that the most significant parameter for determining the emission factor for road vehicles in the average trip speed, that is the average speed between two points on a route, not the average speed measured at a point. This method is used in the CORINAIR emission inventory calculation procedure (EEA 1996). For a given street, a drop in average speed is synonymous with an increase in the variation in the instantaneous speed. That is to say, if a vehicle is driven at an average speed of 25 km/h on a street with a limit of 60 km/h, it is because of traffic limitations, and that driving with a constant speed of 25 km/h on that street would basically never be encountered.

This holds true for all kinds of or urban and rural streets and highways, as shown in Figure 5.8 (next page). This figure shows the standard deviation of the instantaneous vehicle speed as a function of the speed limit of the street or road for traffic in Denmark. Both variables have been normalised by the speed limit, and the figure shows the same relative increase in speed fluctuation when the average speed drops. This is important for emissions, since when the fluctuations increase, there are more high power accelerations, and more time spent at low speed/idle, where engine efficiency is very low and emissions high.

It is important to note that at speeds below the speed limit, constant speed driving is rarely experienced. In practice, it is only on motor ways where nearly constant speeds are observed to a significant extent.

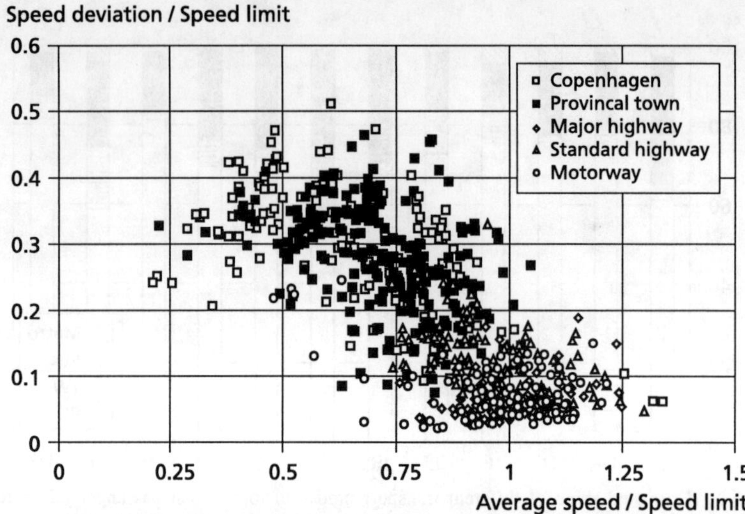

Figure 5.8 Normalised speed variation as a function of normalised trip speed for a variety of driving situations in Denmark (Hansen et al. 1995).

Figure 5.9 shows the variation in the average number of vehicles passing though 4 different streets in Copenhagen during rush hour and during the middle of the day (first two columns in each set) as well at the corresponding average speed of the passenger car traffic on that street (last two columns in each set).

As expected, the traffic slows down during the rush hour, although the effect depends on the street. In the suburban main street, special arrangements had been made to limit the traffic speed, during non-rush hour traffic, and as a result the average trip speeds did not decrease as the number of vehicles increased. The decrease in speed with the increase in traffic has the unfortunate effect of increasing the emissions, not only due to the number of vehicles, but also due to the increase in emission factor, which will be shown in the following sections.

Bus traffic in urban areas is characterised by its highly transient nature. Such driving consists of accelerations, short driving with roughly constant speed, deceleration, followed by stop. A representative cycle for Copenhagen bus traffic is shown in Figure 5.10. This type of driving gives higher emissions than a more steady driving, but is required in order to pick up traffic. It is better suited to diesel engine operation than spark ignition engine, because diesel engines are more efficient at the part and zero load conditions which are so prevalent.

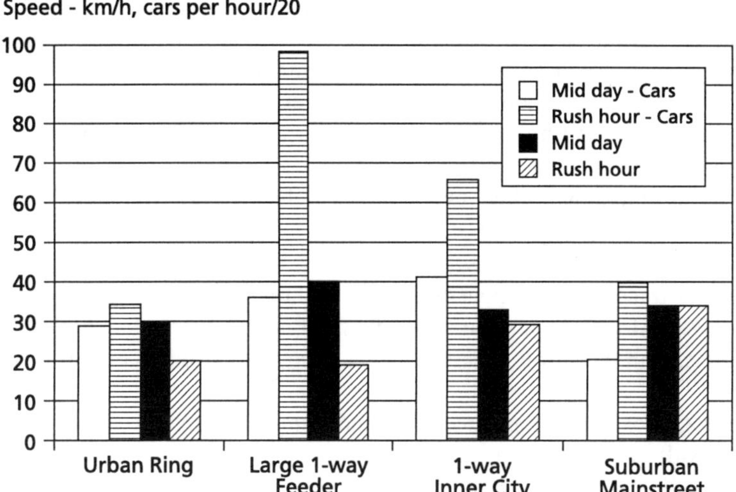

Figure 5.9 The relationship between traffic flow and average speed for different streets in the greater Copenhagen area.

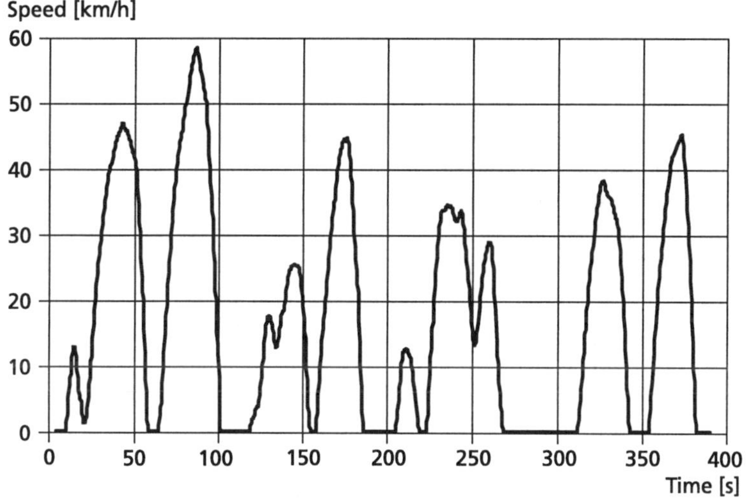

Figure 5.10 Driving pattern for a typical urban bus. The average speed of the trip is 17 km/h, including the stopping time.

5.2 Emission factors for vehicles

5.2.1 REGULATED EMISSIONS

The following discusses emissions of importance from mobile sources. The relative contributions of mobile sources for different cities can be found in Chapter 25.

VOC is often referred to as unburned hydrocarbon (HC or UHC) in the engine field. VOC Emissions from engines result in general from incomplete combustion, which includes phenomena such as flame quenching, misfire, detachment of lubricating oil film etc. While the combustion of fossil fuels is responsible for only one third of total NMVOC emissions, road traffic is responsible for almost 90% of anthropogenic VOC emissions in urban areas. In particular, gasoline passenger cars account for almost two thirds of these emissions, while two wheeled vehicles may also have an important contribution in some cities.

CO from road transport is responsible for over three quarters of CO produced at the European level. At the urban scale, road transport is responsible for more than 90% of CO emissions, with the gasoline passenger car being responsible for almost 75%.

NO_x or nitrogen oxides as regulated for vehicle emissions (NO_x) consist of two gases, nitrogen dioxide (NO_2) and nitric oxide (NO) formed in combustion processes mainly from the oxidation of the nitrogen in the air. Most of the NO_x emissions from engines is emitted as nitric oxide: in the case of spark ignition engines this is effectively the case, while in diesel combustion NO_2 may be up to 30% of total NO_x emissions, depending on the operation mode. Road transport is the main source of urban NO_x emissions. In 1990 at the European scale road vehicles were responsible for about 50% of total NO_x emissions. At the urban scale, road traffic contributes from 50 to 80% depending on the city and its activities.

Particulate Matter (PM) comprises particles emitted from both spark ignition and the diesel engines and they consist of both mineral and organic particles. The mineral based particles originate either from the combustion process (e.g. lead compounds from gasoline engines) or from brake and clutch linings wear. Lead emissions have been greatly reduced by the introduction of unleaded fuel in connection with catalyst technology, and will continue to fall as leaded fuel disappears. The organic particles are mainly associated with the diesel engine, although as diesel particulate emissions are reduced, spark ignition engines will have a larger relative contribution to PM.

Particulate matter emission from diesel vehicles has long been considered as an important hazard for human health, as its organic fraction (i.e. heavy hydrocarbons adsorbed) were identified as primary suspect for carcinogenic and mutagenic effects. However, the results of recent detailed studies have produced clear evidence that the risk comes directly from respirable particles with a diameter less than 2.5 μm. This size range is dominated by combustion related particles, with the diesel engine producing most of the sub-micron particles.

Figure 5.11 presents typical size distributions of PM under urban driving conditions for an indirect injection (IDI) engine. This type of engine has been predominant in light duty urban diesel vehicles, while the direct injection (DI) engine has been predominant in larger diesel engine, such as buses and lorries. DI engines are starting to appear in significant numbers within the group of diesel powered LDV's. The average size of the diesel particulates is around 0.1 to 0.2 μm irrespective of engine combustion type.

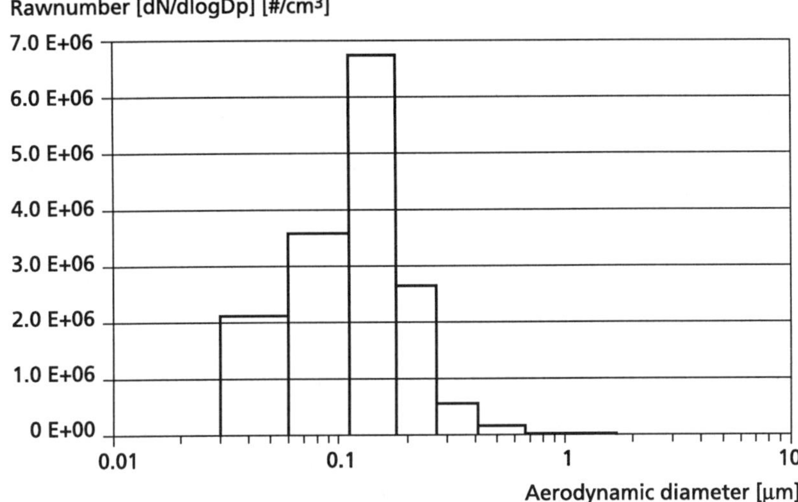

Figure 5.11 Number of particles per unit volume as a function of particle size for PM from an IDI light duty diesel vehicle over the European urban cycle (Pattas et al. 1998).

5.2.2 NON-REGULATED EMISSIONS

In addition to the above substances which are regulated by legislation, (VOC, CO, NO_x, PM), a number of other substances found in the exhaust of mobile sources are of concern.

Benzene represents a special case due to its toxicity and its carcinogenic effect. It is estimated that 80 to 85% of total benzene emissions in the atmosphere comes from the automobile. On a European vehicle fleet fuelled with gasoline containing 3.1% by weight of benzene and 30% by weight of other aromatics, it has been found that about 45% of the emitted benzene has survived combustion, while the rest was created during combustion (Degobert 1995). Benzene can also be found in diesel exhaust, but passenger cars account for about 60-70% of urban benzene emissions.

Polynuclear aromatic hydrocarbons (PAH) have two or more benzene or cyclopentane rings joined in various clustered forms from indane to coronene. The special attention paid to them comes from the fact that they are recognised as carcinogens. Most of them emitted with engine exhaust are found adsorbed on the carbon particles, however, they are also detected in the gaseous exhaust, particularly in gasoline engine exhaust. Benzo(a)pyrene is emitted at practically the same rate from both gasoline and diesel vehicles. Nitrogen containing PAH compounds have also been found in exhaust gases. Although their quantity is small, there mutagenetic activity is very high.

Sulphur compounds (including SO_2, sulphuric trioxide and sulphuric acid) are also emitted by road vehicles and are due to the sulphur content of the automotive fuels.

Their importance is declining due to the gradual installation of desulphurisation equipment in the refineries.

Nitrous oxide (N_2O) is also of concern as it contributes to the greenhouse effect (together with methane CH_4), despite the fact that the automobile is only a minor contributor to the emissions of these pollutants. Improper design and wear in 3-way catalyst equipped cars can lead to N_2O emissions that are a significant fraction of the total NO_x emissions.

5.3 Emission factors for automobiles

5.3.1 HOT EMISSION FACTORS AS A FUNCTION OF SPEED

Emissions of all vehicle categories are typically expressed in gram of pollutant per kilometre travelled. For each individual vehicle category, these representative emissions - usually referred to as emission factors - depend on two main parameters: the pollution control technology and the driving conditions of the vehicle. The average driving speed has been found to be the most important parameter which reflects with sufficient certainty the effect of driving on emissions.

The following figures present characteristic emission factors for most of the vehicle categories discussed in Section 5.1. These factors are valid for fully warm engine operation.

The emission factors for petrol passenger car are much lower for the cars equipped with 3-way catalysts. Figures 5.12 and 5.13 show that the catalyst technology results in a reduction of emission factors of about an order of magnitude.

Figures 5.14 and 5.15 (page 76) show emission factors for diesel powered passenger cars with current technology, and with technology that will be used in the near future. Diesel passenger cars have very low emissions of VOC and CO, even lower than those from petrol powered cars with a 3-way catalyst. NO_x emissions are comparable to those of catalyst cars as well. Further reduction of NO_x from diesel passenger cars will be difficult, however, while there is substantial potential for improvement for petrol powered passenger cars by improving catalyst and engine control strategy. This is indicated by the comparable level of NO_x emission factors for the current and advanced technology engines, while the VOC, CO and PM emission factors will be reduced by about 1/2 to 1/3.

Particulate emissions are a significant problem with diesel powered passenger cars, but legislation will require that they be significantly reduced in the future. Until the early 1990's, very little attention was paid to particulate emissions from petrol engines.

Recent studies indicate that the PM emissions from petrol powered passenger cars with 3-way catalyst are on the order of 0.02 g/km. Advanced technology diesel powered passenger cars are beginning to approach this emission level. There are still few data available to estimate the dependence of petrol powered passenger car particulate emissions on vehicle speed.

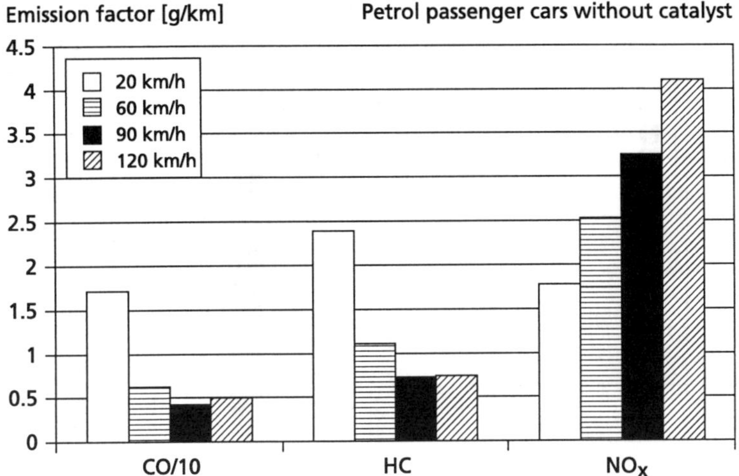

Figure 5.12 Emission factors for warm petrol powered passenger cars without catalyst as a function of trip speed.

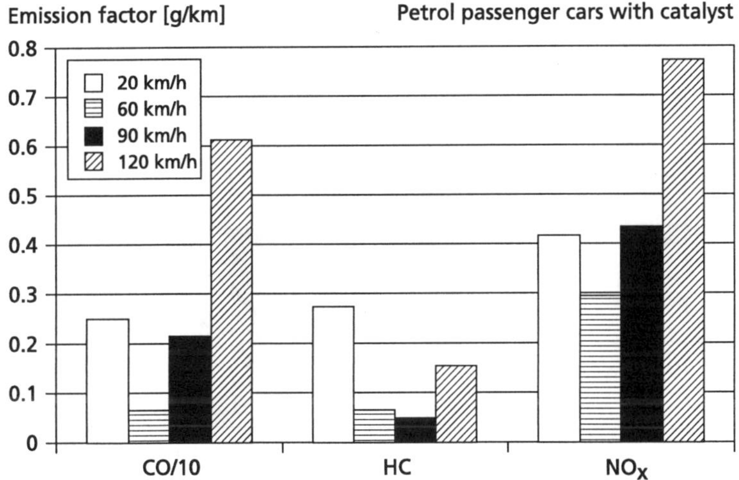

Figure 5.13 Emission factors for warm petrol powered passenger cars with 3-way catalyst as a function of trip speed.

Figure 5.16 (page 77) shows the emission factors for light lorries. This class of vehicles includes delivery vans, and lorries, the type of lorry most commonly found in urban traffic. Emissions of CO, VOC and NO_x are near the levels of those from petrol powered passenger cars without catalyst, and are the NO_x emissions are higher by a factor of about 2. Particulate emission factors are comparable to those of current diesel powered passenger cars.

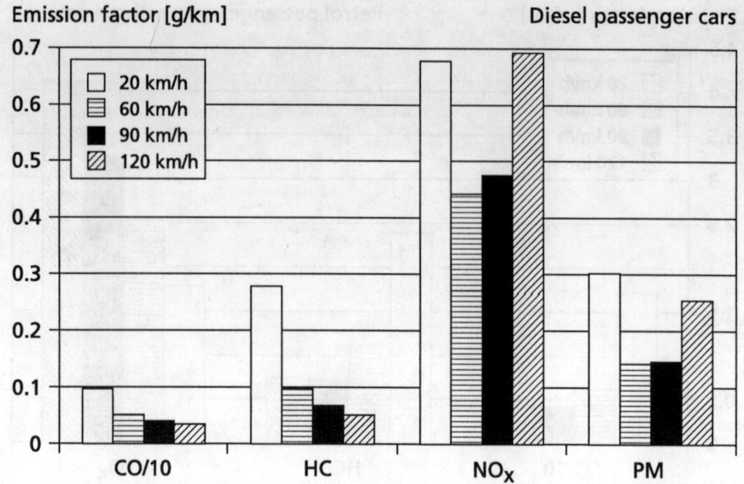

Figure 5.14 Emission factors for current diesel powered passenger cars with conventional technology as a function of trip speed.

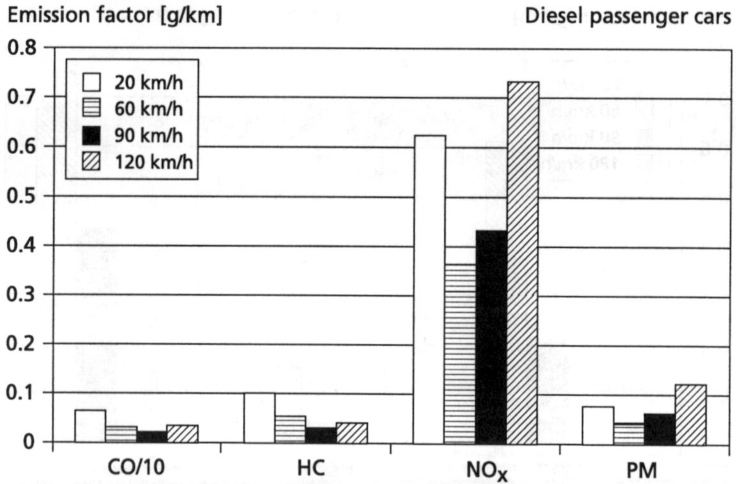

Figure 5.15 Emission factors for warm diesel powered passenger cars with advanced technology as a function of trip speed.

Figure 5.17 shows emission factors for urban buses at two speeds and levels of emission control, for operation in urban areas. They have the largest emission factors for NO_x and particulate matter of all the vehicles in normal urban traffic. As with other diesel powered vehicles, CO and HC emissions are modest. The NO_x emission factors for the newest engines are on the order of 8 to 13 g/km. Depending of the speed of the comparison, this indicates that an average loading factor of about 20-30 persons on a bus is required to give equal NO_x emissions on a person-kilometre basis for catalyst

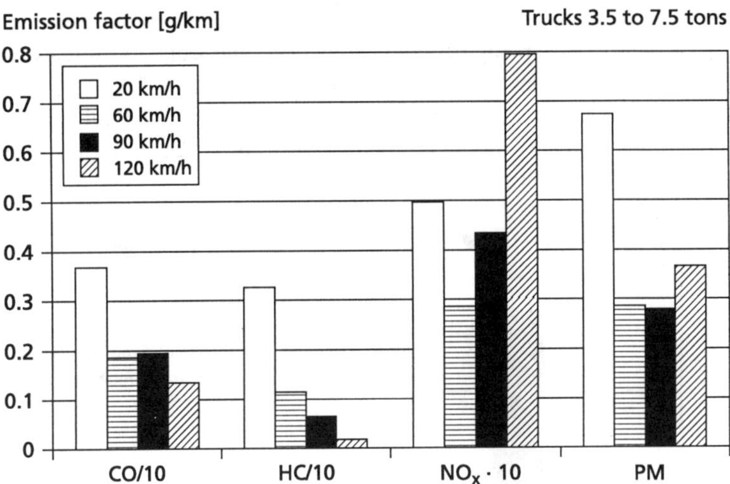

Figure 5.16 Emission factors for smaller lorries as a function of trip speed.

equipped petrol and advanced diesel passenger cars. Advances in diesel technology will improve this picture, but proposed reductions in NO_x emissions from heavy duty diesel engines will be modest, due to technical difficulties and a negative impact on particulate emission. Alternative fuels such as compressed natural gas, liquefied petroleum gas, DME and alcohol are being considered and tested in diesel buses for low-NO_x and low particulate emission buses.

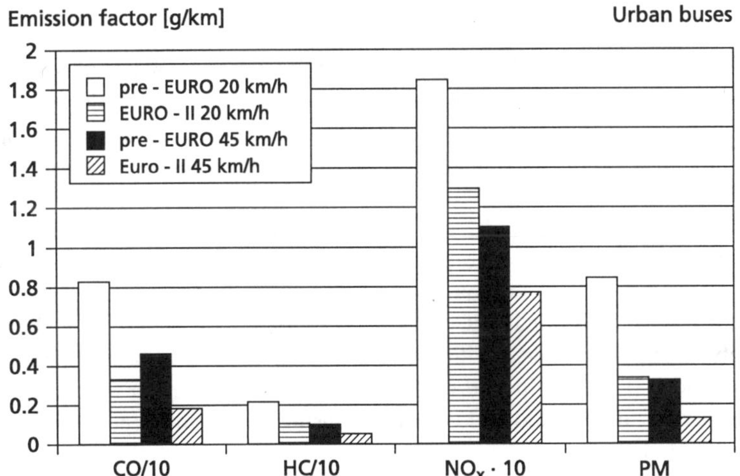

Figure 5.17 Emission factors for urban buses as a function of trip speed for two technology levels.

Figure 5.18 (next page) shows that two wheeled vehicles have extremely high emissions factors for CO and VOC. They are powered by small petrol engines that have not been subject to emissions regulation, and often rely on fuel rich operation to achieve acceptable drivability. Hydrocarbon emissions of two stroke engines are high - up to half the fuel consumption. These engines are very difficult to improve because of strong competitive pressures to keep engine and vehicle price and weight as low as possible.

For nearly all the vehicles and emissions shown in Figures 5.12 - 5.17, the lowest emission factors are found at speeds in the vicinity of 60-90 km/h. They increase considerably when vehicle speed is reduced to 20 km/h, indicating that traffic conditions play a role in determining emission levels.

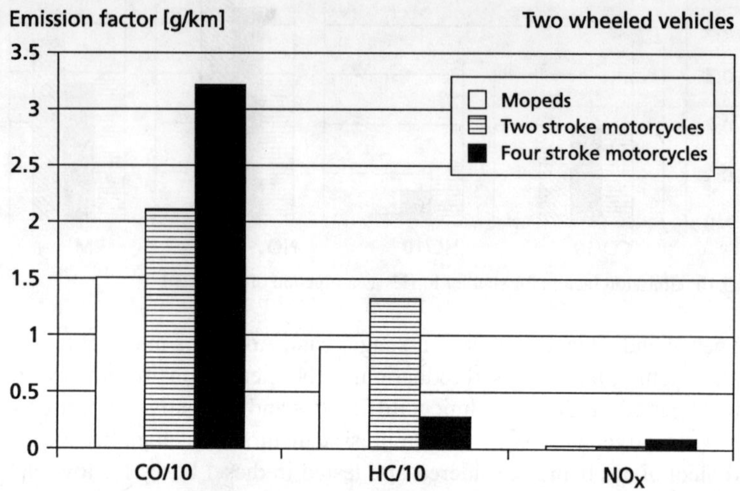

Figure 5.18 Emission factors for two wheeled vehicles as a function of trip speed.

5.3.2 COLD START EXTRA EMISSIONS

The starting temperature of the engine is generally lower than the normal operating temperature, and it can be significantly lower in the colder areas of Europe. The same holds true for the pollution control systems of the passenger cars and the gasoline light duty vehicles. As a result, the emissions produced by the vehicle during this period of engine operation are generally higher from those of the normal operation, the latter calculated as "hot emissions". CO and VOC emissions are the greatest problem with cold starting. Cold starts can take place under all driving conditions, however, they seem to be most likely for urban driving. In principle they occur for all vehicle categories. Typical additional HC and CO emissions of about 5 and 50 grams respectively can be encountered with cold starts at 20°C. At 0°C these amounts increase to about 10 and 100-200. Cold start emissions vary widely from car to car. The duration of the cold period and the cold extra emission during this period depend on a number of parameters, including:

- Engine temperature at the beginning of the trip
- Ambient temperature
- Driving behaviour, which can be evaluated through the average speed
- Vehicle technology

The engine temperature at the beginning of the trip depends again on a number of parameters, such as:

- Temperature at the end of the previous trip
- Parking duration
- Ambient temperature

5.3.3 EVAPORATIVE LOSSES

Evaporative losses refer only to VOCs from gasoline powered vehicles and are distinguished in three separate parts:

- Diurnal losses, mainly coming from the fuel tank through the tank breathing system
- Hot and warm soak losses, coming mainly from the mixture preparation system, due to the effect of the increased engine temperature and the lack of fuel recirculation to the tank
- Running losses, occurring during the vehicle operation

Diesel fuel has a very low vapour pressure, and evaporative emissions from diesel vehicles are minimal. Evaporative losses depend on:

- Trip length distribution
- Ambient temperature
- Fuel volatility (RVP - Reid vapour pressure)
- Car technology

5.3.4 OTHER PARAMETERS AFFECTING EMISSIONS

Apart from engine technology, a number of parameters affect emissions either directly (vehicle mileage, engine temperature, altitude) or by altering the engine mode of operation (road gradient, vehicle load). The vast part of the available experimental data on emission factors refers to "standard" testing conditions, i.e. zero altitude, zero road gradient, empty vehicle, etc. The vehicle mileage is a parameter usually recorded prior to testing, while engine temperature is usually either 20°C (cold start - as it is defined by the legislation) or normal operating temperature (hot start).

Road gradients can increase or decrease the resistance of a vehicle to traction, as the power employed during the driving operation is the decisive parameter for the emissions of a vehicle. Even in the case of large-scale considerations, however, it cannot be assumed that - for example - the extra emission when travelling uphill is compensated by correspondingly reduced emission when travelling downhill.

In principle the emissions and fuel consumption of both light and heavy duty vehicles are affected by road gradient. However, because of their higher masses, the gradient influence is much more significant in the case of heavy duty vehicles.

Vehicle load affects the driving resistance of a vehicle through vehicle mass, i.e. higher vehicle mass requires higher power of the engine during driving especially in acceleration modes. Because of the well known fact that emissions and fuel consumption are proportional to the engine power emission and fuel consumption the emission calculations have to take into account in principle vehicle load. This effect is less important in the case of passenger cars due mainly to the fact that the load range of a passenger car is small, while in the case of heavy duty vehicles the vehicle load has an important influence on emissions and fuel consumption.

Other parameters which are also known to affect the emissions include:
- Maintenance (frequency and quality)
- Ancillaries (e.g. air condition)
- Ambient conditions (temperature, humidity, pressure)

5.4 History of emissions regulations

5.4.1 PASSENGER CARS

Due the large number of passenger cars, this vehicle type was the first to be subjected to emissions standards. *European emissions standards and test procedures* are formulated in connection with ECE, are currently established by the European Union for its member states. In the late 1980's and early 1990's, several countries formed the "Stockholm Group". These countries adopted American equivalent emissions standards at an earlier time than the rest of Europe adopted requirements leading to 3-way catalysts. Since that time, however, several members have joined the EU, whose emission standards now are close to the American requirements.

A historical development of European emissions standards for light duty vehicles is shown in Tables 5.1 and 5.2. The standards are given in terms of grams per kilometre for a standard test cycle. Until 1996, the test was conducted on the Urban Driving Cycle (UDC), which has a maximum speed of 50 km/h and is intended to simulate driving in an urban environment. After 1996, a new driving cycle has been adopted, which used 4 repetitions of the UDC plus a Extra Urban Driving Cycle (NEDC), that has a maximum speed of 120 km/h, in order to include the more common effects of high speed driving.

The original emissions standards (ECE15) were typical of the levels of vehicles produced at the time of their introduction. Table 5.1 shows that emissions factors have been reduced by over 90% for CO and over 80% for the sum of HC and NO_x. These are very large reductions, and required the introduction of the three-way catalyst on 1993 vehicles. The reductions in the emissions standards for light duty diesel powered vehicles have been smaller than for the petrol cars, However, the emissions of VOC, CO and NO_x from original diesel engines were much smaller than those from pre-catalyst petrol cars.

5.4.2 HEAVY DUTY VEHICLES

Emission standards are also in force for heavy duty vehicles. In this case emissions from the engines alone are regulated instead those in connection with actual vehicles, due to the large number of engine-vehicle-transmission combinations with heavy duty vehicles. The emissions are given in terms of gram per hour per kilowatt power produced over an engine test cycle (Table 5.3). The first cycle used was a 13 mode test cycle with the engine operating under steady state conditions. The new emission standards may be applied to a combination of a steady state cycle (OICA) and a transient test cycle (FIGA). The transient cycle is a better simulation of actual driving conditions, and the conditions which produce high levels of particulate emissions.

Table 5.1 Historical Development of European Emission Standards (Light Duty Petrol Vehicle Emissions standards in g/km. The UDC is used up to 1993, the NEDC after 1993. ECE 15-xx emissions are conformity of production values for a 1 tonne car. Euro 3 and 4 emissions standards are estimated from proposal by the EU Commission).

Standard	Year	CO	$HC+NO_x$	HC	NO_x
ECE 15	1971	39.70		3.00	
ECE 15-01	1975	31.80		2.56	
ECE 15-02	1977	31.80		2.56	3.56
ECE 15-03	1979	25.70		2.27	3.01
ECE 15-04	1984	19.80	6.32		
89/76/EEC	1990	11.30	3.80		< 1.50
91/441EEC	1993	2.72	0.97		
94/12/EEC	1997	2.00	0.50		
Euro 3 (proposed)	2000	2.30		0.20	0.15
Euro 4 (proposed)	2005	1.00		0.10	0.08

Table 5.2 Light Duty Diesel Vehicles Emissions standards in g/km (Light Duty Diesel Vehicles Emissions standards in g/km, the UDC is used up to 1993, the NEDC after 1993 Euro 3 and 4 emissions standards are estimated from proposal by the EU Commission).

Standard	Year	CO	$HC+NO_x$	NO_x	PM
91/441EEC	1993	2.72	0.97		1.14
94/12/EEC-indirect injection	1997	1.00	0.70		0.10
94/12/EEC-direct injection	1997	1.00	0.90		0.10
Euro 3 (proposed)	2000	0.64	0.56	0.50	0.05
Euro 4 (proposed)	2005	0.50	0.30	0.25	0.025

Table 5.3 European emission standards for heavy duty transport engines in g/kW-h. (*) UBA proposal from June 1996 over the OICA cycle. If the FIGA cycle is used then: CO: 8.6, HC: 0.197, NO_x: 1.83, PM: 0.115 The European Commission has only published indicative reductions from current standard: NO_x -30% and PM -30%.

Standard	Year	CO	HC	NO_x	PM
EURO 1 < 85 kW	1992	4.5	1.1	8.0	0.5
EURO 1 > 85 kW	1992	4.5	1.1	8.0	0.36
EURO 2	1995	4.0	1.1	7.0	0.15
EURO 3 (*)	2000	3.0	0.154	1.8	0.068

5.5 Emission control technology

As seen by the increasingly stringent emissions legislation, emissions from various mobile sources have been reduced substantially over the past two decades. A variety of technological measures have been used to obtain this reduction. For both light and heavy duty vehicles, a combination of fuel and engine modifications has been utilised to achieve these reductions.

5.5.1 IMPROVEMENT OF CONVENTIONAL FUELS

Fuel, engine, and vehicle technology are all important for determining emissions. There are several reasons for improving fuel quality with respect to emissions reductions. For the first, fuel changes in most cases can be applied to all technologies, and emissions reductions can be achieved for engines with different certified emissions levels. Secondly, fuel changes can be introduced more rapidly that engine/vehicle technology changes. Due to the long life time shown in Figure 5.3 (page 67), changes in vehicles will take over 10 years to reach the great majority of all vehicles on the road. Fuel changes can be accomplished at the refinery, and once made there, can be applied to all type of vehicles, without the phase in time of new vehicle technology.

In addition, certain improvements in fuel quality can be necessary for utilisation of improved emission control technologies. One of the best examples of this is that of *leaded fuels*. For many years, organic lead compounds were added to petrol in order to increase the fuel octane rating and permit more efficient engine operation. However, the lead emitted from these engines constituted a substantial load on the environment. Consequently, unleaded petrol began to appear in the 1980's. It was not until emission regulations became strict enough to require the use of catalysts that a major reduction of the amount of lead petrol produced occurred. The lead in fuel renders the catalyst ineffective after a short time, and therefore cannot be tolerated. Therefore, the stringent requirements on gaseous emissions also have resulted in a major reduction in the amount of lead emitted to the environment.

In order to regain the lost octane characteristics, petrol composition has changed with unleaded petrol. In particular, increasing amounts of aromatics have been added to petrol to maintain octane ratings. Since *aromatics* have some negative environmental impacts, this change has some drawbacks. This is a typical example of technical conflicts which often arise as one attempts to reduce emissions while maintaining engine and vehicle efficiency. Unfortunately, engine operation conditions which lead to high efficiency often lead to high emissions, otherwise emissions problems from vehicles would not be so great today. An alternative to higher aromatic content has been the use of high octane oxygenated additives, such as MTBE (methyl, tertiary-butyl ether), which in addition to high octane ratings have a much lower tendency to produce ozone in the production of photochemical smog. There have recently been reports of MTBE appearing in ground water in areas using MTBE, again indicating the difficulties of finding solutions to environmental problems without undesirable "side-effects".

In the recent years, the role of fuel of determining emissions has been investigated more closely. *The Auto-Oil Programme* in Europe has formed the foundation for new specifications for automotive fuels for the coming decade. Table 5.4 shows the European Commission's estimate of petrol composition for the year 2000 if the specifications in column 3 are adopted, compared to an estimate of the composition which is expected in the year 2000 if no changes were made. Also shown are The European Parliament's proposals for the years 2000 and 2005. At the time of printing, the final requirements have not been established. Much of the discussion concerns the relative cost and benefits of these reductions.

Table 5.4 Proposed petrol specifications for fuels in the European Union.

	Commission Proposal	Commission estimates for market fuels in 2000		Parliament proposals	
	2000	Current	Their proposal	2000	2005
Vapour pressure - kPa	60.0	68.0	58.0	60.0	60.0
Benzene - %	2.0	2.3	1.6	1.0	1.0
Sulphur - ppm	200.0	300.0	150.0	50.0	30.0
Mass % Oxygen	2.3	0.6	1.0	2.7	2.7
Aromatics - vol. %	45.0	40.0	37.0	35.0	30.0

Petrol vapour pressure is important in determining the evaporative emissions from vehicles. Sulphur is important because it can be converted to sulphates (particulates) during warm operation or hydrogen sulphide (odour) in cold operation. In addition, sulphur has a negative effect on catalysts, and very low sulphur levels may be required for future catalyst technologies. Benzene and other aromatics are responsible for a number of toxic and carcinogenic problems. An extra benefit of reducing aromatic composition in fuels is that this also results in slightly lower NO_x emissions. The benefit of oxygen is primarily a reduction in the ozone formation potential of the VOC emissions, as most of the substances used to add oxygen to the fuels have a low photochemical reactivity.

The effects of using a *reformulated petrol* similar to that in the European Parliament proposal for the year 2000 is shown in Figure 5.19 (next page). The reformulated petrol reduces all emissions and the ozone forming potential of the exhaust gas by between 5 and 27%, with the lowest reduction being for the NO_x emissions and the greatest reduction being for the CO emissions.

Proposals for *diesel fuel* are shown in Table 5.5 (next page). The most significant changes proposed for diesel fuel are reduction in the aromatic and sulphur contents. Both of these substances are known to contribute to higher particulate emissions. Sulphur in diesel exhaust reduces the effectiveness of catalysts, of both the oxidation and "lean-NO_x types", and contributes to the long term plugging of particle filters in the exhaust.

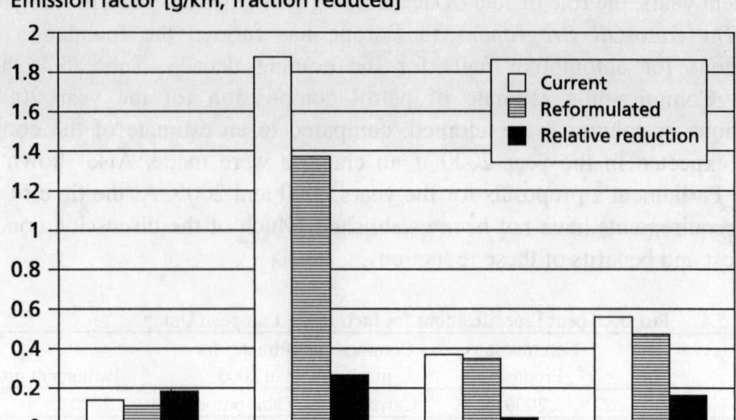

Figure 5.19 The effect of using reformulated petrol on emissions and ozone forming potential on vehicles with 3-way catalysts. The fuel has approximately the characteristics of the European Parliament proposal for the year 2000.

Table 5.5 Proposed diesel specifications for fuels in the European Union.

	Commission 2000	Parliament - 2000	Parliament - 2005
Cetane number - min	51	52	58
Density @ 15°C km/m³ max	845	837	825
Sulphur content - ppm	350	100	50
Polyaromatic Hydrocarbons vol % max	11	6	1
Temperature @ 95% distillation - max	360	350	340

5.5.2 SPARK IGNITION ENGINE EMISSION CONTROL TECHNOLOGY

A large fraction of the engines used for light duty transport consists of spark ignition or petrol engines. Previous to 1992, it was possible to satisfy emissions regulations through engine adjustments, mainly more careful carburettor adjustment and regulation of ignition timing. The 1992 standards, however, were so stringent, that these adjustments were either insufficient, or resulted in an unacceptable deterioration in vehicle performance. Therefore, the use of the *3-way catalysts (TWC)* was adopted for these vehicles. The TWC is a noble metal catalyst in the exhaust systems, which simultaneously reduces CO, HC, and NO_x emissions, hence the name. The TWC requires engine operation very close to stoichiometric fuel air mixtures at all times. Its introduction was therefore accompanied by the introduction of electronic fuel injection, improved ignition systems and computer control of engines. An additional benefit of the TWC was the increase in the amount of unleaded petrol used.

In some instances, the use of a TWC is not adequate for the reduction of NO_x emissions, and the use of *exhaust gas recirculation (EGR)* is required. In this technique, a portion (approx 10-20%) of the exhaust gas is recirculated to the intake system of the engine. This dilutes the intake charge which lowers combustion temperatures and results in lower emissions of oxides of nitrogen.

Another method which has been proposed for emission control has been the *lean burn engine*. In this method, an engine is operated with a excess of air (20-30%). The resulting air in the exhaust system keeps HC and CO emissions low, and the charge dilution with air results in low combustion temperatures and low NO_x emissions. Engine operation under these conditions is much more sensitive than that with the stoichiometric mixtures of the TWC. In addition, NO_x control must be achieved through the combustion process alone, since the TWC will not affect NO_x with oxygen present. A recent development in this area is the direct injection petrol engine, which uses in-cylinder injection of fuel combined with a spark plug to enable lean operation. A major benefit of this type of engine is an improved engine efficiency. However, when the emissions standards become more stringent, the engine modifications necessary to satisfy them substantially reduce these benefits. The newest versions of these engine also switch between lean burn and stoichiometric operation, in order to reduce emissions for conditions where lean burn operation results in too high emissions. The most successful lean burn engines to date have been in stationary applications with natural gas.

5.5.3 DIESEL ENGINE EMISSION CONTROL TECHNOLOGY

There are two types of diesel engine in operation in road vehicles today. One is the *direct injection engine (DI)* in which the fuel in injected into a single combustion chamber. The other is the *indirect injection engine (IDI)*, in which the fuel is injected into a secondary, or pre-chamber where combustion starts and the force of the combustion process is used to force the burning fuel and air into the main combustion chamber for completion. The IDI engine has a lower efficiency due to flow and heat losses associated with the extra chamber. It is only used in applications requiring high speed operation, mainly light duty vehicles. The DI engine has 10-15% higher efficiency than the IDI engine, and is used in applications where fuel consumption is a driving factor, for example buses and heavy duty lorries. In recent years, DI engines have been developed which can operate satisfactorily at the high speeds required for LDV applications.

Because of a mixing controlled combustion process, the diesel engine must always operate with an excess of air, prohibiting the use of the TWC. Thus, *gaseous emissions* reductions to date have been obtained through the improvement of engine design, such as better fuel air mixing, higher fuel injection pressures, and electronic engine control. Exhaust gas after treatment is limited to oxidation catalysts, which help reduce the already very low HC and CO emissions. In addition, oxidation catalysts provide some reduction in particulate emissions by oxidising some of the gas phase fuel components which are absorbed on the particulate matter. Effort is underway to develop a *"lean-NO_x"* catalyst which can reduce the NO_x to nitrogen and water, in spite of the oxidising environment. To date efficiencies of these devices in practice have been under 30%.

Another method which may be used to reduce NO_x emissions in diesel exhaust is that of *selective catalytic reduction (SCR)* of the NO_x. This is a method which has been used with good results on stationary diesel engines. It involves the addition of a reducing agent such as ammonia, urea, or fuel to the exhaust gas, where the reducing agent and the NO_x react on a catalyst to achieve the above reduction. On road application is difficult due to metering problems for rapidly changing loads, but demonstrations of reductions on the order of 60% have been demonstrated for a road vehicle. Due to size and weight restrictions, this technique appears to be only suited to larger diesel vehicles, particularly lorries.

Particulate emissions have been reduced significantly by improving engine combustion conditions. In order to meet the emissions standards shown in Table 5.3 (page 81), engines have been modified to run with higher injection pressures, more carefully optimised combustion chamber geometry and increased use of turbocharging and intake air after-cooling.

An alternate method of reducing particulate emissions is through the use of particulate filters or traps. There are various forms of such traps that introduce a porous barrier in the exhaust system. The exhaust gasses will pass through this barrier, but most of the particles are left behind on the barrier resulting in highly reduced particulate emissions. Filtration efficiencies of well over 90% are possible with reasonable pressure losses. Problems with these traps have been centred on durability and regeneration. The particles accumulate with time and must be burned off periodically, on the order of hours. The exhaust gas temperature is often too low to do this, and so help is needed. This has been in the form of auxiliary burners, fuel additives, catalytic coatings, exhaust gas throttling, oxidation by NO_2 or a combination of these factors. Durability problems have typically been associated with regeneration occurring under inappropriate conditions, resulting in filter destruction, or the lack of regeneration, resulting in filter overload and possible mechanical failure. Significant improvements have been made in this area in recent years, but particle emissions requirements to date do not require the use of such filters, and they are currently mostly used in applications of specially severe local emission problems.

5.5.4 ALTERNATIVE FUELS FOR DIESEL ENGINES

An area which shows good potential for the reduction of emissions from diesel engines is the use of alternative fuels. This area is achieving most interest in the urban environment, where bus engines are of particular interest. Current and near future diesel technology gives NO_x emission levels of about 4-5 g/kWh, with particulate emissions under 0.1g/kWh. Particulate traps can reduce the particulate emissions, but simultaneous reduction of NO_x emissions to under 2 g/kWh and particulate emissions to under 0.03 g/kWh currently requires a change in fuel.

One fuel proposed is *Natural Gas (NG)*, most commonly in the compressed form. This requires the replacement of the diesel engine with a spark ignition gas engine. Gas engines can run either with a TWC or in the lean burn mode, and have very low particulate emissions as well as NO_x emissions levels below those of advanced diesel engines. Lean burn NG engines often have problems with methane emissions at very low NO_x emission levels. Disadvantages with natural gas engines are the more complex refuelling system, and 4 times larger tank size requirement. Engine efficiency in bus operation is also approximately 20% lower than that of the diesel engine. In situations where air quality is bad, this may be accepted to reduce local pollution. LPG has also been used in bus applications with similar benefits as with natural gas, but its efficiency is even lower than that of natural gas engines.

Alcohols such as methanol and ethanol have been used in diesel engines in bus operation. These fuels are not well suited as diesel fuels, and require the use of specially designed engines or the addition of expensive ignition improvers. The oxygen in the fuels has a very positive effect on particulate emissions, and there are few if any particles originating from the combustion process.

Another excellent alternative diesel fuel containing oxygen is *di-methyl ether (DME).* It gives very low particle emissions and NO_x levels of 2 g/kWh or less, without aftertreatment. Contrary to alcohols, DME is an excellent diesel fuel. Like alcohols, DME must be produced from other sources, including natural gas, coal or biomass, although it has been shown possible to produce DME at a lower price than methanol. DME is currently produced in smaller quantities primarily as a cosmetic propellant, and a substantial increase in DME production facilities is needed if it is to be used as a transport fuel.

Fuels based on vegetable oils have been promoted as environmentally friendly, mostly because they are derived from biological sources, although the energy balance/CO_2 emissions for vegetable oils are under discussion. In terms of engine emissions, they are quite similar to diesel engines. One advantage is that the composition of the particles is not a toxic as that of diesel engines, even though the general particulate emission level is near that of diesel fuel.

5.5.5 ALTERNATIVE PROPULSION TECHNOLOGY

Electric vehicles provide the potential for no emissions at the local level, and from that point of view are attractive in a urban application. The problems of the electric vehicle are the well-known difficulties of battery weight and limited capacity. Charging infra structure and time issues also are obstacles to customer acceptance of such vehicles. A compromise between the conventional and electrical vehicle is the hybrid vehicle, which uses a combustion engine to charge batteries on-board, and to provide power directly when battery power is not sufficient. The combustion engine will be smaller than that for a conventional vehicle, and could run at optimum conditions. However, the system is more complex, and the extra electrical components work against the weight reduction from a smaller engine.

Another future possibility is the use of *fuel cells* to generate electricity from directly fuels. One proposal is to generate hydrogen from the decomposition of methanol. Difficulties include the system for the on-board generation of hydrogen, improved materials for the fuel cells themselves and cost.

5.6 Traffic management

5.6.1 DRIVING CONDITIONS

Vehicle driving conditions play a significant role in determining the emission factors for all types of vehicles. Therefore, there is a potential to affect the emissions from mobile sources by some sort of traffic management in order to ensure that vehicles operate at conditions with better emissions characteristics. For a majority of the vehicle classes shown in Figures 5.12 to 5.17 minimum emissions factors are encountered at average trips speeds between 60 and 90 km/h. That is, a smooth traffic flow at a reasonable speed gives the lowest emissions factors. Figure 5.20 shows the effect of a smooth traffic flow for the urban ring street in Copenhagen from Figure 5.9 (page 71). The *"Green Wave"* values represent operation with typical speed variations, but for driving where there are no stops due to traffic lights and other traffic limitations. The "week day" values correspond to driving under conditions where there are more frequent stops.

Figure 5.20 shows that the smooth traffic flow can result in a reduction in VOC and CO emissions reduction on the order of 40% compared to traffic flow with frequent stops. Since emissions factors for all kinds of vehicles in typical urban driving show similar emissions-trip speed dependencies, the "Green Wave" applies to all vehicle types. It is obviously not possible to apply this technique to all streets at all time, but it does show the importance of smooth traffic flow, and indicates that significant emissions reductions can be achieved on major streets.

Another traffic management method which could be used is that of a *bus lane and priority signals for busses*. The idea is to increase the average speed of buses and thereby reduce their emissions factors. A disadvantage of this approach is that it will decrease street capacity for other forms of traffic, reducing their average speed, and hence increasing their emissions factors. Figure 5.21 shows the relative effects of implementing a bus lane and bus priority signals on an urban street in Copenhagen. The overall effects are relatively small, and the total emissions of some substances actually increase.

5.6.2 INSPECTION/MAINTENANCE

An Inspection and Maintenance (I/M) programme aims to ensure that motor vehicle emission control systems are functioning properly throughout the lifetime of the vehicle. It can only be effective if it is technically sound, socially acceptable and not too costly. Among these features, the technical aspects are crucial and have to be considered carefully in order to yield reliable results.

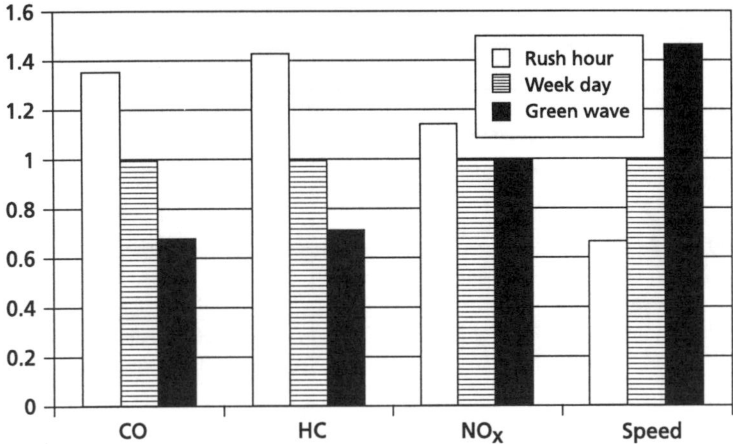

Figure 5.20 Effect of implementing a "Green Wave " on emissions.

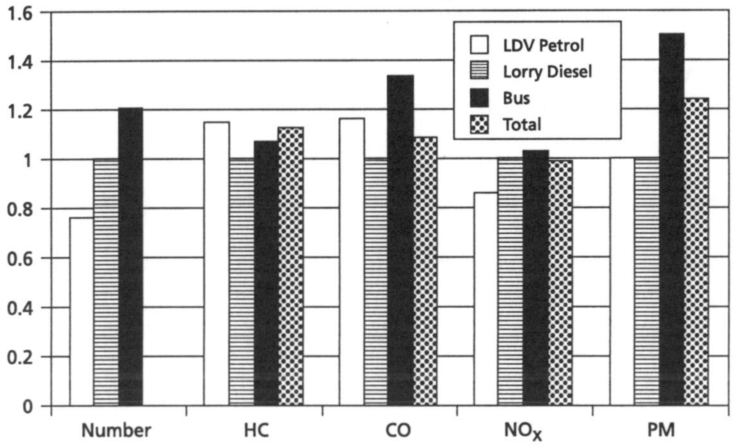

Figure 5.21 Effect of implementing a bus only lane with traffic signal on number of vehicles and emissions.

In the case of catalyst equipped cars, the potential of *maintenance* is illustrated in Figure 5.22 (next page) for CO (Samaras et al. 1997). Trends for HC and NO_x are similar. After thorough maintenance of the high polluters of the sample, total CO emissions are reduced by about 40%, HC emissions by 25%, while the potential in reducing NO_x emissions is lower than 10%. However, these results should be considered as reflecting the maximum potential of I/M in Europe. If the results are compared to the findings of other projects running in parallel (e.g. VROM 1994), then the potential should be closer to 10% for CO, 5% for HC and 0 to 5% for NO_x.

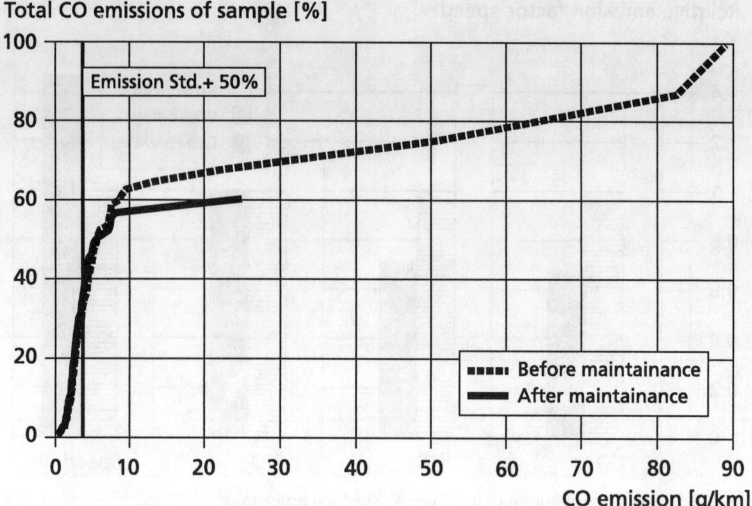

Figure 5.22 The potential of maintenance to reduce the emissions of the catalyst car fleet on the basis of the test results over the NEDC.

Today at the European level only an *idle emissions inspection test* is being implemented in most countries. The effectiveness of this test is under scrutiny in view of the possibility and eventually the need of introducing more sophisticated test procedures to enhance the high polluter identification.

5.7 Summary

Mobile emissions sources have been shown to be major contributors to air pollution in the urban environment. Representative emissions factors for the most important types of mobile sources have been presented, and the effects of vehicle speed and traffic conditions illustrated. Emission factors from most mobile sources have decreased due to legislation, and will continue to decrease in the future. A variety of technical measures are available for future emissions reductions. The discussion here has only focused on technical measures alone. The price of these measures is an important parameter in determining the optimal solution to urban air problems. Consumer choice and traffic management initiatives can also help to reduce urban air pollution. Such measures apply to all technologies.

5.8 References

Degobert, P. (1995) *Automobiles and Air Pollution*, SAE and Editions Technip, Paris.

EEA (1996) Chapter 7: Road Transport, 71 pp, and Chapter 8: Off-Road Transport, 40 pp, in Joint EMEP/CORINAIR *"Atmospheric Emissions Inventory Guidebook"*, Volume 2, Copenhagen.

European Commission (1996) *Air Quality Report of the Auto Oil Programme* - Report of Subgroup 2, Brussels.

ECMT (1995) (European Conference of Ministers of Transport) *1965 - 1990 Statistical trends in Transport,* Paris.

Hansen, J.Q., Winther, M., Sorensen, S.C. (1995) The Influence of driving pattersn on petrol passenger car emissions, *Science of the Total Environment,* **169**, 129-139.

Jørgensen, M.W., Sorensen, S.C. (1997) Estimating Emissions from Rail Traffic, *Proceedings of the 4th International Scientific Symposium, Transport and Air Pollution*, pp. 209-216, Avignon.

Kalivoda, M.T., Feller, R. (1995) ATEMIS A tool for calculating air traffic exhaust emissions and its application, *Science of the Total Environment,* **169**, 241-247.

Krawack, S., Sorensen, S.C. (1993) Traffic Management and Emissions, *Science of the Total Environment,* **134**, 305-314.

MEET (1997) *Methodologies for Estimating Air Pollutant Emissions from Transport. Emission Factors and Traffic Characteristics Data Set Deliverable 21*, Draft Final Report to EC.

Pattas K., Samaras Z. et al. (1998) *Real time measurement of size distribution of diesel particulates*, paper to be presented in SAE international Congress.

Samaras Z., Zierock K.-H. (1994) Assessment of the Effect in EC Member States of the Implementation of Policy Measures for CO_2 Reduction in the Transport Sector, *Commission of the European Communities Document, ISBN 92-826-6505-4,* Luxembourg.

Samaras, Z., Zachariadis, Th., Joumard, R., Vernet, I., Hassel, D., Weber, F.-J., Rijkeboer, R. (1997) Alternative short tests for Inspection & Maintenance of in-use cars with respect to their emissions performance, *Proceedings of the 4th International Symposium Transport and Air Pollution*, pp. 281-288, Avignon, France.

Stanners, D., Bourdeau, P. (eds) (1995) *European Environment Agency Europe's Environment - The Dobris Assessment*, Copenhagen.

Trozzi, C., Vaccaro, R., Nicolo, L. (1995) Air pollutants emissions estimate from maritime traffic in the Italian ports of Venice and Piombino, *Science of the Total Environment,* **169**, 257-263.

VROM (1994) *Project In-Use Compliance - Air Pollution by Cars in Use, Annual Report 1991-1992 including data over the last five years*, Ministry of Housing, Physical Planning and Environment, The Netherlands.

Chapter 6

EMISSION INVENTORIES

RAINER FRIEDRICH and UWE-BERND SCHWARZ
Institut für Energiewirtschaft und Rationelle Energieanwendung
University of Stuttgart, Hessbruehlstr. 49A, D-70565 Stuttgart, Germany

6.1 Requirements and methods

With increasing abatement costs it becomes more and more important to elaborate emission control strategies in order to ensure that the environmental goals are obtained in a cost-efficient way.

The investigation of the potentials and the costs of the different emission reduction measures requires a detailed knowledge of the features of the emission sources - especially the technical processes causing the emissions, the emission factors related to these processes and the amount or degree of utilisation of the technique. Therefore identification of efficient emission reduction measures requires a detailed emission inventory including information about the structure and activity of the emission sources. In the past and to a large extent also at present, such inventories have been based on annual values, spatially divided either into grid cells or administrative units.

However, environmental goals often do not refer to the annual emissions, but to exceedance of certain ambient pollution levels during shorter or longer time periods. In addition, the relations between emission and concentration as well as deposition of pollutants depend on the meteorological conditions, and they are therefore time-dependent (Chapters 10-12). Furthermore, these relation are - even for fixed weather conditions - in many cases highly non-linear. Typical examples are secondary pollutants like ozone, secondary aerosols, and acid deposition. In order to show that emission reduction, measures for this type of pollutants, are efficient and successful, carefully validated atmospheric models, that are able to calculate the expected changes in ambient concentrations and deposition of pollutants due to changes in emissions, must be applied. Such models need emission data with special features, of which the most important are:

- High temporal resolution is required; usually a time resolution of one hour is used, although some models operate with 6-hour intervals.
- Many atmospheric models are nested, i.e. larger scale models provide the boundary conditions for the models analysing the actual investigation area. Therefore emission data must be available on different scales, e.g. beginning with a European data base. This is demonstrated in Figure 6.1, where NO_x emission data for four nested areas are presented. Starting with the whole of Europe in a 54 x 54 km grid, emissions for Germany, East Germany and finally the area of Berlin with a 2 x 2 km grid are shown.
- The pollutants, that have to be covered in the emission inventory, are determined by the chemical mechanisms, which are used in the atmospheric models; e.g. for the RADM2-mechanism VOC emissions must be split into 15 species or groups of species. As an example, Table 6.1 shows a VOC split for emissions from road transport with 25 classes. These 25 classes can then be aggregated again to the classes needed for RADM2, CBMIV or other chemical mechanisms.

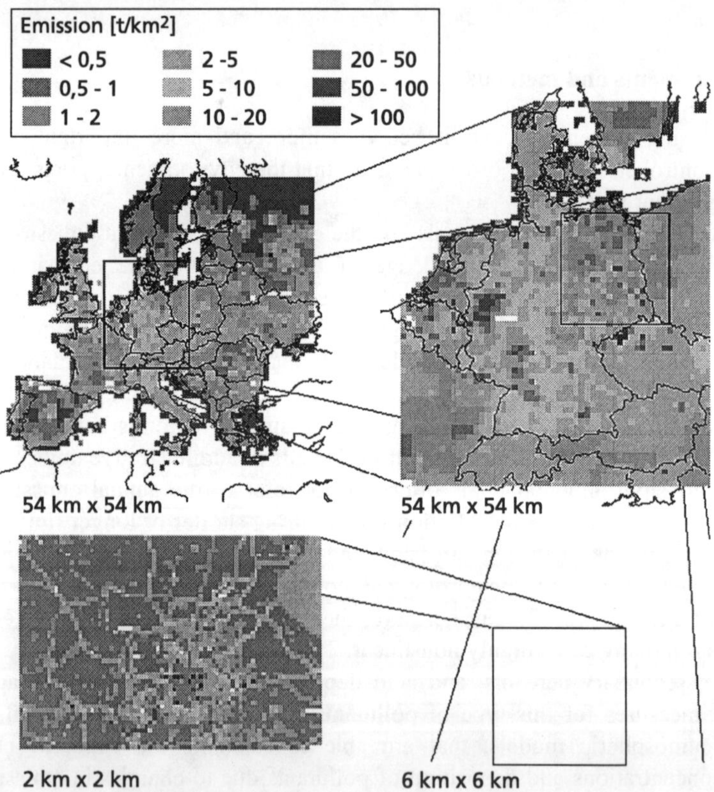

Figure 6.1 Nested grids of NO_x-emissions for 1994 focusing on Berlin (lower left grid).

Table 6.1 VOC-split for road transport (r= rate constant for reaction with OH radicals)

Species	Fraction in %
Methane	5.12
Ethane	0.62
Propane	0.38
Alkanes (r=2,500-5,000 $ppm^{-1}min^{-1}$)	8.60
Alkanes (r=5,000-10,000 $ppm^{-1}min^{-1}$)	24.27
Alkanes (r=10,000-20,000 $ppm^{-1}min^{-1}$)	9.68
Alkanes (r>20,000 $ppm^{-1}min^{-1}$)	8.78
Ethene	5.46
Propene	2.09
1-Alkenes	1.63
x-Alkenes (x>1)	1.22
1-Alkenes/x-Alkenes-mixture	4.55
Benzene, halobenzenes	2.97
Aromatics (r<20,000 $ppm^{-1}min^{-1}$)	6.28
Aromatics (r>20,000 $ppm^{-1}min^{-1}$)	11.00
Styrenes	0.13
Formaldehyde	1.57
Aldehydes >C1	0.90
Acetone	0.35
Organic acids	0.18
Ethine	3.28
Other VOC (r=5,000 $ppm^{-1}min^{-1}$)	0.14
Other VOC (r>10,000 $ppm^{-1}min^{-1}$)	0.09
Not identified VOC	0.52
Not specified VOC	0.38

Such emission inventories are generated by multiplying specific emission factors with the related activities:

$$EM(ijkl) = EF(jkl) \cdot AC(ijk)$$

where:

i = territorial unit or grid cell EM = emission
j = process or class of processes AC = activity
k = time unit EF = emission factor related to AC
l = species

Examples of activity data are:

– The quality and amount of coal burnt and the dry ash extraction in a given power plant during a specific period.
– The amount of paint with a specified solvent content used in a carpentry during a certain time period.
– The number of cars with gasoline engine, a cylinder capacity larger than 2000 cm^3, EURO2 - norm, driving on a specific highway segment at a specific time and with a defined velocity distribution and driving pattern.

An example of an emission factor in the traffic sector is: A passenger car driving with an average velocity of 19 km/h in a city on a flat side road with dense traffic in 1994 in Germany emits 0.86 g NO_x per km.

Obviously such data are usually not available. In some cases, inquiries can be made by asking the operators of emission sources about their energy consumption or, if they have continuous measurement devices, about their measured emissions. On roads, traffic counts can be carried out accompanied by measurement of the velocity distribution and the reading of the car registration plates. Other possibilities are remote sensing measurements. However, regarding the large number of emission sources, in most cases the activity data must be simulated with emission models, that reflect the relations between the activities and the available or known data - e.g. based on statistical or scientific relationships.

6.2 European emission inventories

It is impossible to list all European emission inventories for different urban areas, heavily polluted areas, regions and states here. However, in Europe there are two emission inventories, that are set up officially and on a regular basis: the CORINAIR inventory (CORINAIR 1996) set up for the European Environmental Agency and the EMEP inventory (MSC-W 1996) generated for the EMEP programme of the UN-ECE. These two inventories are harmonised, i.e. they use the same database nomenclature.

6.2.1 EMISSION PARAMETERS

The 1994 EMEP database consists of emission data for 7 pollutants (CO, NO_X, SO_2, CH_4, CO_2, NH_3, NMVOC - Non Methane Volatile Organic Compounds) and 37 European Countries. It contains official national data (anthropogenic and natural) and estimates of land-based emissions over regions within the EMEP modelling area, releases from international shipping, biogenic emissions over sea and land as well as information on the temporal variation of SO_2, NO_X, NMVOC and CO emissions.

The emissions are spatially harmonised on the 50 x 50 km^2 EMEP-grid covering Europe. The CORINAIR data base contains annual emission data for:

- The pollutants CO, NO_X, SO_2, CH_4, CO_2, N_2O, NH_3, and NMVOC; further 9 heavy metals and 10 persistent organic pollutants will be included from 1994 and forward.
- 11 main sectors including one for biogenic emissions, divided into 270 subsectors.
- About 25 European countries, further divided into administrative units.
- Point sources and area sources.

Another important international emission inventory is the IPCC/OECD programme (IPCC 1994), which was initiated in Paris in 1991 by the OECD during a workshop concerning the "Estimation of Greenhouse Gas Emissions and Sinks". Harmonisation of the CORINAIR inventory and the IPCC inventory is envisaged.

6.2.2 CORINAIR DATA

Preliminary CORINAIR emissions for 1994 comprising 20 European countries - the 15 EU member states, Croatia, Iceland, Malta, Norway and Switzerland are:

CO	46,130,595	tons/year
NMVOC	17,423,190	tons/year
CO_2	3,249,477,000	tons/year
SO_2	14,428,813	tons/year
NO_X	12,892,631	tons/year
NH_3	3,658,480	tons/year
CH_4	27,125,886	tons/year
N_2O	1,386,306	tons/year

Not yet included in the CORINAIR '94 inventory are Poland, Romania, Hungary, The Czech Republic, Bulgaria, Slovakia, Lithuania, Latvia, Slovenia and Estonia. Table 6.2 shows the contribution in % of the countries to the overall emission.

The main contributors among the listed countries are France, Germany, Italy and the United Kingdom with large populations. Another important country, Poland, which had a share of 8-14 % in 1990 has not yet been included.

Table 6.2 Relative contribution of the individual European countries (in %) to the overall emission 1994 of the 20 countries (source: CORINAIR).

	CO	NMVOC	CO_2	SO_2	NO_X	NH_3	CH_4	N_2O
Austria	2.56	2.62	1.36	0.38	1.33	2.37	2.33	0.97
Belgium	2.53	2.11	3.50	1.93	2.90	2.16	1.60	2.04
Croatia	1.01	0.82	0.54	0.62	0.45	0.86	1.04	1.24
Denmark	1.55	0.95	1.93	1.10	2.14	2.56	2.89	1.27
Finland	0.95	1.02	2.43	0.77	2.23	1.12	0.87	1.18
France	20.96	15.83	9.48	7.03	13.06	18.23	10.83	16.69
Germany	14.73	14.53	26.91	20.78	17.58	17.02	17.87	15.76
Greece	0.67	6.00	2.29	3.71	1.78	12.16	1.77	12.81
Iceland	0.06	0.04	0.07	0.06	0.17	-	0.08	0.04
Ireland	0.72	1.01	0.86	1.22	0.91	3.41	3.11	1.87
Italy	20.01	15.98	13.14	25.45	16.73	10.63	15.50	9.92
Luxembourg	0.31	0.11	0.29	0.09	0.17	0.19	0.08	0.05
Malta	0.05	0.03	0.07	0.11	0.13	0.17	0.03	0.94
Norway	1.87	2.09	1.16	0.24	1.69	0.68	1.72	1.02
Portugal	2.59	3.60	1.66	1.89	1.94	2.53	1.41	3.98
Spain	10.43	11.31	8.03	14.28	9.52	9.42	11.66	14.12
Sweden	2.86	3.94	2.67	0.51	3.45	1.38	7.24	3.04
Switzerland	1.19	1.85	1.33	0.21	1.11	1.66	1.38	1.56
Netherlands	2.00	2.20	5.32	0.94	4.18	4.72	4.39	4.29
United Kingdom	12.94	13.95	16.96	18.69	18.52	8.75	14.19	7.18

The source categories with the largest relative contributions to the emissions of the individual pollutants are listed in Table 6.3.

Table 6.3 Relative contribution from the main sectors in % of the overall recorded European emissions in 1994 (Source: CORINAIR).

	CO	NMVOC	CO_2	SO_2	NO_x	NH_3	CH_4	N_2O
Public Power/ Heating Plants	1.23	0.86	32.32	50.88	18.93	0.09	0.18	4.57
Comm., etc. Combustion Plants	12.88	3.27	19.40	7.18	4.44	0.04	1.53	2.12
Industrial Combustion	5.76	0.36	18.29	16.41	9.45	0.05	0.19	2.29
Production Process	5.39	5.87	4.71	3.99	1.75	2.83	0.17	22.12
Fuel extraction and distribution	0.23	6.00	0.08	0.22	0.88	-	13.82	-
Solvent Use	-	23.56	0.13	-	-	0.06	-	0.68
Traffic	61.35	27.97	20.74	3.34	47.37	0.96	0.62	3.96
Other mobile Sources	7.55	5.03	3.95	1.89	15.87	-	0.07	1.09
Waste Treatment and Disposal	4.66	1.28	2.09	0.58	0.84	1.09	30.44	1.65
Agriculture	0.74	17.17	-1.80	-	0.29	94.65	39.52	34.45
Nature	0.21	8.64	0.09	15.50	0.17	0.23	13.47	27.06

As it appears, the most important sources among the main sectors in CORINAIR are:
- "Traffic" for the pollutants CO, NO_X and NMVOC.
- "Power plants and industrial combustion" for SO_2 and CO_2.
- "Agriculture" for NH_3, N_2O and CH_4.

6.3 Temporal resolution

Inventories like CORINAIR contain annual emissions. However, as mentioned above, usually a higher time resolution, e.g. hourly emissions, is needed for modelling purposes. First approaches to temporally resolved emissions applied in the past were based on plausible assumptions, and they yield simple patterns of relative emissions in time (Lenhart, Heck, Friedrich 1996). These patterns can be multiplied by annual emissions and result in first order estimations of the seasonal and daily variations of the emissions. National or regional differences of the temporal behaviour of emission sources are in this case neglected. Neither has - with the exception of biogenic emissions and NMVOC evaporation - the strong influence of the temperature on emissions been taken into consideration in the past.

These methods are regarded to be too rough for a reliable estimation of the emissions. Therefore, methods to estimate hourly emissions with higher accuracy and reliability have been developed within the EUROTRAC subproject "GENEMIS" by analysing the temporal behaviour of the processes that cause the emissions. These methods use various available data like e.g. traffic counts, energy consumption, production indices, working hours and user behaviour as well as meteorological data (ambient temperature, degree days) as indicators for the temporal resolution in data for activities relevant for the emissions. Table 6.4 gives an overview of the most important indicators for fourteen emission source sectors (Lenhart, Friedrich 1995).

Some emitters show a very strong temporal variation in pollutant emission, e.g. traffic, whereas the output of others is almost constant (e.g. plants in the primary industry). The emissions of industrial combustion depend on production rates controlling the energy consumption for production processes, outside temperature controlling energy consumption for space heating and production or working hours. Small consumer combustion includes households, institutional and commercial fuel consumers, farms, etc. and is related to fuel consumption, outside temperature, working hours and regional user behaviour. For both sectors equations have been derived, that give the time variation as a function of these parameters.

Table 6.4 Indicators used for the estimation of the temporal variation in emissions.

Sector	Indicators for temporal segregation
Public power	Fuel use, load curve
Refineries	Fuel use, working hours, holidays
Small consumers	Fuel use, degree days, production, user behaviour
Industrial combustion	Fuel use, temperature, degree days, production, working hours, holidays
Production processes	Production, working hours, holidays
Fuel extraction and distribution	Production, working hours
Solvent use	Production, working hours, holidays
Road transport	Traffic counts, road statistics
Gas evaporation	Temperature, traffic counts, road statistics
Air traffic	LTO cycles, passengers, freight
Mobile sources	Working times, user behaviour
Waste treatment	Time factors
Agriculture	Use of fertiliser, animal breeding
Nature	Temperature, land use

The sectors: Industrial, commercial and private solvent use pose some problems for emission inventorying, because they are characterised by a huge number of small and heterogeneous emission sources. A reasonable assumption is, that emissions are related to production, working times, holidays and user behaviour. Hourly emissions of these sectors can be estimated according to working hours. For private solvent use higher emissions are assumed during the summer and weekends.

As an example of the results generated with such methods, hourly NMVOC-emissions of all main sectors are presented in Figure 6.2 (next page) for the Stuttgart area during the episode 29th of July - 5th of August 1990 (Sunday to Sunday). The rush hours in the morning and in the evening can easily be identified. Total day and night emissions range may differ up to one order of magnitude.

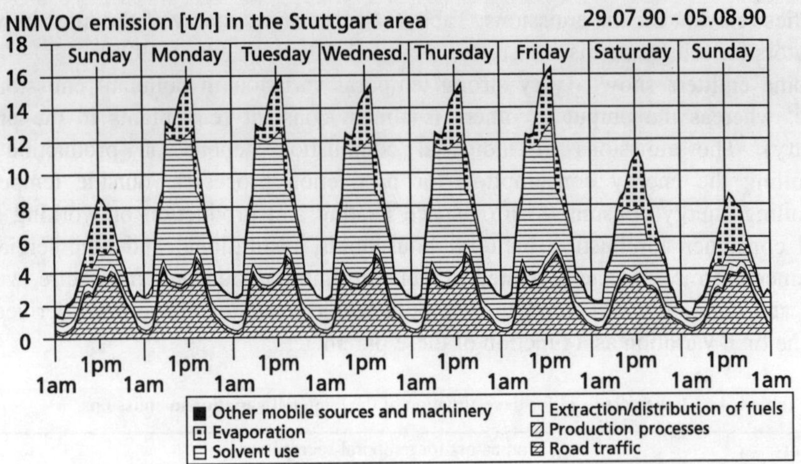

Figure 6.2 Total hourly NMVOC-emissions of all main sectors within 8 days in the Stuttgart area.

As mentioned above the contribution from road traffic to the total NO_x-emission is significant (Figure 6.3). The contributions from other sectors change from region to region, depending on the location of power plants and various industrial facilities, temperature etc. Figure 6.4 clearly demonstrates the difference in VOC evaporation from vehicles in three European cities (Stockholm, Hamburg and Madrid).

Another example for the temporal distribution of emission data is the total daily NO_x-emissions in Germany, France and Greece (Figure 6.5). In France and Germany, the emissions are considerably higher during the winter than during the summer (except during the Christmas holidays). Significantly lower (20-30%) emissions are observed during weekends than on working days.

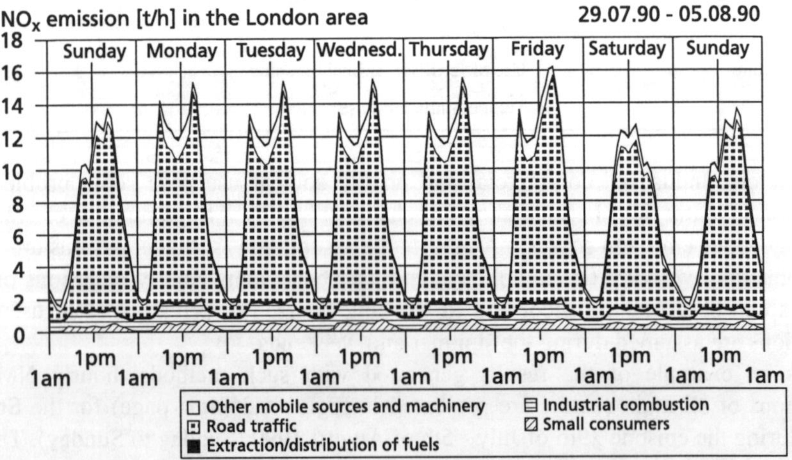

Figure 6.3 Total hourly NO_x-emissions of all main sectors within 8 days in the London area.

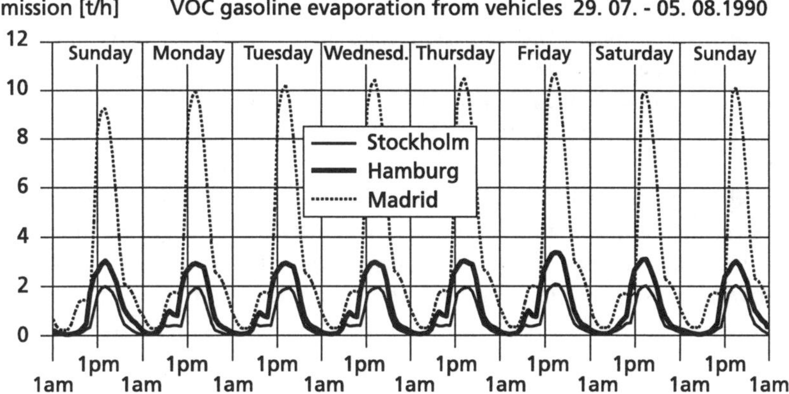

Figure 6.4 VOC gasoline evaporation from vehicles in 3 European cities.

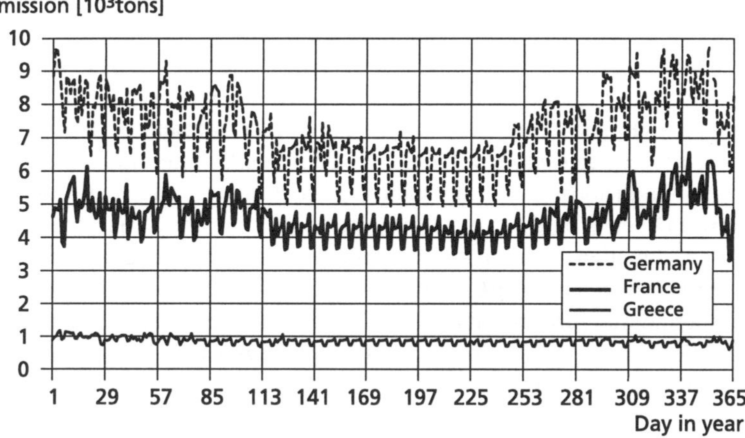

Figure 6.5 Daily anthropogenic NO_x-emissions of Germany, France and Greece in 1990.

These variations are typical for most emissions of air pollution, which show a strong seasonal, weekly and diurnal variation. At night, emissions are typically by factors 5-20 lower than during daytime. During the weekends, the emissions from anthropogenic sources 20-30% lower than during workdays. The ozone relevant VOC/NO_x-ratio is in summer higher than in winter, during the day higher than by night and (at least during the summer) higher during the weekends than during the weekdays (Lenhart et al. 1995).

6.4 Spatial resolution

The basic spatial resolution in emission data is in principle restricted by the spatial resolution of the activity data. For large point sources, the co-ordinates of the stack are known. Line sources, i.e. roads and streets, are usually given as polygons. For area sources (all sources, which are neither point nor line sources) basic data are available for administrative units. The emission data given in these units then must be transformed into the spatial units used by the atmospheric model. This transformation is carried out with a geographical information system (GIS).

Sometimes the spatial resolution achieved with the methods explained above is not sufficient. In such cases, the spatial resolution can be improved by using land use data or other types of geographical information. For example, land use data from satellites may give residential areas with high and low density of building and industrial areas; emission from households or industry can be allocated to these areas. If the co-ordinates of motorways are known, emissions from motorway transport can be assigned to these co-ordinates.

6.4.1 MAP PROJECTIONS

As the Earth is an ellipsoid, geographical locations can only be exactly referenced by a non-Cartesian co-ordinate system - the *Global Reference System.* In this co-ordinate system with co-ordinates longitude and latitude the units of measure (degrees, minutes, seconds - DMS) represent different distances depending on the location on the Earth. For example 1° longitude on the Equator equals about 111,000 km whereas 1° longitude at 60° latitude covers ca. 56,000 km. In order to calculate distances and areas it is easier to project the surface of the Earth on a planar 2-dimensional map.

However, the distortion of geographical properties is an inherent problem of map projections. It depends on the type of projection, which of the basic properties - shape, area, distance, direction - will be more or less distorted. As there are many applications of map calculation, there are just as many different projections. In addition, the Earth is no exact spheroid: the surface is uneven and the South Pole is closer to the Equator than the North Pole. Therefore, some projections are preferably used for the North American continent, whereas others produce less distorted maps for the European continent. It depends on the use of the map, on the part of the Earth to be examined and on the overall size of the map area.

Maps are classified according to the basic property they preserve. Each of them can preserve one basic property, or if all of them are essential all to some degree:

- Equal-area maps preserve all areas
- Conformal maps preserve local shapes
- Equidistant maps preserve certain distances
- Azimuthal or true-direction maps preserve certain directions

As for spatial resolution of emission inventories geographical data are collected from different international sources, the incoming maps are of many different projection types. If the maps contain exploitable information they must first be transformed by map projection into a standard co-ordinate system.

6.4.2 VARIATION OF THE NMVOC/ NO_x-RATIO

The spatial variations of the emissions as well as the NMVOC/NO_x-ratio are large. Table 6.5 shows as an example the influence of the grid size for grids, which include the city of Stuttgart, on the NMVOC/NO_x-ratio. As it can be seen, the NMVOC/NO_x-ratio in a 5 x 5 km^2 grid of the city of Stuttgart is 2.6 higher than the average ratio of the region. Also, the ratio is much higher during summer days at daytime than the annual average. This illustrates the importance of the temporal and spatial variation of emissions and of using atmospheric models with sufficiently small grid size and time steps.

Table 6.5 NMVOC/ NO_x-ratios within selected grid elements in Baden-Württemberg. In brackets: anthropogenic emission only.

	Grid size [km^2]	Annual average	Friday, 03.08.1990 11-12am
Densely populated	60 x 60	1.7 (1.6)	2.8 (2.3)
Rural	60 x 60	1.9 (1.4)	3.7 (1.7)
Baden-Württemberg	150 x 200	1.5 (1.2)	2.5 (1.5)
Stuttgart area	60 x 60	1.7 (1.6)	2.8 (2.3)
Town area	25 x 20	1.9 (1.8)	2.8 (2.6)
City of Stuttgart	5 x 5	3.9 (3.9)	7.7 (7.5)

As an example for the spatial and temporal resolution of emission data Figure 6.6, next page, shows the NMVOC-emissions in Europe for certain hours (1 am, 8 am, 1 pm and 9 pm) of Monday, 1st of August 1994. As it appears the emissions are significantly lower in the night than during the day except for some large emission sources with constantly high activities.

6.5 Uncertainties

To assess the reliability of results obtained from atmospheric models, information concerning the accuracy of the input data, and especially of the emission data, is needed. However, it is difficult to give e.g. standard deviations for emission inventories, since in most cases information about deviations of input parameters for the emission models, i.e. activities, statistical data and emission factors are not available. Nevertheless, the following two procedures to assess the accuracy of emission data can be applied:

- Efforts to determine the probability distribution for the key parameters used to calculate emissions should be made; if no other information is available, expert judgement should be used. Then a classical error analysis should be carried through. Such analyses just have started to be carried out. A first result for the estimation of NO_x-emissions on a specific highway road segment is, that the mean error is 23% and the maximum error 50% of the calculated emissions (John, Friedrich, Kühlwein, Obermeier 1996). This uncertainty mainly comes from the uncertainty in the emission factors, followed by uncertainties in the assumptions concerning technology and velocity distribution.
- Emission data, activity data or emission factors calculated or measured with different methods or for different locations and periods of time can be com

an example, a comparison between CORINAIR NMVOC emissions and emissions that were calculated for Baden-Württemberg using very detailed methods by Obermeier, Friedrich, John, Seier, Vogel, Fiedler and Vogel (1995) revealed a difference of only ca. 5% for Baden-Württemberg as a whole but differences of up to a factor of two when looking at smaller administrative units (Kreise).

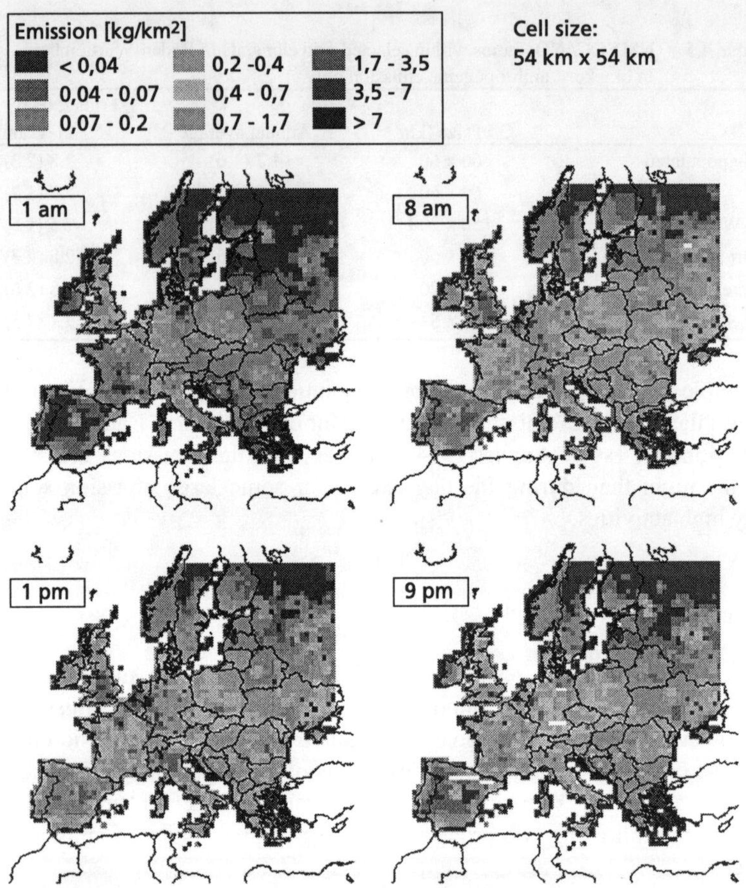

Figure 6.6 NMVOC-emissions on a European scale at different times during Monday, August 1st, 1994.

However, the insight in uncertainty ranges gained with these procedures may be incomplete, as systematic errors may occur (e.g. due to unknown sources, neglecting of diffuse emissions, malfunctioning emission reduction devices, wrongly used emission factors, missing information) and as often similar methods and emission factors are used for the generation of emissions in different inventories. So, the comparison of estimated emission data with independently measured data is indispensable.

For emissions from road transport, a number of tunnel studies have been carried out. By measuring concentrations at the entrance and the exit of the tunnel, emissions inside the tunnel can be estimated and compared to results from emission models. In the

United States, some tunnel studies (Pierson, Gertler, Robinson, Sagebiel, Zielinska, Bishop, Stedman, Zweidinger, Ray 1996) revealed, that so-called high emitters, i.e. cars, where the three way catalyst is not working, may be a cause for underestimation of emissions by models. In an experiment at the Gubrist tunnel in Switzerland, deviations between measured and calculated emissions amounted to between 10 and 35% (John, Friedrich, Staehelin, Schläpfer, Keller, Stahel, Steinemann). As driving pattern and air resistance in a tunnel are different from the situation outside, measurements are also carried out downwind and upwind of roads outside tunnels. First results of a study made on a motorway near Mannheim (Fiedler, Friedrich 1997) showed, that CO emission may be underestimated by emission models.

Future plans to assess the accuracy of emission inventories include measurements of emission fluxes for a whole city by measuring upwind and downwind fluxes using air planes, balloons, air ships and ground stations.

Emission rates can also be determined by using *tracers* (usually SF_6). A known quantity of the tracer is released within emission sources and concentrations of the tracer and the pollutants are measured downwind. From the concentration ratios pollutants/tracer and the tracer release rate, the emission of the other pollutants can be estimated.

The measurement of concentration ratios of individual substances itself can be used to determine the reliability of emission data, as this ratio reflects the ratio of the emissions. For example, measurements of the CO/NO_x and $NMVOC/NO_x$ concentration ratios carried out in several US cities revealed, that the measured ratios are significantly (factor 2 to 7) higher than the ratios calculated with emission models (NRC 1991).

A very large uncertainty is believed to be connected with *biogenic emissions*, the uncertainty range for biogenic emission is estimated to be a factor 2 to 5. Currently a number of experiments as well in Europe (EUROTRAC) as in the US are directed towards reducing this uncertainty range.

Although some results hint on large differences between measured and calculated VOC and CO emissions and VOC/NO_x and CO/NO_x-ratios, these results cannot be generalised, as the different emission inventories use different methods and parameters and so have different accuracy's. The determination of the uncertainties is currently a major research topic, so that new findings will be available within the next years.

6.6 References

CORINAIR (1996) *Atmospheric Emission Inventory Guidebook*, European Environment Agency, DK-1050 Copenhagen, Denmark.

Fiedler, F., Friedrich, R. (1997) *Projekt "Experimentelle Überprüfung von Emissionsdaten für den Kraftfahrzeugverkehr auf Autobahnen".*

IPCC (1994) Intergovernmental Panel on Climate Change - *IPCC, Guidelines for National Greenhouse Gas Inventories* (3 volumes), Cambridge University Press.

John, C., Friedrich, R., Kühlwein, J., Obermeier, A. (1996) Abschätzung und Bewertung der Unsicherheiten hochaufgelöster NO_x und NMVOC- Emissionsdaten im Projekt Europäisches Forschungszentrum für Maßnahmen zur Luftreinhaltung (Hrsg): *13. Statuskolloquium des PEF, Bericht FZKA-PEF*, 221-233.

John, C., Friedrich, R., Staehelin, J., Schläpfer, K., Keller, C., Stahel, W., Steinemann, U. *Comparison of Emission Factors for Road Traffic from a Tunnel Study and from Emission Modelling*, to be published.

Lenhart L., Friedrich R. (1995) European emission data with high temporal and spatial resolution, *Water, Air & Soil Pollution*, **85**, 1897-1902.

Lenhart, L., Heck, T., Friedrich, R. (1996) *The GENEMIS Inventory European emission data with high temporal and spatial resolution*, Institute of Energy Economics and the Rational Use of Energy, University of Stuttgart, Germany.

MSC-W (1996) *MSC-W Status Report 1996 Part One*, the Norwegian Meteorological Institute, N-0313 Oslo 3 Norway.

NRC (1991) National Research Council, *Rethinking the ozone problem in urban and regional air pollution*, Committee on Tropospheric Ozone Formation and Measurement, National Academic Press, Washington, D.C.

Obermeier, A., Friedrich, R., John, C., Seier, J., Vogel, M., Fiedler, F., Vogel, B. (1995) *Photosmog* Ecomed Verlag, Landsberg, Germany.

Pierson, W.R., Gertler, A.W., Robinson, N.F., Sagebiel, J.C., Zielinska, B., Bishop, G.A., Stedman, D.H., Zweidinger, R.B., Ray, W.D. (1996) Real-World Automotive Emissions - Summary of Studies in the Fort Mc. Henry and Tuscarora Mountain Tunnel, *Atmospheric Environment*, **30**, 2233 - 2256.

BASIC ATMOSPHERIC PHENOMENA

The main part of the mass of the earth atmosphere (between 80 and 90%) is located in the lowest 10 km above the surface. However, most of the physical and chemical processes governing transport, dispersion and deposition of air pollution take place in an even thinner layer - the so-called atmospheric boundary layer (ABL). The ABL varies highly in depth, but it is generally confined to the lowest 2 to 3 km of the atmosphere. Pollutants are emitted into this layer and advected with the wind, and dispersion and deposition of air pollution is therefore governed by the mixing ability of the lower atmosphere.

The basic physical processes in the atmosphere are expressed in thermodynamic and hydrodynamic laws. An central parameter is the stability of the atmosphere which is crucial for the prediction of weather as well as pollution conditions and may thus be regarded as a measure of the ability of the lower atmospheres of mixing air pollution. At unstable conditions the turbulent exchange of air masses cause a fast mixing of the lower atmosphere, whereas stable conditions lead to low mixing. Differential surface heating by short wave solar radiation drives the exchange of air masses on as well local as regional scale. The absorption of solar radiation depends on type of surface. The heated surface emits long wave radiation (heat) which may form thermal turbulence and thereby strongly affect the mixing of the atmosphere.

Gases and particles may be removed from the atmosphere by dry and wet deposition. Dry deposition is the direct removal when the compound comes in contact with the surface. This process is depending on chemical properties of the compound, the mixing of the atmosphere and the surface characteristics. Wet deposition consists of in-cloud and below-cloud scavenging. In-cloud scavenging is the uptake in cloud droplets and below-cloud scavenging the uptake in falling rain drops. The In-cloud scavenging is generally the most important due to the much longer lifetime of cloud droplets compared with falling rain drops.

These basic physical processes governing *transport and deposition* of pollutants in the ABL are described in Chapter 7.

Advection, dispersion and deposition govern the concentrations of primary (emitted) air pollutants. However, not all pollutants are emitted in significant quantities; a group of species - secondary pollutants - are formed by chemical reactions. Understanding of the processes forming these compounds is fundamental, since a large fraction of the health impacts, effects on vegetation and material damage is associated with secondary pollution. Examples are photochemical compounds like ozone, nitrogen dioxide, and peroxy acetyl nitrate, and acidifying compounds like sulphuric acid, nitric acid, particulate sulphate, nitrate and ammonium.

The hydroxyl radical (OH) drives the chemical reactions in the atmosphere during daytime and is responsible for a significant part of the chemical conversion of hydrocarbons in the atmosphere. This is despite the very low OH concentration in the background tropospheric boundary layer on the order of 0.05-0.5 ppt (10^6-10^7 molecules $\cdot$ cm^{-3}). The OH oxidation of the various hydrocarbons emitted into the troposphere plays an important role in the formation of photochemical oxidants like ozone and PAN. At night time the nitrate radical (NO_3) takes over from OH as the most important oxidant in the troposphere. Despite the considerably lower reactivity compared with OH, its higher peak concentrations allow NO_3 radicals to play a major role in the chemical transformations of organic compounds.

Nitrogen dioxide episodes have been observed during wintertime in Northern Europe. For the cold and dark periods with low mixing, different chemical mechanisms have been suggested to be responsible for these episodes, but the mechanisms are still not fully explored.

Tropospheric chemistry with relation to urban air quality is reviewed in Chapter 8.

Particles play a major role in tropospheric air pollution. Especially in the urban environment, where the concentration of fine fractions aerosol particles often is significant, they are considered to be a health hazard. An aerosol is defined as a system of a gas and solid or liquid particles, which remain suspended for at least several minutes. In the background troposphere, high aerosol particle concentrations may lead to substantial reductions of visibility.

Particles are directly emitted from e.g. combustion processes or chemically formed in the atmosphere. For the latter, the transformation of sulphur dioxide and nitrogen dioxide into sulphuric acid and nitric acid and the subsequent reactions with ammonia play a major role. Newly formed as well as directly emitted particles are found in the fine fraction, but may grow through nucleation of particles and condensation of water vapour and various gases on the particle surface.

Usually particles are grouped into three so-called modes: nucleation, accumulation and coarse particle modes. Air quality scientists use terms as coarse, fin, PM_{10} and $PM_{2.5}$. PM_{10} and $PM_{2.5}$ refer to the size fractions of the suspended particulate matter with diameters less than 2.5 and 10 µm, respectively. The particles which appear to be most injurious to human health are fine (or accumulation mode) particles. Coarse fraction particles appear not to be so damaging to human health and act more as nuisance through the soiling of surfaces through the accumulation of grit and dust.

An introduction to the *physics of aerosol particles* is given in Chapter 9.

Chapter 7

DYNAMICAL AND THERMAL PROCESSES

KNUT E. GRØNSKEI
Norwegian Institute of Air Research
P.O.Box 100, N-2007 Kjeller, NORWAY

7.1 Introduction

This chapter gives a brief introduction to the physical processes in the atmospheric boundary layer (ABL) which to a large degree governing the relations between emissions and pollution concentrations in the atmosphere. The atmospheric processes may be expressed by the *main thermodynamic and hydrodynamic parameters* i.a.: pressure, temperature, mass, volume, density, water vapour, radiation, wind velocity and turbulence. The relations between these parameters are expressed in the thermodynamic and hydrodynamic laws of the atmosphere, that are formulated in the four conservation equations (of mass, heat, momentum and water/other chemical compound, the gas law (or equation of thermodynamic state) and the radiation laws. A general introduction to the equations is given in the following. For further information see the references given in section 7.7. In Chapter 10 these equations are discussed in the context of the simplifications that must be applied in models describing mesoscale phenomena. It is important to note that the governing atmospheric processes take place on a broad range of temporal and spatial scales (Figure 7.1, next page).

In the processes with general impact on transport and dilution of pollution in the atmosphere, the behaviour of *small scale eddies* is important for dilution of pollution released from individual chimneys and in street canyons.

In the description of turbulence, it is common to apply the Reynolds averaging approach in which the wind velocity is described in time into an average and a fluctuation component: $V = \overline{V} + V'$. This is a useful approach for analysis of observations from monitoring stations. However, for numerical solution of equations with finite difference approximations, it is better to use an averaging procedure with respect to space, and a fluctuating part to account for sub-grid processes. Such a procedure was adopted by Deardorf and others and is used in the following. The fluctuating part is here characterised by statistical parameters.

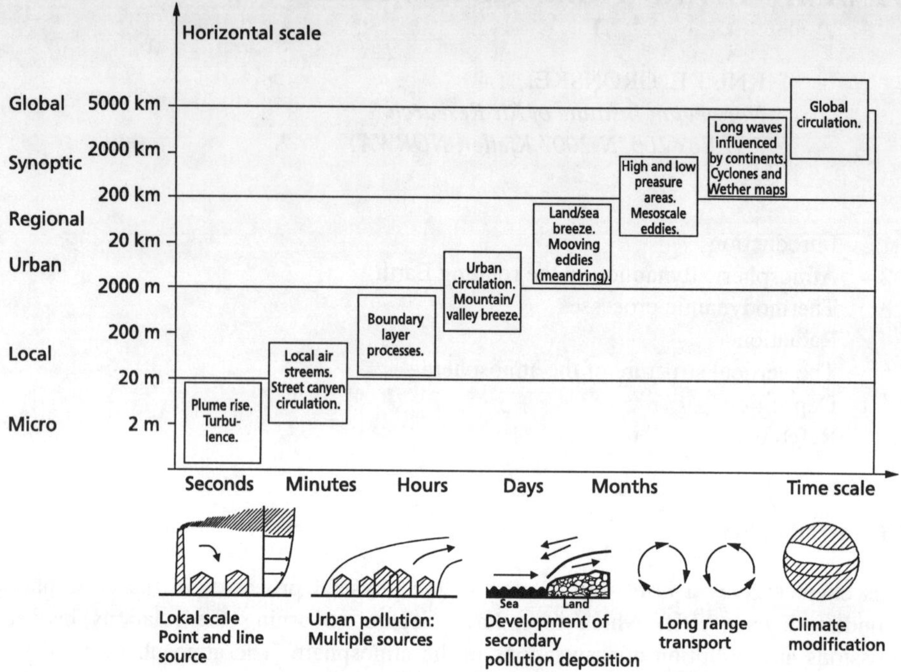

Figure 7.1 Time and spatial scales of atmospheric processes governing the dispersion and deposition of air pollution.

7.1.1 PREDICTABILITY

The predictability of atmospheric motions in time and space is generally constrained by the limited knowledge about the atmospheric state. This applies in particular to the fluctuating components of the atmospheric parameters. The atmospheric conditions may for a certain situation be classified as unstable. In this case, the subsequent development in the conditions is difficult to predict. However, stable conditions are easier to describe. Accordingly, it is important to discriminate between stable and unstable atmospheric conditions in order to predict the development in weather as well as pollution conditions.

All over Europe large resources are allocated for the characterisation of the atmospheric conditions leading to the development and movements of high and low pressure systems. New measuring techniques and the availability of satellite observing platforms have made it possible to predict the variation in weather parameters with reasonable accuracy for time periods of a few days. The fluctuating component of the wind can, however, only be described by statistical parameters and this cause inherent uncertainties in the real time description and prediction of air pollution concentrations.

Regarding the eddies characterised by smaller space and time scales, detailed description and prediction of the urban scale eddies is still beyond the current technical possibilities. However, prediction of quasi-stationary systems of motions may easier be described (Chapter 11).

7.1.2 CLIMATE

The climate in Europe varies considerably from south to north. The variation in climate influences highly the living conditions and the general characteristic of urban areas (Chapter 1). The climate is influenced by parameters like:

- the intensity of the solar radiation which is a function of latitude,
- the surface type of the region (e.g. distribution between land and sea) which due to the differences in heat capacity highly affect the temperature variations, and
- the height above sea level of the terrain.

On local to urban scale the micro-climate is affected by parameters like:

- the local topography, and
- the structure of the surface.

Geological observations show that strong variations in climate have occurred on Earth for many periods. The variations have tentatively been explained by changes in solar radiation and by continental drift on the Earth surface.

During the 19th century rising temperature and increased precipitation have been observed. Based on measurements and results of model calculations is has been argued that the increasing atmospheric content of greenhouse gases: (CO_2, CH_4, H_2O and CFC) already has changed the global energy balance of the Earth resulting atmosphers system, some of the change in climate.

7.2 Atmospheric dynamics on the rotating Earth

The basic equation for conservation of motion as a function of time and location may be written in the following way:

$$\frac{\partial \overline{V}}{\partial t} = -\overline{V} \cdot \vec{\nabla} \overline{V} - \frac{1}{\rho} \nabla p - g\vec{k} - 2\Omega \cdot \overline{V} \quad - \frac{1}{\rho} \vec{\nabla} \left(\rho \, V' \, V' \right) \tag{7.1}$$

Forces: Pressure Gravity Coriolis Friction

where $\overline{V}$ is the average wind velocity, V' the fluctuation around the average value ($\overline{V'} = 0$). These two parameters may also be expressed as

$$V = \overline{V} + V' = (\overline{u} + u')\boldsymbol{i} + (\overline{V} + V')\boldsymbol{j} + (\overline{w} + w')\boldsymbol{k} \tag{7.2}$$

which is the expression for the wind velocity in a rectangular co-ordinate system and ***i, j, k*** are unit vectors along the three axes. $\vec{\nabla}$ is the gradient operator ($i\partial/\partial x + j\partial/\partial y + k\partial/\partial z$) and ρ is the atmospheric density, p the atmospheric pressure, Ω the angular velocity of the Earth. The local value of average wind velocity change as a result of advection of average wind velocity ($\overline{V} \cdot \vec{\nabla}\overline{V}$), the local value of pressure force ($1/\rho\vec{\nabla}p$), gravity acting along the vertical axes ($g\vec{k}$), the Coriolis force ($2\Omega \cdot \overline{V}$) and the friction ($1/\rho\vec{\nabla} \cdot (\rho V'V')$).

The pressure force is a results of the spatial variation of atmospheric pressure. Figure 7.2 illustrates the pressure force acting on a small volume of air: ΔV=Δx·Δy·Δz, *P* is the normal force per unit area on the left side (Δy·Δx) side of the cube. The pressure on the right side equals $\left(p + \frac{\partial p}{\partial x}\Delta x\right)\Delta y \Delta z$. The pressure difference is equal to the net forces acting on the cube in the x-direction.

$$\left(p - \left(p + \frac{\partial p}{\partial x}\Delta x\right)\right)\Delta y \Delta z = -\frac{\partial p}{\partial x}\Delta x \Delta y \Delta z \tag{7.3}$$

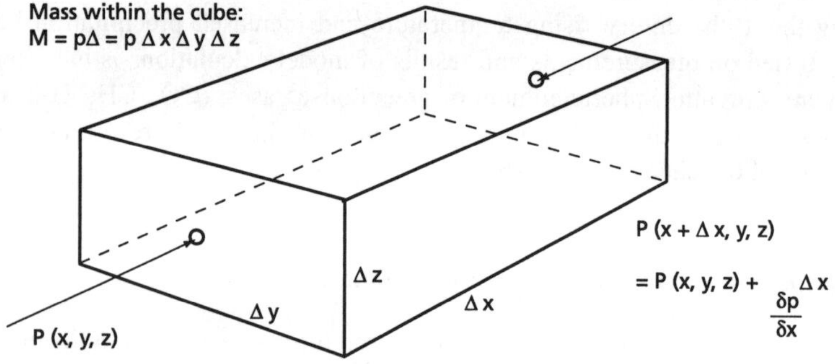

Figure 7.2 The x-component of the pressure force.

The air mass within the cube equals $\Delta x \Delta y \Delta z \cdot \rho$ and the pressure force per unit mass in the x-direction becomes

$$\frac{\partial p}{\partial x} \cdot \frac{\Delta x \Delta y \Delta z}{\rho \Delta x \Delta y \Delta z} = -\frac{1}{\rho}\frac{\partial p}{\partial x} \tag{7.4}$$

The pressure forces acting in other directions are calculated in the same way, and the net pressure force the in three directions follows as:

$$-\left(\frac{1}{\rho}\frac{\partial p}{\partial z}i + \frac{1}{\rho}\frac{\partial p}{\partial y}j + \frac{1}{\rho}\frac{\partial p}{\partial z}k\right) = -\frac{1}{\rho}\nabla p \tag{7.5}$$

Gravity. In most meteorological problems the Earth may be considered as a perfect sphere with radius of 6,371 km. The gravity represents the net force acting on a unit mass rotating with the Earth. At mean sea level, the gravity acts perpendicular to the sea surface, and is often expressed by the potential gradient.

The Coriolis force. The conservation of momentum is expressed by Newton's second law. In atmospheric sciences force normalized by mass is considered and used in the equations. The acceleration represents the change of velocity with time following an object in an inertial coordinate system:

$$\vec{a} = \frac{d_a V_a}{dt} \tag{7.6}$$

The Earth is rotating with constant angular velocity Ω. The velocity V_a of an air parcel may be written as the sum of the velocity relative to the Earth (V) and the velocity resulting from the rotation of the Earth (Ω · R), where R is a position vector of the parcel of air as measured from the centre of the Earth. Figure 7.3 illustrates the component of the angular velocity along the rectangular co-ordinate system following Earth of the surface.

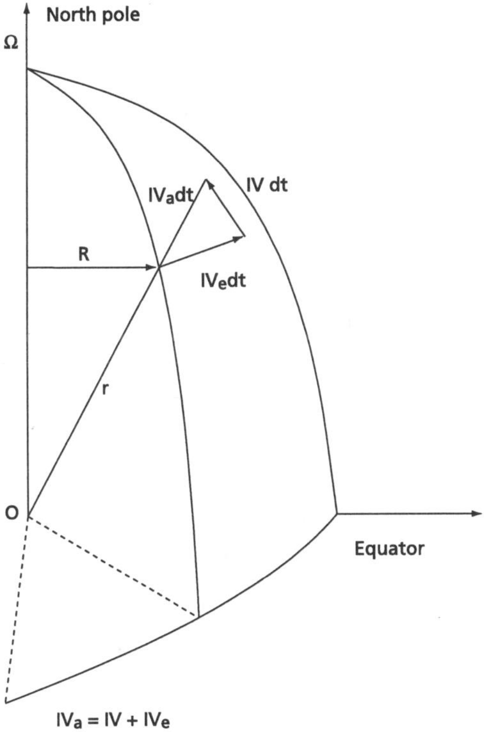

Figure 7.3 Absolute and relative velocity on the rotating Earth.

$$a = \left(\frac{d}{dt} + \Omega\right) V_a = \frac{dV}{dt} + 2\Omega \cdot V + \Omega \cdot (\Omega \cdot R) \tag{7.7}$$

where the first term describe the acceleration for a given a co-ordinate system following the Earth surface. The second term is the Coriolis acceleration or the Coriolis force, proportional and perpendicular to the wind velocity. The third term is the centripetal acceleration which is smaller than the Coriolis term and the effect is often considered in combination with the force of gravity.

The shallow layer close to the surface where *friction* is important, support the well being and even survival of life on Earth. The dispersion of air pollution also occur in this layer. The small scale eddies called turbulence refer to the irregular almost random fluctuations in wind velocity as well as air temperature and humidity. A considerable part of the solar radiation reach the surface. The energy received by the surface is transmitted to the atmosphere by turbulent transport of heat and water vapour as well as by radiation. The turbulent layer close to the surface (the ABL) is characterised by friction and exchange of heat by turbulent motion and has an extend to the height of 500-1000 m.

The friction of fluid flow is caused by turbulent exchange of momentum. The retardation of an air stream due to random motion of gas molecules is referred to as the molecular viscosity. When laminar flow gives way to irregular turbulent motion, which has an effect similar to molecular viscosity, but also include larger eddies in the air stream, the internal friction produced by turbulent whirling eddies is termed eddy viscosity. Near the surface, it is related to the roughness of the surface. As the airstream blows over the surface it breaks into a series of irregular eddies that can influence the air flow for hundreds of meters.

Eddy motions created by roughness elements like houses and trees are termed *mechanical turbulence*. Mechanical turbulence creates a frictional drag on the air streams far greater than that caused by molecular viscosity.

Surface heating influence the air streams and may cause the development of eddy motion. This type of turbulence is called *thermal turbulence*. Turbulent eddies may also develop above the surface as the result of sharp wind gradients. The instability of air streams as a result of sharp wind gradients is termed Helmhotz instability. When turbulent eddies are created in the upper air this is referred to as "clear air turbulence". Their size may vary from a few meters to several hundred meters. From a practical point of view these eddies may influence the dispersion of effluents from tall stacks, and they furthermore influence the comfort and safety of air plane passengers.

The divergence of the stress tensor describes the change of momentum in an air parcel as a result of change in turbulent momentum fluxes (Equations 7.1 and 7.8).

$$\frac{1}{\rho}\nabla \cdot (V'V') = \frac{1}{\rho}\nabla \cdot \begin{pmatrix} u'u'\,\mathbf{ii}, & u'v'\,\mathrm{ij}, & u'w'\,\mathbf{ik} \\ v'u'\,\mathbf{ji}, & v'v'\,\mathrm{jj}, & v'w'\,\mathbf{jk} \\ w'u'\,\mathbf{ki}, & w'v'\,\mathbf{kj}, & w'w'\,\mathrm{kk} \end{pmatrix} \tag{7.8}$$

7.2.1 FRICTION IN THE SURFACE LAYER

When the Reynolds averaging procedure is applied for the flow in the friction layer close to the surface, the fluctuations in air density are of minor importance compared with the fluctuations in wind velocity, temperature and moisture. Accordingly, the density may be approximated by a local constant. Since the friction layer is relatively thin, the changes in density are of minor importance for the description of friction in the equation of motion. Accordingly, it is sufficient to consider the co-variation of fluctuations in wind-components when the friction term, F, is approximated by the divergence of the Reynold stress tensor $\nabla\,\overline{(V'V')}$:

$$F = \frac{1}{\rho}\vec{\nabla}\cdot\left(-\rho\, V'\cdot V'\right)$$

$$= \left[\left(\frac{\partial\overline{u'^2}}{\partial x}+\frac{\partial\overline{u'v'}}{\partial y}+\frac{\partial\overline{u'w'}}{\partial z}\right)+\left(\frac{\partial\overline{u'v'}}{\partial x}+\frac{\partial\overline{v'^2}}{\partial y}+\frac{\partial\overline{v'w'}}{\partial z}\right)\boldsymbol{j} + \left(\frac{\partial\overline{v'w'}}{\partial x}+\frac{\partial\overline{v'w'}}{\partial y}+\frac{\partial\overline{w'^2}}{\partial z}\right)\boldsymbol{k}\right] \quad (7.9)$$

When the horizontal scale of the air stream under consideration is much larger than the vertical scale, it is a reasonable approximation to omit the terms for horizontal variations i.e. all terms described by derivation with respect to x and y. When the average motion is predominantly horizontal, the vertical component of friction may be omitted:

$$F_h = -\rho\left(\frac{\partial(u'w')}{\partial z}\mathrm{i}+\frac{\partial(v'w')}{\partial z}\mathrm{j}\right) \quad (7.10)$$

7.3 Thermodynamic processes

The thermodynamic processes include the description of the relation between mechanical work and heat. Based on two basic laws, relations between thermodynamic parameters are given. The first law of thermodynamics describes the conservation of energy. The second law of thermodynamics describes the principles of transport and transformation of energy. A system of thermodynamics is described by pressure, volume, temperature, mass, entropy and internal energy.

7.3.1 EQUATION OF STATE FOR AN IDEAL GAS

The air temperature is a measure of the translation energy of the molecules in a thermodynamic system. The walls of the thermodynamic system (container) are exposed to bombardment by the gas molecules. The total effect of all these impacts is equivalent to a normal force, acting upon each part of the wall, and the pressure is the force per unit area. For a gas in equilibrium, the molecules move in all directions without preference, and the pressure is independent of the orientation of the surface, which is expressed by:

$$pV = RT \tag{7.11}$$

A mole is defined as a unit of mass numerically equal to the molecular weight of the substance. For a mole of any gas, the relation between p, V and T is the same. Dividing by the mass of the system on both sides of the equation:

$$p\alpha = \frac{R}{M}T \tag{7.12}$$

where $\alpha = V/M = 1/\rho$ is the specific volume, and $R' = R/M$ is the specific gas constant. Even though no ideal gas actually exists, the gases constituting the atmosphere may for all practical purposes be treated as ideal gases. Avogadro found that one mole of any gas occupied 22.4 litre at 0°C and 1 atmosphere. One mole of a gas has therefore a gas constant of: $m \cdot R' = R^* = 8.31$ Joules $mole^{-1}$ K^{-1}, where R^* is the universal gas constant and m is the molecular weight of the considered gas. Table 7.1 show the content in air of the four main constituents.

Table 7.1 Main constituents of the atmosphere.

Gas	Mole weight	Mass (%)
Nitrogen	28.016	75.52
Oxygen	32.000	23.15
Argon	39.444	1.28
Carbon dioxide	44.010	0.05

Using the data from Table 7.1 and applying the universal gas constant, the gas constant for the mixture of gases in ambient air can be calculated assuming that:

- all gases obeys the equation of state, and
- the total pressure arising from the mixture of gases may be obtained as the sum of partial pressures exerted by each of the gases.

$$R'_{air} = \frac{\sum_{j=1}^{4} MjR_j}{\sum_{j=1}^{4} M_j} \tag{7.13}$$

where $R_j = R^*/m_j$ and m_j is the mole weight of component j. When data from Table 7.1 are used in Equation 7.13 to determine the gas constant for air: $R'_{air} = 0.287$ Joule g^{-1} K. According to the first law of thermodynamics, the heat added to a thermodynamic system equals the change in internal energy plus the work done by the system i.e.

$$\delta H = du + pd\alpha \tag{7.14}$$

where δH *is* a small amount of heat, du is the change in internal energy, and $pd\alpha$: is the work done by a unit mass of the system. The specific heat capacity of a gas is defined as the amount of heat that are necessary to increase the temperature by one degree (K). The specific heat capacity may be determined at constant pressure (c_p) and at constant volume (c_v) i.e.

$$c_p = \left(\frac{dH}{dT}\right)_{p=constant} \qquad c_v = \left(\frac{dH}{dT}\right)_{v=constant} \tag{7.15}$$

According to the definitions and the first law of thermodynamics $du = c_v dT$. Differentiation of the equation of state reads

$$pd\alpha + \alpha dp = RdT \tag{7.16}$$

Applying the equation for conservation energy, it is found

$$dH = (c_v + R)dT + \alpha dp \tag{7.17}$$

$$\left(\frac{dH}{dT}\right)_{p=const} = (c_v + R)dT \tag{7.18}$$

Accordingly $c_p = c_v + R$.

7.3.2 ADIABATIC PROCESSES AND POTENTIAL TEMPERATURE

An adiabatic process is a thermodynamic change of state, in which there is no heat exchange between the system and the environment. From the equation for conservation of energy and the equation of state (Equations 7.17 and 7.11) it follows

$$0 = c_p \frac{dT}{T} - R\frac{dp}{p} \tag{7.19}$$

and with integration it follows that

$$\frac{T}{T_o} = \left(\frac{p}{p_0}\right)^{R/c_p} \tag{7.20}$$

This equation is used for the following definition of potential temperature. The potential temperature θ of the atmosphere at temperature T and pressure p is the temperature obtained by adiabatic compression or expansion to $p_o = 1000$ mb.

$$\theta = T\left(\frac{p_o}{p}\right)^{R/c_p} \tag{7.21}$$

7.3.3 MOIST AIR

The atmosphere contains variable amounts of water vapour, liquid water, ice and snow. The phase change between vapour liquid and solid phase involves conversion of considerable amounts of energy that are taken from or given to the surrounding space.

The latent heat of melting, L_m, denotes the amount of heat necessary to transform one gram of ice to one gram of water at the same temperature. The latent heat of vaporisation, L_V, is the amount of heat required to change one gram of ice to vapour at the same temperature and the latent heat of sublimation L_s is the amount of heat necessary to transform one gram of ice to one gram of vapour at the same temperature. In order to preserve energy of reversible processes

$$L_s = L_m + L_v \tag{7.22}$$

At ambient atmospheric pressures the equation of state for water vapour follows the equation of state for a perfect gas i.e.:

$$e = \rho_v R_v T \tag{7.23}$$

where e is the vapour pressure, T is the temperature, ρ_v is the density of water vapour and R_V = 0.461 Joule g^{-1} K^{-1}. Based on the thermodynamic theory, the Clausius-Clapeyrons equation describes the relation between the temperature and the saturation pressure for water vapour.

$$\frac{1}{e_s}\frac{de_s}{dT} = \frac{L_v}{R_v T^2} \tag{7.24}$$

For the saturation pressure over ice the latent heat of vaporisation is exchanged with the latent heat of sublimation L_s .

One measure of the water vapour content in air is given by the *mixing ratio* which is defined as:

$$m = \frac{\rho_v}{\rho_d} \tag{7.25}$$

where ρ_V is the density of the water vapour in the atmosphere and ρ_d is the density of dry air. Based on the equation of state for dry air and water vapour: m = 0.628 e/(p-e).

A second measure is the *specific humidity* which is defined as:

$$q = \frac{\rho_v}{\rho_d + \rho_v} = \frac{0.622\, e}{p - 0.378\, e} \qquad (7.26)$$

For practical purposes, dry adiabatic variation of temperature with pressure applies also for adiabatic expansion or compression for unsaturated moist air.

A third measure of the water vapour in the atmosphere is the *relative humidity* which is given by:

$$R = \frac{e}{e_s} \qquad (7.27)$$

Finally there is the *dew-point temperature*, T_d, which is defined as the temperature that an air mass will obtain when the air is cooled by constant pressure until the saturation pressure is met:

$$E(T_d) = e_s \qquad (7.28)$$

7.3.4 RADIATION

All objects emit energy in the form of electromagnetic waves. The energy flux increase and the wavelength of emitted radiation decreases with increasing temperature. The Earth and the atmosphere receive heat by absorption of electromagnetic radiation from the sun, and simultaneously emit energy to space as electromagnetic waves depending on the temperature of the Earth-atmosphere system. When the radiation is absorbed by the Earth or by the atmosphere, it is transferred to heat.

From a point source energy is emitted in all directions. Considering two concentric spheres equal amount of energy is passing each of the spheres if no absorption occur in the space between the spheres (Figure 7.4, next page). The flux of energy through an area perpendicular to the radiation is proportional to $1/r^2$:

$$F_o = \frac{F}{r^2} \qquad (7.29)$$

The flux per unit area in a direction K is termed the radiation intensity, and differ with the location and direction K:

$$I_K = \frac{F_K}{A}$$

for which a typical unit will be Wm^{-2}. Figure 7.5 (next page) shows the modification of the flux of energy, when the area is oblique to the direction of radiation.

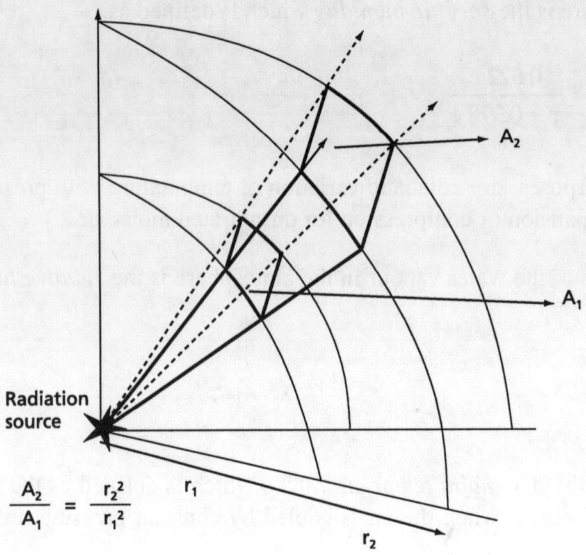

Figure 7.4 Constant flux of energy through spheres.

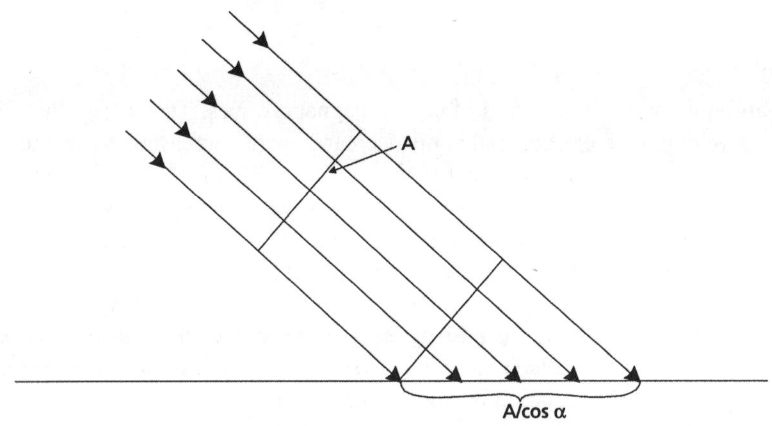

Figure 7.5 The flux of energy through a surface that is not perpendicular to the direction of radiation.

$$E = \frac{|I|}{r^2} \cos\alpha \qquad (7.30)$$

Usually radiation consists of waves with different wavelengths. The behaviour of radiation, regarding for example absorption and scattering, varies with the wavelength. Therefore it is necessary to introduce monochromatic radiation i.e. radiation at a specified wave length λ. The total radiation intensity is the integration of the monochromatic radiation intensity with respect to wavelength.

$$I = \int_0^\infty i \; d\lambda \tag{7.31}$$

7.4 Radiation

7.4.1 RADIATION LAWS

When radiation meet a surface a part is absorbed and another part is reflected with the same wavelength. The absorbed part is transformed to heat leading to an increase of the temperature of the object. For a body which absorb all the received energy regardless of the angle of incidence, the surface is termed a black body. Such a body emits radiation according to *Planck's law*:

$$I_\lambda(T) = \frac{2hc^2}{\lambda^5\left(\exp\left({}^{kc}/_{\lambda\, kT}\right) - 1\right)} \tag{7.32}$$

where h is Planck's constant (usually in Js), c is the speed of light and k is Boltzman's constant. According to Equation 7.32 the radiation intensity varies with respect to temperature and wavelength.

Wien's displacement law. From Equation 7.32 it is found that for a specific temperature the maximum monochromatic radiation intensity occur for

$$\lambda_{max} = \frac{k_4}{T} \tag{7.33}$$

where k_4 equals 2,884 μm·K. Figure 7.6 (next page) shows the conditions of solar radiation and Earth radiation. Equations 7.32 and 7.33 are used to calculate the values. The figure shows the intensity of solar radiation and of Earth radiation as function of wavelength.

Stefan-Boltzman's law. When Equation 7.32 is integrated over all wave lengths, the density of energy flux from a black body (f_b) is obtained:

$$f_B = H\int_0^\infty I_{B,\lambda} d\lambda \tag{7.34}$$

where $f_b = \sigma T^4$, $T = 6.669 \cdot 10^{-8}$ W m^{-2} K^{-4}. The law was established experimentally by Stefan and was expressed theoretically by Boltzman.

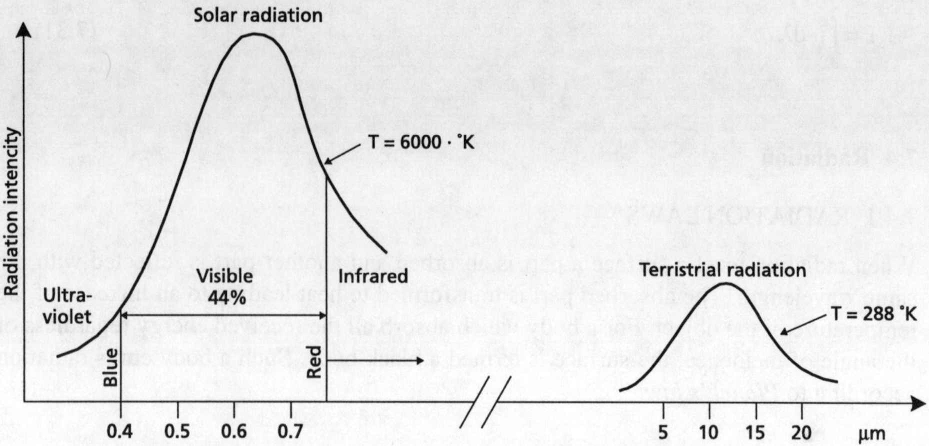

Figure 7.6 Intensity of solar and terrestrial radiation.

Kirchhoff's law. Considering monochromatic radiation reaching a surface under an angle of incidence θ, a fraction a_λ will be absorbed. According to Kirchhoff's law, the surface will emit radiation in the opposite direction according to:

$$I_\lambda = a_\lambda I_{\beta,\lambda}(T) \tag{7.35}$$

Kirchhoff's law relates absorption and emission at a given wavelength with temperature.

Beer's law states that when the radiation pass through mixture of gases and particles constituting the atmosphere, the intensity decreases as a result of the absorption and dispersion taking place:

$$\delta I = -I(k+d)\delta m \tag{7.36}$$

where I is the radiation intensity, k is the coefficient of absorption, d is the coefficient of dispersion and m is the mass. When I_0 is the incoming radiation. This may be expressed as:

$$I = I_0 e^{-(k+d)m} \tag{7.37}$$

7.4.2 SOLAR RADIATION

As the solar radiation travels through space essentially nothing interferes with the process until the radiation enters the Earth atmosphere. The solar constant (having a value of 13,700 W m^{-2}) is the intensity of the energy flux received per unit area perpendicular to the radiation outside the atmosphere, when the Earth is at an average distance from the sun. Only small fluctuations in the solar constant are observed. Upon entering the atmosphere a number of interactions take place. Energy is absorbed by gases such as ozone in the upper atmosphere. Dust particles and air molecules scatter incoming solar radiation in all directions. Shorter wavelength are scattered more effectively than radiation with longer wavelengths. The ratio between outgoing and incoming solar radiation which represent the reflectivity of a surface, is termed the "albedo". Surfaces characterised by a high albedo are white. For water and dark soil, the albedo is small and the absorbed radiation is correspondingly high. Table 7.2 show typical values of the albedo for different surface types. These data demonstrate the large variability between typical surfaces in nature.

Table 7.2 Typical values of the albedo (α) for various types of surfaces.

Surface	α (%)
Fresh snow	75-95
Dark soil	5-40
Sand	15-45
Thick clouds	60-90
Forest	15-20
Water	3-10
Ice	30-45

The scattering of the solar radiation may be divided in scattering caused by air molecules, and by particles and aerosols in the atmosphere.

When the solar radiation pass through the atmosphere it is affected. *The energy of the short wavelengths* (Ultraviolet part of the spectrum) is reduced as a result of absorption by ozone in the upper atmosphere. The shorter wavelengths are scattered more effectively than longer wavelengths. *In the visible range*, the energy is reduced primarily as a result of scattering by molecules and particles. *The energy of the long wavelengths* (infrared part of the spectrum) is reduced primarily as a result of absorption by water vapour. Figure 7.7 (next page) shows the average daily radiation reaching the Earth as a function of latitude and season. The figure shows that in northern Europe (60-70°N) the average daily radiation at summer solstice read 450-500 W m^{-2}, whereas at winter solstice radiation is less than 50 W m^{-2}. In southern Europe (40°) the daily average of radiation varies between 200 W m^{-2} at winter solstice and 450 W m^{-2} at summer solstice. The daily average intensity varies with latitude during winter, but not much during summer.

The cloud cover and the content of water vapour in the atmosphere are particularly important for the perturbation of the solar radiation before reaching the surface; only about 25% reaches the surface in overcast conditions. Accordingly the degree of cloudiness is highly important for the climate and for the heat balance close to the surface. The albedo varies considerably with the structure of the surface (Table 7.2). Furthermore the topography modifies the angle of incidence and the solar intensity at the surface. Parameters like topography, structure of the surface, the water vapour content and the cloudiness influence the radiation intensity and the local heat balance.

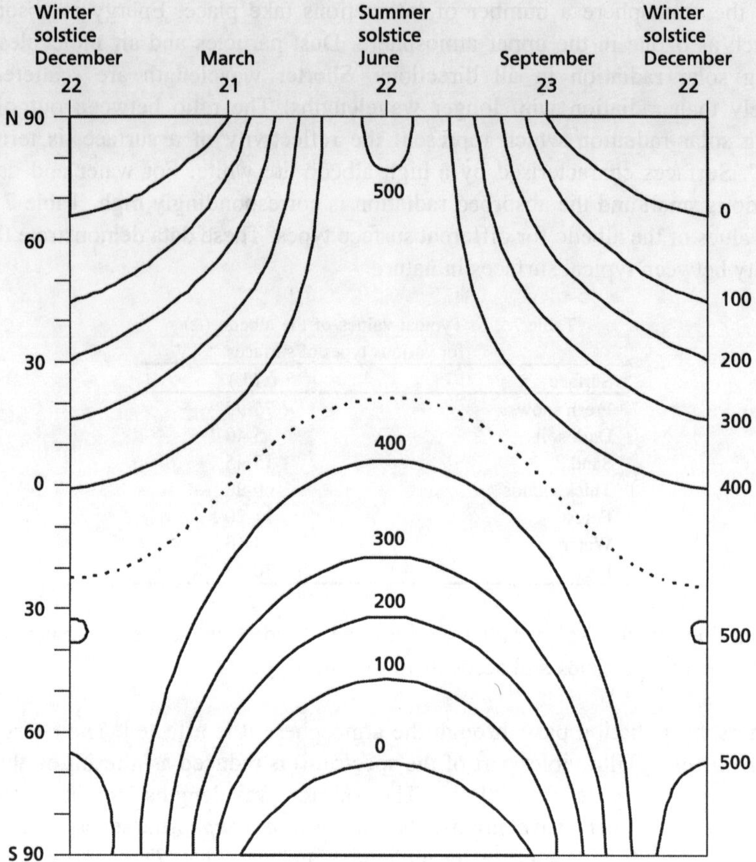

Figure 7.7 Daily insolation as a function of latitude and time of the year.

In high pressure situations, the large scale wind is weak and a general subsidence in the atmosphere cause poor dispersion conditions close to the surface and pollution episodes occur as a result of urban and local scale emissions. For given atmospheric conditions, variations in the local heat balance influence the local dispersion conditions. When dispersion models are applied for studies of local and urban scale pollution episodes resulting from antropogenic sources, the influence of local air streams may be of primary importance.

7.4.3 THE TERRESTRIAL RADIATION

Depending on the temperature the surface emits radiation nearly as a black body. The atmosphere absorbs some of this radiation and emits radiation depending on the temperature distribution in the atmosphere. The effective radiation from the Earth becomes, where for clear sky conditions E_A is about 75% of $\sigma \cdot T^4$:

$$E_{eff} = \sigma T_o^4 - E_A \tag{7.38}$$

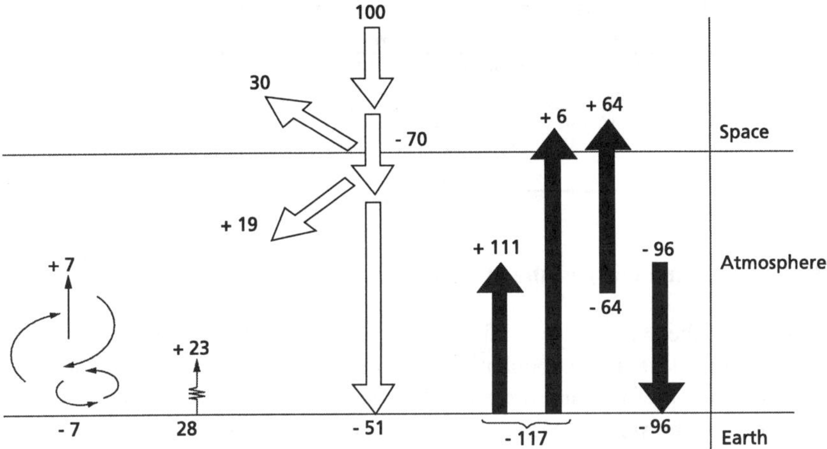

Figure 7.8 Typical energy fluxes in the Earth-atmosphere system.

In order to study the heat balance of the Earth-atmosphere system the following parameters have to be included:

- Solar radiation,
- Earth's albedo,
- Terrestrial radiation,
- Convection and conduction of sensible heat, and
- Evaporation and latent heat.

Table 7.3 (next page) is based on the data given in Figure 7.8, and shows the relative importance of the various processes in each of the regions: space, atmosphere and surface.

- 70% of the total solar radiation is absorbed by the surface and the atmosphere (51% by the surface and 19% by the atmosphere). The energy emitted from the atmosphere is equal to the terrestrial radiation (6% from the Earth and 64% from the atmosphere).
- 19% of the solar radiation is absorbed in the upper atmosphere and 59% in the lower atmosphere. 160 units are emitted from the atmosphere as long wave radiation, 64 units go to space, 96 units to the surface.
- 117 units are emitted from the Earth surface as long wave radiation (111 to the atmosphere, 6 units go to space). The transfer of 23 units as latent heat and 7 units as sensible heat from the Earth to the atmosphere balance the heat budget in both the atmosphere and the Earth. Each of the processes varies horizontally and with a number of parameters. For the Earth-atmosphere system as a thermodynamic system it is likely that several equilibrium situations appear; Long term fluctuations may occur.

Table 7.3 The heat balance of the Earth atmosphere system. Unit: 100: the average solar radiation 338 W/m^2.

	Solar radiation	Terrestrial radiation	Evaporation	Convection	Balance
Space	70 = 100-30	70 = 6+64			0
Atmosphere	19	-49 = -160+111	23	7	0
Surface	51	-21 = -117+96	-23	-7	0

7.5 The vertical structure of the atmosphere

The structure of the atmosphere is influenced by gravity and solar radiation. The density and the pressure decreases exponentially with height above the surface. The vertical distribution of the physical and thermodynamic parametres is shown in Figure 7.9. The pressure close to the surface corresponds to the weight of a ten m thick layer of water. At the surface the density of the atmosphere is 1.3 kg m^{-3} and the temperature between 250 and 310 K. The dry atmosphere behaves to a good approximation as an ideal gas following Equation 7.5 (page 112).

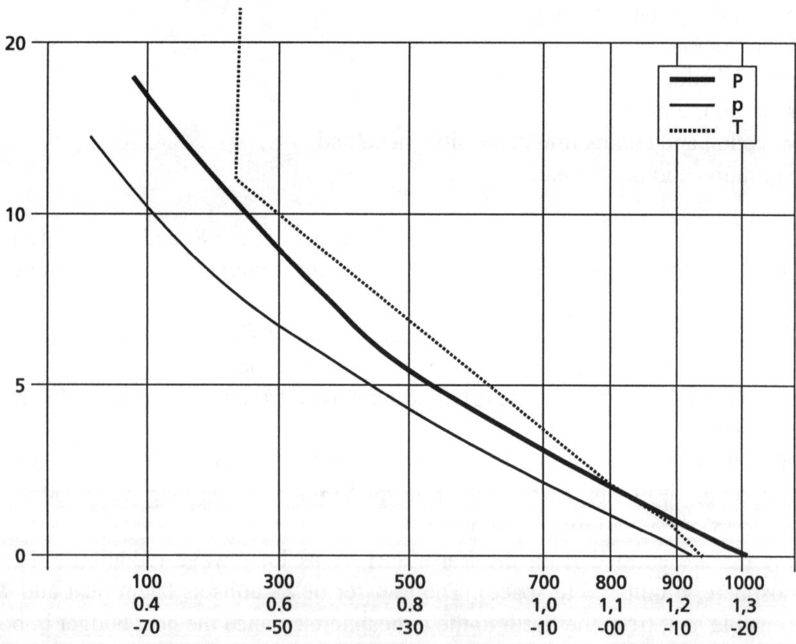

Figure 7.9 Variation of thermodynamic parametres with height above the Earth surface.

The temperature distribution is locally influenced by dynamical processes and by solar radiation. The lowest part of the atmosphere, the troposphere, is limited at the top by the tropopause. The troposphere is characterized by temperature decrease with height and appreciable water vapour content. The weather processes occur in this part of the atmosphere. The tropopause, the boundary between the troposphere and the stratosphere is characterized by an abrupt increase in temperature with height. Usually the vertical wind velocity is small in the atmosphere, and the hydrostatic approximation may be applied.

$$\frac{\partial p}{\partial z} = -g\rho = -g\frac{p}{RT} \tag{7.39}$$

When the temperature distribution is known, the corresponding distribution of pressure and density may be obtained. Emission, dispersion and deposition of pollution primarily occur in the lowest hundred metres of the atmosphere, the planetary boundary layer (PBL), also termed the friction layer. The turbulent exchange is important for:

- the exchange of horizontal momentum in the horizontal equation of motions,
- the turbulent exchange of sensible and latent heat is the thermodynamic energy equation,
- the buoyant force influences the vertical motion and is proportional to the Vaisala-Brundt frequency for small perturbations:

$$\upsilon_s^2 = \frac{g}{T}\left(\frac{dT}{dz} + g / C_p\right) \tag{7.40}$$

7.5.1 STABILITY

In Figure 7.10 (next page) stable and unstable temperature stratifications (lapse rates) are illustrated by comparing with the dry adiabatic temperature variation with height. Plume behaviour in stable and in unstable atmospheric conditions is shown facing the stratification curves.

The static stability of the atmosphere is a simple way to categorize the structure of turbulence. The behaviour of the plumes is important for transport and dilution of pollution concentrations (Figure 7.10). The dispersion of point source releases is furthermore dependent on the horizontal wind which is generated by the wind above the mixing height and local modification by friction and local wind systems, i.e.

- land/sea breezes,
- mountain/valley circulations, and
- urban/rural circulations.

7.5.2 VERTICAL STRUCTURE OF THE ATMOSPHERIC BOUNDARY LAYER

Figure 7.3 (page 113) shows the ABL divided into three sub-layers. The layer from the surface and up to the roughness length (z_0), termed the roughness layer. The height of the roughness layer may be determined from the profile of the horizontal wind velocity. The surface layer from z_0 to h_s, where h_s varies from 10 m to 200 m. In this layer the vertical flux terms of horizontal momentum and heat are the most important terms in the equation of motion and in the equation for vertical transfer of heat. This layer is also termed the *constant flux layer*. The *Ekman layer* from the surface layer to the top of the ABL is a transition layer between the surface boundary layer and the free troposphere.

The location of the mixing height is often characterised by an abrupt change in temperature stratification intensified by large scale vertical motion, and the transfer of radiation. Above the mixing height the temperature stratification changes as a result of variations of turbulent exchange and radiation transfer. In the stable boundary layer the vertical component of wind velocity is reduced as a result of buoyancy. Horizontal motion is not prohibited and interaction with the ground may be very complex under stable conditions.

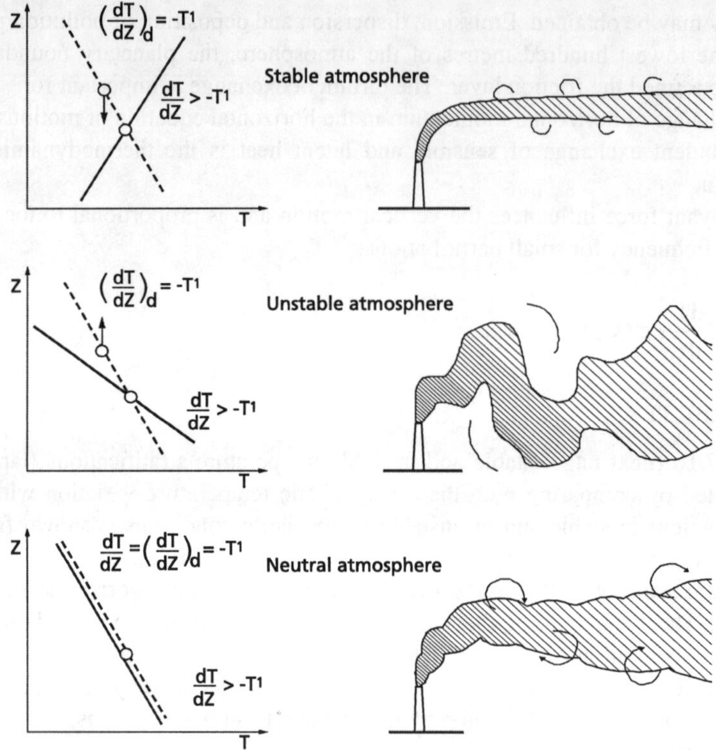

Figure 7.10 Variation of temperature with height above the ground and plume behaviour.

7.6 Deposition

Dry deposition of gases and particles from the atmosphere to the surface may be expressed as:

$$F = V_d(z) \cdot (c(z) - c_s) \tag{7.41}$$

where F is the dry deposition flux, V_d is the dry deposition velocity, c(z) is the concentration at height z and c_s is the concentration at the surface which usually is assumed very small and therefore neglected.

The dry deposition velocity, $V_{d(z)}$ depend on the height above the ground surface, the conditions at the surface as well as the structure of turbulence in the atmosphere. The deposition flux is approximately constant with height, which simplifies the description and combination of processes in the different layers. One way to express these processes mathematically is by the inverse value of the dry deposition velocity:

$$r_d = \frac{1}{v_d} \tag{7.42}$$

r_d is the resistance to dry deposition, which may be divided into three parts:

$$v_d = \frac{1}{r_a + r_b + r_c} \tag{7.43}$$

where r_a is the aerodynamic resistance of the atmospheric surface layer, r_b is the resistance to transport through the quasi-laminar layer close to the depending on on molecular diffusion, and r_s is the surface resistance accounting for uptake at the surface. r_a is dependent on the turbulence in the surface layer of the atmosphere. The wind speed, the heat exchange and the roughness height are important for this transport.

The transfer through the quasi-laminar sub-layer depends on the turbulence in the friction layer, the structure of the ground surface and the behaviour of the pollution component. Based on results of experiments in wind tunnels, the following expression was suggested by Hicks et al. (1987):

$$r_b = \frac{2}{\kappa u_*}\left(\frac{S_c}{P_r}\right)^{2/3} \tag{7.44}$$

where $S_c = \nu / D_i$ is the Smidts number, ν is the kinematic viscosity of air, D_i is the molecular diffusivity of the pollution component, P_r is the Prandtl number for the air motion close to the roughness elements.

The surface resistance r_c. A number of the processes which determine the uptake at the surface are the dependent on the structure of the surface and the type of soil. According to Hicks et al. the surface resistance is determined by:

- plant physiological factors (r_{stom} and r_m),
- the soil conditions (r_{soil}), and
- the external leaf surface conditions (r_{ext}), i.e. water content and chemistry in the water film at the plant.

The variation of surface conditions is complex and the procedure to obtain representative values for surface dry deposition may be difficult. In order to determine the dry deposition, the following land use types are considered:

- urban areas,

– agricultural land,
– range land,
– deciduous forest,
– coniferous forest,
– mixed forest including wetland,
– water surfaces,
– barren land,
– non forested wet land,
– mixed agricultural and range land, and
– rocky open areas with low-growing shrubs.

The role of rainfall is complex since the acidity of the rain influence the acidification. However, the rain is also a cleansing process for a polluted atmosphere. In polluted areas close to the main sources, the effect of rainfall may also be to remove deposited pollution from exposed surfaces.

Far from the sources (long range transport) scavenging of pollution by rain is the most important removal process. Close to the sources local dry deposition to the relatively more important.

7.6.1 STRUCTURE OF THE ATMOSPHERIC BOUNDARY LAYER

The boundary layer of the atmosphere may be divided in zones where various processes are important. *The atmospheric surface layer* is characterised by constant vertical fluxes of heat and momentum. Further the scale of roughness elements influence the scale of turbulence and correspondingly the vertical exchange in the layer. When abrupt horizontal changes occur as a result of changes in the characteristics of the surface a transition zone is found before the stationary profiles are re-established. *The canopy layer* where the roughness elements interrupt horizontal homogeneity. Stationary eddies may develop. The eddies are characterised by wind direction and solar orientation in relation to the surface roughness elements. Figure 7.11 shows a quasi-stationary air stream in a street. The wind above the street canyon governs the circulation (Chapter 12). Solar heating of roughness elements may intensify or diminish the canopy circulation particular in weak wind situations.

7.6.2 PROCESSES AT THE SURFACE

At the surface the processes of importance for deposition include chemical reactions as well as electrostatic forces. The conditions of the surface regarding wetness and biological processes may also govern the deposition. These effects differ from one component to another. Three options are considered for the actual deposition at the surface (Figure 7.12); absorption in the plant tissues after the pollution is transferred through the stomata and the inner parts of the plant. Absorption directly in the soil. Absorption in plant surfaces. The parameterization of deposition is dependent on the wetness of the surface considering many pollution components.

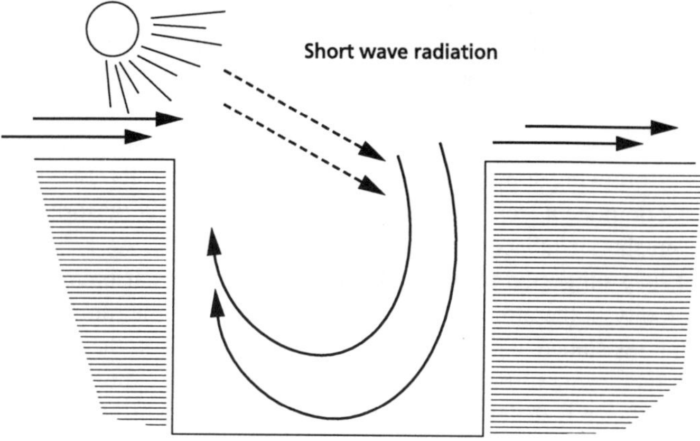

Figure 7.11 Air circulation in a street canyon.

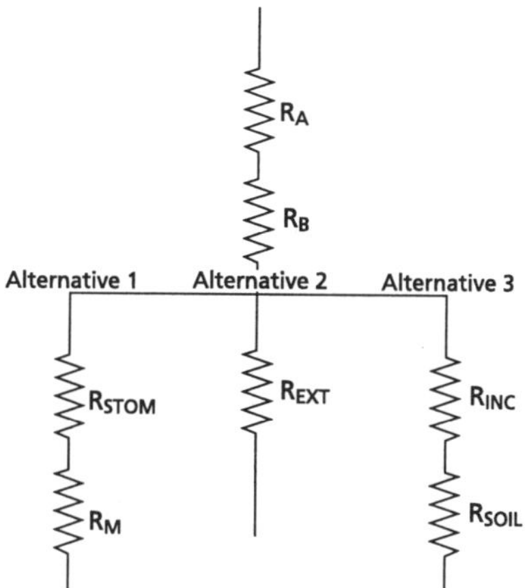

Figure 7.12 Various deposition processes at the ground depending on the ground characteristics. The most effective deposition process is selected.

When a stationary situation is obtained, the transfer processes may be compared to transmission of electricity in an electric circuit with different resistances coupled in series and/or in parallel. The alternative ways of deposition described in the previous paragraph correspond to resistances coupled in parallel. Resistances corresponding to the penetration of one layer after the other corresponds to resistance coupled in series. With reference to Figure 7.12 the following equation may be written for the surface resistance to deposition (R_s)

$$R_s^{-1} = \left[\frac{1}{R_{stom} + R_m} + \frac{1}{R_{inc} + R_{soil}} + \frac{1}{R_{ext}} \right] \qquad (7.45)$$

where R_{stom} is the Stomata resistance, R_m is the resistance to deposition in the inner parts of the plant, R_{ext} is the resistance to deposition at the plant outside stomata, R_{inc} is the canopy resistance to account for plant covered surface, and R_{soil} is the resistance to deposition at the surface.

For most gases in areas covered with vegetations where adverse pollution effects are considered, the following simplification may be applied $R_s >> R_m$. In other words; when the pollution component reaches parts of the plant, it is absorbed fast and the concentration in the air adjacent to the plant becomes small. As a first approximation Hicks et al. (1987) suggested to parameterise the quasi-laminar boundary resistance by the friction velocity in the atmospheric surface layer and of the molecular diffusivity of the gases. The resistance increases with the complexity of the gas molecules and decreases with increasing turbulence (friction velocity in the atmospheric surface layer). The stomata resistance is dependent on global radiation, the surface temperature and the kinematic viscosity of the gas component. The aerodynamic resistance in canopy r_{inc} is defined as

$$R_{inc} = \frac{b \cdot LAI \cdot h}{\upsilon} \tag{7.46}$$

where b is a proportionality factor for different areas, LAI is leaf area index which dependent on vegetation and season, h is the height of vegetation, and u_* is the friction velocity. The typical values of these parameters are shown in Table 7.4.

Table 7.4 The value of parameters for calculation of the canopy resistance for gases over agricultural land and in forested areas. f(month) is a function of the month of the year (Figure 7.13).

Area class	LAI	h (m)	b (m^{-1})
Deciduous forest	$5 \cdot f(month)$	20	14
Coniferous forest	5	20	14
Agricultural land	$5 \cdot f(month)$	1	14
Range land	$5 \cdot f(month)$	1	14

The canopy resistance for gases and small particles is about tem times larger during summertime than during winter (Figure 7.13). For other area classes, the canopy resistance is regarded to be small i.e in:

- urban areas,
- forest and wetland,
- water surface,
- barren land and desert,
- wetland without forest, and
- rocky open areas with low growing shrubs.

For many pollution components it is important to differentiate between water, soil and snow-covered surfaces.

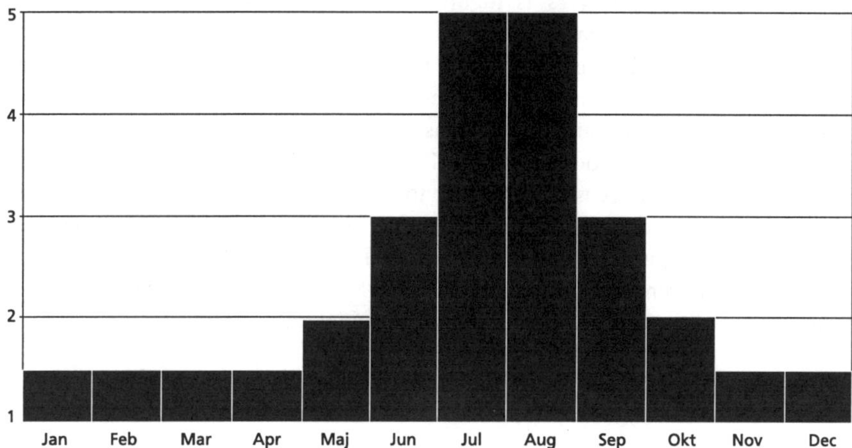

Figure 7.13 The leaf index as a function of time of the year.

7.6.3 SURFACE RESISTANCE FOR VARIOUS POLLUTION COMPONENTS

The dry deposition of *Sulphur dioxide* – SO_2 is enhanced over wet surfaces. However, the functional relationship and the chemical processes at the wet surfaces are not well known. Parameterizations have been developed accounting for the temperature and the relative humidity of the atmosphere. The resistance increases with decreasing temperature and decreasing humidity.

Emission of NH_3 occurs from fertilized soil and pasture. In other areas NH_3 is deposited by stomata uptake and by deposition. In addition to the type of plant cover the deposition is dependent on solar radiation, temperature and humidity. Typical values for the surface resistance are given in Table 7.5.

Table 7.5 Typical resistance values for deposition of NH_3. Unit: s m^{-1}.

Land use		Day		Night	
		Dry	Wet	Dry	Wet
Pasture	summer	1000	1000	1000	1000
	winter	50	20	100	20
Crops	summer	'stom	50	200	50
	winter	'stom	100	300	100
Forest		500	0	1000	0

The main deposition of NO_2 takes place through the stomata of plants. The surface resistance has a typical value about 2000 s m^{-1}.

Natural emissions of *NO* are observed more often than deposition. The processes are complex and not well known, and the resistance may be at least one order of magnitude larger than for NO_2.

Deposition of HNO_3 seems to be limited only by the aerodynamic resistance. The surface resistance has a typical values are 10-50 s m^{-1}.

The dry deposition of *particles* is highly size dependent (Figure 7.14). Particulate material in the atmosphere has a variety of sources and the mass is distributed over a large size range. Vertical transport through the atmospheric surface layer is dominated by atmospheric turbulence; the effect of sedimentation is of minor importance, when the particle size is smaller than 10 µm. Transport through the quasi-laminar sub-layer at the surface depends on various processes. Brownian motions increase with decreasing particle size and cause decreasing resistance to deposition, when the particle diameter is less than 0.1 µm. In the ambient atmosphere the turbulence and the Brownian motion cause coagulation of particles changing the size distribution which again influence the deposition. Particle transport through the quasi-laminar sublayer may be expressed by a formulae similar to Stoke's law describing the terminal fall velocity:

$$V_d = \alpha \frac{\Delta\rho \cdot}{\pi} \tag{7.47}$$

where V_d is the dry deposition velocity, α is a proportionality factor, $\Delta\rho$ is the difference in density between particle and the air, and ν is the dynamic viscosity coefficient of air. In addition to sedimentation effects, particle impaction can be important at the surface when streamlines move close to sharp edges. Processes as sedimentation, impaction and Brownian motion influence the deposition process in addition to the geometry and surface texture. Deposition due to electrical charges of particles may also be important in particular for the small particles. In a moist atmosphere hygroscopic growth modifies the size distribution of particles.

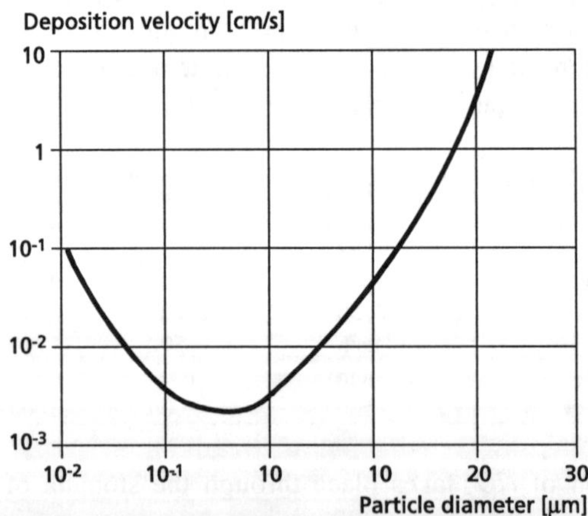

Figure 7.14 Deposition velocity of particles as a function of particle diameter.

Instead of describing all processes, it may be suitable for practical purposes to estimate the dry deposition over the entire particle size range expressed in one resistance value. Empirical estimates of deposition tend to be larger than values obtained from existing theory. From a practical point of view empirical estimates of integrated dry deposition of particulate sulphur and nitrogen as suggested to evaluate pollution effects on ecosystems. Equations to estimate the combined effects of canopy resistance and surface resistance have been suggested by Erisman et al. (1997) for various chemical components in aerosol particles:

$$V_{ds} = \left(\frac{u_*^2}{U_h} \right) E\,(u_*, rh) \tag{7.48}$$

where V_{ds} is the dry deposition velocity, u_* is the friction velocity, U_h is the estimated wind velocity at the canyon height, E is a factor accounting for the interaction between pollution component and the surface. The parameterization of E is suggested for components that influence the acidification and nutrification of ecosystems

$$E = a\, u_*^b \cdot g(rh) \tag{7.49}$$

The values of a and b are vary with pollution component and the wetness of the surface (Table 7.5). The function g(rh) accounts for the growth of particles as function of relative humidity. The increase in particle size leads to an increase in deposition velocity.

Table 7.5 Parameterization of dry deposition of chemical components in particles.

Compound	Wet surface		Dry surface	
	a	b	a	b
NH_4	0.066	0.41	0.05	0.23
SO_4	0.08	0.45	0.05	0.28
NO_3	0.1	0.43	0.063	0.25
Na	0.679	0.56	0.14	0.12

When the surface is wet, the dependence on the relative humidity in the atmosphere follows the same formulae

$$\begin{aligned} &\text{For } rh \le 80\% \qquad && g\,(rh) = 1 \\ &\text{For } rh > 80\% \qquad && g\,(rh) = 1 + 0.37\exp\left(\frac{rh - 80}{20} \right) \end{aligned} \tag{7.50}$$

When the surface is dry, the dependence of dry deposition velocity on the relative humidity follows a different formulae:

$$g(rh) = 1 + 0.18\exp\left(\frac{rh - 80}{20} \right) \tag{7.51}$$

For sodium (Na^+) containing particles, the dry deposition decreases with rh increasing above 80%

$$g(rh) = 1 - 0.09\exp\left(\frac{rh - 80}{20}\right) \tag{7.52}$$

The parameterization shows that the surface resistance to deposition decreases with increasing wetness of the surface and relative humidity in the air. The canopy resistance decreases with increasing friction velocity in the atmosphere.

7.7 References

Arya, S.P. (1988) *Introduction to micro meteorology*, Academic Press Inc., San Diego.

Eliassen, A.E., Pedersen, K. (1977) *Meterology. An Introduction Course. vol. I. Physical Processes and Motion vol. II. Application to weather and weather systems*, Universitetsforlaget, Oslo.

Erisman, J.W., Draaijers, G., Duyzer, J., Hofschreuder, P., Leuwenen, N., Römer, F., Ruijgrook, W., Wyers, R., Gallagher, M.W. (1997) Particle deposition to forests - summary of results and application, *Atmospheric Environment*, **31**, 321-332.

Hicks, B.B., Baldocchi, D.D., Meyers, T.P., Matt, D.R., Hosker, R.P. (1987) A muliple preliminary resistance model for deriving dry deposition velocities from measured quatities, *Water, Air, Soil Pollut.*, **36**, 3111-330.

Nieuwstadt, F.T.M., Van Dop, H. (editors) (1982) *Atmospheric Turbulence and Air Pollution Modelling*, Reidl, Dordrecht.

Wallace, J.M., Hobbs, P.V. (1977) *Atmospheric Science. An Introductory Survey*, Academic Press inc., Orlando.

Chapter 8

TRANSFORMATION OF AIR POLLUTANTS

RICHARD G. DERWENT
Atmospheric Process Research, Meteorological Office
London Road, Bracknell, United Kingdom

OLE HERTEL
National Environmental Research Institute
P.O.Box 358, DK-4000 Roskilde, Denmark

8.1 Introduction

An air pollutant is usually understood to be a substance which between the point of its emission into the atmosphere and its ultimate removal, causes harm to a target whether ecosystems, material, man, or ultimately climate. Urban air pollution arises from the competition between emission processes which force up pollutant concentrations and dispersion, advection and deposition processes which reduce and remove them. However, not all urban air pollution phenomena conform to this simple characterisation; an important category of pollutants are not emitted into the atmosphere themselves in any significant quantity; they are formed there by chemical reactions. This type of pollutants are usually termed secondary pollutants to distinguish them from the precursors: primary or emitted pollutants. The present chapter is about some of the main chemical processes on local as well as regional scale which are considered of importance for present urban air quality in Europe.

Despite the significant improvements made in urban air quality over the last three decades, there are three major urban secondary pollution problems which potentially may still exert some public health impact within Europe. These problem pollutants include: nitrogen dioxide (NO_2), ozone (O_3) and suspended particulate matter (mainly in the particle range less than 10 µm, PM_{10}), and the transformations which produce and degrade them form the main subject of this chapter.

During wintertime pollution episodes, NO_2 concentrations in urban areas may exceed internationally accepted air quality criteria set for the protection of human health. This somewhat unexpected phenomenon has been reported in some large industrial and urban centres in Northern Europe during stagnant wintertime weather conditions. The sole precursor of the elevated NO_2 levels is the nitric oxide (NO) emitted by motor traffic and by stationary combustion sources, such as industrial, commercial and domestic boilers fuelled by coal, gas or oil (Chapters 4-6).

During summertime pollution episodes, photochemical reactions driven by sunlight may lead to the conversion of organic compounds and oxides of nitrogen into photochemical oxidants, in particular ozone. This phenomenon is widespread throughout North-West Europe during most summers. These photochemical reactions also lead to the oxidation of sulphur dioxide into fine haze of sulphuric acid aerosol. This photochemically generated aerosol contributes to suspended particulate matter and gives rise to visibility reduction and the loss of distant horizons. Elevated urban concentrations of ozone and of suspended particulate matter may exceed internationally accepted environmental criteria levels, set to protect human health.

8.2 Nitrogen oxides

The oxides of nitrogen are ubiquitous urban air pollutants whose main sources are road traffic, power plants and industry (Chapters 4-6). Nitric oxide (NO) is on mass basis by far the most important nitrogen compound emitted into the atmosphere. Usually NO constitutes more than 90% of the nitrogen oxide release from combustion processes where the remaining part is emitted as nitrogen dioxide (NO_2). Few adverse environmental impacts are associated directly with NO and most concerns have been associated with its atmospheric transformation products, following its release into the ambient atmosphere. The nitric oxide is formed from atmospheric nitrogen (N_2) at high temperatures in combustion processes. Its main fate in the atmosphere is to react with ozone:

$$NO + O_3 \rightarrow NO_2 + O_2 \tag{8.1}$$

Under typical tropospheric boundary layer conditions, this reaction takes place within a few seconds and either lead to the nearly complete conversion of all the O_3 to form nitrogen dioxide or nearly all of the NO to NO_2 with an excess of unreacted O_3. In a highly polluted atmosphere (like a typical urban area) or close to individual pollution sources, the former behaviour is usually observed because, although ozone is widely distributed in the lower atmosphere, its concentration is not usually high compared with NO in the highly polluted atmosphere and hence ozone concentrations become rapidly depleted.

Nitrogen dioxide is hence the first and most immediate reaction product of the atmospheric oxidation of the NO emitted by human activities. During daylight, most of the formed NO_2 absorb solar ultraviolet radiation (wavelengths λ 200 to 420 nm) and undergo photolysis, reforming NO and O_3:

$$NO_2 + h\nu \rightarrow NO + O(^3P) \tag{8.2}$$

$$O(^3P) + O_2 + M \rightarrow O_3 + M \tag{8.3}$$

where M represents an O_2 or N_2 molecule, which acts as third body that absorbs the excess vibrational energy and thereby stabilises the formed O_3 molecule. Reaction (8.3) is the only production path for ozone in the atmosphere. This reaction is very fast, and in an overall analysis reactions (8.2) and (8.3) may be considered as a one step reaction, where the product of the photodissociation of NO_2 leads to the formation of an ozone molecule. The rate of photolysis of NO_2 is naturally a function of the solar actinic irradiance and this in turn depends on the height of the sun in the sky and hence time-of-day and season as well as on the amount and height of any cloud or haze, which may obscure the sun. For much of the daytime portion of the year, the lifetime of NO_2 is only a matter of minutes before it is photolysed back to NO.

The reaction of NO with O_3 and the photolysis of NO_2 form a cycle, which occurs rapidly over the timescales of minutes in the sunlit atmosphere and ensures that under most tropospheric conditions, NO and NO_2 will coexist as a mixture often termed NO_x ($NO_x = NO + NO_2$). A steady state (Leighton 1961) is rapidly established:

$$\frac{[NO][O_3]}{[NO_2]} = \frac{j_{8.2}}{k_{8.1}} \tag{8.4}$$

where the parentheses indicate concentrations of the given compound, $j_{8.2}$ is the solar radiation dependent photolysis rate coefficient in reaction (8.2) that under summer conditions in the mid-afternoon at mid-latitudes and clear sky have a typical value of about $7 \cdot 10^{-3}$ s^{-1} and $k_{8.1}$ is the temperature dependent reaction rate coefficient in reaction (8.1) with a typical value about $4 \cdot 10^{-4}$ ppb^{-1} s^{-1} (Seinfeld, Pandis 1998). This leads to a typical relationship in Equation 8.4 of about 18, which implies that in the sunlit atmosphere: $[NO_2] = 2 \cdot [NO]$ at about 30 to 40 ppb ozone, and $[NO_2] = 5 \cdot [NO]$ at about 90 ppb ozone.

During *combustion processes* at high temperatures e.g. inside the motor of a petrol or diesel-engined vehicle, NO is formed from atmospheric nitrogen. However, in the very NO rich air inside the exhaust pipe of vehicles and inside chimneys of power plants and industries another oxidation path than reaction (8.1) takes place:

$$NO + NO + O_2 \rightarrow 2NO_2 \tag{8.5}$$

Reaction (8.5) is a third order reaction with a second order dependence of the NO concentration, and has a reaction rate coefficient of $2.3 \cdot 10^{-38}$ $molecule^{-1}$ cm^{-3} s^{-1} (about $6 \cdot 10^{-28}$ ppb^{-1} s^{-1}) (Finlayson-Pitts, Pitts 1986). Table 8.1 shows calculations of the chemical production of NO_2 at various NO concentrations. From these results it is obvious that this reaction is of limited importance under usual ambient tropospheric conditions, even inside urban street canyons where the NO concentration arising from traffic pollution may reach concentrations even up to 1000 ppb but the residence time only is a few minutes (see the description of modelling pollution levels in street canyons in Chapter 12). However, in large conurbations such as London and with persistent accumulation under wintertime inversions, it may still produce NO_2 concentrations of some significance (Bower et al. 1994).

Table 8.1 NO_2 yield from reaction (8.5) at various nitric oxide concentrations. The oxygen concentration is set to 20% (about $5 \cdot 10^{18}$ molecules cm^{-3} for at standard boundary layer atmosphere).

NO concentration (ppb)	NO_2 yield (ppb h^{-1})
1	0.000,019
100	0.19
1000	19
100,000	190,000

8.2.1 WINTERTIME NO_2 EPISODES

During wintertime, NO_2 pollution episodes have frequently been observed in the cities of North-Western Europe. In London during December 1991, hourly peak concentrations of NO_2 as high as 423 ppb were reported in association with NO concentrations in excess of 1000 ppb (Bower et al. 1994), see Figure 8.1.

The typical background ozone concentration in the winter time is in the range of 40 ppb. Assuming that the NO_2 fraction is about 5% of the NO_x emission, the contribution from directly emitted NO_2 will be in the range 70 to 80 ppb. A total chemical conversion of all available ozone to NO_2 would yield an NO_2 concentration of 110 to 120 ppb, which is considerably less than observed. It has been argued that the $NO+NO+O_2$ reaction (reaction 8.5) may contribute to the enhanced NO_2 concentrations in such situations (Bower et al. 1994; Derwent et al. 1995). However, Shi and Harrison (1998) suggested a gas phase reaction mechanism in the dark winter conditions to be responsible for the excess NO_2 concentration. This suggested mechanism is initiated by the cyclic hydrocarbons emitted from petrol vehicles. These cyclic hydrocarbons are oxidised under the formation; and as a result a fast formation of the highly reactive peroxy radicals (Sections 8.4 and 8.5) lead to conversion of NO to NO_2 with a rate on the order of 50 ppb h^{-1} at concentrations of the cyclic hydrocarbons of a few ppm (the suggested reaction path is shortly described in Section 8.11). Such concentrations in the low ppm range are considered improbable for the urban atmosphere under such episodes, and so the mechanism behind these winter time NO_2 episodes is, however, still not fully resolved.

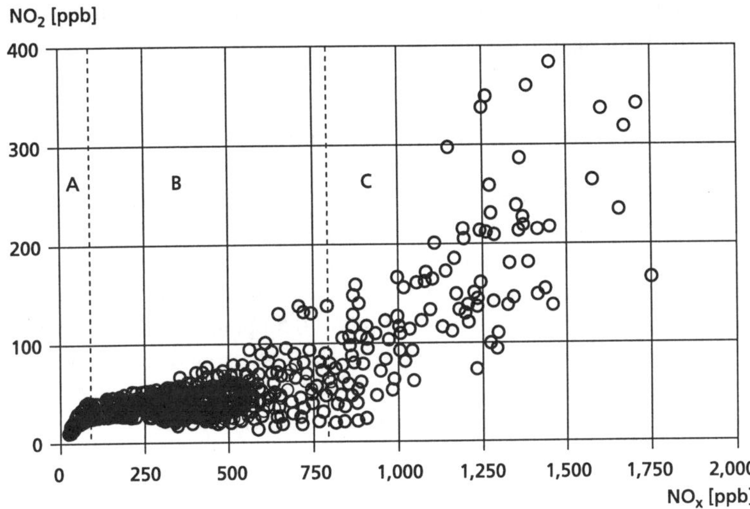

Figure 8.1 Observed hourly mean concentrations of NO_2, as function of NO_X ($NO + NO_2$), at a busy roadside (Cromwell Road) in London in January, February, November and December 1991 (Harrison, Shi 1996).

8.2.2 NO_X REMOVAL

The main chemical removal mechanism for NO_x in the troposphere is the reaction between NO_2 and the OH radical; leading to formation of nitric acid:

$$NO_2 + OH + M \rightarrow HNO_3 + M \tag{8.6}$$

For typical daytime tropospheric background conditions, this reaction has a NO_2 conversion rate of approximately 5% per hour. The produced nitric acid may in turn react with atmospheric ammonia in gas phase or on the surface of aerosols, or it may be taken up by existing particles; in both cases forming particulate nitrate (NO_3^-) (Section 8.10). HNO_3 may photolyse back to NO_2 and OH in the atmosphere, but this reaction pathway is insignificant in the troposphere.

8.2.3 FORMATION OF NITROUS ACID

A reaction similar to (8.6), but less important, is the reaction between NO and OH forming nitrous acid, HONO:

$$NO + OH + M \rightarrow HONO + M \tag{8.7}$$

During daytime HONO photolyses ($\lambda < 400$ nm) rapidly back to the reactants:

$$HONO + h\nu \rightarrow OH + NO \tag{8.8}$$

Therefore, HONO formed in the late evening may serve as night time reservoir of OH and NO, which are subsequently liberated again the following morning, when sunlight starts up reaction (8.8) again.

8.3 Formation and reactions of the hydroxyl radical

The chemistry of the troposphere during daytime is driven largely by the OH radical, which due to its high reactivity leads to oxidation of a significant fraction of the trace constituents with a significantly long lifetime with respect to removal by dry and wet deposition. This is despite the very low OH concentration in the background tropospheric boundary layer on the order of 0.05 - 0.5 ppt ($1\text{-}10\cdot10^6$ molecules cm^{-3}).

The main production mechanism of OH in the background troposphere is initiated by the photolysis of ozone at wavelength < 319 nm to yield electronically excited oxygen atom $O(^1D)$, which subsequently react with water vapour:

$$O_3 + h\nu \rightarrow O_2 + O(^1D) \tag{8.9}$$

$$O(^1D) + H_2O \rightarrow 2OH \tag{8.10}$$

Reaction (8.10) takes place in competition with $O(^1D)$'s reaction with a molecule oxygen or nitrogen acting as a third body to form $O(^3P)$, that in turn (via 8.3) reforms ozone. For each $O(^1D)$ molecule formed under typical water vapour concentrations in the tropospheric background, 0.2 OH radicals are produced (Seinfeld, Pandis 1998).

Ozone is an important photo-oxidant in the atmosphere with serious impact on human health as well as on vegetation (Chapters 18 and 20). As shown in reactions (8.9) and (8.10), ozone initiates the formation of OH radical. In the following sections it is demonstrated how OH radical in turn has an important role in the excess formation of tropospheric photo-oxidants including ozone.

8.3.1 CHEMISTRY OF CARBON MONOXIDE AND NITROGEN OXIDES

As already stated the OH radical is a key compound in the atmosphere, where it interacts with peroxy radicals that are responsible for the formation of excess concentrations of photo-oxidants like ozone. In the background troposphere, carbon monoxide (CO) plays a role in this system:

$$CO + OH \rightarrow CO_2 + H \tag{8.11}$$

The hydrogen atom formed in (8.11) combines quickly with atmospheric oxygen and forms another highly reactive free radical - the hydroperoxy (HO_2) radical:

$$H + O_2 + M \rightarrow HO_2 + M \tag{8.12}$$

This conversion is the only significant reaction of the hydrogen atom. The conversion is so rapid that for all practical purposes reactions (8.11) and (8.12) can be written as a one step reaction:

$$CO + OH + M \xrightarrow{O_2} CO_2 + HO_2 + M \qquad (8.13)$$

Whenever NO is present, the most important ambient atmospheric reaction of the HO_2 radical is the conversion of NO to NO_2:

$$NO + HO_2 \rightarrow NO_2 + OH \qquad (8.14)$$

Note that in the reactions (8.13) and (8.14), the OH radical is recycled. On a global scale about 70% of OH react with CO (reaction (8.11)) and about 30% with methane (Section 8.5). However, on a local scale the reactions of OH with CO and methane are supplemented with reactions with NO_2 and non-methane volatile organic compounds (NMVOCs), which we will return to in the next sections.

The chain of reactions in (8.13) and (8.14) leads to a *net photochemical production of ozone* in the background troposphere, since NO is transformed into NO_2 without consumption of O_3. When NO_2 is photo-dissociated via (8.2), ozone is subsequently formed from the produced $O(^3P)$ radical via reaction (8.3). The photochemical ozone formation can therefore be described as the process by which peroxy radicals shifts the balance in the NO-NO_2-O_3 photo-stationary state - in favour of ozone production. This chain of reactions is terminated, e.g. when NO_2 is transformed into nitric acid in the reaction with OH radical, see reaction (8.6).

In urban areas (8.11) is too slow to have local impact, since the lifetime of CO with respect to this reaction is on the order days to months (depending on the present OH concentration); the reaction rate coefficient has been determined to $2.5 \cdot 10^{-13}$ cm^3 $molecules^{-1}$ s^{-1} (about $6.6 \cdot 10^{-3}$ ppb^{-1} s^{-1}) (Finlayson-Pitts, Pitts 1986).

8.3.2 OXIDATION OF HYDROCARBONS WITH THE HYDROXYL RADICAL

The hydroperoxy radical is only one of many peroxy radicals that take part in ozone production in the troposphere; a wide range of organic peroxy radicals play likewise an important role.

Organic peroxy radicals are mainly formed by the attack of the highly reactive hydroxyl radical on the organic compounds ubiquitously present in the polluted tropospheric boundary layer. These reactions follow a similar path as the previously described CO oxidation, and may in a simplified form be represented as:

$$OH + RH \rightarrow R + H_2O \qquad (8.15)$$

$$R + O_2 + M \rightarrow RO_2 + M \qquad (8.16)$$

where RH represent the organic compound, whereas R is an organic radical such as alkyl radical and RO_2 an alkyl peroxy radical.

The only significant atmospheric reaction pathway of the *alkyl radicals* is (8.16), the reaction with O_2 to form alkyl peroxy radicals (Finlayson-Pitts, Pitts 1986). The detailed mechanisms of the reactions which convert organic compounds into their corresponding peroxy radicals depend naturally on the structure of the individual organic compounds involved. Most organic compounds react with hydroxyl radicals either by H-abstraction or by addition, if they contain carbon-carbon multiple bonds. In the case of alkanes, cycloalkanes, carbonyls and oxygenated hydrocarbons, the main path is that the hydroxyl radical removes a hydrogen atom originally connected to the carbon skeleton of the parent compound, forming a carbon radical and water vapour. These carbon radicals quickly react with oxygen to form the corresponding peroxy radical. For example, if the parent organic compound is methane, the organic radical formed will be methyl radical. This radical rapidly combines with oxygen forming a methyl peroxy radical:

$$OH + CH_4 \rightarrow CH_3 + H_2O \tag{8.17}$$

$$CH_3 + O_2 + M \rightarrow CH_3O_2 + M \tag{8.18}$$

Methane is the globally most abundant organic compound in the troposphere (about 1.7 ppm). The reaction (8.17) is the most important degradation path of methane (lifetime on the order of years) taking place in the background tropospheric boundary layer.

In the case of *alkenes, alkynes and aromatics*, the main reaction path is that the hydroxyl radical adds to the multiple bond, producing a carbon radical which again almost invariably, reacts with oxygen in the analogous process to form a peroxy radical. For example, with ethylene (ethene) the reaction sequence will look like:

$$OH + C_2H_4 + M \rightarrow HOC_2H_4 + M \tag{8.19}$$

$$HOC_2H_4 + O_2 \rightarrow HOC_2H_4O_2 \tag{8.20}$$

Coupling together the OH attack on the parent organic compound with the conversion of the peroxy radicals to alkoxy radicals and of the alkoxy radicals to carbonyls, the ozone production begins to take shape. Considering methane, the steps in (8.17) and (8.18) are followed by conversion of two molecules of NO to NO_2, and subsequently by the conversion of another NO to NO_2 by reaction (8.14):

$$CH_3O_2 + NO \rightarrow CH_3O + NO_2 \tag{8.21}$$

$$CH_3O + O_2 \rightarrow HO_2 + HCHO \tag{8.22}$$

Adding the photolysis of two NO_2 molecules and the reaction of $O(^3P)$ with O_2, this system will produce two molecules of O_3. A similar sequence is found for ethylene (ethene):

$$HOC_2H_4O_2 + NO \rightarrow HOC_2H_4O + NO_2 \quad (8.23)$$

$$HOC_2H_4O + O_2 \rightarrow HO_2 + HCHO + HCHO \quad (8.24)$$

In this sequence of rapid consecutive reactions, the OH radical is recycled, the nitric oxide and nitrogen dioxide are recycled (adding again to these systems the photolysis of NO_2), and the overall system for these two compounds may be written as:

$$CH_4 + 2O_2 \rightarrow HCHO + 2O_3 \quad (8.25)$$

$$C_2H_4 + 2O_2 \rightarrow 2HCHO + 2O_3 \quad (8.26)$$

In this way, a small steady state concentration of the highly reactive OH radical can degrade substantial concentrations of organic compounds, producing ozone as an important reaction product. In order to estimate the rate of ozone production, an understanding of the rate of degradation of the individual organic compounds is needed, and hence the steady state concentrations of hydroxyl radicals need to be estimated (Section 8.8).

Formaldehyde (HCHO) is produced in reactions (8.25) and (8.26), and it is one of the more important photochemically labile organic compounds in the troposphere. The degradation of HCHO may occur by photolysis or by reaction with OH radical:

$$HCHO + h\nu \rightarrow H_2 + CO \quad (8.27)$$

$$HCHO + h\nu \rightarrow H + HCO \quad (8.28)$$

$$HCHO + OH \rightarrow H_2O + HCO \quad (8.29)$$

Subsequently, it react with O_2 and produce an HO_2 radical (reaction 8.10). Similarly, HCO react with O_2 which also leads to the production of an HO_2 radical:

$$HCO + O_2 \rightarrow CO + HO_2 \quad (8.30)$$

The fate of HCHO depends highly on the local conditions, e.g. the sunlight intensity and the concentration of OH. In the tropospheric boundary layer, the main path of HCHO is degradation by photolysis; about 50%, 30% and 20% of HCHO reacts by (8.27), (8.28) and (8.29), respectively. The overall tropospheric lifetime of HCHO is of the order of hours. The degradation of formaldehyde leads to production of ozone, via the produced HO_2 radicals (via the reaction 8.14 followed by reactions (8.2) and (8.3)).

8.4 The role of the nitrate radical

At night the NO_3 radical takes over from the OH radical as being the most important oxidant in the troposphere. Despite the considerably lower reactivity compared to OH, its higher peak concentrations in the night-time troposphere allow the NO_3 radical to play a major role in the chemical transformations of organic compounds. The impact of these two important radicals is complementary, since OH is formed photochemically only during daytime, while NO_3 is quickly photolysed during daytime and hence only can survive during night. The NO_3 radical is formed in the reaction between NO_2 and ozone:

$$NO_2 + O_3 \rightarrow NO_3 + O_2 \qquad (8.31)$$

The typical night time NO_3 radical concentrations in the tropospheric boundary layer are in the order 10^7 to 10^8 molecules cm^{-3} (ppt range).

8.4.1 NITRATE RADICAL OXIDATION OF HYDROCARBONS

The NO_3 radical attacks alkanes by hydrogen abstraction in a similar way as the previously described reactions of the OH radical:

$$RH + NO_3 \rightarrow R + HNO_3 \qquad (8.32)$$

Followed by (8.12) the formation of a peroxy radical (RO_2) that again may oxidise an NO molecule to NO_2. Also for the alkenes, the attack of the NO_3 radical is similar to the reactions of the OH radical; the NO_3 radical adds to the double bond. This reaction is followed by rapid O_2 addition which leads to the production of a peroxy radical. The OH radical reactions are typically 10 to 1000 times faster that the NO_3 radical reactions, but due to the much higher NO_3 radical concentrations, the conversion rates of some of the hydrocarbons are comparable in the background troposphere. This applies especially for the biogenic emitted hydrocarbons - terpenes and isoprenes (see the discussion in Section 8.9).

8.4.2 THE NITRATE RADICAL RESERVOIR SPECIES - N_2O_5

In the tropospheric boundary layer, the nitrate radical has a reservoir species in N_2O_5 (dinitrogen pentoxide):

$$NO_3 + NO_2 + M \rightarrow N_2O_5 + M \qquad (8.33)$$

$$N_2O_5 + M \rightarrow NO_3 + NO_2 + M \qquad (8.34)$$

During night-time in the tropospheric background atmosphere, the system (8.33) and (8.34) will rapidly (in about one minute) reach a steady state. It is therefore common to assume that these two compounds (NO_3 and N_2O_5) are in chemical equilibrium in the nocturnal tropospheric boundary layer.

8.4.3 REMOVAL OF THE NITRATE RADICAL

During daytime, NO_3 photolysis rapidly in the tropospheric boundary layer (with a noontime lifetime of about 5 s) via two different reaction paths (λ represent the wavelength):

$$NO_3 + h\nu(\lambda < 700nm) \rightarrow NO + O_2 \quad (8.35)$$

$$NO_3 + h\nu(\lambda < 580nm) \rightarrow NO_2 + O(^3P) \quad (8.36)$$

During night-time the main removal mechanism for the NO_3 radical in the tropospheric background is the indirect removal through the heterogeneous conversion of N_2O_5:

$$N_2O_5 + H_2O \rightarrow 2HNO_3 \quad (8.37)$$

The lifetime of N_2O_5 with respect to this removal is on the order of minutes in the tropospheric boundary layer.

Close to pollution sources from combustion processes e.g. road traffic or power plants, the NO_3 radical is quickly removed by reaction with NO:

$$NO_3 + NO \rightarrow 2NO_2 \quad (8.38)$$

Reaction (8.38) is so fast that the NO_3 radical and NO cannot coexist in mixing ratios above a few ppt. This means that the hydrocarbon reactions of the NO_3 radical are insignificant inside an urban area, whereas it may be formed in high concentrations downstream of urban areas in so-called urban plumes.

8.5 Peroxy acetyl nitrate (PAN)

Aldehydes are among the most important intermediate products of the atmospheric oxidation of hydrocarbons in the tropospheric boundary layer. When the aldehydes photo-dissociate or react with OH, acyl radicals are formed (Section 8.5). These acyl radicals may in turn produce peroxy acyl nitrates, which in the tropospheric boundary layer serve as important reservoirs of NO_x. The most abundant of these nitrates is peroxy acetyl nitrate (PAN):

$$\begin{aligned} &CH_3CHO + h\nu \rightarrow CH_3C(O) \\ &CH_3CHO + OH \rightarrow CH_3C(O) + H_2O \end{aligned} \quad (8.39)$$

$$CH_3C(O) + O_2 \rightarrow CH_3C(O)OO \tag{8.40}$$

$$CH_3C(O)OO + NO_2 + M \rightarrow CH_3C(O)OONO_2 + M \tag{8.41}$$

Reaction (8.40) is very fast, and (8.39) and (8.40) may therefore for many practical purposes be regarded as taking place in a one step reaction.

High concentrations of PAN have often been observed along with high ozone concentrations during photochemical smog episodes and in such cases PAN may be a harmful compound especially to plants but also to humans and animals.

PAN is thermally unstable in the atmosphere and an equilibrium between the peroxy acetyl radical and NO_2 on one side and PAN on the other side is established in the tropospheric boundary layer. In a cold atmosphere these compounds have a long lifetime, whereas they rapidly degrade at higher temperatures. The thermal degradation of PAN gives it a lifetime of about 1.7 h at 273 K and 50 h at 263 K.

The formation of PAN in (8.41) is competing with the NO degradation of peroxy acetyl radicals:

$$CH_3C(O)OO + NO \rightarrow CO_2 + NO_2 + CH_3O_2 \tag{8.42}$$

Reaction (8.42) dominates for NO concentrations at ppb levels, which means that PAN is formed in the background atmosphere, and not inside urban areas. However, substantial PAN concentrations may still be observed in urban areas, especially at relatively low temperatures. Moreover, PAN and similar peroxy acyl nitrates act as reservoirs of NO-NO_2 e.g. they temporarily reduce the amount of free NO-NO_2 by tying up these in less reactive compounds. The other peroxy alkyl nitrates include compounds produced via similar pathways as PAN, but are generated from biogenic isoprene emissions. These compounds may be of importance in southern Europe and have health impacts, and they have thermal degradation pathways similar to PAN.

8.6 Nitrogen oxides - main reaction paths

The tropospheric boundary layer contains a number of nitrogen oxide compounds - the primary pollutants NO and NO_2 and their reaction products. The main reaction pathways of these compounds have been discussed in the previous sections. Figure 8.2 illustrates the most important interactions between the various nitrogen oxide compounds. In addition to the gas phase reactions illustrated in the figure, a number of heterogeneous processes are taking place on the surface of or inside tropospheric aerosol particles and/or cloud droplets; processes that have major impact on the chemical conversions and atmospheric fate of nitrogen oxide compounds in the troposphere. A brief description of these heterogeneous processes is given in the Sections 8.12 and 8.13, and a more detailed description of the physical processes of tropospheric particles may be found in Chapter 9.

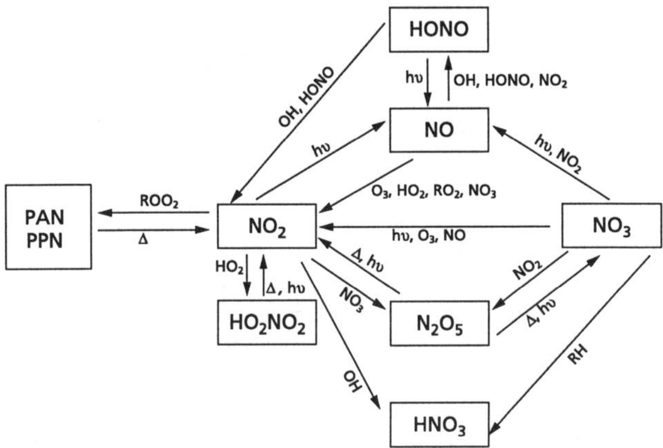

Figure 8.2 Illustration of the interaction between the various nitrogen oxides compounds in the tropospheric boundary layer. Δ represents energy leading to thermal degration, hν solar radiation leading to photo-dissociation and RH a hydrocarbon reacting with the specie in question. PPN is a notation for other peroxy nitrates than PAN.

8.7 Photochemical balance in the sunlit atmosphere

The OH radical concentrations in the sunlit tropospheric boundary layer are established by the fast photochemical balance reactions, which link together each of the sources and sinks of the major free radical species. There are two main pools of free radical species in the sunlit boundary layer, the pool of hydroxyl radical (OH) and the pool of peroxy radicals, both hydroperoxy (HO_2) and organic peroxy radicals (RO_2). In addition, there are six major categories of free radical reactions which together make up the fast photochemical balance. These six categories comprise reaction which:

- serve as sources of OH,
- serve as sources of HO_2 and RO_2,
- interconvert OH into HO_2 and RO_2,
- interconvert HO_2 and RO_2 into OH,
- serve as sinks for OH,
- serve as sinks for HO_2 and RO_2 radicals.

The resultant of all these processes is the formation of a steady state concentration of the free radical species in the sunlit tropospheric boundary layer. In this system hydrogen peroxide (H_2O_2) serves as a reservoir species for free radicals. A simplified diagram of the conversion processes of free radicals in the boundary layer (the above categories 1 to 6) is given in Figure 8.3 (next page).

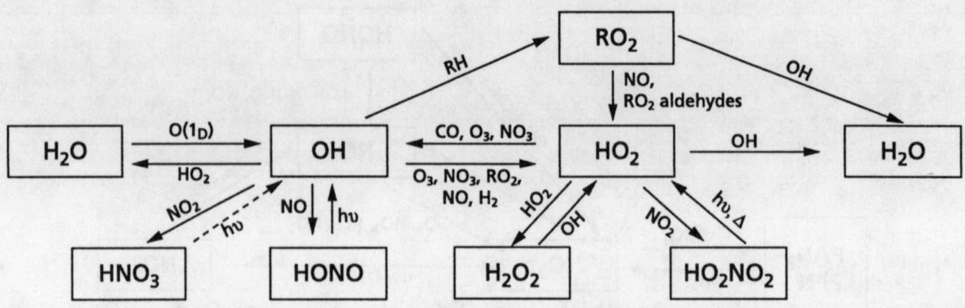

Figure 8.3 Illustration of the conversion processes involving OH, HO_2, and RO_2 compounds in the tropospheric boundary layer. Δ represents energy leading to thermal degradation, hν solar radiation leading to photo-dissociation and RH a hydrocarbon reacting with the specie in question.

Typically, close to noon, under moderately polluted conditions, these steady state concentrations may approach 0.2 ppt OH, 4.5 ppt HO_2 and 1.1 ppt CH_3O_2. Applying these steady state concentrations, the rate of ozone production in the background troposphere may be estimated:

$$\frac{d[O_3]}{dt} = k_{NO+HO_2}[NO][HO_2] + k_{NO+CH_3O_2}[NO][CH_3O_2] + \text{other} \tag{8.43}$$

$$\frac{d[O_3]}{dt} = 1 + 0.2 + 1.2 \text{ ppt s}^{-1} = 8 \text{ ppb h}^{-1} \tag{8.44}$$

This analysis implies that to reach the ozone concentrations typically found in regional scale ozone episodes of about 100 ppb, an elevation of about 50-70 ppb is required above the northern hemispheric background level. On the basis of the above estimates, this requires 6 to 9 hours of intense, sustained photochemical activity which might imply 2 days total reaction time and of the order of 500 km of travel. Long range transport is therefore anticipated to be an important dimension to regional scale photochemical episodes.

8.8 Photochemical ozone creation potential

Measurements suggest that ground level concentrations of ozone over Europe have more than doubled in the past century, brought about largely because of increased emissions of nitrogen oxides and volatile organic compounds (VOCs) from anthropogenic sources (Logan 1985). However, the chemical characteristics are varying greatly between the different VOC species. Table 8.1 (page 140) presents the chemical lifetime of a number of key photochemically active VOC species with respect to typical concentrations of OH and NO_3 radicals.

Table 8.2 Timescales for tropospheric reactions of daytime OH and night time NO_3 radicals with various hydrocarbons. The examples are for typical radical concentrations; OH radical concentration of 0.2 ppt and NO_3 radical concentration of 11 ppt. The lifetimes of the hydrocarbons are derived from reaction rate coefficients given in various articles in Wayne (1991). Figures in parentheses give the percentage of non-methane emissions by mass in European hydrocarbon emissions, according to Derwent and Jenkin (1991).

VOC class	Species	OH lifetime	NO_3 lifetime
Methane	CH_4	306 d	-
Alkanes (31%)	Ethane	8 d	-
	n-butane	18 h	-
Alkenes (4.2%)	Ethene	5 h	198 d
	1-butene	2 h	3 d
	Propene	2 h	4 d
Aromatics (19.5%)	Toluene	8 h	-
	o-xylene	3 h	104 d
	Benzene	37 h	-
Alcohols (11.6%)	Ethanol	47 h	>19 d
Biogenic (NA)	Isoprene	0.5 h	1.3 h
	α-Terpene	8 min	6 min
Aldehydes	Formaldehyde	5 h	57 d
	Acetaldehyde	3 h	14 d
Organic sulphur	DMS	9 h	1 h

The photochemical ozone creation potentials (POCPs) of the various VOC compounds may be used in future reduction strategies, to account for the differences in their chemical characteristics. The POCP concept may be used in European ozone studies to express the ozone forming ability of a specific VOC compound relative to that of ethene (Derwent, Jenkin 1991):

$$POCP_i = 100 \cdot \frac{\Delta\, \mathrm{ozone}_i}{\Delta\, \mathrm{ozone}_{ethene}} \tag{8.45}$$

where the change in ozone production for ethene and species i is expressed for identical changes in emissions by mass of the two compounds. Thus the POCP of ethene is 100 by definition. The application of the POCP concept may serve as a useful tool in future VOC emissions reduction strategies. An example of calculated POCPs of various hydrocarbons following their reaction with OH radicals is given in Table 8.3 (next page).

8.9 Cyclic hydrocarbons

A large fraction of the NMVOCs in the troposphere consists of aromatic compounds such as benzene, toluene and xylene, which may account for up to 10% of unleaded gasoline. High levels of these compounds are therefore observed in many urban streets over Europe and benzene has been seen to exceed present guideline values. The only significant degradation pathway for these compounds is by reaction with the hydroxyl radical. The OH radical may add to the double bonds of the aromatic ring and in cases like toluene abstract a hydrogen atom from the methyl group (Figure 8.4).

Table 8.3 Theoretical chemical potential for ozone formation by several hydrocarbons following their reaction with OH radicals. It is assumed that all the NO_2 molecules formed in the oxidation lead to formation of ozone molecules. The number of NO molecules converted to NO_2 per hydrocarbon is computed for two stages of the oxidation: From initial stage leading only to the formation of aldehydes and ketones, and the further oxidation of these carbonyl compounds. Numbers given are rounded to nearest integer. MEK = methyl ethyl ketone ($CH_3COCH_2CH_3$) (based on Simpson 1995; Warneck 1988).

Compound	Intermediate	Number of O_3 molecules produced		
	Aldehydes and ketones	Initial	From carbonyl compounds	Total
Ethane	CH_3CHO	2	4	6
Ethene	2HCHO	2	2	4
Propene	CH_3CHO, HCHO	2	5	7
Ethanol	CH_3CHO	1	4	5
n-butane	CH_3CHO (35%), MEK (65%)	2	7	10
o-xylene	CH_3COCHO, $CH_3COCHCHCHO$	2	11	13

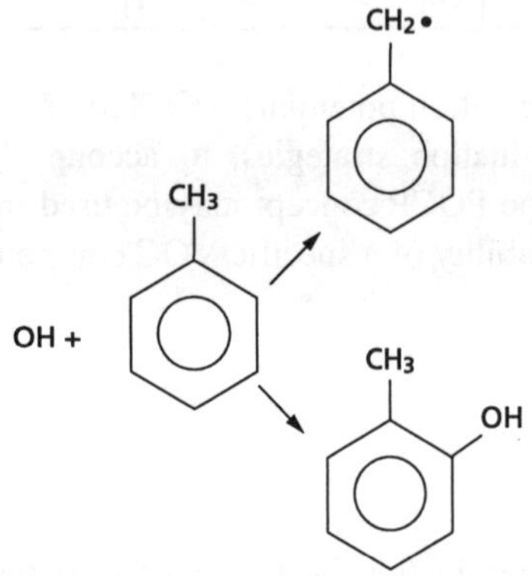

Figure 8.4 An illustration of the pathways of OH reactions with toluene in the tropospheric boundary layer.

The drawing of the adduct in Figure 8.4 symbolises that the free electron is shared by five of the carbon atoms in the ring. OH may actually add to any of the four different carbon positions of the ring, but the shown structure is the most important. The further reactions of these produced radicals involves reactions with O_2, NO, NO_2, and fragmentation of the cyclic structure. The products are a variety of aromatic compounds and smaller organic compounds. All these compounds will subsequently be degraded in the troposphere by their own reaction pathways and/or, as for most of the acids be removed from the troposphere by wet deposition.

In general the reactions of aromatic hydrocarbons with NO_3 radical are too slow to be of importance in the troposphere. However, phenols react fast with NO_3 radical and the lifetime is shorter with respect to reaction with NO_3 than with OH radical. The reaction proceed mostly via hydrogen abstraction of the phenolic hydrogen atom (Figure 8.5).

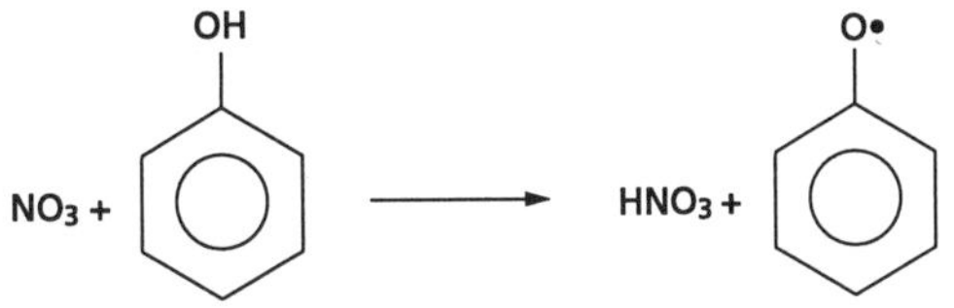

Figure 8.5 Illustration of the NO_3 radical reaction with phenol in the tropospheric boundary layer.

The chemical degradation of most of the cyclic hydrocarbon compounds is too slow to be of importance in the urban atmosphere. However, as earlier mentioned there may be exceptions. Harrison and Shi (1996) suggested compounds such as 1-methyl 1,3 cyclo-pentadiene to be responsible for the formation of high winter time NO_2 concentrations in London. The reaction pathway may be as illustrated in Figure 8.6.

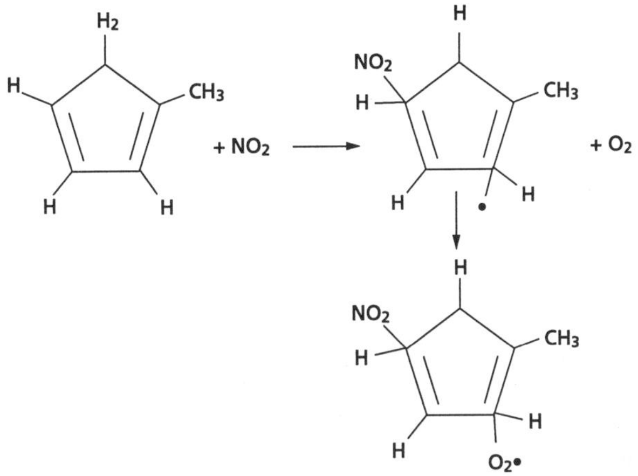

Figure 8.6 Illustration of the by Harrison and Shi (1996) suggested formation of peroxy radicals from 1-methyl 1,3 cyclo-pentadiene during dark winter time NO_2 episodes in London.

The figure shows how NO_2 adds to the cyclic hydrocarbon which in turn leads to the formation of a peroxy radical. The formed peroxy radical may then convert an NO molecule to NO_2. The suggested cyclic compounds have been found in exhaust gases from gasoline cars, but whether the concentration levels are sufficient to account for the needed conversion rate is still unclear.

8.10 Secondary suspended particulate matter

The first clear and unambiguous evidence that the heat hazes, which are frequently seen in fine, sunny weather in Europe, are man-made in origin became available nearly three decades ago. Lovelock (1972) showed how regionally polluted and photochemically reacted air masses were advected from continental Europe to the remote Atlantic coast of Ireland. The occurrence of simultaneous elevations in turbidity (a measure of reduced visibility) and CFC-11 (one of the freons), which is a unique man-made halocarbon tracer, showed conclusively that the summertime heat haze which is invariably associated with the large European anticyclonic weather systems is man-made. Subsequently, Cox et al. (1975) showed that these turbid photochemically reacted air masses also contained elevated ozone concentrations. Long range transport can bring elevated concentrations of ozone and suspended particulate matter to the most remote regions of Europe.

Airborne suspended particulate matter is responsible for the turbid nature of summertime anticyclonic air masses and hence for the visibility reductions associated with them (Chapter 21). The suspended particulate matter in these turbid air masses consists of a wide range of different chemical substances with a wide range of physical properties. Suspended particulate matter can only be described relative to some measurement method and cannot be described in an absolute way. There are therefore many ways of characterising these polluted turbid air masses.

8.10.1 PARTICLE SIZE DISTRIBUTION AND COMPOSITION

Aerosol and cloud physicists usually define three particle size ranges or particle modes in their characterisation of boundary layer aerosol particles:

- nucleation,
- accumulation, and
- coarse particle modes

Air quality scientists use terms such as coarse, fine, PM_{10} and $PM_{2.5}$. Generally speaking, the particles which are thought to be most injurious to human health (Chapter 18) are those that air quality scientists term as fine particles and aerosol physicists term accumulation mode particles. Coarse fraction particles are not thought to be so damaging to human health and act more as nuisance through the soiling of surfaces through the accumulation of grit and dust. The terms PM_{10} and $PM_{2.5}$ refer to the size fractions of the suspended particulate matter with diameters less than 2.5 or 10 μm, respectively, and both are usually reported in units of μg m^{-3}. The physics of aerosol particles is treated in Chapter 9. In the following is given a description of the formation of new aerosols from gas phase reactions and the chemical composition of tropospheric aerosol particles.

Particles, with a size range of less than 2.5 μm, $PM_{2.5}$, have a variable composition in time and space. During the summertime regional pollution episodes associated with heat haze and visibility-reduction, the main particle components appear to be:

- ammonium bisulphate,
- ammonium sulphate,
- sulphuric acid,
- sodium nitrate,
- elemental carbon,
- ammonium chloride, and
- sodium chloride

of which, ammonium sulphate appear to account for the largest fraction of the fine particulate mass. Elemental carbon is a primary pollutant and as such is given no further consideration in this chapter. The remainder of this section gives a description of the transformation processes by which SO_2, NO_x and ammonia emissions are converted into aerosol ammonium sulphate, nitrate and chloride and sulphuric acid.

8.10.2 FORMATION OF SECONDARY AEROSOL PARTICLES

The only significant source of fine fraction particulate ammonium sulphate in the troposphere is the chemical conversion process involving gas phase SO_2, since the direct emission of these particles is insignificant. The formation process begins typically with the emission of sulphur dioxide from coal- and oil-burning in stationary and mobile sources (Chapter 6). During regional pollution episodes in the summertime, sunlight-driven photochemical reactions, driven by hydrocarbons and oxides of nitrogen emitted by human activities, lead as already mentioned to elevated concentrations of the extremely reactive hydroxyl radical. Hydroxyl radicals oxidise sulphur dioxide to sulphur trioxide in two reactions rather similar to the oxidation path of CO:

$$OH + SO_2 + M \rightarrow HOSO_2 + M \tag{8.46}$$

$$HOSO_2 + O_2 \rightarrow HO_2 + SO_3 \tag{8.47}$$

Sulphur trioxide reacts very quickly with water vapour to form sulphuric acid vapour:

$$SO_3 + H_2O \rightarrow H_2SO_4 \tag{8.48}$$

The HO_2 radical produced in (8.47) may follow the already described reaction path in which NO is transformed to NO_2 which in turn photolyse and the chain of reactions eventually leads to production of ozone. This is why there is a close relationship between haze production, visibility reduction and photochemical ozone production.

The hydroxyl radical is recycled in the photochemical oxidation of sulphur dioxide in the sunlit tropospheric boundary layer, through the hydroperoxy radical and its subsequent reaction with NO. In this way, a small concentration of hydroxyl radical on the order of one hundredth to one tenth of a ppt, can lead to a substantial SO_2 oxidation rate, approaching a few percent per hour. This once more illustrates the importance of the OH radical in the sunlit atmosphere. In the overall conversion of SO_2 to sulphate on the global scale, the OH radical path accounts for approximately half of the transformation. The other half takes place inside cloud droplets where dissolved SO_2 is oxidised by H_2O_2 and O_3.

Sulphuric acid vapour readily nucleates on its own or with water molecules to form a fine aerosol of sulphuric acid droplets in the nanometre (nm) size range of the nucleation mode. These exceedingly small droplets will then grow by coagulation and coalescence with other sulphuric acid droplets or with the pre-existing suspended particulates and droplets in the sub-micron particle size range. The end-prod aerosol nucleation and growth processes is a dynamic distribution of sulphuric acid droplets and particles with varying sizes from nm to µm. Freshly oxidised material produced by these gas-to-particle conversion processes, is generally in the smallest size ranges and aged material in the larger sub-micron size range. By far the largest number of particles are in the nm range and by far the largest contribution to the particle mass in the sub-micron range.

8.10.3 SECONDARY AMMONIUM PARTICLE FORMATION

Ammonia is the only alkaline gas of any significance in the atmosphere. It is emitted mainly from agriculture through the disposal of animal wastes and the use of nitrogenous fertilisers. This take-up can be exceedingly rapid which make it capable of competing with similarly very rapid dry deposition of ammonia. These processes gives in general ammonia a very short tropospheric lifetime; Asman and Janssen (1987) found from model fit to measurements, that ammonia in Europe has an average lifetime in the tropospheric boundary layer of 3-4 hours. The end-point of the irreversible uptake of ammonia onto the surface of the sulphuric acid aerosol, is a mixture of particles and droplets containing sulphuric acid and ammonium sulphate.

Ammonia also reacts with other acidic gases, such as nitric acid and hydrogen chloride, present in the atmosphere to produce neutral and low volatile ammonium compounds:

$$NH_3 + HNO_3 \rightleftarrows NH_4NO_3 \tag{8.49}$$

$$NH_3 + HCl \rightleftarrows NH_4Cl \tag{8.50}$$

As indicated, these reactions are reversible reactions, in contrast to the reaction between ammonia and sulphuric acid which is irreversible. Both ammonium nitrate and ammonium chloride may dissolve in pre-existing aerosol droplets or may adsorb onto the surface of any pre-existing aerosol particles. In this way, nitrate, chloride and ammonium species become incorporated as secondary pollutants into suspended particulate matter in the size range less than 2.5 µm.

Whereas hydrogen chloride (HCl) is a primary pollutant emitted by coal burning and incineration, nitric acid is the main secondary pollutant from oxidation of NO_x emissions (reaction 8.6). A major part of the nitrate present in suspended particulate matter is formed by the sea-spray displacement reaction, which occurs on the surface of sea-spray particles (Wall et al. 1988):

$$NaCl(s) + HNO_3(g) \rightarrow NaNO_3(s) + HCl(g) \tag{8.51}$$

In the acid rain models such as the EMEP model this reaction is accounted for by a first order decay of HNO_3 of $10^{-5}\ s^{-1}$, and a reverse reaction rate coefficient which is half this size (Hov et al. 1994). Measurements by Wall et al. (1988) in California showed that nitrate in the coarse-particle mode is primarily associated with high sodium levels in maritime air and that the coarse nitrate particles have a peak in their distribution at the 3 μm diameter, where the product of the sodium surface and mass distribution also peaks.

Tropospheric boundary layer concentrations of HCl in the ppt range and in some case even ppb range have been reported for rural areas, which may indicate that despite often being neglected in transport-chemistry models, HCl may be of importance for the transformation of ammonia. Ammonium nitrate and ammonium chloride may evaporate back to the gas phase compounds, where these two systems may find into an equilibrium. Stelson and Seinfeld (1982a,b) determined the equilibrium product between the gas phase concentrations of NH_3 and HNO_3 as a function of humidity and temperature; a concept which is adopted in many transport-chemistry models of today.

Similar equilibrium expressions as for NH_3 and HNO_3 were derived by Pio and Harrison (1987) for the equilibrium products of NH_3 and HCl. Since sulphate associate more easily with NH_3 than does HNO_3 and HCl; sulphate is neutralised before NH_4NO_3 and NH_4Cl is formed.

In many cases it may be a reasonable assumption that there will be equilibrium between NH_3 - HNO_3 and NH_4NO_3, and similarly between NH_3 - HCl and NH_4Cl. However, observations of the particle size distribution of inorganic nitrogen, sulphur, and chlorine species in maritime air over the North Sea have shown that the observed products of partial pressures of $[NH_3][HNO_3]$ and $[NH_3][HCl]$ often fall below the theoretical lines of equilibrium (Ottley, Harrison 1992), and cautions must therefore be taken when applying the assumption of equilibrium.

8.10.4 IMPACT OF PARTICLES AND CLOUD DROPLETS ON TRANSFORMATION OF GASES

The chemical reaction chains in the atmosphere are drastically altered once an air parcel is cooled, condensation occurs, and a cloud develops. Although the volume fraction of liquid water is very small (usually between 10^{-7} to 10^{-6}), some atmospheric gases are "concentrated" in a relatively small volume, which can enhance reaction rates substantially. Moreover, the "cage" effect that water molecules exert on reactants increase the reaction probability after collision of reactants. It is generally assumed that about half of the conversion of SO_2 to sulphate takes place inside cloud droplet. The main oxidant is in this case H_2O_2, but also ozone contributes to this conversion in an pH dependent oxidation.

Lelieveld and Crutzen (1991) discussed *the role of clouds* in tropospheric photochemistry, and they defined six processes that are potentially rate limiting for chemistry in cloud droplets:

- Transfer of gases from the gas phase to the droplet surface
- Transfer across the gas-liquid interface
- Volatilisation of gases
- Dispersion of gases throughout the droplet
- Attainment of aqueous-phase equilibria
- Aqueous-phase reactions

Based on simple parameterizations of these processes they developed a model for the some of the most important chemical conversion processes in cloud droplets. From the model simulations Lelieveld and Crutzen (1991) showed that clouds substantially reduce the concentrations of NO_x, HCHO, OH, HO_2 and H_2O_2 in the tropospheric boundary layer.

Also aerosol particles in general play a role in the chemical conversion in the atmosphere. The already discussed reactions of HNO_3 and HCl with NH_3 are believed to take place on the surface of aerosol particles (continental or sea spray particles). Notholt et al. (1992) observed an interrelation between simultaneous peaks in NO_x concentrations and aerosol particle surfaces during foggy periods, and peaks in HONO concentrations in the highly polluted Po Valley in Northern Italy. This was taken as an evidence for heterogeneous conversion on aerosol surfaces through either of the reactions:

$$NO + NO_2 + H_2O \rightarrow 2HNO_2 \quad (8.54)$$

$$2NO_2 + H_2O \rightarrow HNO_2 + HNO_3 \quad (8.55)$$

Probably this type of chemical conversion plays an important role also in many urban areas over Europe, but so far only few studies have been carried out.

8.11 Concluding remarks

Our knowledge concerning the chemical transformation processes in the tropospheric boundary layer has increased substantially during the last few decades. However, there are still phenomena that are not fully understood and where investigations are needed.

Major efforts in air pollution research will in the coming years be given to the study of the processes governing aerosol particle concentrations in urban streets and on the role of these particles in the conversion of gas phase compounds on local as well as mesoscale. Some of the particulate material is derived from the chemical conversion during long range transport of the primary pollutants sulphur dioxide and nitric oxide emitted from anthropogenic sources over Europe. The implemented and future reduction strategies in Europe should therefore lead to reductions in these levels in the coming years but the extent of any reductions are uncertain at present.

Another area of concern is the winter time NO_2 episodes in northern Europe, and the summertime photochemical smog episodes in southern Europe. Significant effort should be given to the understanding of the chemical processes governing these episodes.

8.12 References

Asman, W.A.H., Janssen, A.J. (1987) A long range transport model for ammonia and ammonium for Europe, *Atmospheric Environment*, **21**, 2099-2119.

Bower, J.S., Boughton, G.F.J., Stedman, J.R., William, M.L. (1994) A winter NO_2 smog episode in the UK, *Atmospheric Environment*, **28**, 461-475.

Cox, R.A., Eggleton, A.E.J., Derwent, R.G., Lovelock, J.E., Pack, D.H. (1975) Long-range transport of photochemical ozone in north-western Europe, *Nature*, **218**, 118-221.

Derwent, R.E., Middleton, D.R., Field, R.A., Goldstone, M.E., Lester, J.N., Perry, R. (1995) Analysis and interpretation of air quality data from urban roadside location in Central London over the period from July 1991 to July 1992, *Atmospheric Environment*, **29**, 923-946.

Lelieveld, J., Crutzen, P.J. (1991) The role of Clouds in Tropospheric Photochemistry, *J. Atmos. Chem.*, **12**, 229-267.

Derwent, R.G., Jenkin, M.E. (1991) Hydrocarbons and the long range transport of ozone and PAN across Europe, *Atmospheric Environment*, **25A**, 1661-1678.

Harrison, R.M., Shi, J.P. (1996) Sources of nitrogen dioxide in winter smog episodes, *Science of the Total Environment*, **189/190**, 391-399.

Hov, Ø., Hjøllo, B.Aa., Eliassen, A. (1994) Transport distance of ammonia and ammonium in Northern Europe, 1. Model description, *J. Geophys. Res.*, **D9**, 18,735-18,748.

Finlayson-Pitts, B., Pitts, J.N. (1986) *Atmospheric Chemistry: Fundamentals and Experimental Techniques*, John Wiley & sons, New York.

Leighton, P.A. (1961) *Photochemistry of Air Pollution.* InterScience Publishers, New York.

Logan, J.A. (1985) Tropospheric ozone: Seasonal behaviour, trends, and anthropogenic influence, *J. Geophys. Res.*, **D18**, 10463-10484.

Lovelock, J.E. (1972) Atmospheric Turbidity and CCl_3F concentrations in Rural Southern England and Southern Ireland, *Atmospheric Environment*, **6**, 917-925.

Notholt, J., Hjorth, J., Raes, F. (1992) Formation of HNO_2 on aerosol surfaces during foggy periods in the presence of NO and NO_2, *Atmospheric Environment*, **26A**, 2111-217.

Ottley, C.J., Harrison, R.M. (1992) The spatial distribution and particle size of some inorganic nitrogen, sulphur and chloride species over the North Sea, *Atmospheric Environment*, **26A**, 1689-1699.

Pio, C.A., Harrison, R.M. (1989) The equilibrium of ammonium chloride aerosol with gaseous hydrochloric acid and ammonia under tropospheric conditions, *Atmospheric Environment*, **21**, 1243-1246.

Shi, J.P., Harrison, R.M. (1997) Rapid NO_2 formation in diluted petrol fuelled engine exhaust - A source of NO_2 in winter smog episodes, *Atmospheric Environment*, **31**, 3857-3866.

Seinfeld, J.H., Pandis, S.N. (1998) *Atmospheric Chemistry and Physics. From Air Pollution to Global Change*, John Wiley & Sons Inc., New York.

Simpson, D. (1995) Hydrocarbon reactivity and ozone formation in Europe, *J. Atmos. Chem.*, **20**, 163-177.

Stelson, A.W., Seinfeld, J.H. (1982a) Relative humidity and temperature dependence of the ammonium nitrate dissociation constant, *Atmospheric Environment*, **16**, 983-992.

Stelson, A.W., Seinfeld, J.H. (1982b) Relative humidity and pH dependence of the vapour pressure of ammonium nitrate - nitric acid solutions at 25°C, *Atmospheric Environment*, **16**, 993-1000.

Wall, S.M., John, W., Ondo, J.L. (1988) Measurements of aerosol size distributions for nitrate and major ionic species, *Atmospheric Environment*, **22**, 1649-1656.

Warneck, P. (1988) Chemistry of the Natural Atmosphere, Academic Press, San Diego, California.

Wayne, R.P. (1991) The Nitrate Radical: Physics, Chemistry, and the Atmosphere - special issue, *Atmospheric Environment*, **25A**, 1-206.

[illegible]

[illegible] 19[illegible] 1[illegible]

[illegible]

Chapter 9

PARTICLES

HELMUTH HORVATH
Institute of Experimental Physics, University of Vienna
Boltzmanngasse 5, A-1090 Vienna, Austria

9.1 Particle formation processes

The atmosphere contains particles of sizes ranging from slightly larger than molecules up to several mm like hailstones or lapilli emitted during volcanic eruptions. They consist of a variety of chemical compounds. Particles are always present in the atmosphere, however in highly variable concentrations. A system consisting of a gas and solid or liquid particles, which remain suspended for at least several minutes, is called an aerosol, and by this definition the whole atmosphere is an aerosol. The aerosol particles vary in sizes from a few nm to fractions of a mm in diameter, thus covering a huge size range of 5 orders of magnitude. Obviously the particles have different origins and properties. A few examples are listed in Table 9.1.

Table 9.1 Typical sizes of particles in the urban atmosphere.

Diameter	Substance	Origin
≈ 50 μm	Rubber, mineral material cement dust, insect fragments	Tire wear, erosion of pavement and buildings
≈ 10 μm	Fog droplets, pollen, bacteria fly ash, soil material	Condensation of water, biologic origin combustion, wind erosion, resuspended road dust
≈ 5 μm	Fly ash, soil material	
≈ 1 μm	"Wet particles", sea salt	Sulfate, nitrate grown to solution droplets by uptake of water in a humid environment
≈ 0.5 μm	Sulfates, nitrates, organics	Various combustion processes, end product of condensation on existing particles
	"Mixed particles"	Re-evaporated and processed cloud droplets
≈ 0.1 μm	Soot, cigarette tar and ash	Internal combustion engine, smoking
≈ 30 nm	Metallurgic fumes and condensation	Metal processing, primary particles in Diesel engine
≈ 10 nm	Gas to particle conversion	Reactions of precursor gases
≈ 3 nm	Original nuclei of photooxidation	Aitken particles

The mass of particles in 1 m^3 air is very small; in a very clean urban environment it can be about 10 μg/m^3, whereas a typical urban background concentration is about 20 to 50 μg/m^3 and in polluted city centers values > 100 μg/m^3 can be observed. Particles in the air are essential, without particles no clouds and fog could form and no precipitation would occur.

Particles can be of local origin or be the product of long range transport of pollutants. Due to their life time of about a week, particles in the size range between 0.1 and 1 μm, observed in the urban environment, can at least in part be from sources outside the urban region. Comparative measurements in a town and upwind give a quick possibility to determine whether the particles are mainly of local or distant origin.

9.1.1 DIRECT EMISSIONS

A major urban source for particles is the emissions from all kinds of vehicles (Chapter 5). Typical for European towns is the large number of Diesel powered cars and lorries (50% of the newly registered light vehicles and ≈ 100% of the heavy duty vehicles). Modern Diesel motors emit soot particles mainly in the submicrometer range with a maximum in the size distribution close to 0.1 μm, the particles consist of black carbon and hydrocarbons. Depending on the sulphur content of the fuel the emitted particles may also contain sulphates. The emission factor is about 1 g/l, thus e.g. in Vienna with a total diesel fuel consumption of 520 tons/day, the diesel powered vehicles emit 650 kg/day of soot. The emission from vehicles powered by gasoline engines is less than half of that. These particles contain hydrocarbons and lead, if leaded fuel is used, which still is the case in some European cities.

Additional sources of directly emitted particles are other combustion processes. Most larger power plants and industries in Europe have very efficient filtration, thus these sources only emit small amounts compared to the many smaller sources e.g. local space heating (Chapter 4). Especially during winter, heat production causes an additional load of particles. There is a large difference depending on the age of the heating system. The older systems use small stoves or fireplaces heating one room, use coal, coke or wood and have considerable emissions, since the air supply is not well controlled. The London smog episode in 1952 was in part caused by this (Brimblecombe 1987). The emitted particles contain ashes, e.g. potassium, black carbon and organic carbon. Nowadays either large central heating units, district heating or co-generation of electricity and heat with very small emissions are in use. A typical tracer for the combustion of residue oil is Vanadium. As an alternative small heating units powered by natural gas are in use many places. Depending on the degree of modernisation, space heating is a considerable source of particles in cities like Prague and Budapest, but only of little importance in Scandinavian towns, where district heating is the most common type of space heating. However, due to additional sources, the particle concentration is in general higher during winter.

Any combustion process leads to the formation of substances which have a very low vapour pressure at ambient temperatures. During cooling, a high supersaturation occurs. Homogeneous nucleation form nm-sized particles that usually coagulate rapidly forming larger particles, and thus reducing the particle number. The coagulation gets slower with decreasing number of particles. Usually particles with sizes between 50 nm and 1 μm are emitted, but also nano-particles are emitted in large numbers (due to their small sizes) e.g. by cars. This type of particles has only been studied recently. A similar process takes place during the formation of metallurgic fumes.

Due to limited space for landfills, household refuse is burned in many European towns. Modern incineration is quite advanced and careful cleaning of the flue gases gives little emission (Chapter 4). But no filtration process is perfect, thus small amounts of elements ranging from zinc (e.g. cans) to calcium (e.g. paper) are emitted.

The movement of vehicles produces turbulence near the surface of the roads and particles deposited on the road such as wear of brake linings or tires are brought in the airborne state. Abrasion of road material (gravel) and tires produce a reservoir of coarse particles which are resuspended by traffic or by high wind speed. In many European towns a serious dust problem can often be observed towards the end of the winter since (1) gravel, which is put on the roads to facilitate driving in snow and ice, produces enormous amounts of coarse particles under dry conditions and (2) studded tires are very abrasive on dry road surfaces. Similarly, wind erosion from soil or building surfaces produces coarse particles. These particles are of natural or quasi natural origin, and have sizes above a few μm, and are mainly of local importance, especially in arid regions. The particles contain the chemical elements present in the Earth's crust and consist mainly of Fe, Ca, Si and Al. Especially near the sea, also sea salt particles are found and thus also Na and Cl is contained in the super μm particles in these regions.

9.1.2 GASEOUS EMISSIONS WITH SUBSEQUENT GAS TO PARTICLE CONVERSION

Most fossil fuels (coal, oil) contain sulphur which during the combustion process is oxidized mainly to sulphur dioxide which is then emitted. In the hot flames atmospheric nitrogen and oxygen react, forming nitric oxides, which are therefore also emitted by all combustion processes. The mass concentrations of sulphur dioxide and nitrogen dioxide in an urban environment are usually higher than the mass concentration of the particles. In the atmosphere both gases are further oxidized, and eventually sulphates and nitrates emerge. The oxidation is a process which takes some time, with SO_2 oxidation rates between 0.2 and 15%h^{-1} with the lower values in the arid regions. Some SO_2 is oxidized in the gas phase by reacting with OH or other free radicals forming gaseous sulphuric acid. This will acquire water and nucleate to sulphuric acid droplets. Most SO_2 is oxidized in the aqueous phase, by being dissolved in wet particles. The substances contained in the particles can catalyse the oxidation, e.g. Fe, Mn, C or dissolved HNO_2. Thus with continuing oxidation process the particle contains more and more sulphates. Depending on the ammonia present in the atmosphere the sulphuric acid can be neutralized, and particles which mainly consist of ammonium sulphate will be formed. The oxidation of nitrogen oxides to nitric acid is similar. Since it takes time to oxidize sulphur dioxide, the particles can appear at locations far away from the source of the precursor gas.

Sulphur dioxide and nitric oxides are not the only substances which can be converted in the atmosphere to substances of very low vapour pressure, which subsequently form particles. This also occurs naturally: Plants, especially conifers emit organic vapours, e.g. terpenes; they are converted in the atmosphere to organic substances which form particles. The haze seen in forested regions can have this natural origin. Since organic vapors move with the air mass, this can also occur in towns.

9.1.3 PHOTOCHEMICAL SMOG

If hydrocarbons and nitric oxides occur together, the presence of sunlight triggers a chain reaction, which leads to the oxidation of NO to NO_2, oxidation of hydrocarbons and the formation of ozone (Chapter 8). In this process also a vast amount of particles is generated, leading to a foggy appearance of the atmosphere. It was first observed in California, but similar incidences are now also documented for European towns. With the strict use of catalysts the automotive emissions contain less nitrates and organics, therefore photochemical smog is not as dramatic, as it used to be. It takes some time for the reactions to take place, thus with air movement the products of photochemistry, e.g. ozone, are found at several tens of kilometers away from the source, and an area where no sources exist may be heavily polluted.

9.2 Size distributions

A simple characterisation of the suspended particles is the mass concentration, M. It is the total mass contained in one cubic meter of air. It can e.g. be determined by drawing the air through a filter, which has a deposition close to 100% for all sizes, which among others is the case for a glass fibre filter. Since the particles have various origins, many chemical species are contained in the particles deposited on the filter and a chemical analysis can give the mass concentration M_i for chemical species i. Another simple characterisation of the aerosol is the total particles number, N per cubic meter or number concentration, as it can be measured e.g. with a condensation nucleus counter.

Since the range of particle sizes is about five orders of magnitude, neither the mass concentration, M nor the number concentration, N sufficiently characterises the aerosol. The particles can better be described by the size distribution, number and mass size distributions are in use. Let us consider particles which are in the size range [r, r + dr]. The number of the particles in this size interval be dN, the mass be dM. We call dN/dr = n(r) the number size distribution and dM/dr = m(r) the mass size distribution. Obviously dN = n(r)·dr and dM = m(r)·dr. The total particles number is obtained by integration between the smallest and largest radius:

$$N = \int_{r_{min}}^{r_{max}} n(r)dr \text{, and similarly:} \tag{9.1}$$

$$M = \int_{r_{min}}^{r_{max}} m(r)dr \tag{9.2}$$

Sometimes a cumulative size distribution is used, which gives the particle number or particle mass smaller than or larger than a given radius.

For the large range of particle sizes it is useful to use logarithmic intervals [ln r, ln r + dln r] and in this case the number and the mass size distribution is defined as dN/dln r and dM/dln r. Since dln r = (1/r)dr the following relations hold: dN/dln r = r·dN/dr and dM/dln r = r·dM/dr. If the particles are spherical and have the density ρ, one can convert the number/size distribution to a mass/size distribution and vice versa by:

$$\frac{\mathrm{dM}}{\mathrm{dr}} = \frac{4}{3}\pi\,\rho\,\mathrm{r}^3\,\frac{\mathrm{dN}}{\mathrm{dr}} \tag{9.3}$$

Size distributions measured in the atmosphere have been observed to follow certain rules and thus model size distributions are in in use. A commonly used model is the lognormal distribution. For most aerosol processes the generated particles are of variable size due to a certain degree of randomness during generation. In many cases the size distribution can be described by a log-normal distribution:

$$\frac{\mathrm{dN}}{\mathrm{dln(r)}} = \frac{\mathrm{N}}{\sqrt{2\pi}\,\sigma_g}\exp\left[\frac{\left(\ln(r)-\ln(r_g)\right)^2}{2\ln^2(\sigma_g)}\right] = \mathrm{N}\cdot\mathrm{LND}\,(r_g,\sigma_g) \tag{9.4}$$

Also the mass size distribution can be described by a similar distribution. The radius r_g determines the maximum of the distribution curve and is called the number geometric mean diameter and mass geometric mean diameter respectively and σ_g determines the width of the curve and is called the geometric standard deviation. Particles which originate from one single process usually have a small σ_g, e.g. 1.1 for well controlled condensation and about 1.7 for less controlled processes. Particles which originate from various processes have distributions which could be approximated by wide lognormal distributions with $\sigma_g > 2$, and it is questionable, whether the distribution then is a good approximation.

Several other functions are used to approximate the size distribution of the atmospheric aerosol particles - at least in a certain size range: The power law distribution $dN/dr = a \cdot r^{-\nu}$, with $\nu \approx 3$, is a simple approximation for sizes between 0.1 and 10 μm, modified gamma functions in the form of:

$$\mathrm{n(r)} = \mathrm{a}\cdot\mathrm{r}^{\alpha}\exp\left(-\mathrm{b}\cdot\mathrm{r}^{\gamma}\right) \tag{9.5}$$

with a, b, α and γ being constants are also in use - especially for atmospheric optics.

Whitby (1978) has suggested to characterise the urban aerosol by a sum of three log-normal distributions, which are called modes:

- *The nucleation mode*, originating from condensation of supersaturated vapours. The primary particles are a few nm in size, but due to the high concentrations they coagulate until the number is too small for further growth.

- The *accumulation mode*, consisting of long lived particles of sizes of a few tenths of a micrometer. They stay in the atmosphere for approximately a week and compete for condensation and coagulation with the particles of the nucleation mode. Due to the larger surface of the accumulation mode particles, the heterogeneous coagulation with them exceeds homogeneous coagulation.
- The *coarse mode*, the particles of which are generated by mechanical processes such as sea spray, erosion, and resuspension. It must be noted, that the sampling of coarse particles can be flawed by the inlet system, which may make sampling of particles above 10 μm difficult.

Values for the three modes are given in Table 9.3. For comparison the urban model given by Jaenicke (1988, p. 408), there is little difference between the two models, except for the coarse mode, where sampling is crucial for losses of large particles. The model size distributions are for the whole collective of the particles without differentiating the various species.

Table 9.3 Characteristics of the three modes of the atmospheric aerosol. Log-normal volume size distributions are used. A value for the volume of 10^{-12} can be interpreted as 0.001 mm^3/m^3 and with particles having the density of water this corresponds to $\mu g/m^3$.

	d_g	d_g ($\mu g/m^3$)	volume (0.001 mm^3/m^3)
Atmospheric average (Whitby 1978)			
Nucleation mode	15-40 nm	1.6	0.0005-9
Accumulation mode	0.15-0.5 μm	1.6-2.2	1-300
Coarse mode	5-30 μm	2-3	2-1000
Urban average (Whitby 1978)			
Nucleation mode	38 nm	1.8	0.62
Accumulation mode	0.25 μm	2.16	38.4
Coarse mode	5.7 μm	2.21	30.8
Urban model (Jaenicke 1988)			
Nucleation mode	29 nm	1.67	0.48
Accumulation mode	0.3 μm	2.17	35
Coarse mode	16.5 μm	4.63	64.3

9.2.1 MEASURED URBAN AEROSOL SIZE DISTRIBUTIONS

Size distributions are obtained with size selective instruments. Optical, mechanical and electrical principles for size segregation are used. The data reported in this example were taken with a cascade impactor. By pointing a jet of particle laden air on a plate, particles which are unable so make a sharp turn are deposited. By increasing the speed and decreasing the distance to the deposition plate on the subsequent stages, the impactor produces deposits of the different size classes of the aerosol. An example (Horvath et al. 1997) for a mass size distribution is shown in Figure 9.1a. The models discussed above have a similar shape, but obviously do not agree with the example, e.g. the mean diameter of the accumulation mode is larger than postulated by the models. This has frequently been observed and can be a hint that European towns are different to others, since Europe is densely populated and the accumulation mode particles have

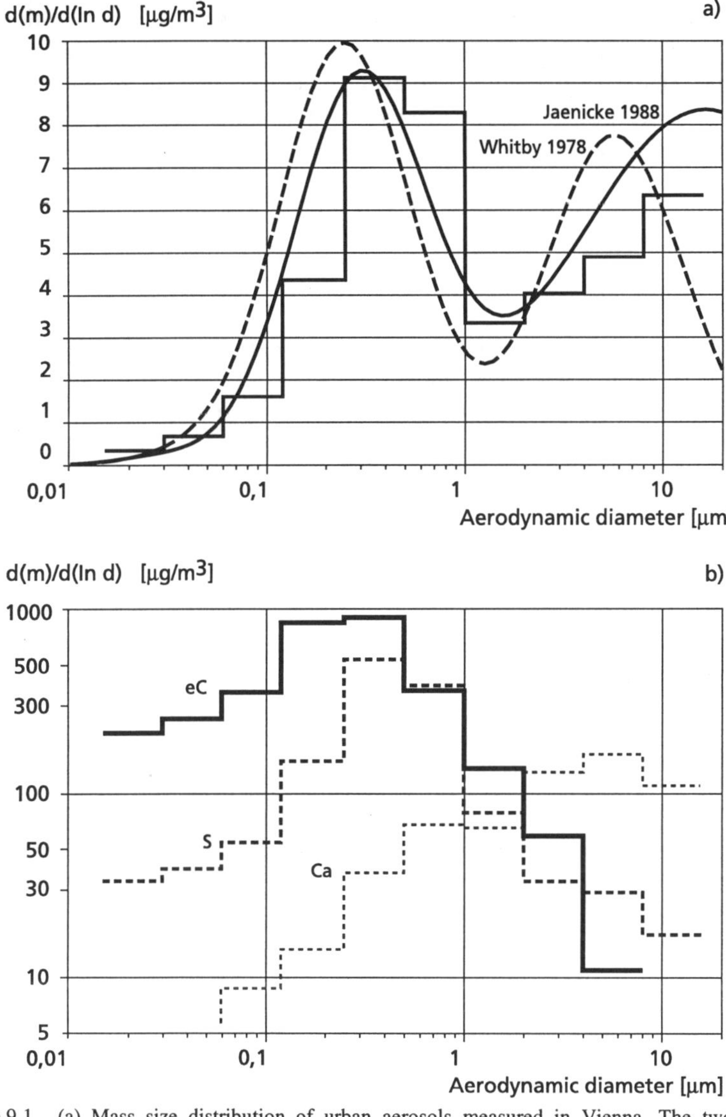

Figure 9.1 (a) Mass size distribution of urban aerosols measured in Vienna. The two urban models (taken from table 9.3) are shown for comparison, the model values have been halved to simplify comparison. (b) Measured mass size distribution for sulfur (S), calcium (Ca) and black carbon (eC).

sufficient possibilities to grow by coagulation and condensation. The different species forming the aerosol can be determined by analysing the deposited particles, in this example by PIXE. Three elements are shown for illustration in Figure 9.1b: Sulphur has only one mode with a mean diameter of $\approx$ 0.4 μm. There is no difference, neither in mass nor in size to sulphur data measured upwind, thus from the previous considerations one can conclude, that the sulphur aerosol has its origin outside the town. Calcium mainly exists in the coarse mode, and the upwind concentration is about 55%, thus erosion and resuspension is a possible urban source. Black carbon has a peak at ≈ 0.2 μm, thus it is smaller than the average aerosol, since diesel motors, the main source for black carbon in Vienna, mainly emit particles with a mean diameter of this size. The upwind concentration is 54%, thus the major contribution is local.

9.3 Dynamics of aerosol particles

9.3.1 SEDIMENTATION

All particles settle with a velocity v_s, which can be determined by equating the gravitational force ($F_g = 4\pi \cdot r^3 \cdot \rho \cdot g/3$ for spherical particles) with the drag force obtained from the Stokes law ($F_d = 6\pi \cdot \eta \cdot v_s / C_s$ with ρ the density of the particles, η the viscosity of the air, and C_s the slip correction factor given as:

$$C_s = 1 + \frac{\lambda}{r} + \left[1.257 + 0.4 + \exp\left(-1.1\, r\lambda\right)\right] \tag{9.6}$$

were λ is the mean free path of the molecules). A few settling velocities are listed in Table 9.2. The settling velocity is important for particles larger than ≈1 μm, e.g. a 3 μm particle takes 9 hours to settle a distance of 200 m.

Table 9.2 Distance an aerosol particle moves in 1 s by sedimentation or diffusion, residence time in the atmosphere and distance travelled with a wind speed of 2 m/s. Air of 20°C and 1 bar, particles with density of 1000 kg/m³ is assumed.

Diameter	10 nm	30 nm	0.1 μm	0.3 μm	1 μm	3 μm	10 μm	30 μm
Sedimentation (m)	$6.7\cdot10^{-8}$	$2.1\cdot10^{-7}$	$8.7\cdot10^{-7}$	$4.2\cdot10^{-6}$	$3.5\cdot10^{-5}$	$3.1\cdot10^{-3}$	0.027	0.248
Diffusion (m)	$2.6\cdot10^{-4}$	$1.1\cdot10^{-4}$	$3.7\cdot10^{-5}$	$1.6\cdot10^{-5}$	$7.4\cdot10^{-6}$	$4.1\cdot10^{-6}$	$2.2\cdot10^{-6}$	$1.3\cdot10^{-6}$
Residence time (d)	0.3	1.7	5	7.6	7.6	4.6	2.7	0.5
Distance traveled (km)	60	300	860	1320	1320	800	460	86

9.3.2 BROWNIAN MOTION, DIFFUSION

The gas molecules perform a random motion and also impact on the particles suspended in the gas. If the number of impacts is large, the net momentum transferred to the particles by the molecules is zero. But with decreasing particle size the number of impacts decreases and the net momentum will vary randomly, causing a random motion of the particles in the gas, which is known as Brownian motion. The mean distance a particle moves in a given direction in time Δt is given by:

$$\sqrt{\frac{2kTC_s\Delta t}{6\pi\eta r}}\,. \tag{9.7}$$

Some values are given in Table 9.2. One can see that for submicrometer particles the displacement due to Brownian motion is orders of magnitude larger than the distance settled, thus diffusion is the main mechanism to transport particles with diameters below 1 μm through a gas.

9.3.3 COAGULATION

During Brownian motion there is the possibility for the particles to collide with other particles. The collision usually results in a sticking of the particles on each other thus forming a new, larger, and mostly irregular particle. The number of particles with number concentration, N colliding per second and thus disappearing is given by:

$$\frac{dN}{dt} = KN^2 \text{, with the coagulation coefficient being } K = \frac{4kTC_s}{3\eta}. \qquad (9.8)$$

The coagulation coefficient exhibits little variation with size; coagulation is a second order process and is therefore mainly influenced by the square of the particle number. Therefore coagulation is rapid, if the particle number is high, e.g. with a concentration of $N = 10^{14} m^{-3}$ it takes 20 seconds to half the particle number and 140 seconds to double the particle size. For a concentration two orders of magnitude less the corresponding values are 55 hours and 16 days. Concentrations this high rarely occur in the atmosphere. Only particles a few nanometers in size occur at concentrations where coagulation is important. Thus at the lower end of the size spectrum atmospheric particles are reduced in number by coagulation.

9.3.4 HUMIDITY GROWTH

It is well known that e.g. rock salt, when left in a humid environment will become humid and eventually a solution of salt in water will be formed. A substance which has this property is called deliquescent. This means, that the substance takes up a certain quantity of water, when a minimum humidity is exceeded. The amount of water uptake depends on the humidity and the substance. For a given humidity an equilibrium is reached, when the saturation vapor pressure over the solution equals the vapour pressure in the air. The relation between the concentration of the solute, usually given as molality (moles of solute per kg of solvent), and the water activity (ratio of the equilibrium vapour pressure over the solution and the one over pure water, equals the considered relative humidity) have been determined for many substances and are available in the literature (e.g. Robinson, Stokes 1959). For pure substances the phase transition is abrupt and large hysteresis effects occur. The atmospheric aerosol contains both particles consisting of pure substances, which show a well defined phase transition with hysteresis effects, and particles, which are a mixture of several substances where no exact phase transition can be found. A summary of intense investigations is given by Hänel (1976). For the different types of atmospheric aerosols the increase in mass is e.g. tabulated by Jaenicke (1988, p. 427). When the humidity increases from below 60% to 80%, the rural aerosol takes up about its own mass of water, the urban aerosol about half and the maritime aerosol about twice its mass.

9.3.5 CONDENSATION AND EVAPORATION

A supersaturated vapour (i.e. a vapor with the density ρ_v which is larger than the saturation density ρ_s) stimulates the production of particles by condensation. If condensation nuclei are present, the vapour mainly condenses on the particles available and considerable growth of the particles by covering them with the condensing substance occurs. This happens e.g. in cloud or fog formation, where the supersaturation of water vapor mainly is produced by adiabatic expansion. The supersaturation ($\rho_v = \rho_s$-1) in clouds is usually just a few per cent. The vapour pressure ρ over a curved droplet is smaller than over a flat surface (the saturation pressure ρ_s) and is given by the Kelvin equation:

$$\rho = \rho_s \cdot \exp\left(\frac{2\gamma M}{\rho RTr}\right) \tag{9.9}$$

with γ, M, ρ, R, T and r being the surface tension, molecular weight and density of the liquid, the gas constant, absolute temperature and droplet radius. For a given supersaturation one can solve the Kelvin equation with respect to the droplet radius. Droplets smaller than this size will evaporate, droplets larger will grow by condensation. Due to the positive exponential smaller droplets require higher supersaturations to grow. If the condensation occurs onto a salt particle, the needed supersaturation for growth is less than for a pure water droplet, its value depends on the chemical composition of the salt and the size of the particle. During cloud or fog formation only these particles will act as cloud condensation nuclei and form a droplet which will grow at the existing humidity. In an urban environment this is only a small fraction. The non-activated particles will still increase their size by taking up of water and forming a solution droplet and can collide with cloud or fog droplets, and after evaporation a particle will be formed being an agglomerate of several previously separate particles and this agglomerate may also contain substances dissolved in the cloud/fog water. Since nine out of ten clouds evaporate, clouds considerably contribute to the transformation of particles.

Gas phase photochemical reactions in an urban atmosphere can yield low vapour pressure products. Due to the low vapor pressure of the reaction product a supersaturation occurs. At supersaturation of several hundred percent agglomerates of vapour molecules are continuously formed and disintegrate. If an agglomerate has reached the critical diameter, immediately condensation of more vapor molecules will occur, thus a condensation nucleus is formed by homogeneous nucleation. These nuclei occur in high numbers and after a rapid first growth sub-micrometer particles of considerable concentration are formed. The number of particles formed per time and volume can easily exceed $10^8/s^{-1}m^{-3}$ for high supersaturations. These particles can coagulate with each other, thus slowly forming larger particles, or they can attach to already existing particles.

9.4 Life time of aerosol particles and variability

9.4.1 DEPOSITION

A particle which is suspended in the air will eventually be removed from the atmosphere. Depending on the size, different processes are important:

- For particles larger than a few micrometers the sedimentation limits the lifetime of the particles considerably. But also the inertia of the particles can cause a rapid removal from the gas: In a cloud a falling raindrop can collide with a particle too large follow the air stream around the droplet. Similary, when air flows around an obstacle, e.g. a trunk of a tree, the particles can impact onto it and thus be removed from the air.
- Particles less than 100 nm essentially move by diffusion. Although the distance traveled in a second is small, it can be sufficient for the particle to move to a surface through the laminar layer surrounding it, or to collide with other suspended particles and coagulate to an agglomerate. By this the particle will stop existing as such and will form a new particle. This process is effective for nm, or tens of nm, sized particles.
- For sizes between tenths of μm to μm's both above mentioned processes are ineffective, and the removal of particles is slow and mainly due to in-cloud incorporation and precipitation.

Using all possible deposition mechanisms average residence times of particles in the lower troposphere have been estimated (Jaenicke 1988); values are shown in Table 9.2 (page 168). The particles can travel with the air movement an average distance of 50 km for tens of nm or super micrometer sizes and > 1000 km for particles sized a few tenths of μm. Thus in an urban environment, depending on the size, the particles can be mainly of local origin or transported over large distances.

9.4.2 TIME PATTERN

The urban pollution can undergo dramatic changes within a few hours. A hazy day can be followed by a day with crystal clear vision for no obvious reasons. In general the cause is meteorological conditions, but also changing sources can be responsible. We can use a simple box model for demonstrating possible influences on particle concentration: Let us consider a source which emits particles (directly or by gas to particle conversion). Since we consider a large area, the source may be an area source emitting m particles per unit time and unit area. The emitted particles are carried with the air having a wind velocity v_w and are eventually turbulently mixed (we assume homogeneously) to a height h. Let us assume that some particles are removed, which is given by a filter function f, where f = 1 means no removal, f = 0 means all removed. Let us now consider a strip of land, which has a length Δx and a width of w perpendicular to the wind direction. Consider the air passing over this strip in time $\Delta t = \Delta x/v_w$. All the particles emitted on the considered strip having a mass of $m \cdot w \cdot \Delta x \cdot \Delta t$, become airborne and are eventually mixed to an altitude h and are thus contained in a volume of $h \cdot w \cdot \Delta x$. The mass concentration ΔM in this volume, also considering removal is thus:

$$\Delta M = \frac{m \cdot \Delta x \cdot f}{h \cdot v_w}. \tag{9.10}$$

If the air mass moves a distance, x over a source region we obtain:

$$M = \frac{m \cdot x \cdot f}{h \cdot v_w}. \tag{9.11}$$

The (mass) concentration of the aerosol is influenced by the distance x, the air mass travels over a source region, the source strength, m, the wind velocity, v_w, the mixing height, h, and the filter function, f. All five can be of importance:

- The longer the distance over a source region, the higher the concentration.
- Source dominated aerosols vary with the source strength, the most typical example is the correlation of black carbon peaks with traffic peaks or lower pollution on the weekends. But even with no traffic, the concentration does not go down to zero, since both particles reach the region from outside and particles will be airborne for some time. The effect of emissions of certain industries can occasionally be estimated, if operations are stopped e.g. due to a strike. Due to the lifetime of at least several days the emissions of far away sources also influence the particle concentration. Thus the wind direction is important, and air pollution roses permit the identification of sources. Typical examples have been observed several years ago in Berlin, where the particulate pollution was high, when the air came from the then polluted East European countries, or the strong dependence of pollution on the wind direction at locations on the Atlantic coast. Another example is the selective radioactive fallout in certain regions of Europe after the Chernobyl nuclear reactor accident, depending on the air mass passing over the source. Occasional transport of Saharan dust can considerably increase the particle load also in European cities.
- With higher wind velocity emissions are distributed in a larger volume of air. Thus in all cities pollution decreases with increasing wind speed. Wind speed is the single most important factor for the pollution level. Heavy pollution such as the London killer fog 1952 (Brimblecombe 1987), occurred when the air was stagnant for almost a week. An interesting example was reported by Vignati et al. (1996), who compared the pollution in Copenhagen and Milan, two towns having similar size and emissions. The wind speed in Copenhagen usually is between 2 and 10 m/s whereas in Milan it is between 0 and 3 m/s causing a difference in the pollution level by a factor of three. A graph representing the wind speed dependent pollution level of both towns is following the same curve, with Milan at the low wind speed and high pollution side of the graph and Copenhagen at the low pollution and high wind speed side.
- Depending on the meteorological conditions, the height up to which the emissions from ground level are dispersed, can vary. Inversions are frequent in winter and the mixing height is low, causing increased pollution levels. With increasing insolation during the day the mixing height increases and a decrease in pollution levels in the middle of the day is partly caused by the increased mixing.

- In Europe, with an average wind speed of ≈ 3 m/s, it takes about 4 days for an air mass to move 1000 km. Removal of particles at this time scale is not very efficient (see the half life times of Table 9.2, page 168). On the other hand low pollution is frequently observed after rain, which is often attributed to the cleaning action of rain. However, this is not the case, since a cold front, which can cause heavy rain, is usually associated with high wind speeds, and due to the shorter residence time the pollution goes down. There maybe one exception: Foehn (Föhn) occurs when moist air masses pass over a high mountain range. Due to the adiabatic cooling of the rising air the water vapor condenses and condensing clouds are the most efficient removal mechanism for particles, which leave the clouds with the rain drops. When descending on the other side of the mountain the remainder of the clouds evaporate and the air is almost particles free. At towns located on the slope of the mountain (e.g. Innsbruck or Munich) a crystal clear view can be observed during Föhn.

9.5 References

Brimblecombe, B. (1987) *The big smoke*, Methuen, London.

Hänel, G. (1976) The properties of atmospheric aerosol particles as a function of the relative humidity at thermodynamic equilibrium with the surrounding moist air, *Advances in Geophysics*, **19**, 73-188, Academic Press, New York, USA.

Horvath, H., Kasahara, M., Pesava, P. (1996) The size distribution and composition of the atmospheric aerosol at a rural and nearby urban location, *J. Aerosol Science*, **27**, 417-435.

Jaenicke, R. (1988) Aerosol physics and chemistry, in: Helwange, K.H., Madelung, O. (editors) Chapter 9 of *Landolt Börnstein, Numerical Data and Functional Relationships in Science and Technology, Vol. 4 Meteorology, Subvolume b. Physical and chemical properties of the ai,* pp. 402-457, Springer, Berlin.

Robinson, R.A., Stokes, R.H. (1959) *Electrolyte solutions*. Butterworth, London.

Vignati, E., Berkowicz, R., Hertel, O. (1996) Comparison of the air quality in streets of Copenhagen and Milan, in view of the climatological conditions, *Sci. Total Environm.*, **189/190,** 467-473.

Whitby, K.T. (1978) The physical characteristics of sulfur aerosols, *Atmospheric Environment*, **12**, 135-159.

In Europe with an average wind speed of ~ 5 m/s, it takes about 4 days for an air mass to move 1000 km. Removal of moisture at this time scale is not very efficient [illegible] the lifetime in Table 9.2, page [illegible]. On the other hand low cell [illegible] frequently observed [illegible] cleaning [illegible] rain. However, this is not the case since a cold front, which can cause heavy rain, is usually associated with high wind speeds, and due to the shorter residence time the pollutant [illegible]. One exception [illegible] air masses pass over high mountain ridges [illegible] of the [illegible] the most efficient removal [illegible] the clouds with the rain [illegible] on the other side of the mountain [illegible] the clouds [illegible] Alps [illegible] Munich [illegible] crystal clear view can be observed [illegible].

References

[illegible]

Hales, J. M. (19[illegible]) [illegible]

[illegible]

IV

AIR POLLUTION MODELLING

The wide spectrum of atmospheric phenomena governing air pollution concentrations take place on various temporal (from seconds to month and years) as well as spatial scales (from few meters to thousands of km) (Chapter 7). Generally it is common to divide the processes into local (micro), regional (meso) and large (global) scale phenomena. Understanding of the general features of the processes on the various scales is crucial when mathematical models for air pollution simulations are formulated. Assumptions that may be valid on one scale may on the other hand be highly violated when regarding another scale.

Many air pollution problems are associated with mesoscale features like land-sea breezes, internal boundary layers, mountain-valley flows as well as flow systems generated by urban heat islands. Pielke (1984) has formulated a definition where "mesoscale applies to those atmospheric systems that have a horizontal extend large enough for the hydrostatic approximation to the vertical pressure distribution to be valid, yet small enough for the geostrophic and gradient winds to be inappropriate as approximations to the actual wind circulation above the planetary boundary layer". The mesoscale phenomena take place on spatial scales from a few km and up to about 2000 km. These processes are crucial for the regional transport of air pollution which also has strong effects on the urban pollution levels of many compounds. In order to describe the phenomena in mathematical terms special model descriptions are needed and simplifications applying to the specific scale are crucial for the development of operational tools. The general outlines of *mesoscale modelling* is presented in Chapter 10.

The weather and air pollution conditions in urban areas are governed by physical processes on scales between mesoscale and microscale. The urban scale processes cannot be considered as isolated microscale processes, since city structures may generate motions at scales as large as the whole city area itself. The vertical structure of the urban atmosphere is more complex than over the surrounding rural areas and the lowest parts of the boundary layer, the so-called surface boundary layer (SBL), may be considered as composed by two layers: the canopy and the roughness sub-layers. The canopy-layer is composed of a number of individual building canyons, whereas the roughness sub-layer is a non-equilibrium transition layer in which vertical fluxes of momentum, energy, moisture and pollution from individual canyons blend together. In recent years, models specifically developed for scribing the complex processes on this scale have been developed. A review of the current knowledge and present *modelling activities of urban scale processes* is presented in Chapter 11.

In urban areas, the highest air pollution levels occur in street canyons where the dilution of car exhaust gases is limited by the presence of buildings flanking the street. These hot spots are important since they represent the locations of highest human exposures to air pollution. The main features of *wind flow and dispersion conditions in urban streets* are well known and they are shortly outlined in Chapter 12 together with a general introduction to mathematical modelling of pollution phenomena on this scale. Several models, currently in use in Europe are briefly presented, and a more detailed discussion is given for the Danish model, the Operational Street Pollution Model (OSPM). The short residence time for pollution in urban streets leave only time for very fast chemical transformations. An example is the NO, NO_2 and O_3 system and the treatment of this system in the OSPM is described.

Diagnostic models demand a detailed knowledge about all involved processes. In some cases e.g. in connection with forecasting of air pollution episodes, a *stochastic modelling* approach provides a simple and good alternative. This type of modelling is described in Chapter 13.

Physical models represent an important tool to investigate the flows in urban areas. The physical modelling is usually aimed at studies on the scales of street canyons and individual buildings. This type of studies have been very important for the development of mathematical models and a general introduction to *wind tunnel experiments* is therefore given in Chapter 14.

Chapter 10

REGIONAL/MESOSCALE MODELS

GEORGE KALLOS
University of Athens, Department of Applied Physics
Panepistimioupolis, PHYS-5, GR-15784 Athens, Greece

10.1 Introduction

It is well known that many air quality problems are associated with various so-called mesoscale features. When mesoscale circulations are developed in a given region, they influence the ventilation and thereby the air quality in the area. The most important mesoscale features are circulations exhibiting a diurnal cycle. To such circulation types belong land-sea breezes, internal boundary layers, up-slopes, down-slopes and drainage flows, as well as flow systems created by urban heat islands. They result in the development of sharp flow gradients and vertical stratification which affect the air quality in the considered region. These mesoscale circulations are mainly due to differential heating of the surface in the region which may i.e. be due to variations in surface type. Another category of mesoscale circulations is related to mechanical effects such as blocking and channelling forced by the topography of the region.

The discipline of mesoscale atmospheric modelling started back in the 1970's, and is generally associated with numerical simulations describing atmospheric disturbances exhibiting a diurnal cycle. The wide spectrum of the atmospheric disturbances from large Rossby waves to molecular motions may e.g. be divided into four scales: large, regional, mesoscale (or local) and turbulence. Mesoscale has a horizontal scale from a few kilometres to several hundreds of kilometres, and a vertical scale from a few tens of meters to at least the depth of the boundary layer. The temporal scale is ranging from 1 to 12 h or even higher. In the meteorological community, the same range of motions is divided into six scales: global, synoptic, meso-alpha, meso-beta, meso-gamma and turbulence. The relationship between the spatial and temporal ranges of these scales is shown in Table 10.1 (next page). It should be noted, however, that these scales and ranges are more or less arbitrary and may not apply in all cases. A more formal definition of the mesoscale was given by Pielke (1984), where "mesoscale applies to those atmospheric systems that have a horizontal extend large enough for the hydrostatic approximation to the vertical pressure distribution to be valid, yet small enough for the geostrophic and gradient winds to be inappropriate as approximations to the actual wind circulation above the planetary boundary layer".

Table 10.1 Scales of atmospheric motion used in air quality applications.

Scale		Spatial scale	Temporal scale	Practical use connected with Air Quality
Large Scale	Global	4,000 - 40,000 km	2 - 10 days	Long-range transport; significant for "background" quantities in upper troposphere and in the stratosphere
	Synoptic	2,000 - 4,000 km	1 - 3 days	Long-range transport; significant for continental scale transport, large scale removal processes
Regional	Meso-alpha	200 - 2,000 km	12 - 48 h	Long-range transport; significant for large physiographic variations, convective systems; plume integration
	Meso-beta	10 - 30 km	2 - 12 h	Transport and plume integrations, thermal circulations and convective effects, orographic eddies
Local scale	meso-gamma	1 - 20 km	0.1 - 3 h	Transport and mixing due to local thermal circulations and landscape characteristics
Turbulence	Micro-scale	0.00001 - 2 km	1 - 1800 s	Vertical transport, dispersion, mixing and deposition. Transition zones in inhomogeneous landscapes become very important

A wide spectrum of atmospheric phenomena are falling within the range of mesoscale, and therefore a wide range of applications of mesoscale modelling have been carried out. Among these, application of mesoscale models for air pollution studies is widely used, and is caused by the necessity of providing accurate descriptions of physical parameters like 3-D meteorological fields and the depth of the atmospheric boundary layer (ABL). This is especially important in areas with significant variations in the physiographic characteristics like orography, land-water distribution, landscape cover etc. In such complex areas a variety of phenomena including land-sea breezes, convergence zones, drainage flows from surrounding topography, and the development of thermal internal boundary layers can be developed. All these mesoscale phenomena in association with the larger-scale flow make the dispersion processes very complex (e.g. McKendry 1989; Pielke et al. 1983). In almost all cases where standard dispersion modelling approaches are applied, some crude simplifications are made e.g. for the spatial homogeneity of the ABL height and simplified diurnal variations of the same which is usually estimated from indirect methods based on observations and empirical formulations. However, these assumptions may be violated in coastal regions when marine air is advected towards land (e.g. Mckendry 1989; Pielke et al. 1983). Naturally, significant variations in the wind field are also observed in these regions. These variations are more significant when marine air intrusion occurs over irregular terrain with physiographic variations. In such cases, even if a dense observational network is established, it is impossible to accurately resolve the detailed spatial and temporal variations in wind field as well as ABL depth (Pielke et al. 1989). These parameters are crucial for detailed air quality investigations. In such cases application of a high quality mesoscale model capable of accurately resolving the 3-D wind fields, ABL depth and turbulence characteristics and their temporal variations is necessary. In addition, applying meteorological mesoscale models in association with dispersion-chemistry models, it is possible to accurately describe transport and transformation of air pollutants arising from emissions from various sources over a large area. A brief description of such models and their applications is provided in the following chapters.

10.2 The basic conservation principles

The meteorological processes in the atmosphere may generally be expressed through the main thermodynamic and hydrodynamic parameters, i.e.: pressure, temperature, mass, volume and density, water vapour, radiation, wind velocity and turbulence.

The interrelations between these parameters may be expressed through the thermodynamic and hydrodynamic laws formulated in the equations for the basic conservation principles; the conservation of: mass, heat, motion, water, and various gaseous and particulate compounds. In the following each of these basic conservation equations will be given a brief description.

10.2.1 CONSERVATION OF MASS

The conservation of mass or continuity equation states that in the atmosphere there are no sources or sinks of mass. Mathematically this conservation principle is expressed as:

$$\frac{\partial \rho}{\partial t} + \left(\vec{\nabla} \cdot \rho \cdot \vec{V}\right) = 0 \tag{10.1}$$

where ρ is defined as the density of the air and $\vec{V}$ is the velocity of the elementary volume of air with its three components $\vec{V}_i$, i = 1, 2, 3 along the three axes of the Cartesian co-ordinate system x, y, z respectively. The Cartesian co-ordinate system in a position over the earth is defined with the x axis along the tangent of the parallel cycle passing from this point and with positive direction towards East, y axis the tangent of the mercator passing from the point with positive direction towards North and z axis defined as the perpendicular to the plane defined from x and y axis, which is passing from the centre of the spherical earth and positive direction upwards.

10.2.2 CONSERVATION OF HEAT

The conservation of heat equation is an expression of the first law of thermodynamics for the atmosphere, which states that the differential changes in heat content dQ are equal to the sum of differential work dW performed by an object and differential increases in internal energy dI. The mathematical expression for this conservation law is:

$$\frac{\partial \theta}{\partial t} + \vec{V} \cdot \vec{\nabla} \theta = S_\theta \tag{10.2}$$

where θ denotes the potential temperature which is expressed as:

$$\theta = T_V \cdot \left(\frac{1000}{p} \right)^{\frac{R_d}{C_p}} \tag{10.3}$$

T_V denotes the virtual temperature in degrees K, p the pressure in hPa, R_d is the dry gas constant and C_p the specific heat at constant pressure. T_V is expressed as:

$$T_V = T \cdot (1 + 0.619 \cdot q_3) \tag{10.4}$$

which is the temperature required in a dry atmosphere to have the same value of Pα as in an atmosphere with a specific humidity q of water vapour (the subscript 3 denotes vapour or gas phase). α is the specific volume (i.e., volume per unit mass) which is also expressed as the inverse density of air, p. S_θ denotes the sources and sinks of heat as they are expressed by changes in potential temperature. The most important of the processes contributing to the source-sink term are the condensation-evaporation, freezing-melting, deposition-sublimation, dissipation of kinetic energy by molecular motions, net radiative flux convergence, chemical reactions etc.

10.2.3 CONSERVATION OF MOTION

The conservation of motion equation states that a force exerted on an object causes acceleration which is expressed in Newton's second law. The motions of the air masses in the atmosphere are taking place over the Earth which is rotating with constant angular velocity Ω. Therefore, the velocity of an elementary parcel of air may be considered as the sum of the velocity relative to the Earth and the velocity resulting from the rotation. An apparent force which is termed the Coriolis force, arises from considering this rotating system as a reference. This force may be expressed as $-2\vec{\Omega} \cdot \vec{V}$. The remaining forces exerted on an elementary air parcel may be divided in internal and external forces. Internal forces account for the dissipation of momentum by molecular motions. The effects of these forces are related to the viscosity of the air (or the liquid) and the deformation of the momentum field. External forces are defined as the pressure gradient and gravitational forces. The pressure gradient force acts in all three directions, whereas the gravitational force acts only in the vertical. The conservation of motion is expressed in the Navier-Stokes equations, which given in vector annotation may be expressed mathematically as:

$$\frac{\partial \vec{V}}{\partial t} = -\vec{V} \cdot \vec{\nabla}\vec{V} - \frac{1}{\rho}\vec{\nabla}p - g \cdot \vec{\kappa} - 2\vec{\Omega} \cdot \vec{V} \tag{10.5}$$

where the first term at the right hand side of this equation is the advection term, the second the pressure gradient force, the third the gravitational and the fourth the Coriolis force. $\vec{\kappa}$ is the unit vector with the components (0, 0, 1). The Equation (10.5) is also known as the conservation of momentum equation.

10.2.4 CONSERVATION OF WATER

Water is found in all three phases in the atmosphere: as vapour, liquid and ice, and its phase changes plays an important role in the atmospheric energy balance. When considering the water conservation law, the phase changes and transport in the atmosphere has to be accounted for:

$$\frac{\partial q_i}{\partial t} = -\vec{V} \cdot \vec{\nabla} q_i + S \cdot q_i \qquad i = 1,2,3 \tag{10.6}$$

where q_i denotes the mixing ratio of water in its three phases (vapour, liquid, ice). The first term at the right hand side of the equation accounts for advection, the second for sources and sinks. The source-sink term refers to the changes due to phase transfer, chemical reactions and fluxes from the ground. For most atmospheric applications the changes in water mass due to chemical reactions may be neglected. The remaining processes in the source-sink terms are related to condensation-evaporation, freezing-melting, deposition-sublimation, fallout.

10.2.5 CONSERVATION OF GASEOUS AND PARTICULATE MATERIAL

The atmosphere is composed by a variety of gases. In addition a significant amount of mass is found in particles of various sizes. The conservation law for water in the atmosphere applies also for these gaseous and particle substances. The conservation law applies in these cases to the mixing ratio by mass of the considered substance, defined as the mass of the substance to the mass of the mass of air in a given volume.

10.2.6 THE IDEAL GAS LAW OR EQUATION OF STATE

This equation expresses the relation between atmospheric pressure, density and temperature. In order to account for the impact of water vapour, the virtual temperature is used:

$$P = \rho \cdot R_d \cdot T_v \tag{10.7}$$

The above equation represent a set of non-linear partial differential equations for the 11 dependent variables $u_i(i = 1, 2, 3)$, p, $q_i(i = 1, 2, 3)$, T, T_V, θ and ρ. When a number of other gaseous or aerosol materials in the atmosphere are considered, this system must be treated by a similar number of conservation equations. The presented sequence of equations apply to a wide range of motions (from Brownian to planetary scale) and must be solved simultaneously for the independent variables and taking into account their space and time variation. In general analytical solutions cannot be derived, and even numerical solutions are highly difficult to obtain for the complete form of these equations. Therefore various simplifications have to be applied when solving these sets of equations in what is usually termed numerical models.

10.2.7 SIMPLIFICATIONS OF THE BASIC EQUATIONS

In order to obtain solutions or approximate descriptions of specific events, several simplifications of the basic equations are needed. An example is to consider only those atmospheric motions that are taking place on characteristic scales of the phenomenon under consideration. For describing motions on a specific scale, exact solutions to the differential equations are anticipated for the given scale, whereas for the remaining scales of motion more crude descriptions are applied.

For simplifying the basic equations, scale analysis is needed. This is carried out by determining the order of magnitude of the various terms by applying typical values of the dependent variables and parameters for the various phenomena on the scale under consideration. In the following a short description is given of some of the common simplifications made in mesoscale modelling.

The most important assumptions regarding treatment of *the conservation of mass equation* in mesoscale models are related to the fact that the density variations at synoptic scale are much slower than at mesoscale. The horizontal gradients are likewise much smaller at synoptic scale compared with mesoscale. In the vertical, the comparison of the vertical extend of the disturbances with the density-scaled height of the atmosphere (approximately 8 km) defines the final form of the continuity equation, which may be for either deep or shallow convection. In the case of deep convection, the vertical extend of the atmospheric disturbances is comparable to the characteristic scale of the atmosphere. In this case, the time derivative is small, while the terms including the density are considered only on the synoptic scale and neglected on mesoscale. In the case of shallow convection, the resulting equation is termed anelastic or soundproof because it filters out the acoustic sound waves. This is an important assumption since it makes it possible to use longer time steps without violating the stability criteria. In the case of shallow convection, the spatial variations in density are considered negligible and the resulting equation is termed incompressible. Both approximations are used as replacement of the full equation in application of the conservation of mass equations for studies of a large variety of mesoscale phenomena. The advantage of this is naturally higher computational efficiency.

The most commonly applied simplifications of the *conservation of heat equation* are related to the source-sink term. In the case of adiabatic processes, this term is zero and therefore the equation simplifies accordingly. This assumption is more appropriate in cases with the atmosphere considered to be dry with no phase change in water, the time of integration is small and therefore no significant radiative heating or cooling effects are evident in the atmospheric layers near the surface.

For three components of the *conservation of motion equation* (Navier-Stokes equations) several assumptions and simplifications are usually made. The Navier-Stokes equations may be simplified by applying the Boussinesq (1977) approximation which states that the variation of pertubation density may be neglected in the continuity and momentum equations, except through its influence on the buoyancy term. However, the Boussinesq approximation has been interpreted in different ways, which has let to different formulations of the continuity equation (e.g. Thunis, Bornstein 1996). The anelastic approximation considers $\partial\rho/\partial t=0$, yielding $\vec{\nabla}\rho\vec{V}=0$. The approximation of incompressible air considers ρ to be constant in time and space ($\partial\rho/\partial t=0$ and $\vec{\nabla}\rho=0$), yielding the non-divergent flow equation: $\vec{\nabla}\vec{V}=0$. However, the most known simplification applies to the vertical component and is termed the hydrostatic approximation. This approximation implies that the vertical acceleration is much less than the magnitude of the pressure gradient force. In this case the hydrostatic equation is applied:

$$\frac{\partial p}{\partial z}=-\rho\cdot g \qquad (10.8)$$

This assumption is considered valid in cases where the horizontal scale of an atmospheric disturbance is on the order or less of the density scaled height, which is approximately 8 km.

In the two horizontal components of the equation of motion, it is in general difficult to simplify or omit various terms in the case of mesoscale phenomena. This is due to the large variety of mesoscale phenomena where the various terms dominate the each other. For some phenomena with relatively small horizontal extension, the Coriolis term may be omitted whereas in slow moving systems this term becomes important. One oversimplification is the consideration of balance between the Coriolis and the gradient term, the so-called geostrophic balance. Due to the nature of the mesoscale phenomena, this cannot generally be considered as a realistic assumption. Turbulence processes are considered to have importance on mesoscale, and therefore the viscous force cannot be neglected.

Considering the equation for *conservation of water, gaseous and particulate material*, the source-sink term may be simplified for water and in some cases also for other compounds. This is especially the case when water does not undergo phase transitions. The amount of water generated in chemical reactions is generally negligible compared to other terms (e.g. advection). However, for the studies of air quality, the source-sink term must in general be described in detail.

As a summary of the simplifications which may be applied to the basic equations in mesoscale studies, one may characterise the flow as anelastic, hydrostatic or non-divergent, according to the assumptions adopted. In general, following the definition of Pielke (1984) the atmospheric processes may be characterised as mesoscale processes, when the horizontal scale is large enough to make the hydrostatic assumption valid at the same time as it is small enough to made the Coriolis term small relative to the advective and pressure gradient forces, and therefore to have a flow field significantly different from the gradient wind even with the friction term neglected. This definition does not mean necessarily that all the mesoscale processes are hydrostatic and that other effects are negligible. A more detailed scale analysis related to mesoscale processes is provided in Pielke (1984).

10.3 Application of atmospheric models for generation of meteorological fields

The basic conservation equations presented in Section 10.2 are applied for construction of numerical algorithms for calculation of meteorological fields. These algorithms are usually constructed considering the Eulerian concept of motion, and the models are mainly divided in diagnostic and prognostic models.

10.3.1 DIAGNOSTIC MODELS

The diagnostic models are used to analyse available observations at discrete points and at predefined time intervals. They are based on a selected subset of the conservation equations described in Section 10.2, and apply various numerical optimisation techniques for selected parameters like divergence. The main drawback of this methodology is the limitations enforced by the usually low number of measuring points. Some of these models take into account topographic effects, but for all models of this type consideration of balancing and the misrepresentation of local thermal effects put severe constrains which may cause considerable deviations between the generated fields and the reality (Pielke et al. 1991).

10.3.2 PROGNOSTIC MODELS

The prognostic or dynamic models are based on the complete set of basic conservation equations given in Section 10.2. Most of the dynamic models used for air quality applications are derived for application to a limited area and in general, only the type of models used for climatic studies are of global coverage.

The modes (wavelengths) represented by the dynamic equations of the limited area model (LAM) covers a certain part of the energy spectrum (Figure 10.1). The LAM has respective overlaps with global coverage model data (GL) and sub-grid area data (SG) for the two remaining parts of the energy spectrum. The overlap with the sub-grid area is due to the filtering procedures usually applied for the smaller wavelength of the model. The overlap with the global coverage model is generally an advantage, since the dynamic model then will require less time after the initialisation to reproduce the modes with wavelengths smaller than the cut-off of the global model (the sub-grid representation of the modes is not resolved by the global model).

The initial conditions are usually gridded fields prepared from available observations at a specific time. For many air quality applications, where limited area models are used, especially for coverage of urban areas, this initialisation is performed with a "horizontally uniform" technique. In this case a single sounding of the vertical structure is applied to the whole model domain. This vertical structure is assumed to be constant through the integration time. These initial conditions are assumed to describe the larger scale circulation during all the model simulation. This assumption is crude also for air pollution applications, and may lead to erroneous conclusions.

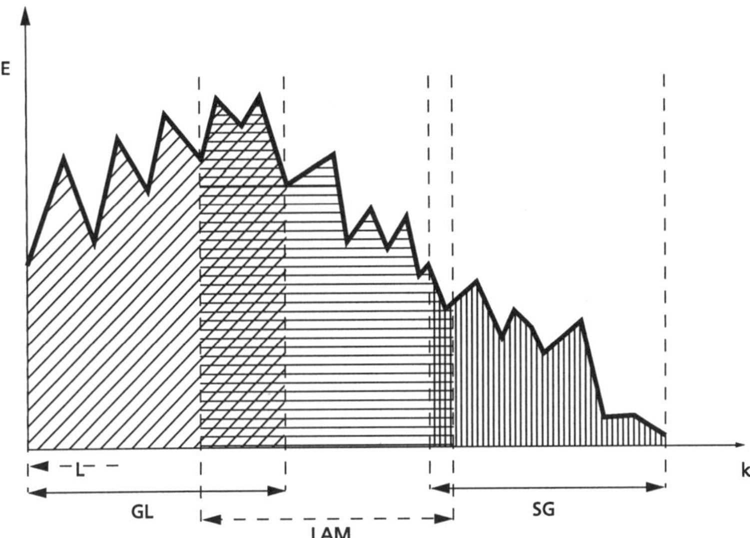

Figure 10.1 Schematic representation of the energy spectrum of the atmosphere and its splitting according to various-scales modelling. GL denotes the portion of the spectrum covered by the global meteorological models, LAM by the limited area meteorological models and SG by the sub-grid parameterization equations.

10.4 Air quality models

Air quality models are usually based on the conservation of mass equation. In order to describe transport and diffusion these models utilise meteorological variables that are typically generated by a meteorological model. Often these meteorological models include implicitly the equations governing the air pollutants in the form of a module to simulate chemical transformations, and other modules for description of dry and wet depositions. The inclusion of a certain number of species into a dispersion problem requires a similar number of continuity equations to be solved. This set of non-linear equations may be solved together with the equations for atmospheric motion or separately by utilising results from the latter. In cases where several chemical species are considered, the number of equations becomes high and the numerical treatment requires substantial computer resources. The concentrations of the chemical species vary many orders of magnitude, and for some of the compounds the concentration changes in time are also highly variable. The system of ordinary differential equations describing the chemical mechanisms is usually a “stiff” system and requires appropriate numerical techniques that are computationally expensive.

When air quality modelling started in the 1970ties, the computer resources were limited and crude simplifications had to be applied in order to solve the basic equations. With the increasing computer power available more detailed models have now replaced the early ones. In the following, a brief description of the various types of air quality models is given.

10.4.1 BOX MODELS

In a box model, the pollutants are usually assumed to be uniformly mixed from ground to the depth of the boundary layer and source release rates and winds are assumed constant over the considered model domain. These assumptions lead to a considerably simplified set of mass conservation equations for the local pollutant concentrations. An example of such a model was given by Gifford and Hanna (1973). Models of this type are considered to describe reasonably well cases where the dominating terms are advection and horizontal mixing of releases from area sources, but it cannot resolve large horizontal concentration gradients. This will e.g. take place at low wind speeds and when the wind field exhibits considerable variations in direction. This type of models is not applicable for mesoscale problems.

10.4.2 GAUSSIAN PLUME MODELS

This type of models describes the vertical and cross wind dispersion of a single plume under steady-state conditions. Time averaged concentrations (e.g. 1 h mean or less) downwind of a single source are obtained as function of mean winds and boundary layer stability (e.g. Turner 1970). Several sources may be treated in such models, but then each of them is treated as individual Gaussian plumes. This type of models are world-widely used for regulatory purposes of releases from e.g. power production plants and industries using stacks. Some of the basic shortcomings of the methodology in these models are the lack of spatial and temporal variations in the meteorological parameters which impose limitations in the combined use with mesoscale meteorological models. In general, these dispersion models may be applied to describe concentrations in an area up to 10-20 km away from the source, in an area with only small topographic and landscape variations.

10.4.3 LAGRANGIAN PUFF MODELS

In this type of models a single puff of contaminated air is followed over large distances and long time periods after release from the source. Lagrangian puff models are used e.g. in cases where the spatial and temporal scales are too large for the application of Gaussian plume models. The puff represented with the centre of its mass is advected along a trajectory calculated from 3-D wind fields. The dilution of the pollution is computed as the puff moves along the trajectory. Details concerning this type of modelling may be found in the description of the CALPUFF/CALGRID models (Yamartino et al. 1993; Scire et al. 1996).

10.4.4 EULERIAN MODELS

In the Eulerian models, the conservation of mass equation is solved for a set of receptor points in a 2-D or 3-D grid space. The production and loss terms include exchange with the surrounding grid elements as well as emissions, chemical transformations and scavenging by dry and wet deposition. In some cases, the meteorology and transport-chemistry is solved in one integrated model system, where the governing equations for emissions, transformations, advection and dispersion of the air pollution are build into a mesoscale meteorological model. Especially for chemically inert species this is a practical approach, since for reacting pollutants the treatment of the chemical reactions typically is the most computer demanding part (in many cases up to 90% of the computer time).

In the Eulerian models, emissions are considered evenly distributed over the grid, which has the impact that sharp concentration gradients close to the sources tend to vanish in the model. This effect is of course highly dependent on the actual horizontal and vertical grid resolution in the model. Especially in models with high resolution, an undesired oscillation (Gibbs phenomenon) may appear in cases of strong gradients in the emissions. Use of smoothing and/or filtering procedures on the emission data can ease the numerical treatment, but this is an unphysical solution to the problem and may cause misleading results.

Eulerian models are considered appropriate for describing long-range transport and apply well when the horizontal and vertical scales of the pollutants are larger or at least equal to the horizontal and vertical increments in the model. Furthermore the spatial scale of the air pollutant must be much larger than the scale of turbulence (Pielke et al. 1983; Pielke 1984).

The system of conservation of mass equations for air pollutants are usually treated numerically by use of splitting techniques (e.g. Zlatev 1995). This means that the terms for advection, diffusion, chemical transformations etc. are treated separately according to the characteristic time scales of the various processes; a procedure that ease the numerical treatment in the model. The obtained solution will contain a certain numerical error, however with an appropriate splitting operation this error can be kept small.

10.4.5 LAGRANGIAN PARTICLE DISPERSION MODELS

The Lagrangian particle dispersion models (LPDM) are also known as *Lagrangian statistical or stochastic models, Markov chain models, random walk models, random flight models or Lagrangian Monte Carlo models*. In the LPD models, the turbulent dispersion is modelled by tracking the release of a large number of particles as they are advected with the flow, which is typically generated by a meteorological model and represented by mean flow and turbulent fluctuations. The release of particles may be either sequential (as in a plume) or simultaneous (as a puff). Concentration fields are determined from the position of the spatial distribution in the set of predicted particle positions.

The LPDMs have some advantages compared to the Eulerian models; the turbulent diffusion is in reality a Lagrangian phenomenon. The Lagrangian model do not have problems with sharp gradients in the emissions, and may easily be applied for point and line sources. Eulerian models may in some cases have problems with negative concentrations due to numerical errors, which cannot appear in Lagrangian models. Finally Lagrangian models are generally flexible, computational inexpensive and easy to apply compared with Eulerian models. A good description of the two dispersion modelling techniques, the Eulerian and the Lagrangian, may be found in (Moran, Pielke 1996a,b).

10.4.6 LAGRANGIAN TRANSPORT-CHEMISTRY MODELS

In Lagrangian transport-chemistry models, using generated wind fields (e.g. from a mesoscale meteorological model) trajectories are computed backwards in time (typically on the order of 4 days) from selected receptor points for selected arrival times. After applying initial concentrations for the pollutants, the air parcel is advected along the trajectory to the receptor point. During the transport, the air parcel receives emissions from the sources at ground, chemical transformation and dry and wet deposition take place. Examples of such models are the various versions of the EMEP Lagrangian models (Iversen et al. 1990; Simpson 1993; Hov et al. 1994). In the EMEP Lagrangian models the pollutants are assumed evenly distributed from ground to the depth of the boundary layer and a simplified exchange with the free troposphere is applied.

10.4.7 HYBRID DISPERSION MODELS

This type of models has been developed in order to surpass some of the disadvantages in Eulerian and Lagrangian models. This is obtained by combining a Lagrangian particle model with an Eulerian transport model (Moran, Pielke 1996a,b and references herein). The Lagrangian model is applied as an attempt to deal with the sub-grid scale aspects of a pollutant release while the Eulerian model takes over when the pollutant is dispersed to a degree that it is adequately resolved on the applied computational grid. The source configuration of such models is always very flexible. Any number of sources may be specified anywhere in the model domain and configured as a point, line area or volume source and the emissions may be instantaneous, intermittent or continuous. The pollutants can be treated as gases or aerosol particles and a radioactive half-life or a settling velocity may be specified. An example of such a model is the Hybrid Particle Concentrations Dispersion Model (HYPACT) by Tremback et al. (1993). This model utilises meteorological fields generated by the Regional Atmospheric Modelling System (RAMS) (Pielke et al. 1992).

10.4.8 RECEPTOR ORIENTED MODELING

In the previous sections, the dispersion models discussed are based on the source oriented approach. The equations are solved forward in time for specified pollutant sources. For several cases, the concentration at a specified point (receptor) is of great interest. In such cases the modeling technique is called receptor oriented approach. In this case, instead of calculating the concentration fields with the classical source oriented method, an integral of pollution concentration over the modeling domain and time of stimulation is calculated. The integral can be defined in accordance with the aim of the study by using the receptor function. In this case, an influence function is calculated instead of concentration. The influence function provides information on contributions from different sources to the specific receptor point. The influence function is usually calculated from backward trajectories of Lagrangian particle releases. In cases of Eulerian models, the partical differential equations are formulated in a variational framework and then the influence function is obtained by solving the adjoint equations backward in time with the receptor function as a source term.

The receptor-oriented models take into account processes like deposition, transformation and of course meteorology but the influence functions calculated are independent of emission sources. The air quality integral can be calculated by averaging the emission field with the influence function taken by weight.

The receptor-oriented models are in general less computationally expensive compared with the source oriented ones. For more information about this type of modeling the reader should look at Uliasz and Pielke (1991).

10.5 Coupling meteorological models and transport-chemistry models

Most of the previously described types of air pollution models require 3-D meteorological fields (winds, temperature, humidity, stability, turbulence) and their evolution in time. The coupling between the meteorology and air pollution processes may be performed on-line or off-line. For the on-line coupling, the meteorological model is run with the air pollution model as a build-in module. This type of coupling is often used for inert (non-reacting) pollutants and usually these models are Eulerian. The off-line coupling is performed by running the meteorological model, store the meteorological fields with selected time intervals and then run the air pollution model that in this case may be of any type (Eulerian, Lagrangian, puff or plume model).

The *on-line and off-line couplings* both have their advantages and disadvantages. In on-line coupling, processes like advection and diffusion are generally considered to be more accurately described, e.g. steeper concentration gradients may in some cases be obtained with this type of coupling compared with off-line coupling. The disadvantages of on-line coupling are mainly related to computer demands. This is especially the case for test and validation, where the simulations must be repeated many times, since the meteorological model often is a highly computer demanding part of the system. A major disadvantage of the off-line coupling is the substantial I/O handling that is usually required. Another problem is the representation of gradients that in some cases may be perturbed due to the relatively coarse time resolution by which data are stored. In Lagrangian dispersion, the particle distribution over the domain may in some cases show artificial sharp gradients due to this problem. Shorter time intervals between storing of the meteorological data decreases this problem, but this is at the expense of larger amount of data which needs first to be stored and afterwards handled by the air pollution model.

Most meteorological models use equal grid resolution in the two horizontal directions; often polar stereographic projection. However, in many air pollution models and especially in many photochemical models other projections are often applied. UTM and longitude - latitude co-ordinates are often considered easier for the handling of input data e.g. the emission inventories. The meteorological models have generally non-equidistant grids in the vertical direction with more layers close to the ground than aloft. This resolution need not necessarily to be the same in the air pollution model. For these cases, interpolation of the meteorological data must be performed before they are used in the air pollution model. Such procedures are not easy to apply and considerable errors may be introduced - especially for model domains with complex terrain.

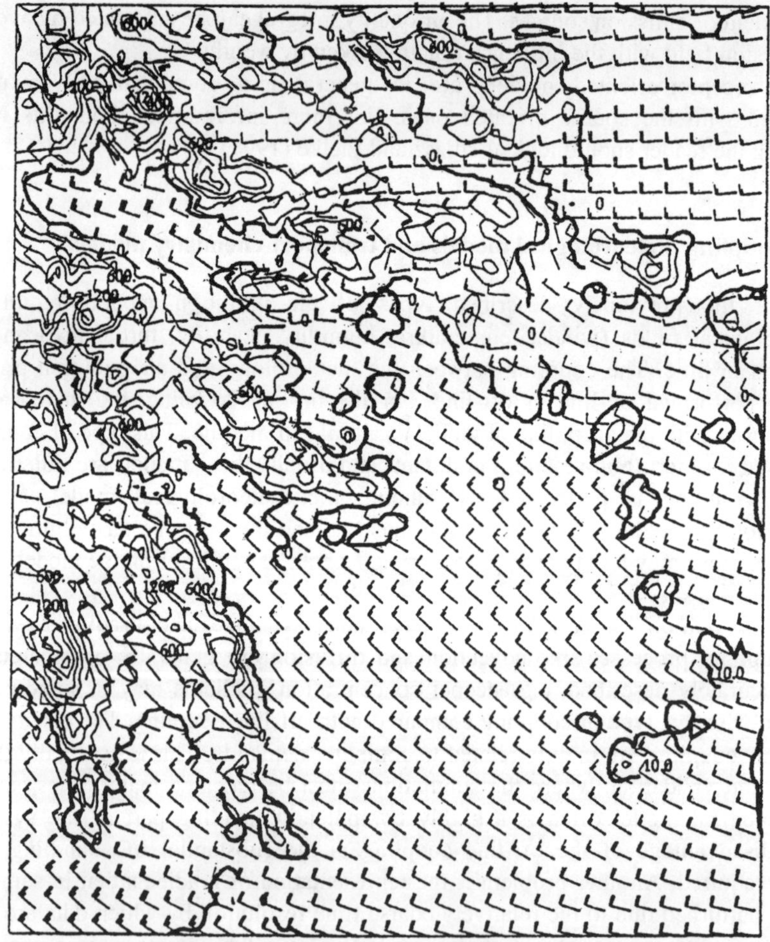

Figure 10.2a Example of the blocking effect of the Peloponnese at wind flows from North-West.

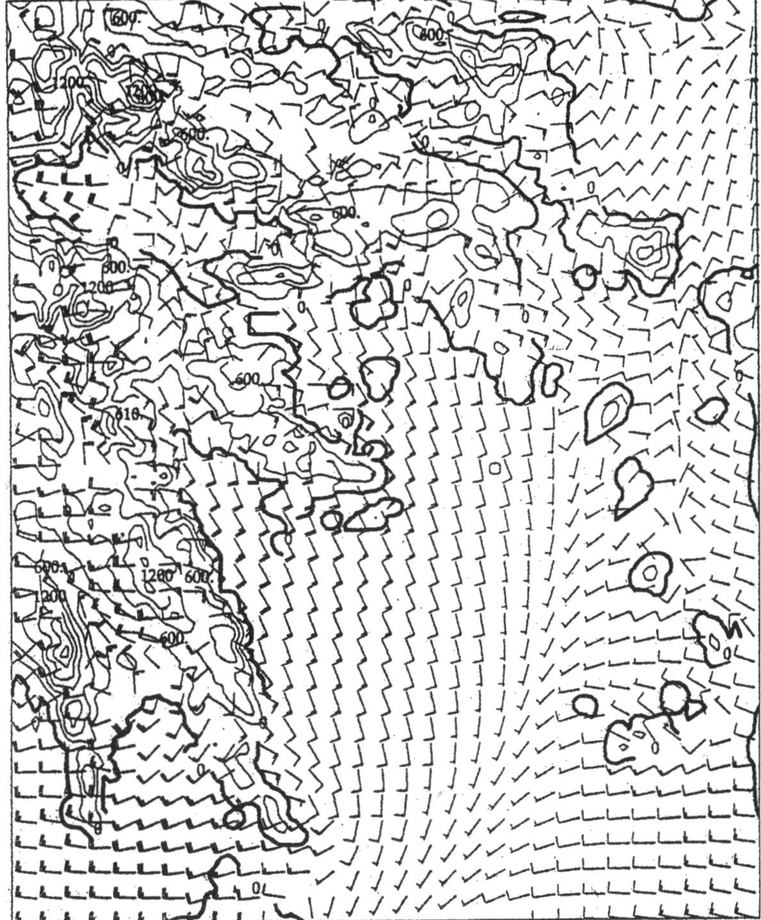

Figure 10.2b Example of the blocking effect of the Peloponnese at wind flows from North.

10.6 Mesoscale features effecting the flow in the Athens area

The mesoscale circulations due to differential heating of the surface in a given region have been the subject of several research projects in the past; the most classic being related to land-sea breezes in coastal areas. As topographic data have been incorporated in the more recent mesoscale meteorological models, similarly important circulation systems have been found over land (Segal et al. 1989a,b). Recent work has furthermore shown that effects governed by position and orientation of the topography of a region may affect significantly the air quality of an urban area. A typical example is found in Athens, Greece, where it has been shown that the topographic blocking effect of the Peloponnese plays an important role for the formation of air pollution episodes (Kallos et al. 1993; 1997). The flow is typically from North-west over all the area during night hours (Figure 10.2), whereas during daytime the wind direction turn in South over the sea east of the Peloponnese. This significant veering of the wind field is the result of the mesoscale circulation developed due to the landscape characteristics of the area around Athens (approximately 100 km x 100 km) and corner and blocking effects of the Peloponnese. Due to the orientation of the Greek Peninsula (and thereby also of the Peloponnese) and the stable conditions due to the anticyclonic circulation, the flows on larger scale (the so-called meso-alpha scale, Table 10.1, page 178) is deflected around Peloponnese and becomes supportive to the local (termed meso-beta, Table 10.1) circulation over and around Athens (Kallos et al. 1993). This condition is necessary in order to overcome the relatively strong trade winds blowing along the Aegean sea during summertime (the so-called Etesians). In such cases, simulations of the flow conditions must be performed in such a way that topographic variations located far from the area of interest are taken into account. Even when the flow is resolved adequately in mesoscale flow simulations over an area around the urban area of interest, the mesoscale features of a larger area around it has also to be considered.

10.6.1 SELECTION OF MODEL DOMAIN

Problems like the one presented in the previous paragraph are associated with more general considerations like - what is the appropriate size of the model domain around the given area of interest, and what is the appropriate vertical and horizontal grid resolution? Furthermore - what is the impact of topography and their representation in the domain and its boundaries? These questions are not easily answered and are of course strongly related to the area and the air pollution problem that is under consideration. The maximum and minimum of modes resolved in the limited area meteorological models are bound by the domain size (maximum wavelength resolved) and the grid spacing (minimum wavelength resolved), see Figure 10.1 (page 185).

The remaining modes are resolved in a less exact way, they are "parameterized". Before a flow simulation is initiated, the local circulations must be identified together with characteristic scales of transport and transformation of the investigated air pollutants. The model domain must have a size that ensures that the local circulation cells developed over the region take place within the boundaries of the domain in order to avoid reflections and similar unwanted and unphysical effects. The horizontal grid spacing must be appropriate in order to resolve the local circulations in an accurate way. Good vertical representation is needed in cases of mountainous barriers in order to resolve effects like blocking and strong updrafts. The latter may in some cases inject pollutants into the free troposphere or trap air masses with elevated pollution levels in the boundary layer through drainage flow at night. The second group of questions is related to the effects of topographic variations and the lateral boundary effects. The representation of the topographic features near the lateral boundaries may have significant impact on the description of the large-scale (synoptic or meso-alpha, Table 10.1) flow within the considered domain (e.g. Kallos et al. 1993).

The air pollution dispersion is affected by the variability of landscape parameters such as soil moisture and vegetation cover, but also by their horizontal distribution. Pronounced variations of horizontal landscape characteristics, generally lead to production of more turbulent flow which affects the pollutant dispersion in the area (e.g. Dalu et al. 1992).

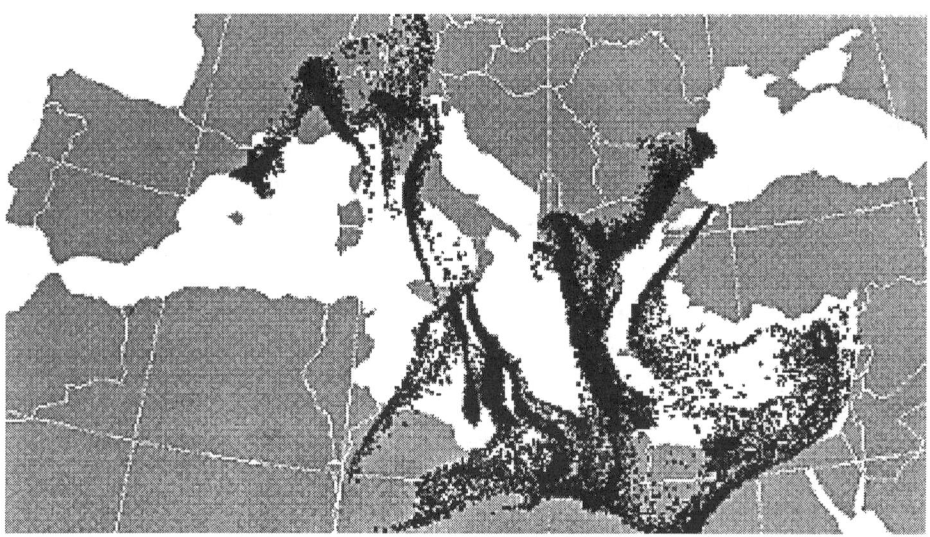

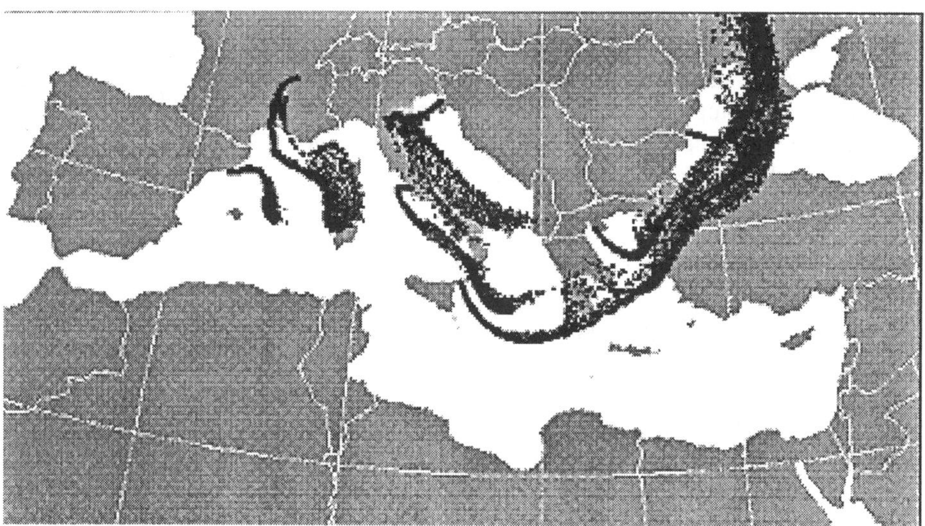

Figure 10.3 Particle projection at the first model level from the HYPACT dispersion model. A) at 1500 UTC, 4 July 1994 (after 60 h of particle release). B) at 1500 UTC, 6 September 1993 (after 30 h of particle release).

10.7 Long range transport and its impact on urban air quality

During recent years, significant research effort has been devoted to the impact of long range transport to the air quality in rural as well as urban areas. The long range transport may affect the urban air quality in various ways. During the last decade, emphasis was put on acidic rain and visibility degradation (e.g. Moran, Pielke 1996a,b and references herein). Recent applications of mesoscale meteorological models in combination with dispersion and photochemical models have shown that the regional component is important, especially for the oxidant formation. This is pronounced in areas with complex landscape variability as e.g. coastal regions during summertime. Studies in the Mediterranean Region and Southern Europe have indicated that in certain periods the urban areas may be significantly affected by sources located several hundreds of kilometres away (Millan et al. 1997; Kallos et al. 1998). This research has shown that air pollution problems like the oxidant formation is a multi-scale problem. In areas with similar climatological characteristics, dispersion and photochemical processes may vary significantly, as it was shown for the Mediterranean Region e.g. by Kallos et al. (1997; 1998). For example, in the Western Mediterranean and the Iberian Peninsula the vertical component of advection is most important whereas in the Eastern part of the Mediterranean the horizontal component dominates. Urban plumes from various locations in Southern Europe may be transported over the Mediterranean maintaining most of their characteristics (Figure 10.3, previous page).

The time scale for such transport is 2-4 days which is sufficient for important physical-chemical transformations (e.g. Luria et al. 1996). While there is not important mixing over the Mediterranean waters there is considerable mixing over the European part of the land due to topographic features. The mixing over the African land area is considerable during day-hours reaching hights at the range of 3-5 km. as it is shown in Figure 10.4 where a North-South vertical cross-section of the Turbulent-Kinetic Energy (TKE) is plotted.

As it is shown in Figure 10.3, areas (both urban and rural) located even at the other side of the Mediterranean sea are affected by emissions (both urban and industrial) in a significant way. Furthermore, this transport mechanism can transfer air pollutants to the middle and upper troposphere on an efficient way. In the specific case, aged air pollutants can enter into the Intertropical Converzence Zone (ITCZ) and be transfered furthermore, or effect the sensitive rain cycle of the region (by supplying CCN's) as it is described in Kallos et al. 1997 and 1998.

The appropriate modelling of such long range transport phenomena requires the coverage of a broad part of the spectrum of atmospheric modes, and the meteorology and transport-chemistry models must be applied for an area covering at least a few thousands of kilometres and vertically up to the top of the troposphere with a horizontal resolution of a few kilometres (less than 5 km) in order to accurately describe the physical and chemical processes governing such phenomena. This type of simulations are in practice impossible to carry out with the available computer power of today. A good alternative is the two-way interactive nesting which allows the use of nested grids with variable resolution (e.g. Tremback et al. 1993). With this technique, fine resolution is used in the area(s) of interest and coarser over a large area of broad consideration. One-way nesting in dispersion and especially photochemical calculations is not recommended, mainly due to the problems related to the double consideration of emissions which may cause serious errors.

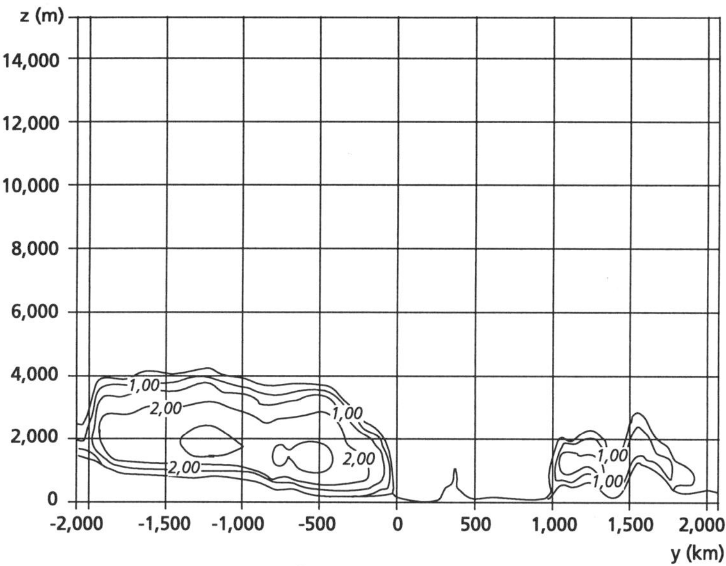

Figure 10.4 Cross section of the turbulent kinetic energy field (at 0.5 m^2s^{-2} interval) following the 25° E meridian, at 1200 UTC, 3 July 1994.

10.8 References

Dalu, G.A., Pielke, R.A., Avissar, R., Kallos, G., Guerrini, A. (1991) Linear impact of subgrid-scale thermal inhomogeneities on mesoscale atmospheric flow with zero synoptic wind. *Annals Geophysicae.*, **9**, 641-647.

Gifford, F.A., Hanna, S.R. (1973) Modelling urban air pollution. *Atmospheric Environment*, **7**, 131-136.

Hanna, S.R. (1987) An empirical formula for the height of the coastal internal boundary layer. *Boundary-Layer Meteorol.*, **40**, 205-207.

Hov, Ø., Hjøllo, B.Aa., Eliassen, A. (1994) Transport distance of ammonia and ammonium in Northern Europe. I. Model description. *J. Geophys. Res.*, **99**, 18,735-18,748.

Iversen, T., Halvorsen, N.E., Saltbones, J., Sandnes, H. (1990) Calculated budgets for airborne sulphur and nitrogen in Europe. *EMEP MSC-W Report 2/90.* Norwegian Meteorological Institute, P.O.Box 43, N-0313 Oslo, Norway.

Kallos, G., Kassomenos, P., Pielke, R.A. (1993) Synoptic and mesoscale weather conditions during air pollution episodes in Athens, Greece. *Boundary-Layer Meteorol.*, **62**, 163-184.

Kallos, G., Kotroni, V., Lagouvardos, K., Papadopoulos, A. (1998) On the transport of air pollutants from Europe to North Africa. *Geophys. Res. Let.*, **25**, 5, 619-622.

Kallos, G., Kotroni, V., Lagouvardos, K., Varinou, M., Uliasz, M., Papadopoulos, A., Luria, M., Peleg, M., Sharf, G., Matveev, V., Tov-Alper, D.S., Vanger, A., Tuncel, G., Tuncel, S., Aras, N., Gullu, G., Al-Momani, I. (1997) "Transport and Transformation of Air Pollutants from Europe to the East

Mediterranean Region", T-TRAPEM. Environmental Research Program AVICENNE, EU, DGXII contract AVI*-CT92-0005. ISBN 960-8468-07-8, University of Athens.

Luria, M., Peleg, M., Sharf, G., Tov-Alper, D.S., Schpitz, N., Ben Ami, Y., Gawi, Z., Lifschitz, B., Yitzchaki, A., Setter, I. (1996) Atmospheric sulfur over the last Mediterranean Region, *J. Geophys. Res.*, **101**, 25917-25930.

McKendry, I.G. (1989) Numerical simulation of sea breezes over the Auckland Region, New-Zealand - Air quality implications. *Boundary-Layer Meteorology*, **49**, 7-22.

Millan, M.M., Salvador, R., Mantilla, E., Kallos, G. (1997) "Photooxidant Dynamics in the Mediterranean Basin in Summer: Results from European Research Projects" *J. Geophys. Res-Atmospheres*, **102**, D7, 8811-8823.

Moran, M.D., Pielke, R.A. (1996a) Evaluation of a mesoscale atmospheric dispersion modelling system with observations from the 1980 Great Plains Mesoscale tracer field experiment. Part I: Datasets and meteorological simulations. *J. Appl. Meteorol.*, **35**, 281-307.

Moran, M.D., Pielke, R.A. (1996b) Evaluation of a mesoscale atmospheric dispersion modelling system with observations from the 1980 Great Plains Mesoscale tracer field experiment. Part II: Dispersion simulations. *J. Appl. Meteorol.*, **35**, 308-329.

Pielke, R.A. (1984) Mesoscale meteorological modelling. Academic Press, Orlando, Florida 32887, USA.

Pielke, R.A., Cotton, W.R., Walko, R.L., Tremback, C.J., Lyons, W.A., Grasso, L.D., Nicholls, M.E., Moran, M.D., Wesley, D.A., Lee, T.J., Copeland, J.H. (1992) A comprehensive meteorological modelling system - RAMS. *Meteorol. Atmos. Phys.*, **49**, 69-91.

Pielke, R.A., Kallos, G., Segal, M. (1989) Horizontal resolution needs for adequate lower tropospheric profiling involved with atmospheric systems forced by horizontal gradients in surface heating. Source of Atmos. and Ocean Techn., **6**, 741-758.

Pielke, R.A., Lyons, W.A., McNider, R.T., Moran, M.D., Moon, D.A., Stocker, R.A., Walko, R.L., Vliasz, M. (1991) Regional and mesoscale meteorological modelling as applied to air quality studies. Air Pollution Modelling and its Application VIII. Edited by H. van Dop and D.G. Steyn, Plenum Press, New York ISBN 0-306-43828-3. 259-289.

Pielke, R.A., McNider, R.T., Segal, M., Mahrer, Y. (1983) The use of a mesoscale numerical model for evaluations of pollutant transport and diffusion in coastal regions and over irregular terrain. *Bull. American Meteor. Soc.*, **64**, 243-249.

Scire, J.S., Strimaitis, D.G., Fernau, M.E. (1996) New developments in the CALPUFF non-steady-state modeling system. Air pollution modelling and its application. Edited by S.-E. Gryning and F. Schiermeier, Plenum Press, New York ISBN 0-306-45381-9. 709.

Segal, M., Garratt, J.R., Kallos, G. and Pielke, R.A. (1989a) The Impact of Wet Soil and Canopy Temperatures on Daytime Boundary - Layer Growth. *J. Atmos. Sci.*, **46**, 3673-3684.

Segal, M., Schreiber, W., Kallos, G., Pielke, R.A., Garratt, J.R., Weaver, J., Rodi, A. and Wilson, J. (1989b) "The Impact of Crop Areas in Northeast Colorado on the Mid-summer Atmospheric Thermal Circulations". *Mon. Weather Rev.*, **117**, 809-825.

Simpson, D. (1993) Photochemical model calculations over Europe for two extended summer periods: 1985 and 1989. Model results and comparison with observations. *Atmospheric Environment*, **27A**, 921-943.

Thunis, P., Bomstein, R. (1996) Hierarchy of mesoscale flow assumptions and regulations, *J. Atmos. Sci.*, **53**, 380-397.

Tremback, C.J., Lyons, W.A., Thorson, W.P., Walko, R.L. (1993) An emergency response and local weather forecasting software system. *Proceedings of the 20th ITM on Air Pollution and its Application*, November 29 - December 3, 1993. Valencia, Spain. 18, 423-429.

Turner, D.B. (1970) *Workbook of Atmospheric Dispersion Estimates*. Environmental Protection Agency, Research Triangle Park, North Carolina, USA.

Uliasz, M., Pielke, R.A. (1991) Application of the receptor oriented approach in mesoscale dispersion modeling, in: Van Pop, H. Steyn, D.G. (editors), *Air pollution modeling and its Application VIII*, Plenum Press, New York ISBN 0-306-43828-3, 399-407.

Yamartino, R.J., Scire, J.S., Carmichael, G.R., Chang, Y.S. (1992) The CALGRID mesoscale Photochemical Grid Model I. Model formulation. *Atmospheric Environment*, **26A**, 1493-1512.

Zlatev, Z. (1995) *Computer treatment of large air pollution models*. Dordrecht, Boston, London: Kluwer Academic Publishers.

Chapter 11

URBAN SCALE MODELS

PATRICE G. MESTAYER
Centre National de la Recherche Scientifique & Ecole Centrale de Nantes
1. rue de la Noe, F-44321 Nantes Cedex 3, France

11.1 Introduction

The atmospheric processes governing the conditions in urban areas take place on scales between mesoscale and microscale. Although the main interest bears primarily on the lowest part of the atmosphere on scales as small as 50 m, the urban scale processes cannot in general be considered as isolated microscale processes. This is mainly because the city structures generate motions at scales as large as the whole city area itself. Recently Thunis and Bornstein (1996) defined a new scale called meso-δ scale (200 m to 2 km), which fits the scales of the urban atmospheric boundary layer.

The present chapter gives a general introduction to the characteristics of the urban atmosphere with focus on the specific features and difficulties of modelling processes on this scale. Then follows discussions concerning development of specific urban scale models, and a brief introduction to some of the existing operational models in Europe.

11.2 The urban atmosphere

The urban atmosphere is highly inhomogeneous and strongly influenced by the canopy. The vertical structure is more complex than over the surrounding rural areas and the lowest parts of the boundary layer, the so-called surface boundary layer (SBL), may be considered to consist of two layers: the canopy and the roughness sub-layers. The canopy sub-layer is composed of a number of individual building canyons, whereas the roughness sub-layer is a non-equilibrium transition layer in which vertical fluxes of momentum, energy, moisture and pollution from individual canyons blend together (Rotach 1993; Mestayer, Anquetin 1994). The canopy sub-layer is a patchwork of quarters with very different structures, which increases the complexity of this sub-layer, and the strong transitions in roughness and heat fluxes lead to the development of successive internal boundary layers in high wind conditions and of quarter thermal breezes and slope winds in low wind conditions (Figure 11.1).

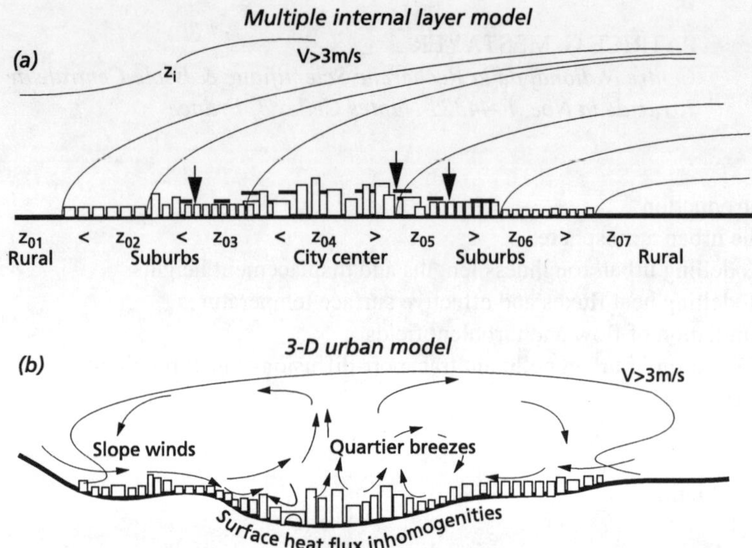

Figure 11.1 During high wind conditions (a) multiple internal boundary layers may be developed in the canopy layer, whereas slope winds and quarter breezes may be developed at low wind conditions.

In most cases the observed conditions in a given urban area will be some kind of a combination of the two situations sketched in Figure 11.1. The development of successive internal boundary layers is observed only for strong wind conditions over flat terrain, in which case the structure of the internal layers in average appears to be quite 2-D with only little lateral cross-influence (Costes 1996). In the cases where the orography or topography of the area is not uniform, e.g. due to the presence of valleys, coastlines and other terrain features, the presence of these features make the structure of the urban atmosphere fully 3-D. Most urban air pollution episodes take place at low wind condition where the structure of the atmosphere just above rooftop has strong impact on transport and dispersion of pollutants in the urban atmosphere and thereby affects also the photochemical pollution transformation processes.

11.2.1 VERTICAL STRUCTURE OF THE URBAN ATMOSPHERE

The urban atmospheric surface layer is dramatically perturbed by the presence of buildings and other constructions which act as large roughness elements (Figures 11.2 and 11.3). The urban canopy is defined as the layer in which the air flows between buildings and other obstacles with the consequent development of many turbulent wakes and recirculation zones.

In densely built city centres, the air flow is canalised in the narrow street canyons and trapped in the courtyards; this layer has some similarities with the dense forest canopy layer (as well as it has similarities to flow in porous media) interacting with the upper atmosphere essentially through vertical fluxes. These vertical fluxes are strongly depending on the thermal characteristics of the surfaces of buildings, roads etc., as well as it depends on sunlight trapping and anthropogenic heating.

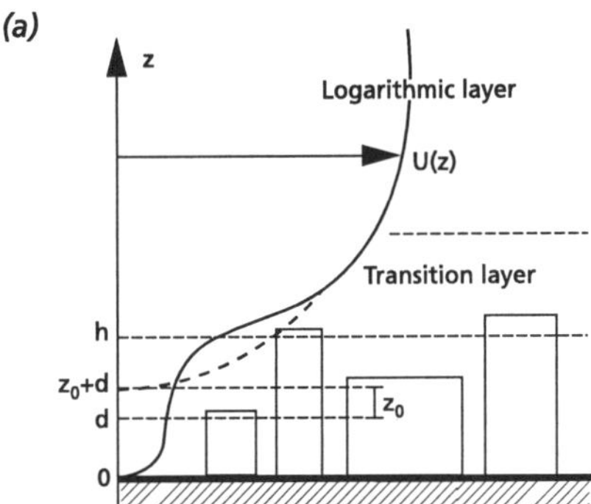

Figure 11.2 Schematic representation of the vertical structure of the atmospheric surface layer over city buildings.

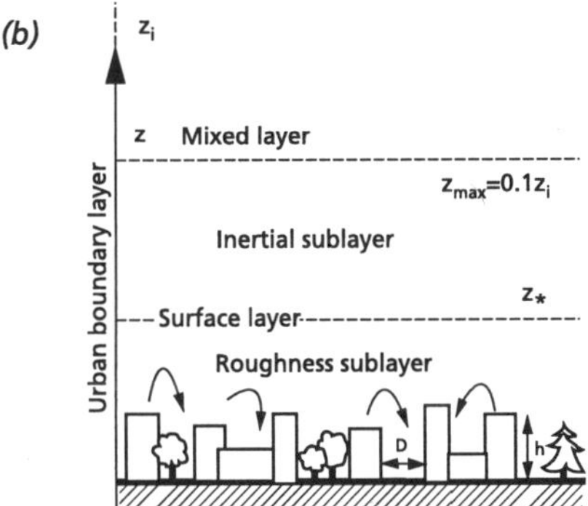

Figure 11.3 Schematic representation of the vertical structure of the urban atmospheric boundary layer.

The less densely built city quarters form a mixture of canalising streets with closed building facades, open areas with aligned or isolated buildings of varying height, as well as man made and natural surfaces. In these areas, the canopy layer becomes more complex, since the air skirts round and above the constructions and trees, and additional turbulence is generated e.g. by sparse elevated buildings.

In the outer quarters, the canopy layer looses its integrity to give place to flow over relatively isolated and randomly placed obstacles in the form of individual buildings as well as clusters of houses, trees, road and freeway constructions. Still the surfaces in the urban out-skirt have their own characteristics including a significant fraction of man made surfaces, which often appear in large patches of commercial, industrial, harbour and airport zones.

As a rule of thumb, the urban roughness length, z_0, is one tenth, and the displacement height, z_d, is two third of the mean building height, h (Oke 1987), although the actual ratios are strongly depending on the building density; in the dense city centres the roughness length is lower and the displacement height higher, while in the low density outer quarters, the roughness length is also lower whereas the displacement height decreases. Assimilating the canopy layer to the displacement height as a first approximation shows that it varies from less than a metre in the outer quarters to being close to the mean building height in the city centre.

Internal boundary layers (IBLs) develop over the transitions between quarters with different roughness lengths and heat fluxes. Their height increases approximately as $x^{4/5}$, where x is the downstream distance from the transition (e.g. Mestayer, Anquetin 1994). Such an IBL is composed of two layers. A lower internal layer with a thickness of about one tenth of the IBL in which the vertical momentum flux is independent of height and in equilibrium with the local wind speed and underlying roughness length. In this layer, the momentum flux decreases towards an asymptotic value with increasing downstream distance. Above the lower internal layer is a non-equilibrium transition layer, where the momentum flux reaches an extremum (maximum or $z_{02} > z_{01}$, minimum $z_{02} < z_{01}$) and then tends towards the value of the upper constant flux layer. Therefore the sketch of Figure 11.1a (page 198) must rather be seen as composed of non-equilibrium layers, perturbing each other, where only thin layers between 2 h and H(x)/10 have some chances to be in local equilibrium, i.e., where Monin-Obukhov similarity theory is verified. Let's consider z_i = 400 m as a typical ASL thickness in weakly unstable conditions, and an urban roughness lengths ranging from 0.5 to 3 m, corresponding roughly to building heights from 5 to 30 m: the IBL upper boundaries reach $z \sim z_i$ for fetches x between 2 and 1.3 km. Since 1 to 2 km is the typical size of European city quarters, i.e., the typical length between two quarter transitions, this means that the maximum height reached by the internal lower layers in local equilibrium may be observed in thin layers over constructions lower than 15 to 20 m (5 to 7 store buildings) but not over higher buildings. In general, most of the lower urban atmosphere is out of equilibrium (Figure 11.4).

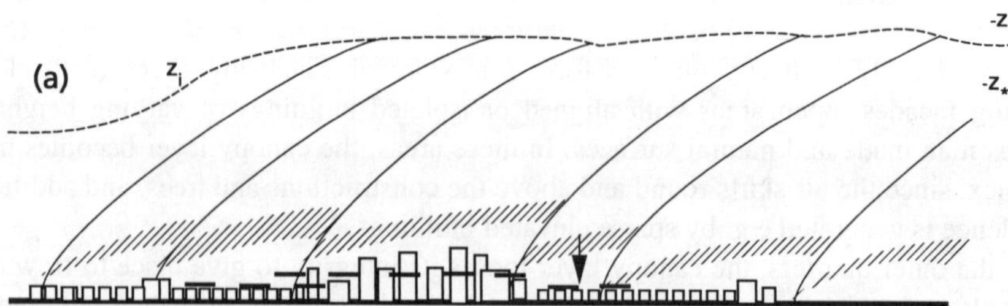

Figure 11.4 In the development of internal boundary layers over successive city quarters, only a small fraction of the atmosphere (in grey) may be expected in equilibrium due to the thick roughness sublayers.

The upper boundary of the ASL is generally higher over urban areas than in the surrounding rural areas; this has at least been observed when an inversion layer has been clearly identified. How the non-equilibrium IBLs blend together in the upper part of the ASL is still not fully explored; They may form a quasi-homogeneous, quasi-equilibrium layer or the perturbations may reach the inversion. This is an important uncertainty in mesoscale models which assume the Monin-Obukhov ASL similarity theories to be valid in the lower 1 or 2 computational layers in order to define lower boundary conditions. These uncertainties are the main reasons for using refined scale PBL models for simulation of the urban atmosphere. The boundary conditions are defined at the PBL top where they can be well expressed. Therefore, although they are mainly confined to the lowest layers, urban scale models need also to include the mixing layer and the entrainement layer. The top of the PBL is believed to be influenced by the presence of the urban area only at a regional scale, except for the case of purely convective conditions.

In moderately unstable conditions, the structure of the IBL is perturbed by the difference in thermal fluxes over the adjacent quarters due to large difference in albedo, heat storage capacity, presence of anthropogenic heat sources, etc. These differences generate local thermal "breezes" and vertical motions at intermediate scales between PBL eddies and small scale turbulence. The local convective motions enhance mixing with pollutants from the nearby quarters. In strongly unstable conditions, the local convective motions interfere with the general urban convective regime.

In stable stratified conditions when the ASL is less than 100 m, the dynamic equilibrium vanishes everywhere and the influence of the canopy appears mainly in thermal flux variability over the various quarters. This generates an inhomogeneity of thermal stability, hence of turbulent intensity and diffusivity.

The variability of sensible and latent heat fluxes therefore play an important role in the local transport, dispersion and mixing of pollutants. Due to the non-equilibrium conditions, determination of these fluxes requires revision of the Monin-Obukhov ASL theory.

11.2.2 URBAN TERRAIN INHOMOGENEITY

In an urban area, the roughness length of the various quarters range from less than 1 mm (e.g. lake and water surfaces) to more than 10 m (e.g. towering business centres). The surface temperatures vary typically more than 30 degrees during a sunny day (considering averages over a few tens of meters). A radiance temperature variance of 54 degrees has been reported for the half a million inhabitant District of Nantes, France (Robin 1995). The humidity fluxes also show considerable quarter scale inhomogeneity, the soils varying from water surfaces to non-permeable surfaces The ratio between natural and man made soil and the vegetation coverage fraction both range from zero to unity: in sunny conditions this inhomogeneity generates strong spatial variations in the latent heat flux. This inhomogeneity also applies to the pollution sources and furthermore to the deposition of pollutants as a result of the variety of surface types.

11.3 Modelling urban roughness lengths and displacement heights

Over the dense canopy of the city centres, determination of the momentum flux requires parameterisations that account for local geometric characteristics of the roughness elements and for the quarter geographical extension and its thermal structure influencing the small scale flow between the buildings. The usual wall laws have furthermore to be extended from the natural surfaces to the urban surfaces by assuming equilibrium constant flux theories to be good approximations at the level z_s where the wall law is applied (usually the boundary between first and second grid mesh, however, the optimal choice of z_s is still subject to research). The wall law is usually drawn from the log profile relationship:

$$U_h(z_s) = \frac{u_{*s}}{\kappa}\left[\ln\left(\frac{z_s - z_d}{z_0}\right) - \psi_m\left(\frac{z_s}{L}\right)\right] \tag{11.1}$$

where U_h is the horizontal velocity component and $u_{*s} = \langle -u'w'(z_s)\rangle^{1/2}$ (the brackets indicate ensemble averages), $\psi_m(z/L)$ is a stability correction term, and L is the Monin-Obukhov length. $U_h(z_s)$ and $\langle u'w'(z_s)\rangle$ are obtained from the Navier-Stokes equation (Chapter 10) computation at the previous time step. The computation implies knowledge of the values of z_0 and z_d for the given grid cell.

Two groups of methods may be employed for determination of the roughness length z_0 and the displacement height z_d in urban areas (Grimmond, Oke 1997): *wind based and land use based methods*. The first group of methods is empirical and may take into account the influence of wind direction. However, these methods require huge numbers of measuring points in order to be applied to a whole urban area. The presently available land use methods take into account only the sizes and arrangement of roughness elements, not the dimensions and other characteristics of the quarters; therefore they provide estimations of the "intrinsic" roughness parameters.

Bottema (1995) defined a notation for building dimensions and interspaces of regular building arrays, which is applied in the following. Wind direction 0^0 is here defined as wind flow perpendicular to the longer building facade or in direction of the largest interspaces. Regular building arrays can be "normal" (Figure 11.5) or "staggered" when the buildings of one row face the building separations of the next row. In typical European urban quarters, individual buildings are not uniform in size, but for simplicity the same notation is used for uniform and non-uniform building arrays:

$$\sum \beta \cdot \gamma \text{ for } \sum_{i=1}^{Nb} \beta_i \cdot \gamma_i \tag{11.2}$$

where β_i and γ_i are individual building sizes and Nb the number of buildings in the considered area.

One type of models for determination of z_0 and z_d, consider only the mean obstacle height (Grimmond, Oke 1997): $z_0 = \alpha_0 h$; $z_d = \alpha_d h$, where the α coefficients are constant (Table 11.1).

Table 11.1 Literature values of α_0 and α_d.

α_0	α_d	Reference
0.1	0.667	Oke 1987
0.142		Paeschke 1938
0.111		Stull 1988
	0.75	Thom 1971
	0.75	Rotach 1993

In another type of models the α coefficients are defined as functions of building density or the built up area fraction λ_p defined as (Kutzbach 1961; Counihan 1971):

$$\lambda_p = \frac{\sum w \cdot l_x}{\sum d_x \cdot d_y} \tag{11.3}$$

The notation is explained in Figure 11.5.

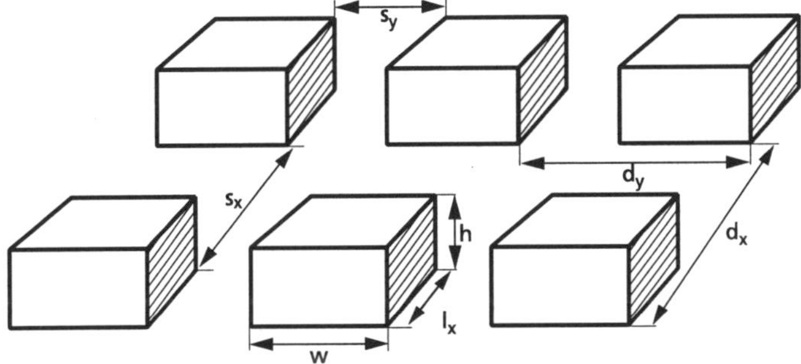

Figure 11.5 Definition of geometrical parameters in regular building arrays.

The empirical determinations of the α-λ_p relationships, from wind tunnel measurements, are not in perfect agreement with each other and Gimmond and Oke (1997) consider that "the Kutzbach method provides reasonable estimates only for low λ_p ratios (< 0.2), the Counihan (1971) method is more applicable for higher values (> 0.15)". Kondo and Yamazawa (1986) propose a quite similar relationship where the average building height is area weighted:

$$\lambda_p = 0.25 \frac{\sum h \cdot w \cdot l_x}{\sum d_x \cdot d_y} \tag{11.4}$$

Grimmond and Oke (1997) note that this model predicts continuously increasing values of z_0/h with λ_p. This is in contrast to the results of Counihan (1971) showing a decrease of z_0/h for $\lambda_p > 30\%$.

Only few values for λ_p have been published, but Soliman (1976) reports a range of λ_p between 5 and 50%. Densities below 10% often coincide with tall buildings (>10 stores) whereas for higher densities there usually cannot be observed any clear relationship between building height and building density.

For determining the roughness height, z_0, it is better to use another parameter; the frontal density ratio, λ_f, which refers to the surface area facing the wind:

$$\lambda_f = \frac{\sum w \cdot h}{\sum d_x \cdot d_y} \tag{11.5}$$

The frontal density ratio is thus a function of wind direction. Lettau (1969) derived first a very simple expression based on λ_f from wind profiles over bushel baskets on the frozen lake Mendota:

$$\frac{z_0}{h} = 0.5\lambda_f \tag{11.6}$$

The roughness length is a measure of the wind drag integral over the ground and obstacles. Rewriting the equilibrium equation (11.1) by relating z_0 to the drag coefficient C_d at a given reference height and neglecting the stability correction, $\psi_m(z/L)$, one obtains (Bottema 1997):

$$C_d(z_{ref}) = \frac{\tau}{½\rho_a U^2(z_{ref})} = \frac{2u_*^2}{U^2(z_{ref})} = \frac{2\kappa^2}{\ln^2\left(\dfrac{z_{ref} - z_d}{z_0}\right)} \tag{11.7}$$

Bottema (1997) assumed that the drag essentially is due to building wakes, neglecting the interspace surface stress. He furthermore replaced the reference height with the mean building height, h, and assumed equilibrium relationships at z = h:

$$\frac{z_0}{h} = \frac{h - z_{d,pl}}{h} \exp\left(\frac{-\kappa}{\sqrt{0.5C_{dh}\lambda_f}}\right) \tag{11.8}$$

where z_d is replaced by $z_{d,pl}$, the *displacement height* in the building median plane (i.e. the theoretical value of z_d if the buildings were of infinite length). This intermediate parameter is a measure of the sheltering effect of one building on another building, considering their relative positions but not their arrangement in the whole array, while $z_d/z_{d,pl}$ takes into account the density influence. A series of $z_{d,pl}$ formulas are derived from the observations of geometry of the recirculation zones and the displacement heights obtained from wind profile measurements in wind tunnels (see the sketches in Figure 11.6).

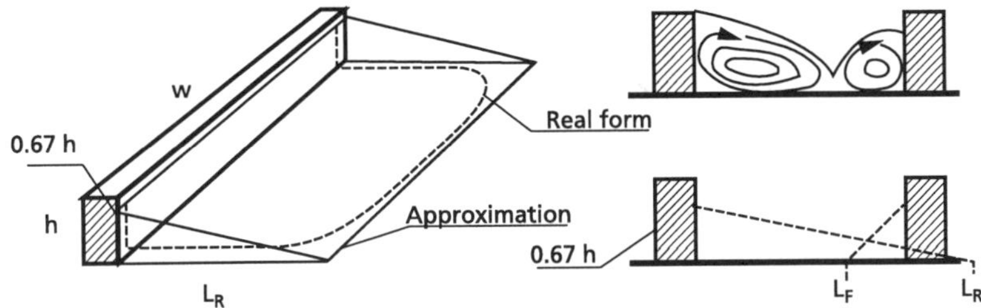

Figure 11.6 Model of flow recirculation between successive buildings and mutual sheltering parameters.

Raupach (1994) assumed the equilibrium Equation (11.1) to be fulfilled at z = h, and expressed z_d by:

$$1-\frac{z_d}{h}=\frac{1-\exp(2c_{dl}\lambda_f)}{\sqrt{2c_{dl}\lambda_f}} \tag{11.9}$$

He then solved (11.1) from:

$$\frac{u_*}{U_h}=\min\left[\left(C_{ds}+0.5\lambda_f C_{dh}\right)^{0.5},\left(\frac{u_*}{U_h}\right)_{max}\right] \tag{11.10}$$

where C_{ds} and C_{dh} are respectively the substrate surface and the unit obstacle drag coefficients estimated at level z. For the constants, Raupach (1994) recommends C_{ds} = 0.003, C_{dh} = 0.6, $(u^*/U_h)_{max}$ = 0.3, ψ_m = 0.193 and c_{d1} = 7.5. This choice and further corrections (Raupach 1995) are discussed in Bottema and Mestayer (1997).

Since the regularity of building arrangements is hard to determine when urban maps are constructed (e.g. Mestayer et al. 1996), Bottema (1995) proposed a *simplified and synthetic power law model*, which fits reasonably well with more detailed model studies:

$$\frac{z_d}{h}=\lambda_p^{0.6} \tag{11.11}$$

He furthermore proposed to account for presence of trees, weighting their influence by their area density and porosity. Such a procedure was given by Grimmond and Oke (1997).

11.4 Modelling heat fluxes and effective surface temperatures

Applying finite difference methods to solve the thermodynamic equations, one needs to define values for the lower boundary conditions, i.e. air temperature and moisture just above the surface, sensible and latent heat fluxes to/from the surface. The surface refers here to the real ground in the city outskirts and the canopy-atmosphere interface in city centres and densely planted quarters. Such data may be obtained from measurements, as well as they may be estimated or modelled.

Due to the high spatial variability, direct measurements of the lower boundary data are seldom possible from ground, whereas remote sensing techniques from satellites may be a more feasible approach. Both methods are shortly described in the following.

Methods for *mapping of surface thermodynamic parameters (STPs)* rely on analysis of the land use data for the area, where the precision in land use classification must be carefully adapted to the resolution in the simulation.

The simplest modelling approach for this purpose is based on the regional STPs and fixed land use dependent values representing the heat island effect. The most detailed STP models have been developed for mesoscale studies of the humidity budgets in complex ecosystems. In such models, the governing partial differential equations for water and heat diffusion processes are solved on a fine vertical grid, and the exchange between different types of vegetation, soil and air are obtained by applying the resistance method for the transfer in the plant elements.

The simplified STP models make in general use of time independent parameterisation. Thielen et al. (1996) presented such computations for short time periods (4 h) with a horizontal resolution of 500 m. The surface sensible (Q_0) and latent (E_0) heat fluxes are expressed as:

$$Q_0 = \mu(x,y) \cdot S_0 \cdot f(z, \varphi, \delta, s_E, s_N) \qquad (11.12)$$

$$E_0 = \frac{Q_0}{B_0(x,y)}$$

where $\mu(x,y)$ is a factor describing the conversion of incoming solar radiation to a sensible heat flux; $B_0(x,y)$ is the Bowen ratio, and the geometric factor, f, describing the reduction in solar energy (S_0 = 1395 W m^{-2}) is a function of the Zenith angle, latitude φ, declination angle δ, and topography slopes in North (s_N) and East (s_E) directions, respectively. The conversion function μ, combining albedo and emissivity, and the Bowen ratio B_0 are assumed to be only varying with the surface type. Thielen et al. (1996) considered four types of land uses: city centres with tall buildings, other urbanised areas, vegetated surfaces and water surfaces.

A good compromise between multi-layer soil models and models applying the assumption of constant surface parameters, is the *two layer or force-restore models* (Deardorff 1978). In these models, equilibrium assumptions similar to (11.1) are assumed for the lowest atmospheric layers of the model domain:

$$T(z_s) - T_s = \frac{Pr_t \theta_*}{\kappa}\left[\ln\left(\frac{z_s - z_{dT}}{z_{oT}}\right) - \psi_h\left(\frac{z_s}{L}\right)\right] \tag{11.13}$$

$$q(z_s) - q_s = \frac{S_{ct} q_*}{\kappa}\left[\ln\left(\frac{z_s - z_{dq}}{z_{oq}}\right) - \psi_v\left(\frac{z_s}{L}\right)\right] \tag{11.14}$$

where P_{rt} is the Prandtl number and S_{ct} is the Schmidt number. z_{dT}, z_{dq} and z_{0T}, z_{0q} are the displacement heights and roughness lengths for T and q, respectively, ψ_h and ψ_v are the stability correction functions for heat and water vapour, respectively. The Monin-Obukhov scaling parameters are defined: $Q_0 = -\rho_a C_p u_* \theta_*$ and $E_0 = -u_* q_*$. The constants z_{0T} and z_{oq} represent the levels where the extension of the logarithmic profiles (11.13) and (11.14) would yield $T(z) = T_s$ and $q(z) = q_s$. It should be noted that the profiles are not logarithmic within the roughness sub-layer, $z < z_*$. In mesoscale models the parameters z_{dT}, z_{dq}, z_{0T} and z_{0q} are usually set constant; an assumption that may easily be violated in the urban canopy (see the following discussion).

Since the $T(z_s)$ and $q(z_s)$ are computed by the atmospheric model, the soil model has to supply values of T_s, q_s, Q_0 and E_0 as function of atmospheric forcing and time. Various versions of force-restore models describe in more detail the heat transfers, the humidity transfers, or the role of vegetation (e.g. Deardorf 1978; Guilbaud 1996). In the following some of the common features of these models are presented and the necessary improvements needed for applications for the urban canopy are discussed.

The force-restore model is based on an equation for the evolution of T_s associated with the instantaneous surface energy budget, and an equation for the evolution of water content in two ground reservoirs yielding q_s. The model accounts for the partial coverage of the soil by vegetation which affects transfers of rain, heat, liquid and vapour water, either by retardation or by root-leave short-circuits (Figure 11.7).

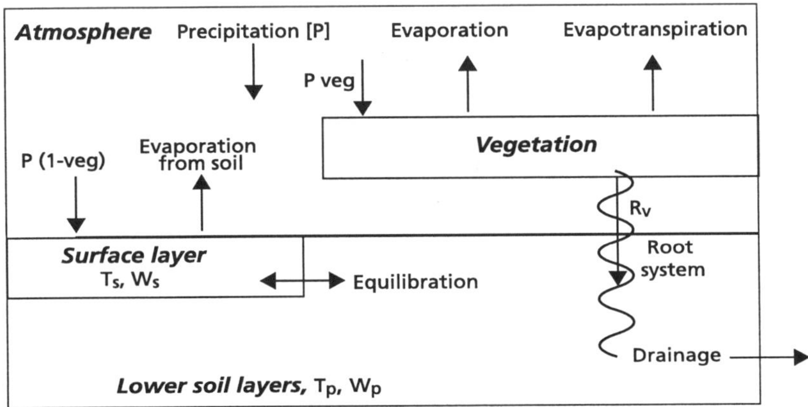

Figure 11.7 Scheme of atmosphere-soil exchange mechanisms in force restore models.

The equation for T_s expresses the diurnal forcing towards a solution of the type $T_s = \langle T \rangle + \delta T \sin(2\pi t / \tau_d)$, <T> is the deep soil temperature.

$$\frac{\partial T_s}{\partial t} = C_T H_{at} - \Delta T \cdot \frac{2\pi}{\tau_d} \tag{11.15}$$

where τ_d = 86400 s (= 24 hours), H_{at} is the total energy flux received from the atmosphere, ΔT is either defined as T_s - <T> or T_s- T_p where the deep layer temperature T_p is computed from $\partial T_p/\partial t = (T_s\text{-}T_p) / \tau_d$, the surface specific humidity q_s is a function of saturation specific humidity, $q_{sat}(T_s)$, at the given surface temperature and pressure, vegetation coverage Veg, vegetation fraction covered by dew, and water content in the surface layer W_s. The later may be expressed in a similar way as T_s:

$$\frac{\partial W_s}{\partial t} = C_1(P_s - E_s) - \frac{C_2 \Delta W}{\tau_d} \tag{11.16}$$

where P_s and E_s are precipitation reaching the soil and evaporation from the soil, accounting for vegetation coverage. They are functions of P and E_0 respectively; ΔW_s is defined as W_s - W_p. The equation for W_p describes the water balance in the deep layer taking into account rain and evaporation through the surface layer and the vegetation, with an eventual draining sink at saturation.

H_{at} is here obtained from the energy budget at the canopy surface:

$$H_{at} = R_n + Q_0 - \rho_0 L E_0 + H_{an} \tag{11.17}$$

where R_n is the net radiative budget between received and emitted, solar and IR, radiation (expressed for a plan surface, and H_{an} is the sum of all additional fluxes including anthropogenic heat sources. Q_0 and E_0 may be obtained from various algorithms based on the bulk aerodynamic formulas:

$$Q_0 = \rho_a C_p C_h U(z_s)[T_s - T(z_s)]C_3 \tag{11.18}$$

$$E_0 = C_v U(z_s)[q_s - q(z_s)]C_4 \tag{11.19}$$

where C_h and C_v are the aerodynamic heat and water vapour transfer coefficients, computed from z_s/L, z_0 and z_{0T} or z_{0q}. Since the Monin-Obukhov length, L, is a function of Q_0, E_0 and u_*, all fluxes must be computed simultaneously with iterative algorithms.

The coefficients C_T, C_1, C_2, C_3 and C_4 are secondary coupling functions representing the vegetation transfer processes, such as shadowing, heat and precipitation retardation, leave resistance, dew fractional coverage and the changes in water transfers, when the soil layers approach saturation. These functions, which differ notably from one model to another, include empirical parameters representing vegetation and soil characteristics.

11.4.1 APPLICATION TO AN URBAN CANOPY

The thermodynamics of the urban canopy have been described previously (e.g. Oke 1987; Mestayer, Anquetin 1994). It includes building shadowing/sky view factors, radiation multiple reflections due to vertical surfaces, artificial diversity, building wall heat fluxes, local heat and moisture anthropogenic sources, draining surfaces. Adapting a soil model to the urban canopy, i.e. creating an urban thermodynamic model, requires a revisit to several of its assumptions and an extension of several parameterisations.

- Although it is probable, there is no evidence that $z_{dT} = z_{dq} = z_d$.
- The model relies heavily on z_0, z_{oT}, z_{0q}. Over natural surfaces it is usually assumed that:

$$z_{0T} = z_{0q} = r_0 \cdot z_0 \tag{11.20}$$

where r_0 is a constant factor, sometimes given the value 1, but more often set to 0.1. Hignett's (1994) experimental results show that over discontinuous canopies, r_0 varies by several orders of magnitude, depending on the terrain structure. The low temperature-humidity correlations observed by Roth and Oke (1993; 1995) hint that the urban heat and humidity sources are being differently dispersed, z_{0T} and z_{0q} should therefore not be equal. Low values of r_0 have been shown to have a strong influence on both the momentum and scalar flux computations (Guilloteau 1997).

- The apportionment between bare soils and soils covered by vegetation can be extended to artificial permeable surfaces (like roads) and waterproof draining surfaces (like building roofs), with their own transfer parameterisation.
- The soil parameters must account for the fact that most urban grounds have been dug in, generating implicit ducts and draining systems.
- The anthropogenic energy sources must be modelled to account separately for the permanent, semi-permanent, and time dependent sources (e.g. domestic heating, industrial releases, traffic).
- Finally, the canopy geometry must be considered for its effects through sky view factors, radiative trap, and canopy internal ventilation. This last process, being due to a mixed wind and thermal convection, introduces a new level of coupling.
- In addition, the radiative model should account for aerosols in the lower atmosphere and horizontal divergence generated by inhomogeneous pollutant concentrations.

Obviously the creation of urban canopy thermodynamic models requires specific model developments, but also, perhaps mainly, radiative and thermodynamic process studies extending the pioneering works of Oke and colleagues (e.g. Oke 1978; 1982; Oke et al. 1991) and of Terjung and colleagues (e.g. Todhunter, Terjung 1988), and of experimental works allowing to determine the numerous empirical constants in the models.

11.4.2 MAPPING OF THE URBAN SURFACE

Urban canopy thermodynamic models and urban roughness models naturally require *local geographic data*. The currently available European land use databases, e.g. CORINE Land Cover, have in general insufficient horizontal resolution for generation of roughness and STP maps for urban scale simulations. Furthermore the surface parameter averaging strategy adopted in mesoscale simulations must be replaced by a new strategy based on the aerodynamic quarter concept.

For generation of STP maps, the main problem is the large variety of surface characteristics within the same land use category. CORINE Land Cover is divided into 44 categories, of which in practise 12 categories appear in urban areas and 2 in built-up areas. It should be noted that presently available mesoscale models can handle even less urban categories; e.g. the soil model of MERCURE (Riou 1987; Carissimo 1997) which is already applied to urban areas, contains parameters for 7 soil categories of which only 3 correspond to built-up areas. STP maps may be refined by combining the CORINE Land Cover database with more local cadastral databases. This task is not trivial due to administrative, geographical, legal and proficiency limits as well as co-ordinate matching. Splitting the urban land cover for the district of Nantes, France in 25 land use categories, Robin (1993) has shown the commonly observed high variability in land use - surface temperature correlations (Figure 11.8). This work illustrates that several land use categories have to be strongly refined in order to reduce the errors in estimated surface temperature.

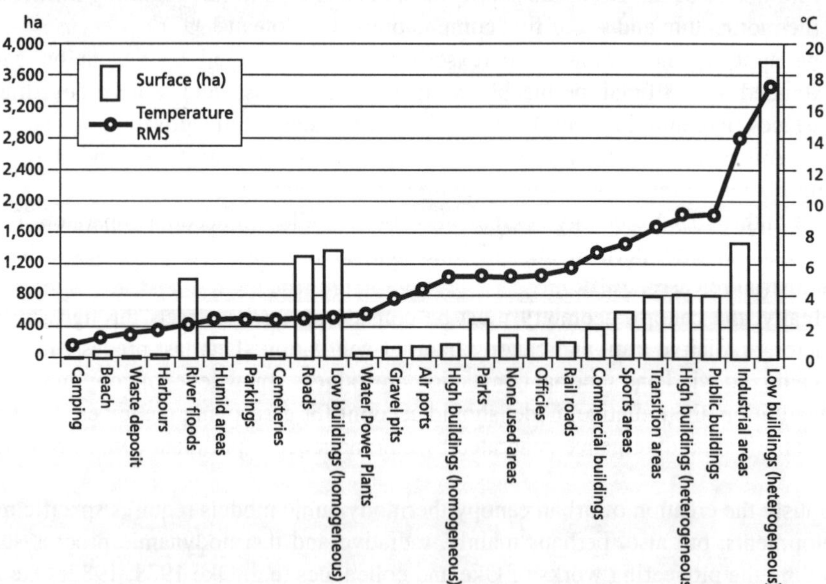

Figure 11.8 Variability in urban surface temperatures: the District of Nantes, France, is split in 25 land-use categories, each bar indicates the area covered by a category and the dot shows the Lansat TM radiative temperature RMS over this area (Robin 1993).

An alternative approach to land use data analysis, is the *direct measurement of the surface temperature* by remote sensing techniques with satellite Infra-Red channels. In practise the high spatial and temporal variability in urban albedo, emissivity, and radiative trapping makes it difficult to accurately transform visible and IR images into surface temperatures. Furthermore this method provides radiance temperatures while the actual temperature in the air just above the surface has to be represented in the models. This method is therefore still subject to research for the coming years.

In order to generate values for z_0 and z_d model input data are required not only for the land use which may be provided by geographic maps, but also for elevation of the various obstacles (data for mean height and frontal density). Such data are not generally available for European cities. In France, the National Geographic Institute has established databases describing location, area and elevation of all buildings and structural elements of most larger cities. These TRAPU databases have been created on the basis of airborne high resolution stereo photographs. In several other European countries similar databases are being constructed for the purpose of cellular telecommunication networks or advanced driving help/electronic route maps. Roughness maps may be generated by combining land use analysis (e.g. from CORINE Land Cover) with TRAPU data statistical analysis on the basis of Raupach and Bottema's roughness models (e.g. Bottema, Mestayer 1997). A difficulty of the method is the huge amount of data that has to be treated.

In principle the roughness parameters may also be obtained directly from satellite radar data since, due to the tilt of the radar beam, the obtained signal is a function of the apparent height of buildings and other elements. Two approaches are in development, using ERS synthetic aperture radar (SAR) data:

- SAR signal empirical correlation with land use and/or roughness parameters.
- SAR interferometry (InSAR) where the phase difference between two frames depends on the scattering of the objects.

In practise, both methods require much signal processing, since SAR response depends on view angle, object shape and orientation relative to the spacecraft orbit, electromagnetic properties of the detected surfaces, presence of humidity in soil, dew on surfaces, wind over water surfaces etc., so that their accuracy is still somewhat uncertain.

Detailed 3-D databases like TRAPU are resource demanding and it will last a while before similar databases are available for all European urban areas. It is therefore desirable to develop methods allowing to retrieve the needed information from 2-D databases like maps and satellite images, e.g. by establishing correlations between surface arrangements and volumic structures of building ensembles.

In mesoscale simulations, the land use resolution is usually finer than the grid mesh while the atmospheric layers of interest are higher than the blending height: an averaging procedure is therefore most often applied, to smooth down effects of surface inhomogeneity. Such procedures often include an area weighted average of fluxes and $\ln(z_0)$ within each grid mesh and a moving average over neighbouring meshes. In contrast, at the urban scale the flow structures generated in the atmospheric lower layers by the terrain inhomogeneities are an important part of the simulation, while the grid is finer than the land use resolution. The physical and statistical representativity in STPs and roughness parameters at the scale of one grid mesh may of course be questioned. The strategy consists of replacing a city map by a patchwork where each patch is an aerodynamic quarter. This concept corresponds to an area over which the roughness parameters and fluxes vary sufficiently little to make their averages representative of the whole area. Extension of the area is determined by obvious transitions in the values of these parameters. The size of the aerodynamic quarter is a compromise between the statistical representativity of the averaged parameters and the generation of sharp transitions possibly causing instabilities in the models.

11.5 Simulation of flow and turbulent fields

The first urban flow models were designed for the purpose of modelling urban comfort. They were based on correlations of the maximum gust and turbulence intensity at a given place within the city to their counterparts at a reference meteorological mast (e.g. Bottema et al. 1992). Unfortunately such engineering rules are insufficient for prediction of the whole PBL turbulent field over an urban area.

Applied modelling efforts include the development of models for thermal and dynamic internal boundary layers based on the energy balance equation in the PBL and the jump in potential temperature at its top (Gryning, Batchvarova 1997). Due to the complexity of the vertical structure of the urban roughness layer, this type of semi-analytical models is still not suited for prediction of the structure of the urban flow fields, especially at low wind conditions where the IBL structure vanishes. They represent, however, an interesting development basis for urban meteorological pre-processors.

The domain of the urban scale is too small for the independent development of the meteorological processes and too large for the PBL structure to be considered unchanged from inflow to outflow. In order to adequately simulate the passage of large scale events and their interaction with the urban area, three basic methods may be distinquished:

- Model nesting where dynamic flow and turbulent fields at the urban domain boundaries are provided by a mesoscale model.
- Meteorological pre-processing that provides diagnostic vertical profiles of the dynamic and thermodynamic variables at the inflow and lateral boundaries using either predicted or measured series of surface or PBL top variables.
- Applied meteorological models using continuous measured variables or mesoscale model output to derive coherent sets of variables at each point of the inflow and lateral boundary planes.

The flow fields are obtained from solving the momentum and continuity equations - Navier-Stokes equations (Chapter 10) on the Eulerian grid. Two types of approaches are in use here*: Large Eddy Simulation (LES) and Reynolds averaging simulation (RAS).*

In LES the variables are split into two components: a grid scale component and a sub-grid scale component. The grid acts as a small-scale filter and the computed fields represent the large eddies. The influence of the smaller scales on the resolved scales is computed by a sub-grid scale model. The various approximations applied for the equations and the methods of solution are discussed in Chapter 10.

In the Reynolds averaging approach, each variable is split in its mean and turbulent component: $a = \langle a \rangle + a'$. Due to the non-linearity of the Navier-Stoke (NS) equations, the turbulent contributions to the mean momentum equations appear in the form of non-zero correlation terms, e.g. $\langle u_i u_j \rangle = \langle u_i \rangle \langle u_j \rangle + \langle u_i' u_j' \rangle$. These terms are computed as functions of the mean variables using turbulence closure models.

In *PBL models* NS equations are simplified by use of Boussinesq approximation (Chapter 10) through different formulations of the continuity equation: either the full compressible equation, the anelastic, the incompressible, or the hydrostatic approximations (Thunis, Bornstein 1996). Considering the scales of interest and the conditions in the urban PBL, the hydrostatic approximation is not applicable, but which of the other three approximations that is the most adequate is still not clear. However, it is the authors point of view that the approximation of incompressible air is too restrictive for the simulation of the local scale convective motions in the urban area. In principle the equation for fully compressible air is the most general, but in practise its discrete form tends to generate spurious pressure waves that must be removed by filtering techniques whose influence on small scale motions is yet not fully controlled.

Urban scale simulations are often performed with main emphasis on a certain part of the urban area, as e.g. the city centre or an area close to certain sources or target groups. For this reason zooming facilities are useful tools in urban scale models. Such facilities may be implemented by various grid refinement and grid nesting techniques (Chapter 10).

11.5.1 LOCAL SCALING MODELS

The non-zero turbulent correlation terms $\langle u_i' u_j' \rangle$ in the mean momentum equations raise the problem of turbulent equation closure. One approach to this problem is to derive the budget equations for the unknown second order correlations. The new equations contain higher order moments which are also unknown. The higher order closure methods parameterise the third and fourth order moments as functions of the mean variables and second moments. Mellor and Yamada (1982) have produced several levels of closure schemes for the PBL, which vary in complexity and in number of adjustable constants. Although powerful in principle, this approach is higher computer demanding due to the high number of equations that has to be solved in parallel (up to 11), and it is delicate to apply due to the high number of constants which need to be determined.

Another approach is the *first gradient closure* based on Boussinesq's turbulent viscosity assumption, which may be written in the form:

$$-\langle u'_i u'_j \rangle = \nu_t \left(\frac{\partial U_i}{x_j} + \frac{\partial U_j}{\partial x_i} \right) - \frac{2}{3} \delta_{ij} \langle u'_i u'_j \rangle \tag{11.21}$$

where $u_j = U_j + u'_j$. The turbulent viscosity ν_t has the dimension of the product of a mixing length scale by a velocity scale: $\nu_t = C_\mu$ L U. In the constant flux layer it has been shown by Monin and Obhukov (1954) that:

$$\nu_t = \kappa z u_* \tag{11.22}$$

Several attempts to extend this relation in the PBL by diagnostic model of L as a function of z have not been successful. For the urban PBL where the turbulent structure is constantly perturbed by the ground inhomogeneities and often out of equilibrium, it seems that models based solely on u_* and z should not be retained.

Local scaling models are based on local characteristics of the turbulent field. The most appropriate models of today are the *k-L and k-ε models*. Both model types use the turbulent kinetic energy (tke) for defining the turbulent mixing velocity scale: $U^2 = k = \langle u_1\rangle^2 + \langle u_2\rangle^2 + \langle u_3\rangle^2$, which yields:

$$\nu_t = C_\mu L k^{1/2} \tag{11.23}$$

where k is obtained from the budget equation:

$$\frac{\partial k}{\partial t} + \vec{V}\vec{\nabla}k = D_k + P_k + S_k - \varepsilon \tag{11.24}$$

where D_k, P_k, S_k are the diffusion, shear production, buoyancy production or dissipation of k, respectively, and ε the tke dissipation rate by viscosity.

For PBL simulations the most effective local mixing length model is that of Bougeault and Lacarrere (1989), based on the assumption that tke depends on thermal stability only. For urban simulations it may present weaknesses, especially in stably stratified conditions. In the k-ε models the mixing length is determined by considering Kolmogorov energy cascade from the large tke producing scales to the small energy consuming scales:

$$L = \frac{k^{2/3}}{\varepsilon}\ ;\ \nu_t = \frac{C_\mu k^2}{\varepsilon} \tag{11.25}$$

A budget equation similar to (11.24) is solved for ε:

$$\frac{\partial \varepsilon}{\partial t} + \vec{V}\vec{\nabla}\varepsilon = D_\varepsilon + P_\varepsilon + S_\varepsilon - \varepsilon_\varepsilon \tag{11.26}$$

with $D_\varepsilon = \alpha_e \vec{\nabla}\left(\nu_t \vec{\nabla}\varepsilon\right)$, $P_\varepsilon = C_{\varepsilon 1} P_k \frac{\varepsilon}{k}$, $B_\varepsilon = C_{\varepsilon 1} B_k \frac{\varepsilon}{k}$, $\varepsilon_\varepsilon = C_{\varepsilon 2}\varepsilon \frac{\varepsilon}{k}$.

Launder and Spalding (1974) original set of constants is known as the "standard" model. It was optimised for reference wind tunnel turbulent fields such as homogeneous grid turbulence and fully developed boundary layer over a flat plane. The standard model has several well identified weaknesses, as e.g. the strong overestimation of tke production in regions where the flow impinges on a wall. Several versions of the model have been proposed to remedy this weakness. Furthermore, the standard model does not take into account two characteristics of the lower atmosphere: some large turbulent scales do not effectively participate in the diffusion (while they contribute to k and ε); the turbulent structure is strongly dependent on stratification. This led Duynkerque (1988) to propose an atmospheric version of the k-ε model, with a C_μ value fitting the observed values of k, ε and ν_t in the neutral constant flux layer:

$$\varepsilon = \frac{u_*^3}{\kappa z}, \quad k \approx \frac{11 u_*^2}{2} \rightarrow C_\mu = \left(\frac{2}{11}\right)^2 \tag{11.27}$$

and an additional term in the ε equation, depending on thermal stratification.

In the lower urban roughness and canopy layers where the relations (11.22) and (11.27) are not observed, it is still unclear which version of the k-ε model which is the most effective, due to the conflicting flow structures in the surface and canopy layers, corresponding to conflicting C_μ values and conflicting buoyancy models (e.g. Delaunay et al. 1997). The experience from mesoscale modelling is of little help to model turbulence in the stably stratified urban PBL, since in these models it is usually just considered negligible. Yet, the observations of plume dispersion in (moderately) stable conditions show that turbulent diffusion is not zero and differs in the vertical and horizontal directions. An interesting approach extending the k-ε model consists of separating the vertical and horizontal diffusivities according to the anisotropic ratios, i.e.

$$\frac{\nu_{tv}}{\nu_{th}} = \left(\frac{\langle w'^2 \rangle}{\langle u_h'^2 \rangle}\right)^{1/2} \tag{11.28}$$

and to relate this ratio to the local thermal stability (thermal lapse rate or Richardson number) either through empirical relationships or with the help of $\langle w'^2 \rangle$ and $\langle \theta'^2 \rangle$ budget equations (Abart, Sini 1997). The development of such models is crucial for simulation of the pollution diffusion during periods where the PBL is stably stratified while the canopy layer is usually unstable due to the anthropogenic heating of the building walls and street surfaces.

No specific urban sub-grid scale model has yet been developed for LES at the urban scale. Specific *urban sub-grid scale models* should account for the presence of inhomogeneous terrain and the fact that turbulence is often out of equilibrium. The most promising model in this context seems to be the dynamic model of Zang et al. (1993), since its grid filtering is performed at each grid cell on the basis of the local turbulent structure; it especially accounts well for the non-homogeneous grids which allows refinements close to the ground. Other promising models in development include solving the sub-grid scale tke transport equation and sub-grid scale variance transport equations for the scalars like sensible heat and pollutant concentrations.

In parallel to the continuity, momentum and turbulence equations, the dynamic solvers are applied to solve the thermodynamic equations for sensible and latent heat, and budget equations for the water phases. These last equations depend on the selected micro-physics model (Chapter 10). These models have not been developed specifically for the urban scale. Specific urban micro-physics studies could require models accounting for chemistry-microphysics interaction in all phases, and non-cloud micro-physics such as fog and smog.

11.6 Simulation of urban pollutant transport-diffusion-transformation

Various types of models may be applied to pollutant dispersion (Chapter 10). The chemistry at the urban scale is still not fully explored (Chapter 8) and involve hundreds of constituents and reactions. Detailed process studies are still needed in order to define reduced but yet efficient reaction mechanisms. For this purpose detailed emission inventories (Chapter 6) on a resolution of hectometre are needed for the urban scale models.

Naturally no models are general and specifically for the urban conditions different model types should be applied for solving various problems. Some of the general constrains on the various model types are given here.

The *Gaussian plume dispersion models* are not well adapted to inhomogeneous terrain and ground base releases in the inhomogeneous canopy. Furthermore they cannot be applied for time varying simulations like Large Eddy Simulations (LES).

Lagrangian particle models are well adapted to describe dispersion in unsteady and strongly inhomogeneous turbulent fields, but in that case they require velocity moments of third order and higher, and especially for the canopy such data are not easily available.

Lagrangian puff or puff-particle models represent an alternative to particle models with most of the advantages, and disadvantages such as the requirement for higher order moments as input for describing inhomogeneous turbulence. While it is still unclear how well they can simulate dispersion within the canopy due to the multiple flow separations and recirculations, they can adequately be applied to the roughness and surface layers.

These three plume-type approaches have been developed for point source dispersion and cannot be used for second order chemistry without strong simplifications or loss of efficiency.

11.6.1 URBAN SCALE MODEL SYSTEMS

Urban scale models can take the form of integrated model systems serving a number of different purposes. These systems may include one set of solvers for the dynamics-turbulence and thermodynamics-microphysics equations and another set of solvers for the transport-chemistry equations. In the following some examples are given of the currently available European mesoscale models with zooming facilities for urban areas.

The *French communal model SUBMESO* (Mestayer 1996) contains a dynamic solver, originally developed for the mesoscale model ARPS (Droegemeier et al. 1995) and is based on the Boussinesq continuity equation for compressible air. The system is solved by finite difference method for an inhomogeneous terrain-induced Cartesian grid. The transport-chemistry module contains the reduced chemical mechanism from the tropospheric chemistry model MoCA (Aumont et al. 1996). The solver is a QSSA predictor-corrector scheme. The system includes two microphysic schemes, and advanced turbulence, sub-gridscale, urban roughness and force-restore soil models, specially created for meso-δ urban scales (Mestayer et al. 1995; Mestayer 1996). The developments by several French groups in co-operation include domain nesting, meteo pre-processing, urban chemistry, dry deposition and canopy thermodynamics.

The *EUMAC Zooming Model (EZM)* (Moussiopoulos 1994) was developed within the framework of the EUROTRAC sub-project EUMAC. The dynamics and transport-chemistry solvers are the regional mesoscale model MEMO and the chemistry-transport-diffusion solver MARS (Graf, Moussiopoulos 1991). The systems uses terrain-induced grids and the Sundqvist scheme for cloud and rain water processes. An adiabatic mass-consistent model is used as a meteorological pre-processor (or as alternative to MEMO for simple conditions) and the Lagrangian particle model is used for computation of dispersion of passive scalars. In a co-operation between 4 European teams, this system is being further developed. The developments include an on-line dynamic-chemistry solver using full 3-D or multi-layer approaches, plume dispersion, canopy processes and micro-physics chemistry interaction (Moussiopoulos 1994).

The *German MITRAS project* has its emphasis on the model nesting concept. A canopy layer, building resolving, model is being nested into the mesoscale model METRAS. Turbulence and chemical schemes are adapted to the canopy scales with focus on fast chemical reactions involving gaseous pollutants and soot aerosols (Schlüenzen et al. 1995).

In a co-operation between several groups (in Switzerland and the European Joint Research Centre) using the TVM-CHEM system, a non-hydrostatic mesoscale model is under development using the finite element discretization technique as an alternative to the vorticity mode model TVM (Schayes et al. 1996). This technique should allow high variations in the computational grid resolution in order to refine areas of interest like city centres. This system includes on-line chemical computations (Clappier et al. 1996). Several *mesoscale models* are being equipped with either specific urban chemistry schemes and/or small scale zooming facilities in order to study the governing processes of urban pollution episodes. The GRAMM model (Almbauer et al. 1995) with a two-way multiple nesting from regional to city scale is used with a simplified version of the tropospheric chemistry model RACM (Stockwell et al. 1990). A nested system has been constructed on the basis of the MEMO/MARS models and the meteorological model NUATMOS; a system which include point source and non-local turbulence sub-models and the RPM advanced photochemical model (Borrego 1998). Another example is the MESO NH-C (Borrel et al. 1996), a non-hydrostatic mesoscale model with multiple nesting, advanced dynamic and turbulence schemes and an urban aerosol module with emphasis on microphysics-radiation-chemistry interactions. Finally the French AZUR project is the assembly of the mesoscale model MERCURE (Riou 1987; Carissimo 1997), the chemical model MoCA, the emission inventory model MIEL (1 km and 1 h resolution) and the AIRQUAL chemistry-transport model (Jaecker-Voirol et al. 1997). This system has been applied for the Paris urban region over a domain of 120 km · 120 km with a resolution as fine as 2 km in the horizontal and 25 m in the vertical near the ground.

Besides a large number of specialised atmospheric dispersion models, there are air quality systems in operation on the urban scale in several European countries.

The Norwegian AirQUIS dispersion modelling system (Bøhler 1998) has been in operation with gradual development for almost 20 years. The system contains 30 different modules and the basic units are two Gaussian dispersion models for long term ground level concentrations over an area with point and area sources. It contains furthermore modules for handling of data for meteorology, road net and traffic, population, domestic, small and large industrial sources and for calculation of population exposures. At present this system is not handling chemical transformations.

A Finnish modelling system combines an urban dispersion modelling system and a dispersion model for pollution from road networks, *UDM-FMI* and *CAR-FMI*, respectively. This system includes emission modules for stationary and traffic sources. A meteorological pre-processor provides hourly time series of atmospheric stability parameters based on synoptic and local meteorological data. The UDM and CAR models are a multiple source Gaussian plume model and a Gaussian finite-line source model, respectively. Each of the models have their own emission and meteorological pre-processing modules. CAR-FMI computes the chemical transformations with a discrete parcel method and includes the chemistry of the NO-NO_2-O_3 system (Karpinen et al. 1997).

The English urban air quality management system *ADMS Urban* (McHugh et al. 1997) is based on a GIS with extensive interactive interface and containing a large number of different modules. It computes dispersion from point, line, area, volume and road sources with an adapted pre-processing of meteorological data based on semi-empirical relationships, and airflow calculations over complex terrain with the FLOWSTAR multi-layer model (Carruthers et al. 1997). It handles dispersion in street canyons and building influence and include a chemical transformation module for the $NO-NO_2-O_3$ system.

These operational systems may be handled on PCs and are generally more modelling toolboxes than real integrated models. Although their individual modules have generally been validated, their overall validation is usually limited and the results may therefore be rather crude. As for most other model simulations, the results from these models depend highly on the user skills. Provided these warnings, these models have the advantage of being operational.

11.7 References

Abart, B., Sini, J.-F. (1997) New first-order closure models for stably stratified flows, *11th Symp. On Turbulent Shear Flows*, 8-11 Sep., Grenoble, France.

Almbauer, R., Pucher, K., Sturm, P.J. (1995) Air quality modeling for the city of graz, *Meteorology and Atmospheric Physics*, **57**, 31-42.

Aumont, B., Jaecker-Voirol, A., Martin, B., Toupance, G. (1996) Tests of some reduction hypothesis in photochemical mechanisms: application to air quality modeling in the Paris area, *Atmospheric Environment*, **30**, 2061-2077.

Borrego, C., Lemos, S., Carvalho, A.C., Coutinho, M. (1998) A modelling system for air quality management, *5th Int. Conf. on Harmonisation within atmospheric dispersion modelling for regulatory purposes*, 18-21 May 1998, Rhodes, Greece, Proc. 641-648.

Borrel, L., Perrier, M., Rosset, R., Joumard, R. (1996) Modelisation de la pollution chronique a l'echelle d'une agglomeration. Tendances Nouvelles en modelisations pour l'environnement, *Actes des journees*, **15-17**, Janvier 1996, CNRS, Paris.

Bottema, M. (1995) Parameterization of aerodynamic roughness parameters in relation with air pollutant removal efficiency of streets, *Air Pollution III*, Computational Mechanics Publication **2**, 235-242.

Bottema, M. (1997) Urban roughness modelling in relation to pollution dispersion, *Atmospheric Environment*, **31**, 3059-3075.

Bottema, M., Leene, J.A., Wisse, J.A. (1992) Towards forecasting of wind comfort, *J. Wind Engin. Ind. Aerodyn.*, **41-44**, 2365-2376.

Bottema, M., Mestayer, P.G. (1997) Urban roughness mapping - validation techniques and some first results, *2nd European and African conference on wind engineering*, Genova, Italy, June.

Bougeault, P., Lacarrere, P. (1989) Parameterization of orographic-induced turbulence in a mesobeta-scale model, *Monthly Weather Review*, **117**, 1872-1890.

Bøhler, T. (1998) AirQUIS - a modern air quality management system, *5th Int. Conf. on Harmonisation within atmospheric dispersion modelling for regulatory purposes*, 18-21 May 1998, Rhodes, Greece, Proc. 655-659.

Carissimo, B. (1997) Numerical simulation of meteorological conditions for peak pollution in Paris, *22nd NATO/CCMS Int. Techn. Meeting on Air Poll. Modelling and its Appl.*, June 2-6, Clermont-Ferrand, France.

Carruthers et al. (1997) ADMS-Urban - an integrated air quality modelling system for local government, *Air Pollution V*, Computational Mechanics Publications.

Clappier, A.P., Perrochet, P., Martilli, A., Müller, F., Krüger, B.C. (1996) A new non-hydrostatic mesoscale model using a CVFE (Control Volume Finite Element) discretisation technique, in: Borell, P.M., Borell,

P., Cvitas, T., Kelly, K., Seiler, W. (editors), Proc. *EUROTRAC Symposium 96*, Computational Mechanics Public., Southhampton, 527-531.

Costes, J.P. (1996) *Simulations númeriques des ecoulements atmospheriques sur sols fortement hétérogénes*, Doctoral thesis, University of Nantes & Ecole Centrale de Nantes.

Counihan, J. (1971) Wind tunnel determination of the roughness length as a function of the fetch and the roughness density of three-dimensional roughness elements, *Atmospheric Environment*, **5**, 637-642.

Deardorff, J.W. (1978) Efficient prediction of ground surface temperature and moisture, with inclusion of a layer of vegetation, *J. Geophys. Res.*, **83**, 1889-1903.

Delaunay, D., Flori, J.P., Sacré, C. (1996) Numerical modelling of gas dispersion from road tunnels in urban environments: comparison with field experimental data, *4th Workshop on Harmonisation within Atmospheric Dispersion Modelling for Regulatory purposes*, Oostende, Belgium, 361-368.

Droegemeier, K.K., Xue, M., Johnson, K., O'Keefe, M., Sawdey, A., Sabot, G., Wholey, S., Lin, N.T., Mills, K. (1995) *Wheather prediction: A scalable storm-scale model.* Chapter 3, Sabot, G. (editor), Addison-Wesley, Reading, Massachussets.

Duynkerque, P.G. (1988) An application of the K-ε model to the neutral and stable atmospheric Boundary layer, J. Atm. Sci., urban roughness parameters: merphometric 45, 865-880. *Analysis 12th symposium on boundary layer and turbulence*, 28 july - 1 August 1997, Vancouver, BC., American Met, soc., 457-458.

Graf, J., Moussiopoulos, N. (1991) Intercomparison of two models for the dispersion of chemically reacting pollutants, *Contr. Phys. Atmos.*, **64**, 13-25.

Grimmond, C.S.B., Oke, T.R. (1997) An intercomparison of methods to determine roughness and displacement heights in urban aread.

Gryning, S.E., Batchvarova, E. (1997) A model for the height of the internal boundary layer over an area with an irregular coastline, *Boundary Layer Meteorology*, **78**, 405-413.

Guilbaud, C. (1996) *Etude des inversions thermiques: application aux écoulements atmosphériques dans des vallées encaissées*, Doctoral thesis, Univ. J. Fourier, Grenoble, France.

Jaecker-Voirol, A., Lipphardt, M., Martinm, B., Quandalle, Ph., Salles, J., Carissimo, B., Dupont, E., Musson-Genon, L., Riboud, P.M., Aumont, B., Bergametti, G., Bey, I., Toupance, G. (1997) A 3D regional scale photochemical air quality model - application to a 3 day summertime episode over Paris, *4th Int. Sci. Symp. Transport and Air Pollution*, 9-13 June, Avignon, France.

Karpinen, A., Kukkonen, J., Konttinen, M., Härkönen, J., Valkonen, E., Koskentalo, T., Elolähde, T. (1997) Development and verification of a modelling system for predicting urban NO_2 concentrations, *22nd NATO/CCMS Int. Techn. Meeting on Air Poll. Modelling and its Appl.*, Jun 2-6, Clermond-Ferrand, France.

Launder, S.P., Spalding, D.B. (1974) The numerical computation of turbulent flow, *Comp. Methods in Applied Mech. and Eng.*, **3**, 269-289.

Lettau, H. (1969) Note on aerodynamic roughness-parameter estimation on the basis of roughness-element description, *J. Applied Met.*, **8**, 828-832.

McHugh, C.A., Carruthers, D.J., Edmunds, H.A. (1997) ADMS-Urban: a model of traffic, domestic and industrial pollution. International Journal of Environment and Pollution, **8, 3-6**, 666-674.

Mellor, G.L, Yamada, T. (1982) Development of a turbulent model for geophysical fluid problem, *Review of Geophysics and Space Physics*, **20**, 851-875.

Mestayer, P.G. (1996) Development of the French communal model SUBMESO for simulating dynamics, physics and photochemistry of the urban atmosphere, in: Borrel, P.M., Borrell, P., Cvitas, T., Kelly, K., Seiler, W. (editors), *Proc. EUROTRAC Symposium '96*, Computational Mechanics Public., Southampton, 539-544.

Mestayer, P.G., Anquetin, S. (1994) *Climatology of cities, Diffusion and Transport of Pollutants in Atmospheric Mesoscale flow Fields*, Gyr, A., Rys, F.S. (editors), *ERCOFTAC Series*, Kluwer Academic Publishers, 165-189.

Mestayer, P.G., Chollet, J.-P., Coppalle, A., George, J., Chaumerliac, N., Ayrault, M., Toupance, G., Martin, B., Sacré, C., Carissimo, B., Courty, J.-C. (1995) Modélisation de la Dynamique, la Physique et la Photo-chimie de l'atmosphère urbaine - Le projet SUB-MESO, in: Elichegaray, C., Fontan, J.,

Laterasse, J., Muller, M. (editors), *Pollution atmosphérique à l'échelle locale et régionale*, Min. Environnement, 89-98.

Monin, A.S., Obukhov, A.M. (1954) Basic laws of turbulent mixing in the ground layer of the atmosphere, *Trans. Geophys. Inst. Akad.*, NAUK USSR, **151**, 163-187.

Moussiopoulos, N. (1994) The EUMAC Zooming Model (EZM): An introduction. *The EUMAC Zooming Model: structure and applications*, Moussiopoulos, N. (editor), EUROTRAC, Int. Sci. Secr., Garmisch-Partenkirchen, March 1994, 7-21.

Oke, T.R. (1982) The energetic basis of the urban heat island, *Quart. J.R. Met. Soc.*, **108**, 1-24.

Oke, T.R. (1987) *Boundary layer climates*. Methuen and Co., Ltd, 2nd edition, New York, USA.

Oke, T.R., Johnson, G.T., Steyn, D.G., Watson, I.D. (1991) Simulation of surface urban heat islands under "ideal" conditions at night, Part 2: Diagnosis of causation, *Boundary-Layer Meteorology*, **56**, 339-358.

Paeschke, W. (1938) Experimentelle untersuchungen zum rauhigkeits- und stabilitetsproblem in der bodennahen luftschicht, *Beitr. Phys. Fur Atmos.*, **24**, 163-189.

Raupach, M.R. (1994) Simplified expressions for vegetation roughness length and zero-plane displacement height as function of canopy height and area index, *Boundary layer Meteorology*, **71**, 211-216.

Riou, Y. (1987) Comparison between MERCURE-GL code calculations, wind tunnel measurements and Thorney Island field Trials, *J. Hazard. Mat.*, **16**, 247-265.

Robin, M. (1993) Apport de L'imagerie satellitaire basse résolution NOAA-AVHRR pour l'étude des ilots de chaleur urbains: exemple des villes de Loire Atlantique, *Cahier Nantais*, **40**, 47-60 (Publi. IGARUN, Nantes).

Robin, M. (1995) *La télédétection*, Editions Nathan, Paris

Rotach, M.W. (1993) Turbulence transfer relationships over an urban surface. I: Spectral characteristics, *Q. J. Roy, Meteorol. Soc.*, **119**, 1071-1104.

Schayes, G., Thunis, P., Bornstein, R. (1996) Topographic vorticity-mode mesoscale-β (TVM) model. Part I: Formulation. *J. Appl. Meteor.*, **35**, 1815-1823.

Schlüenzen, H., Bigalke, K., Niemeier, U., Von Salzen, K. (1995) The mesoscale transport- and fluid-model METRAS - model concept, model realization, *METRAS Tech. Rep. No 1*, Meteorologische Institute, Universität Hamburg, 173 p.

Soliman, B.F. (1976) *A study of wind pressure forces acting on groups of buildings*, PhD thesis, University of Sheffield, UK.

Stockwell, W.R., Middleton, P., Chang, J.S., Tang, X. (1990) *J. Geophys. Res.*, **95**, 16343-16367.

Stull, R.B. (1988) *An introduction to boundary layer meteorology*, Kluwer Academic Publishers.

Thielen, J., Wobrock, W., Mestayer, P.G., Creutin, J.-D. (1996) *The influence of surface parameters on rainfall development in meso-γ scale models: a sensitivity study*. 12th Int. Conference On Clouds and Precipitation, 19-23 Aug., Zurich, Switzerland.

Thom, A.S. (1971) Momentum absorption by vegetation, *Q. J. Roy. Meteorol. Soc.*, **97**, 414-428.

Thunis, P., Bornstein, R. (1996) Hierachy of mesoscale flow assumptions and equations, *J. Atmospheric Sciences*, **53**, 380-397.

Todhunter, P.E., Terjung, W.H. (1988) Intercomparison of three urban climate models, *Boundary-Layer Meteorology*, **42**, 181-205.

Zang, Y., Street, R.L., Kosef, J.R. (1993) A dynamic mixed subgrid scale model and its application to turbulent recirculation flows, *Physics of Fluids*, **A5**, 3186-3196.

[illegible] The [illegible] scheme [illegible]

[illegible]

[illegible]

Zhang, [illegible] (19[illegible]) [illegible] between [illegible] scale [illegible] and [illegible] Theoretical and Applied Climatology [illegible]

[illegible]

[illegible]

[illegible]

[illegible]

Chapter 12

STREET SCALE MODELS

RUWIM BERKOWICZ
National Environmental Research Institute
P.O.Box 358, DK-4000 Roskilde, Denmark

12.1 Introduction

Urban areas cannot be considered as homogeneous entities; the largest pollution levels occur in street canyons, where the dilution of car exhaust gases is limited by the presence of buildings flanking the street. Moreover, we are dealing here with pollution phenomena that take place in the immediate vicinity of the source, the car traffic. Therefore modelling of this pollution must account for micro-scale processes, i.e. a scale that is much smaller than the meso-scale discussed in Chapter 10.

The main features of wind flow and pollution dispersion in street canyons are discussed in this chapter. The mathematical modelling methods that are usually used for these processes are presented. Special consideration is given to applied methods, which can be used for routine evaluation and analyses of air pollution from traffic in streets. Several models, currently used in Europe are briefly presented, and a more detailed discussion is given for the Danish model, the Operational Street Pollution Model (OSPM). Principles of physical modelling, (i.e. modelling in wind tunnels) are discussed in Chapter 14.

Available measurements, either from field experiments or wind tunnel studies, are used to elucidate the street pollution processes. A rather comprehensive dataset of pollution and meteorological measurements in Copenhagen is used here. The urban structure of Copenhagen is typical for many medium-large European cities.

12.2 Wind flow in street canyons

The most characteristic feature of the street canyon wind flow is the formation of a vortex, where the direction of the wind at street level is opposite to the flow above the roofs. Although this phenomena has been known for long (Albrecht 1933) only few field measurements are available. Most of our knowledge about the structure of wind flow in street canyons comes from wind tunnel studies (Chapter 14).

12.2.1 CLASSIFICATION OF FLOW REGIMES

Using the available measurements, Oke (1988) provided a systematic classification of flow regimes in urban street canyons. The flow types are characterised by three regimes depending on the canyon geometry: isolated roughness flow, wake interference flow, and skimming flow. The canyon geometry is defined mainly by the ratio H/W, where H is the average height of the canyon walls, and W is the canyon width. The three flow regimes are illustrated in Figure 12.1. For widely spaced buildings (H/W < 0.3) the flow fields associated with the individual buildings do not interact, which results in the isolated roughness flow regime. At closer spacing (0.3 < H/W < 0.7) the wake created by an upwind building is disturbed by a downwind building, creating a downward flow along the windward face of this building. This is the wake interference flow. Even closer spacing results in the skimming flow regime. In this case a stable circulatory vortex is established in the canyon, and the ambient flow is de-coupled from the street flow. This classification of flow regimes should be considered only as qualitative. Caution must be taken, when wind tunnel results are extrapolated to ambient conditions.

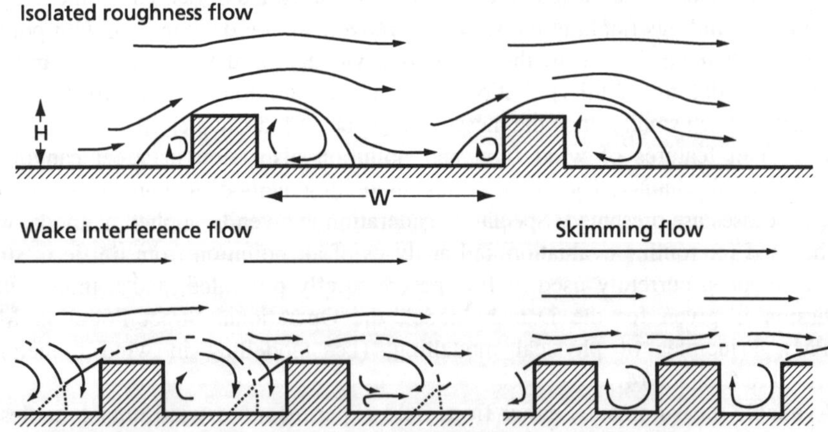

Figure 12.1 Flow regimes associated with different canyon H/W ratios (Oke 1988).

12.2.2 MEASUREMENTS AND MODELS

In field experiments, only few point measurements of wind are usually available and they may even be significantly influenced by very local structures. Therefore it is difficult to use such measurements for determination of a full three dimensional structure of the wind pattern. Instead more elaborated flow visualisation techniques are applied. An example of such an experiment is the work of DePaul and Sheih (1986). The mean wind velocities in a street canyon were determined by analysis of trajectories of tracer balloons that were released in the canyon and photographed in rapid sequence. The balloon trajectories showed the formation of a vortex cell within the canyon, provided the ambient wind velocity exceeds 1.5-2.0 ms^{-1}. An important feature of the flow pattern, demonstrated in this experiment, is that the vertical extent of the cell does not seem to extend beyond the roof level. Velocity vectors at roof level appeared to be nearly parallel to the ambient wind.

Equations describing the mean flow and turbulence in street canyons are basically the same as used in meso-scale models (Chapter 10). The main difference is that the presence of the buildings along the street imposes special boundary conditions on the vertical building walls. Another significant difference is the scale of the modelling domain. In order to resolve the structure of the complex wind flow in a street canyon, the computational mesh must be very fine, in the order of few meters only. This results in a tremendous increase of the computational time, and presently numerical models dealing with street canyon flows are therefore mostly used for research purposes or very specialised studies.

12.2.3 *k*-ε MODELS

The method, which has found the most wide application in numerical modelling of flow conditions in street canyons and also in the case of other obstacles, is the so called *k*-ε method, discussed previously in Chapter 11. The *k*-ε model has been extensively tested and calibrated for industrial flows around bluff bodies and structures. The tests have focused on regions where flow separation can occur, which is also an important property of the flow in and around street canyons. The constants in the models have been determined from experimental data, considering basic universal flows, and afterwards improved by experimental modelling. Several comparisons of model results with data from mainly wind tunnel experiments have proven the usefulness of the *k*-ε method for street canyon modelling (e.g. Johnson, Hunter 1995; Mestayer, Anquetin 1994).

12.2.4 HOTCHKISS AND HARLOW MODEL

A simplified description of wind flow in a street canyon was proposed by Hotchkiss and Harlow (1973). They use the equation of vorticity to describe a two dimensional flow occurring, when wind blows perpendicularly to an infinitely long canyon of width W and height H. From the definition of vorticity:

$$\omega = \frac{\partial u}{\partial z} - \frac{\partial w}{\partial x} \tag{12.1}$$

and the momentum conservation equations (Chapter 7), the vorticity conservation equation is derived:

$$-u\frac{\partial\omega}{\partial x}-w\frac{\partial\omega}{\partial z}+\nu_t\left(\frac{\partial^2\omega}{\partial x^2}+\frac{\partial^2\omega}{\partial z^2}\right)=0 \tag{12.2}$$

The cross canyon (x-direction) velocity component is denoted by u, the vertical (z-direction) component by w, and ν_t is the turbulent viscosity. For vortex-like flows in street canyons it is reasonable to assume that advection of vorticity is small and in this case (12.2) is reduced to:

$$\frac{\partial^2\omega}{\partial x^2}+\frac{\partial^2\omega}{\partial z^2}=0 \tag{12.3}$$

One of the solutions to this equation, satisfying the boundary conditions of vanishing vorticity on the canyon vertical walls, is:

$$\omega=\omega_o\left(e^{ky}+\beta e^{-ky}\right)\sin(kx) \tag{12.4}$$

The expressions for the street canyon velocity components satisfying free-slip boundary conditions for along-walls and canyon bottom components and vanishing normal components, reads:

$$u=\frac{A}{k}\left[e^{ky}(1+ky)-\beta e^{-ky}(1-ky)\right]\sin(kx) \tag{12.5}$$

$$w=-Ay\left[e^{ky}-\beta e^{-ky}\right]\cos(kx) \tag{12.6}$$

where: $k=\frac{\pi}{W}$, $A=\frac{ku_o}{1-\beta}$, $y=z-H$, $\beta=e^{-2kH}$, and u_o is the wind speed above the canyon (the point x = W/2, z = H).

The wind field calculated by (12.5) and (12.6) is shown in Figure 12.2 for a canyon with the ratio H/W = 1. Wind velocities are normalised with respect to u_o.

The Hotchkiss and Harlow model, although being very simplified, reflects the basic properties of wind circulation in street canyons. Comparison with wind measurements in a street with H/W ratio close to one, presented by Yamartino and Wiegand (1986), shows some reasonable agreement. Caution must be taken when the Hotchkiss and Harlow model is used for other canyon configurations. Referring to the aforementioned work by Oke (1988), the wind flow model by Hotchkiss and Harlow describes only the skimming flow regime and is not suitable for canyons with H/W ratio significantly different from one.

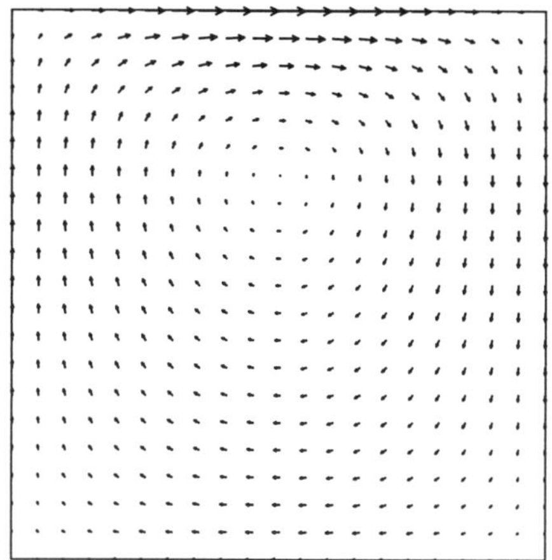

Figure 12.2 Wind flow in a street canyon with H/W = 1 according to the analytical model by Hotchkiss and Harlow (1973).

12.3 Dispersion modelling

Modelling dispersion of pollutants in streets is inevitably connected with wind flow modelling. The mathematical principles are basically the same, i.e. the governing equation is the steady state mass conservation equation for a scalar:

$$u_j \frac{\partial c}{\partial x_j} = -\frac{\partial}{\partial x_j} \overline{c' u'_j} + S \tag{12.7}$$

where c denotes the mean concentration, and c' is the deviation from the mean value. S represents all source and sink terms, e.g. emissions and chemical reactions. The index j denotes here the x, y and z direction component, respectively. The usual convention of summation over repeating indexes is implied.

The key problem is again the determination of the parameterization of the turbulent flux term $\overline{c' u'_j}$ and the most common approach is based on the eddy diffusivity concept:

$$\overline{c' u'_j} = -K_t \frac{\partial c}{\partial x_j} \tag{12.8}$$

where K_t is the eddy diffusivity coefficient, usually assumed to be equal to the eddy viscosity ν_t.

The mean wind fields and diffusivity coefficients can be supplied by a particular flow model and Equation (12.7) solved numerically subject to appropriate boundary conditions. Modern numerical methods and availability of powerful computers have resulted in several such models which have been developed. The diffusivity coefficients are estimated either making use of the mixing length concept (Moriguchi, Uehara 1993; Lee, Park 1994; Kamenetsky, Vieru 1995) or the k-ε, method (Johnson, Hunter 1995; Mestayer, Anquetin 1994).

Another approach is based on the *stochastic Lagrangian trajectory model* (Lamb et al. 1979; Lanzani, Tamponi 1995). Concentrations of pollutants are calculated tracing the movement of particles representing air parcels. The trajectories are calculated using the mean flow fields on which a random fluctuation component is superimposed. The statistics of the fluctuating component depend on the turbulence characteristics of the flow field and can be supplied by the flow model. The stochastic Lagrangian approach permits one to avoid using the diffusivity "closure" (12.8) for modelling dispersion. The eddy diffusivity approach is known to be applicable only when the scale of the turbulent motions is much smaller than the scale of pollution distribution. This might not be the case in street canyons where both scales are of a comparable size and limited by the canyon dimensions.

Numerical models based on solution of the diffusivity equation (12.7) or *stochastic models with the corresponding wind flow models* are still too complex for practical applications, e.g. in support of air pollution management. As research tools they can, however, provide significant insight into the essential processes and results used for constructing more simple models. The models in question here are basically parameterized semi-empirical models making use of a priori assumptions about the flow and dispersion conditions and limitations of such an approach must certainly be recognised. Anyhow, it is this kind of modelling that until now has found the broadest application. Some of the more commonly used models will be discussed in the following.

12.4 Applied street pollution models

Basically all the applied street pollution models make use of the schematic picture of flow and dispersion depicted in Figure 12.3.

The contribution from the emissions from the local street traffic (street contribution c_s) is added to the pollution present in the air that enters from roof level (background contribution c_b).

$$c = c_b + c_s \tag{12.9}$$

The pollution plume from the traffic in the street is advected by the wind vortex towards the leeward side, while on the windward side, the impact is only from the air that has recirculated in the street. Description of the two regimes: the direct advection and recirculation, are the main building elements of many street pollution models.

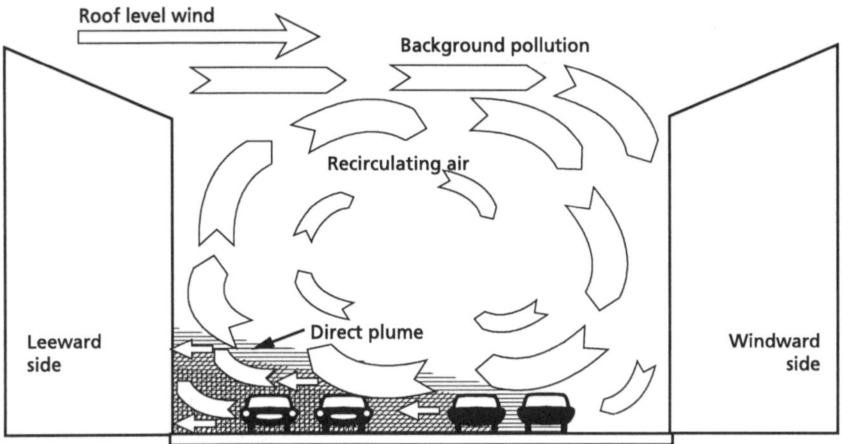

Figure 12.3 Schematic illustration of flow and dispersion in a street canyon.

12.4.1 THE STREET MODEL

One of the earliest street pollution models is the STREET model by Johnson et al. (1973). The model is empirically derived from analysis of measurements in streets of San Jose and St. Louis.

The street contribution is assumed to be proportional to the local street emissions Q ($gm^{-1}s^{-1}$) and inversely proportional to the roof-level wind speed u. For winds blowing at an angle of more than 30° to the street direction, two formulas are derived:

one for the leeward side:

$$c_S = \frac{K}{u + u_o} \sum_i \frac{Q_i}{[(x_i^2 + z^2)^{1/2} + h_o]} \tag{12.10}$$

and one for the windward side:

$$c_S = \frac{K}{(u + u_o)} \frac{H - z}{H} \sum_i \frac{Q_i}{W} \tag{12.11}$$

where:
K is an empirically determined constant ($K \approx 7$),
u_o accounts for the mechanically induced air movement caused by traffic ($u_o \approx 0.5 ms^{-1}$),
h_o accounts for initial mixing of pollutants ($h_o \approx 2$ m),
x_i and z are the horizontal and the vertical distances from the i-th traffic lane to the receptor point,
Q_i is the emission strength of the i-th traffic lane,
H and W are the height and the width of the canyon, respectively.

For wind directions at angles less than 30° to the street direction, the average of (12.10) and (12.11) is recommended, but actually, the model is not designed for this condition.

The model predicts that the concentrations decrease with increasing wind speed and that pollution levels in wide streets are lower than in narrow streets (with the same amount of traffic). Furthermore, concentrations on the leeward side of the street are higher than on the windward side. These are the most essential features of pollutant dispersion in street canyons and therefore the STREET model, with some minor modifications is still widely used, especially for engineering applications. The more detailed features of pollution dispersion in street canyons can, however, not be described by such a simplified model as STREET. An essential drawback of the model is the very crude parameterization of wind direction dependence. Furthermore, at reduced ambient wind speeds (calm conditions), a uniform concentration distribution is expected across the street-canyon. The STREET-model does not describe this feature and actually it is not recommended for ambient wind speeds less than 1 ms^{-1} (although it is frequently used for evaluation of "worst-case" pollution levels, which usually are associated with low wind speed conditions). In spite of these limitations, the model is a useful tool for a "first-order" evaluation of air quality in street-canyons.

Comparison of results from the STREET model with wind tunnel data of Hoydysh and Dabberdt (1988) is shown in Figures 12.4 and 12.5.

Due to a somewhat unclear definition of the source strength in Hoydysh and Dabberdt (1988), not much attention should be given to the absolute concentration values, but it is noticeable that the STREET model does not reproduce the large difference between the leeward and windward concentrations shown by the wind tunnel data. On the contrary, the model predicts much stronger vertical gradients than observed from the wind tunnel measurements. Measurements show very little variation of concentrations with height.

12.4.2 THE "CPBM" MODEL

An innovative approach was introduced by Yamartino and Wiegand (1986) in their Canyon Plume-Box Model (CPBM). The concentrations are calculated combining a plume model for the direct impact of vehicle emitted pollutants with a box model that enables computation of the additional impact due to pollutants recirculated within the street by the vortex flow. The wind flow in the canyon is defined using the model by Hotchkiss and Harlow (1973) for the transverse components, u and w, while a simple logarithmic profile is anticipated for the longitudinal component, v.

An empirical turbulence model is used for the turbulence parameters σ_u, σ_v and σ_w, representing the standard deviations of flow velocities about the mean flow. Variables used to estimate the turbulence include the mechanical component driven by the roof top wind and the thermal component, dependent on the global solar radiation and the number of vehicles. The last accounts for the heat released by the vehicles in the street.

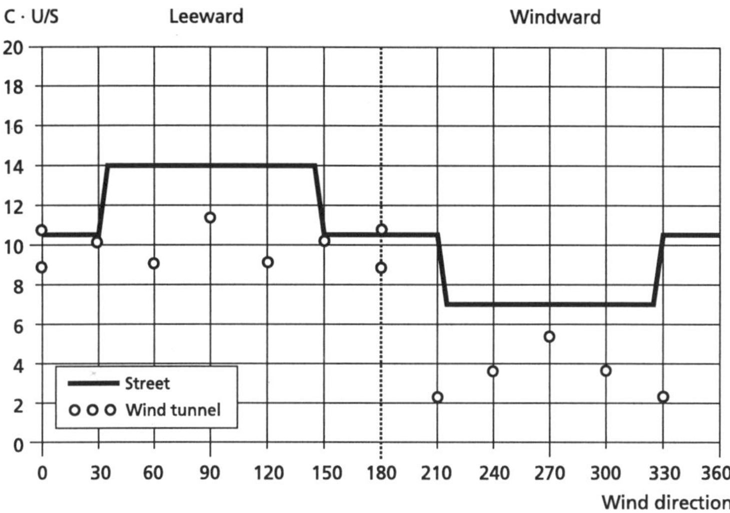

Figure 12.4 Modelled and measured street level concentrations versus wind direction. 0° and 180° correspond to wind parallel to the street axis. Experimental data are from the wind tunnel measurements of Hoydysh and Dabberdt (1998). Concentrations are normalised with wind speed (U) and emission density (S).

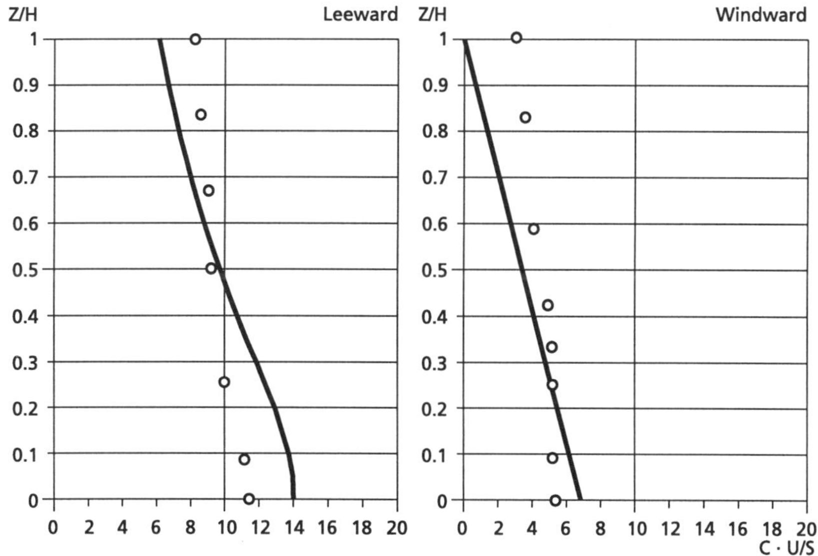

Figure 12.5 Modelled and measured vertical concentration profiles. Wind direction perpendicular to the street. Symbols as in Figure 12.4.

The plume is divided into three segments, which are assumed to follow straight line trajectories and disperse according to Gaussian plume formulas. The vertical dispersion parameter, σ_z, is given by:

$$\sigma_z(t) = \frac{H_l}{\sqrt{2\pi}} + \sigma_w \cdot t \tag{12.12}$$

where H_l is the initial plume dispersion, dependent on the size and the speed of vehicles. σ_w is calculated at the effective source height, being equal to the half of the vehicle height. The transport time, t, is equal to x/u_b, where x is the distance from the source (a vehicle line) to the receptor point and u_b is the cross canyon velocity calculated from the Hotchkiss and Harlow model. "Across canyon" average values, calculated at the effective source height, are used for u_b.

The contribution to concentrations resulting from the *recirculation component* is calculated from the considerations of the mass budget within the canyon, and is given by:

$$C_R = \frac{Q}{u_b\,(W/2)}\frac{F}{(1-F)} \tag{12.13}$$

where Q is the emission rate per street length, W is the width of the canyon and F is the fraction of material that is recirculated. This fraction is calculated using considerations about the vortex recirculation time and the canyon ventilation speed.

For all receptors on the lee side of the canyon, the recirculation concentration C_R is added to concentrations calculated by the direct plume model. For the wind side, where the only contribution arises from the recirculating air, the dilution of the concentrations due to entrainment of the clean air is calculated assuming a linear growth of the size of the jet, with the growth rate proportional to σ_u, the standard deviation of the cross canyon velocity.

The plume-recirculation model is used only for the case when the vortex advection dominates over diffusion. In the opposite case, that usually correspond to situations with the wind direction close to the street axis or in the case of very light winds, a simpler, non-vortex dispersion model is used. Concentrations are then computed by assuming a plume diluted with velocity v (along-street component) and travelling parallel to the canyon axis. The Gaussian plume dispersion parameters are again calculated assuming a linear growth proportional to the turbulence parameters and initial plume size depending on the size and speed of vehicles.

The CPBM model was tested on data from the Bonner Strasse experiment in Cologne, Germany (Yamartino, Wiegand 1986). A subset of the data was used for optimisation of the model. The performance of CPBM was shown to be significantly better than the empirical STREET model, especially considering the broad range of meteorological conditions for which the STREET model was not specifically designed.

12.4.3 THE "CAR" MODEL

An empirical approach was used in development of the Dutch traffic pollution model CAR (Calculation of Air pollution from Road traffic) (Eerens et al. 1993). Based mainly on wind tunnel experiments, a set of empirical relationships was established between wind direction and concentrations for various street configurations. The wind tunnel experiments covered 49 configuration sets which differed with respect to dimensions, distances and shapes of streets and its buildings. Also the influence of trees along streets was investigated.

The results were incorporated in a plume type model, called TNO traffic model, which was the basis for finally development of the more operational CAR model in which few most distinguish street configurations with respect to dispersion conditions were categorised. For each street category a source-receptor relationship is specified as a function of the distance between the receptor point and street axis. Only annual average concentrations are calculated and other statistical means are estimated based on empirical relationships are derived from measurements in the national pollution network. The model is now applied as a regulatory model in Dutch cities, and an international version, CAR International is also available.

12.4.4 THE "OSPM" MODEL

The Danish Operational Street Pollution Model (OSPM) is based on similar principles as the CPBM model by Yamartino and Wiegand (1986). Concentrations of exhaust gases are calculated using a combination of a plume model for the direct contribution and a box model for the recirculating part of the pollutants in the street (Figure 12.3, page 229). OSPM makes use of a simplified parameterization of flow and dispersion conditions in a street canyon. This parameterization was deduced from extensive analysis of experimental data and model tests (Hertel, Berkowicz 1989a; Berkowicz et al. 1997). Results of these tests were used to improve the model performance, especially with regard to different street configurations and a variety of meteorological conditions. The model is described in more details in the following section.

12.5 The Operational Street Pollution Model (OSPM)

The formulas for concentration of pollutants in the street are derived with the constrain that a continuous transition must be obtained between the different flow regimes. This concerns especially the transition from the recirculating regime at higher wind speeds to the non-recirculating regime at lower wind speeds. The same applies for the case of vanishing vortex when the wind direction changes from perpendicular to parallel with the street. Expressions used for the direct contribution and the recirculation contribution are constructed to fulfil these requirements.

12.5.1 THE DIRECT CONTRIBUTION

The direct contribution is calculated using a simple plume model. It is assumed that both the traffic emissions are uniformly distributed across the canyon. The emission field is treated as a number of infinitesimal line sources aligned perpendicular to the wind direction at the street level. The cross wind diffusion is disregarded and the sources are treated as infinite line sources. The direct contribution is thus calculated as:

$$C_d = \sqrt{\frac{2}{\pi}} \frac{Q}{u_b} \int_0^{L_{max}} \frac{dx}{\sigma_z(x)} \tag{12.14}$$

where Q is the emission in the street (g m^{-1} s^{-1}), W is the width of the street canyon, u_b is the wind speed at the street level and $\sigma_z(x)$ is the vertical dispersion parameter at a downwind distance x.

The integration in Equation (12.14) is performed along the wind path at the street level. The integration path depends on wind direction, extension of the recirculation zone and the street length.

The main principles for estimation of the integration path for Equation (12.14) are illustrated in Figure 12.6. If the roof level wind direction is at angle Φ with respect to the street axis, then the street level wind in the recirculation zone forms also an angle Φ with the street axis, but the transverse component is mirror reflected. Outside the recirculation zone the wind direction is the same as at roof level.

The length of the vortex, L_{vortex}, is assumed to be twice the height of the upwind building, H_{upwind}. For wind speeds (roof level) less than 2 m/s, a linear decrease of L_{vortex} with wind speed is assumed. This is consistent with the observations of e.g. DePaul and Sheih (1986) indicating disappearance of vortex circulation at low wind speeds.

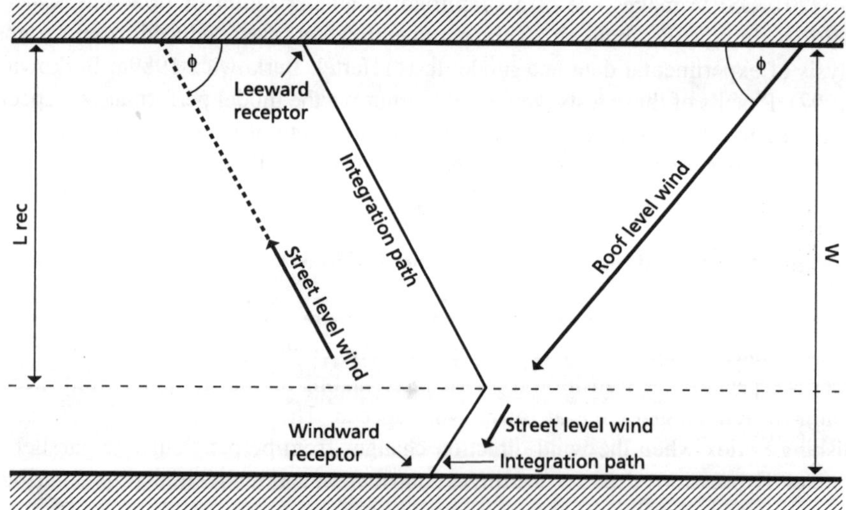

Figure 12.6 Illustration of the wind flow and formation of the recirculation zone in a street canyon. See the text for explanation of the symbols.

H_{upwind} depends on canyon geometry and wind direction at roof level. Openings between buildings will result in H_{upwind} = 0, depressing thereby formation of the recirculating vortex.

The maximum extension of the recirculation zone is given by the width of the street W, or by the length of the vortex L_{vortex}:

$$L_{rec} = \min(W, L_{vortex} \cdot \sin\Phi) \tag{12.15}$$

The vertical dispersion parameter, σ_z, is modelled assuming that the dispersion of the plume is solely governed by the mechanical turbulence. The turbulence due to thermal stratification is disregarded, as it usually is small at street level. The mechanical turbulence is assumed to be generated by two mechanisms: by the wind and by the traffic in the street:

$$\sigma_w = ((\alpha u_b)^2 + \sigma_{wo}^2)^{½} \tag{12.16}$$

where σ_w is the vertical turbulent velocity fluctuation, α is a constant and σ_{wo} is the traffic created turbulence. The proportionality constant, α, is given a value of 0.1, corresponding to typical levels of mechanically induced turbulence. For the vertical dispersion coefficient, σ_z, linear dependence on the travel distance is anticipated:

$$\sigma_z(x) = \sigma_w \frac{x}{u_b} + h_o \tag{12.17}$$

Here h_o is the initial (immediate) dispersion in the wakes of the vehicles and is set equal to the average height of the cars ($h_o = 2$ m).

For a receptor on the leeward side, *the direct contribution* is calculated considering the emissions from traffic in the recirculation zone only. For a receptor on the windward side, only contributions from the emissions outside the recirculation zone are taken into account. If the recirculation zone extends through the whole canyon, no direct contribution is given for the receptor on the windward side.

When the angle between wind direction and the street axis is small (near parallel flow) the recirculation zone may occupy only a small portion of the canyon. As the flow patterns inside and outside the recirculation zone do not differ much in this case (according to the concept used in OSPM, the angle between the respective wind vectors is 2Φ), emissions from outside the recirculation zone may contribute to the concentrations at the leeward receptor. This is accounted for in OSPM by extending the integration path for the leeward receptor to the whole canyon, but the contribution from outside the recirculation zone is weighted by an angle- and wind speed dependent factor R, which is given by:

$$R = \max(0, \cos(2r\Phi)) \tag{12.18}$$

where r is a function of wind speed; r = 1 for roof top wind speed greater than 2 m/s and decreases linearly to zero for wind speed less than 2 m/s. When Φ tends to zero, the weighting factor R tends to 1. For wind speeds larger than 2 m/s the contribution from outside of the recirculation zone is zero for $\Phi > 45°$. The procedure outlined here is solely based on intuitive reasoning, but the main purpose of this approach is to get a smooth transition from a recirculation regime to a parallel flow.

When the integration path is long (as it usually is the case for a near parallel flow) the pollutants may be dispersed so much in the vertical direction that they escape from the canyon. In OSPM it is assumed that this takes place, when $\sigma_z > H$ contributions from sources further upwind are computed assuming an *exponential decay* with the rate given by:

$$\kappa = \frac{\sigma_{wt}}{H} \tag{12.19}$$

where σ_{wt} is the canyon ventilation velocity determined by the turbulence at the top of the canyon:

$$\sigma_{wt} = ((\lambda u_t)^2 + 0.4\sigma_{wo}^2)^{½} \tag{12.20}$$

where u_t is the wind speed at the top of the canyon, and the term with σ_{wo} is added in order to account for the traffic created turbulence. The proportionality constant, λ, is given the same value as α, i.e. 0.1. This is the presently used approach, but contrary to the street level turbulence, some dependence on ambient air stability conditions cannot be excluded; further research on this subject is needed.

The analytical expressions for the dependence of *the direct contribution* on wind direction are given by Hertel and Berkowicz (1989a). These expressions were derived for infinite street canyons, while in the present version of OSPM, some modifications were introduced, so the integration path can be limited by a finite street length if e.g. a broad intersection exists at a shorter distance from the receptor, or the street becomes broader or open.

For the special case of wind direction perpendicular to the street axis the expression for the direct contribution will read:

$$C_d = \sqrt{\frac{2}{\pi}} \frac{Q}{W\sigma_w} \ln\left[\frac{h_o + (\sigma_w / u_b) W}{h_o}\right] \tag{12.21}$$

12.5.2 THE RECIRCULATION CONTRIBUTION

The contribution from the recirculation part is calculated assuming a simple box model. The box model is illustrated in Figure 12.7.

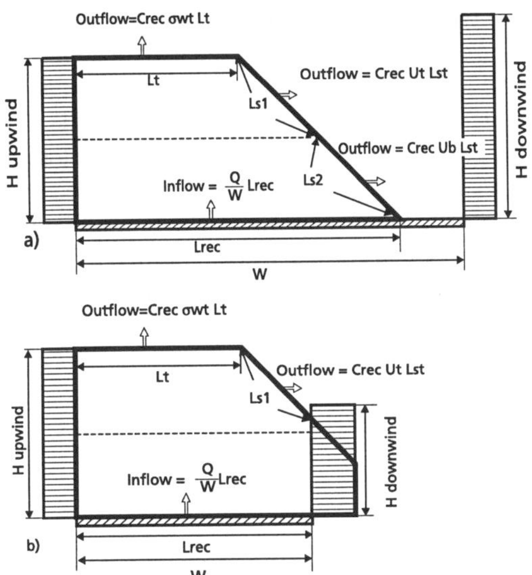

Figure 12.7 Geometry of the recirculation zone. Top: the recirculation zone totally inside the canyon. Bottom: the downwind building intercepts the recirculation zone. See the text for explanation of the symbols.

It is assumed that the canyon vortex has the shape of a trapeze, with the maximum length of the upper edge being half of the vortex length L_{vortex}. The ventilation of the recirculation zone takes place through the edges of the trapeze but the ventilation can be limited by the presence of a downwind building if the building intercepts one of the edges.

The inflow rate per unit length is given by:

$$\mathrm{INFLOW} = \frac{Q}{W} L_{rec} \tag{12.22}$$

where L_{rec} is the width of the recirculation zone. For narrow streets L_{rec} can be determined by the distance between buildings, W, (Figure 12.7, bottom).

The outflow rate through the top and side edges is calculated with flux velocities given by: σ_{wt} - the top edge, u_t - the upper half of the side edge and u_b - the lower half of the side edge.

$$\mathrm{OUTFLOW} = C_{rec} \left(\sigma_{wt} L_t + u_t L_{s1} + u_b L_{s2} \right) \tag{12.23}$$

L_t, L_{s1} and L_{s2} are calculated taking into account the canyon geometry and the extension of the recirculation zone (Figure 12.7).

The concentration in the recirculation zone is calculated assuming that the inflow rate of the pollutants into the recirculation zone is equal to the outflow rate and that the pollutants are well mixed inside the zone. When the wind vortex extends through the whole canyon, the direct contribution at the windward side is zero, and the only contribution is from the recirculation component. The concentration at the leeward side is always computed as a sum of the direct contribution and the recirculation component. The direct contribution is usually much larger than the recirculation component.

Considering the simple case when the vortex is totally immersed inside the canyon (according to assumptions made in the OSPM, this is the case when W/H ≤ 1), the recirculation contribution reads:

$$C_{rec} = \frac{Q}{\sigma_{wt} W} \quad (12.24)$$

12.5.3 WIND SPEED

The wind speed at street level, u_b, is calculated assuming a logarithmic reduction of the wind speed at roof top towards the bottom of the street. A simplified dependence on the angle between wind and the street axis is also introduced:

$$u_b = u_t \frac{\ln(h_o)}{\ln(H)} (1 - 0.2 \cdot p \cdot \sin\Phi) \quad (12.25)$$

where H is the average depth of the canyon, while $p = H_{upwind}/H$; this ratio is not allowed to exceed 1. h_o is the initial dispersion height.

For a street with 15m high buildings, roughness length z_o = 0.60m and a parallel flow (Φ = 0), we find that $u_b = 0.37u_t$. For a perpendicular flow ($\Phi = 90^o$), the reduction is 20% larger. In the case of openings between buildings on the upwind side (H_{upwind} = 0) the wind speed is the same as for a parallel flow.

12.5.4 TRAFFIC GENERATED TURBULENCE

The traffic generated turbulence is modelled by considering the vehicles in the street as a moving flow distortion elements creating additional turbulence in the air (Hertel, Berkowicz 1989c):

$$\sigma_{wo}^2 = b^2 V^2 D \quad (12.26)$$

where V is the average vehicle speed, D is the density of the moving elements (cars), and b is an empirical constant related to the aerodynamic drag coefficient.

The density of the traffic in the street is given by the relative area occupied by the moving vehicles with respect to the street area:

$$D = \frac{N_{veh} \cdot S^2}{V \cdot W} \tag{12.27}$$

where N_{veh} is the number of cars passing the street per time unit, S^2 is the horizontal area occupied by a single car, and W is the width of the street. Substituting (12.27) into (12.26) we obtain:

$$\sigma_{wo} = b\left(\frac{N_{veh} \cdot V \cdot S^2}{W}\right)^{1/2} \tag{12.28}$$

Expression (12.28) tells thus that the traffic created turbulence increases with the square root of the traffic flow ($N_{veh} \cdot V$) and that this turbulence decreases with increasing canyon width. The empirical constant, b = 0.3, as used in the present version of OSPM.

The traffic induced turbulence plays a crucial role in determination of pollution levels in street canyons. During windless conditions the ambient turbulence vanishes and the only dispersion mechanism is due to the turbulence created by traffic. Thereby, the traffic created turbulence becomes the critical factor determining the highest pollution levels in a street canyon.

Some experimental results concerning the contribution of the traffic to the street turbulence are presented in Figure 12.8. Here are shown turbulence measurements from a sonic anemometer, placed at a height of 6m on a mast in the street Jagtvej, Copenhagen. Only observations with the free wind speed U < 1.5m/s are selected and the diurnal variation of the vertical velocity turbulence is shown for working days, Saturdays and Sundays, separately. Additionally, the traffic created turbulence, as calculated by Equation (12.28), is shown by continuous lines in the figures.

It is seen that the turbulence in the street has an significant diurnal variation which follows quite well the traffic pattern. Even the difference in the traffic pattern for working days and week-ends is more or less reproduced in the diurnal variation of the turbulence. The night time values of σ_w are, however, somewhat higher then one would expect from the traffic induced turbulence only. Some other mechanisms must be of importance here too, as e.g. wind circulation induced by across the street temperature differences (Sini et al. 1996) or other local wind effects.

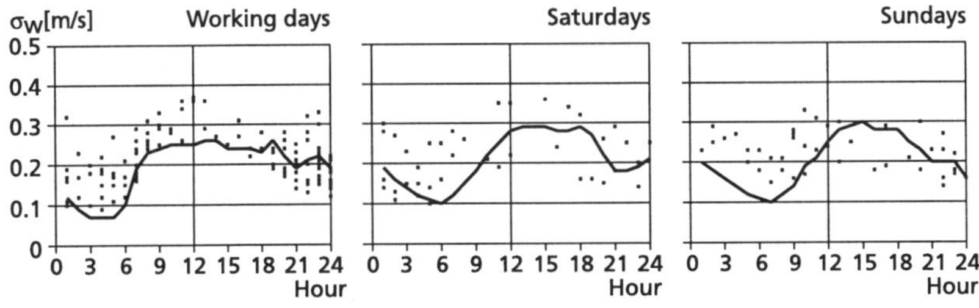

Figure 12.8 Diurnal variation of σ_w for U < 1.5 m/s. The traffic created turbulence as calculated by OSPM is shown by continuous lines.

12.5.5 WIND DIRECTION AVERAGING

The wind direction in urban areas is seldom constant over a time period of an hour or even less, in particular for low wind speeds. Large fluctuations occur and thereby, the dependence of concentrations in a street canyon, when averaged over an hour or so, is significantly smoothed. In order to account for this effect an averaging of the calculated concentrations with respect to wind direction was introduced (Hertel, Berkowicz 1989c). The averaging interval is given by:

$$\begin{aligned} \Delta\Phi &= \pm\frac{\sigma_{vc}}{u} \quad \text{for } u < 1\,\text{m/s} \\ \Delta\Phi &= \pm 0.5 \quad \text{for } u \geq 1\,\text{m/s} \end{aligned} \tag{12.29}$$

where $\sigma_{vc} = 0.5$ m/s, $\Delta\Phi$ is in radians.

12.5.6 RESULTS

Model results are summarised in Figure 12.9. Ground level wall-site concentrations are calculated for two artificial East-West oriented street-canyons with H = 20 m and W = 20 m. A constant traffic flow consisting of 900 light and 100 heavy vehicles travelling with a speed of 40 km/h is assumed. The emission density in the street is set to 1000 units $m^{-1} s^{-1}$.

In the first example, the buildings on both sides of the street are assumed to be of equal height and densely spaced. Calculations are performed for seven different values of roof level wind speed: u_t = 8, 6, 4, 2, 1, 0.5 and 0 ms^{-1}. The dependence of the calculated concentrations on wind direction is shown for receptors on both side of the street. Due to the symmetry of the considered street, the wind direction dependence of concentrations on the respective sides of the street are just shifted by 180° with respect to each other.
The variation of concentrations with wind direction is very pronounced. The leeward concentrations (southerly winds for the south side and northerly winds for the north side) are much higher than the windward concentrations. Maximum concentrations are calculated for wind directions close to parallel with the street (90° and 270°). This result is valid for long street canyons - the case anticipated in this example. Increasing concentrations for winds approaching the parallel direction were also observed in wind tunnel experiments by Hoydysh and Dabberdt (1988) (Figure 12.4, page 231).

For perpendicular winds, a local maximum on the leeward side is seen too, but only for wind speeds larger than 2 m s^{-1}. For wind speeds smaller than 2 m s^{-1} the extension of the vortex starts to decrease. According to assumptions made in OSPM and for the canyon geometry considered here, i.e. for H/W = 1, this will permit additional ventilation of the recirculation zone through the side edges (Figure 12.7, page 237). This results in a small decrease of concentrations on the windward side, where the only contribution is the recirculating component. For the windward side the direct contribution starts to become important for wind speeds less than 1 m s^{-1}.

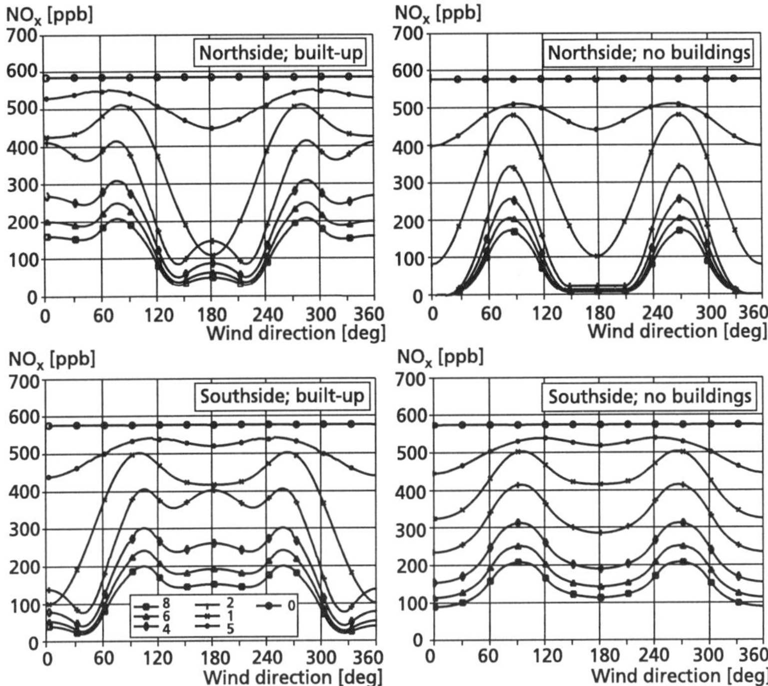

Figure 12.9 Modelled concentrations as function of wind direction and wind speed for two East-West oriented street canyons with: 20 m high buildings on both sides (left panel), 20 m high buildings on the south side only (right panel). Wind speed is given in the legend.

At still lower wind speeds, when the street vortex disappears, the wind direction dependence disappears too. We recall that at vanishing ambient winds the concentration levels are determined by the traffic created turbulence only.

The second example is for a street with 20 m high buildings only on the south side of the street. Considering higher wind speeds, the concentrations calculated for the north side receptor are in general very low, except when the winds are close to parallel. This is due to the absence of a recirculation vortex for northerly winds ($H_{upwind} = 0$), while for the southerly winds, when the vortex is created, the recirculation component (the only contribution to the windward receptor) is strongly diluted due to increased ventilation of the recirculation zone For vanishing wind speeds the concentrations become similar to that in the case of a symmetrical canyon.

The most striking feature of the wind direction dependence of the concentrations on the south side of the street is that the strong minimum observed for the symmetrical canyon is much less pronounced. The difference between the modelled concentrations for southerly and westerly winds is small. The maximum observed for parallel winds is due to the assumption about an infinite street length. Southerly winds, for which the receptor point is leeward, result in concentrations only slightly smaller than in the corresponding situation in the case of the symmetrical canyon. The only difference is here the strength of the recirculation contribution. The receptor point on the southern side receives direct contributions from the traffic in the street regardless of the wind direction. This is the consequence of the absence of recirculation vortex in the case of northerly winds.

12.6 Measurements and comparison with model results

Measurements from the monitoring stations in a few streets in Copenhagen are used here for comparison and analysis of model calculations with OSPM.

Comparisons of measured and modelled hourly concentrations of NO_x in Jagtvej are shown in Figure 12.10. Jagtvej is a busy street with about 22,000 vehicles/day. The street is 25 m wide and is flanked on both sides with about 18 m high buildings (4-5 stories). Orientation of the street is 30° with respect to North. The pollution monitoring station is situated on the east side of the street where the building facades are closed. On the opposite side, narrow side-streets make openings between the building facades. Data from working days only are used in calculations presented here.

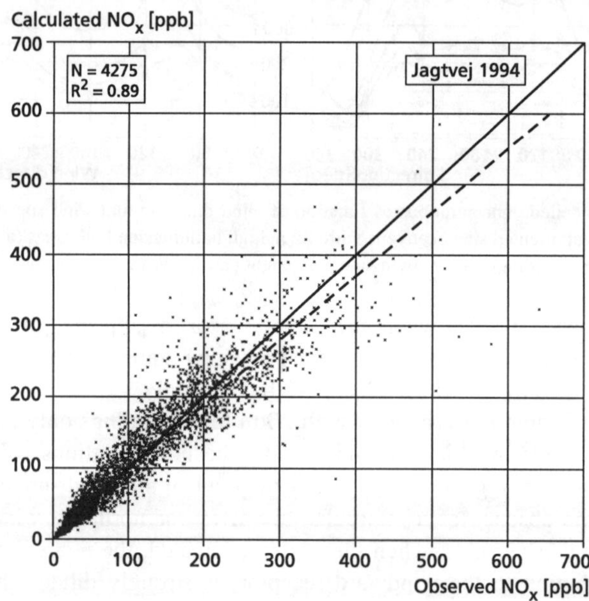

Figure 12.10 Comparison between observed and calculated by OSPM concentrations of NO_x in Jagtvej, Copenhagen.

The good correlation (R^2 = 0.89) between the modelled and the measured concentrations is evident from the results presented in Figure 12.10. Somewhat larger scatter is, however, observed especially at the high-end concentrations. Uncertainties in emission estimations have undoubtedly contributed to this scatter, but the main contribution may be attributed to the uncertainties in the meteorological data and oversimplifications of the flow parameterization in OSPM.

12.6.1 DEPENDENCE ON WIND DIRECTION

A more detailed evaluation of a model performance requires an analysis of the behaviour of model results with respect to the meteorological conditions in comparison with experimental data. Such an analysis is presented in Figure 12.11 (next page), where the measured and modelled NO_x concentrations are shown as a function of wind direction (roof mast measurements). Additionally to Jagtvej, comparison is also made for two other streets in Copenhagen: Bredgade and H.C. Andersens Boulevard.

Bredgade is a narrow street canyon (W = 15 m and H = 15 m) with closed building facades on both sides. The street is oriented 21° with respect to North and the monitoring station is on the west side of the street. The traffic intensity in Bredgade is approximately the same as in Jagtvej (ca. 20,000 vehicles/day).

H.C. Andersens Boulevard is a broad avenue (W = 60 m and H = 25 m) oriented 130° with respect to North. The monitoring station is on the north-east side where the street is flanked by somewhat irregular, but on average 25 m tall buildings. On the opposite side the street is facing the Copenhagen Amusement Park Tivoli, and is thus practically open. H.C. Andersens Boulevard is one of the most heavily trafficked streets in Copenhagen, with about 60,000 vehicles/day.

The measuring data presented in Figure 12.11 are from daytime hours only (from 8 to 18 hour) and additionally divided into two groups: wind speed between 1 m/s and 3 m/s and wind speed between 4 m/s and 6 m/s.

The well known street canyon effect is clearly evident from the results shown for Jagtvej and Bredgade. When the wind direction is such, that the measuring point is on the windward side (westerly winds for Jagtvej and easterly winds for Bredgade), the concentrations are much smaller than in the case of a leeward position. The windward wind sectors are shadowed in the figures. The difference is slightly larger for Jagtvej than for Bredgade, especially in the case of higher wind speeds (4 to 6 m/s). The dependence on wind direction is less pronounced in the case of low wind speeds.

The street Bredgade is narrower than Jagtvej, and the dilution of pollution by the recirculating vortex is smaller. Therefore the concentrations observed at the Bredgade monitoring station for the windward sector are slightly larger than the corresponding concentrations in the case of Jagtvej.

The dependence on the wind direction observed in H.C. Andersens Boulevard is quite different from the two other streets. The south-westerly wind sector, for which the measuring point is on the windward side, is here relatively unobstructed. No concentration minimum is observed, due to absence of the recirculating vortex. On the contrary, there is a pronounced leeward maximum, which occurs for the north-easterly winds. This indicates formation of a vortex in the leeward wind sector.

Model calculations reproduce the observed behaviour very well. Somewhat larger discrepancies observed especially for lower wind speeds are probably attributed to the models inability to adequately describe the flow conditions influenced by the complex street geometry.

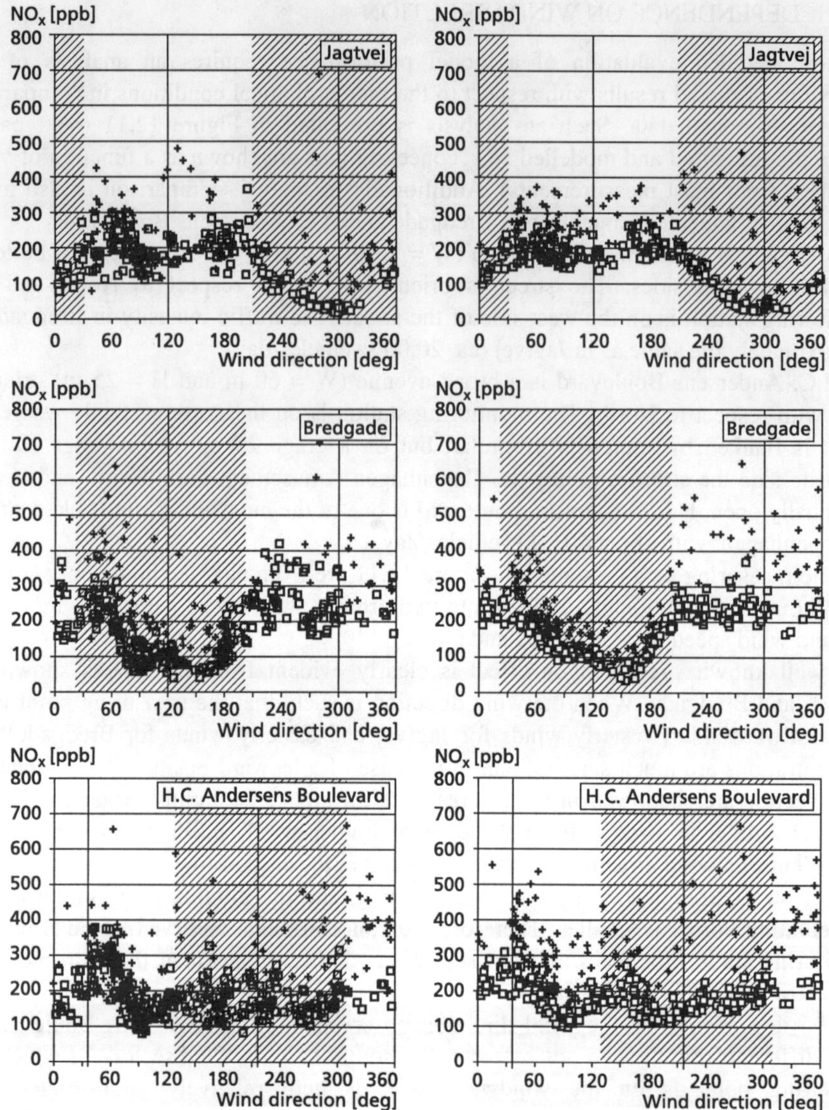

Figure 12.11 Wind direction dependence of measured (left) and modelled (right) NO_x concentrations. The wind sector for which the monitoring station is windward is shadowed. + - 1 m/s < u < 3 m/s; □ - 4 m/s < u < 6 m/s

12.6.2 DEPENDENCE ON WIND SPEED

Dependence of the measured and modelled concentrations of NO_x on wind speed is shown in Figure 12.12. Here, for each of the studied streets, only the lee-ward wind sector is selected, what minimises the dependence on wind direction.

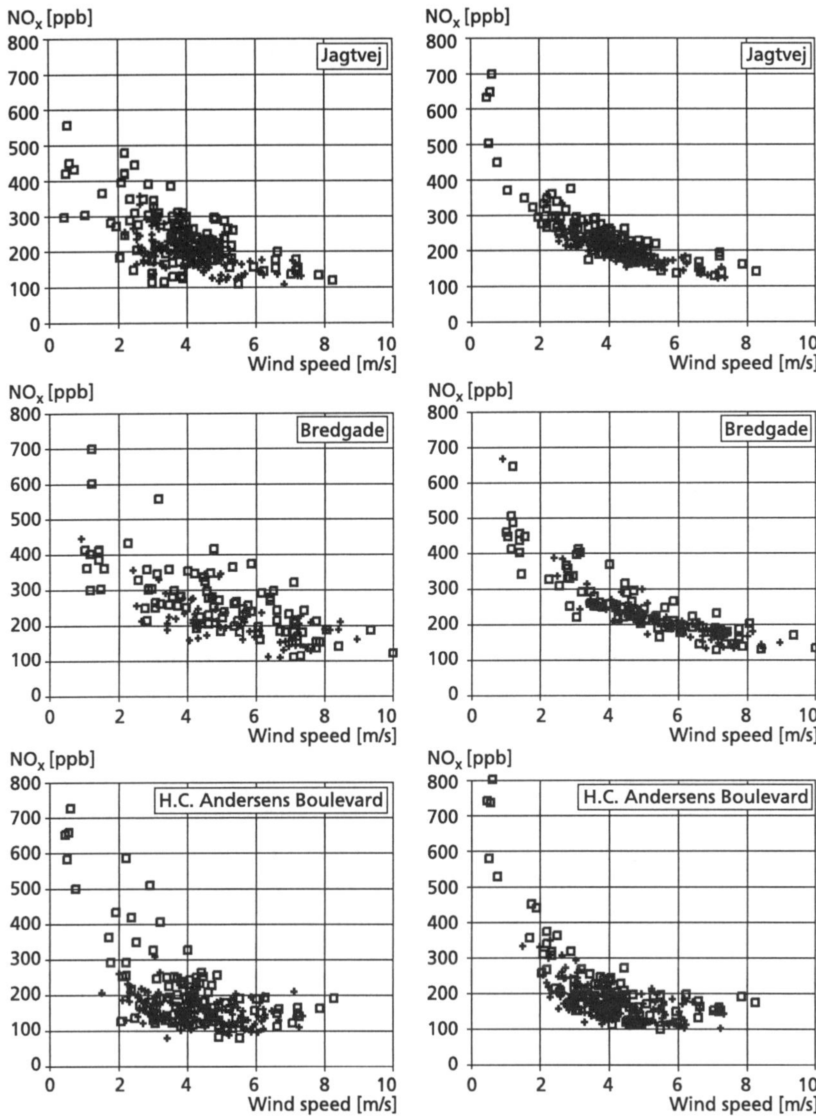

Figure 12.12 Wind speed dependence of measured (left) and modelled (right) NO_x concentrations.
+ - Global radiation > 400 W/m^2 □ - Global radiation < 400 W/m^2

The data are furthermore divided into two groups: one group for which the global radiation, is less than 400 W/m^2, and one for which the global radiation is greater than 400 W/m^2. Only daytime hours are considered. The separation with respect to global radiation is made in order to highlight the possible influence of the thermal stratification of the air on dispersion conditions in streets. The dependence on wind speed is obvious, and as expected, shows decreasing concentrations with increasing wind speed. Model calculations agree in general well with the measuring results. A somewhat larger discrepancy is observed for the lowest wind speeds. No notable dependence is observed on global radiation. This indicates that the thermal stratification does not play any significant role in dispersion of pollutants in these streets. The main mechanism is mechanically created turbulence. In the case of low wind speeds, when the thermal stratification might be important, the traffic created turbulence appears to dominate.

12.7 Chemical processes in street canyons

Transport and dispersion processes are not the only factors determining relationships between emissions and ambient concentrations. Chemistry plays a crucial role in transformation of pollutants resulting in degradation of some species and formation of others. Considering transport of pollution on a larger scale, when the transport times involved are of the order of hours or even days, often hundreds of chemical reactions must be considered in order to account for the chemical composition of the air (Chapter 8).

The situation is quite different when dealing with processes in street canyons. Due to the very short distances between the sources and receptors, only the fastest chemical reactions can have any significant influence on the transformation processes in the street canyon air. It means that most of the pollutants emitted from traffic can be considered as inert components for which chemical transformations inside the street canyon are unimportant. Such inert compounds on these time scales are CO and hydrocarbons which actually constitute the main composition of car exhaust gases.

The *nitrogen oxide gases* are, however, subject to fast chemical reactions (Chapter 8). Of special interest for us are the reactions involving NO - NO_2 - O_3. The time scales characterising these reactions are of the order of tens of seconds and thus comparable with residence time of pollutants in a street canyon. Consequently, the chemical transformations and exchange of street canyon air with the ambient air are of importance for processes leading to NO_2 formation. Taking this into account, the rate of change of the concentrations of NO, NO_2 and O_3 in the street can be approximated by the following equations:

$$\frac{d\,[NO]}{dt} = -k\,[NO][O_3] + J\,[NO_2] + \frac{[NO]_v}{\tau} + \frac{[NO]_b - [NO]}{\tau} \tag{12.30}$$

$$\frac{d\,[NO_2]}{dt} = k\,[NO][O_3] - J\,[NO_2] + \frac{[NO_2]_v}{\tau} + \frac{[NO_2]_b - [NO_2]}{\tau} \tag{12.31}$$

$$\frac{d\,[O_3]}{dt} = -k\,[NO][O_3] + J\,[NO_2] + \frac{[O_3]_b - [O_3]}{\tau} \tag{12.32}$$

where k is the rate coefficient of the reaction between NO and O_3, J is the photodissociation coefficient for NO_2, and τ is the residence time of pollutants in the street.

The first two terms on the right hand side of Equations (12.30) - (12.32) account for the chemical reactions, thereafter follow the contribution from the direct emission in the street ($[NO]_v$, $[NO_2]_v$; naturally there is no direct emission of ozone) and the exchange rate between the street and the background air. The terms with index b are the background air concentrations. The exchange rate is governed by the time constant τ, the residence time.

Assuming that a *steady state* is achieved (time derivatives become zero), equations (12.30 - 12.32) can be solved analytically giving concentrations of NO_2 in the street canyon air:

$$[NO_2] = 0.5\,(B - (B^2 - 4\,([NO_x]\cdot[NO_2]_o + [NO_2]_n \cdot D))^{1/2}) \tag{12.33}$$

where:

$$[NO_2]_n = [NO_2]_v + [NO_2]_b$$

$$[NO_2]_o = [NO_2]_n + [O_3]_b$$

$$B = [NO_X] + [NO_2]_o + R + D$$

The photochemical equilibrium coefficient is given by R = J/k (ppb), while $D = (k\tau)^{-1}$ is the exchange rate coefficient (ppb).

Equation (12.33) is implemented in OSPM (Hertel, Berkowicz 1989b) and has successfully been used for estimation of NO_2 concentrations in Danish streets (Palmgren et al. 1996).

In the model, the direct emission term, $[NO_2]_v$, is set to a constant fraction (= 5%) of $[NO_X]_v$, which is calculated by the dispersion part of OSPM ($[NO_X] = [NO_X]_v + [NO_X]_b$). The residence time, τ, is given by H/σ_{wt}. The photodissociation coefficient, J, is calculated as function of global radiation using an empirical formula derived from measurements of NO, NO_2 and O_3 in rural areas (Hertel, Berkowicz 1989b).

12.7.1 IMPLEMENTATION OF THE NO_X-CHEMISTRY

The NO_2/NO_X relationship computed according to Equation (12.33) is shown in Figure 12.13a (next page). Here calculations are made for three values of background ozone concentrations: $[O_3]_b$ = 5, 40 and 80 ppb. For simplicity, the background concentrations of NO_X are set to zero. The photodissociation coefficient corresponds to a global radiation of 500 Wm^{-2} and the canyon ventilation velocity is set to $\sigma_{wt} = 0.5\ ms^{-1}$ which results in the residence time of 40 s for a 20 m deep canyon. The direct emission of NO_2 is set to 5% of NO_X.

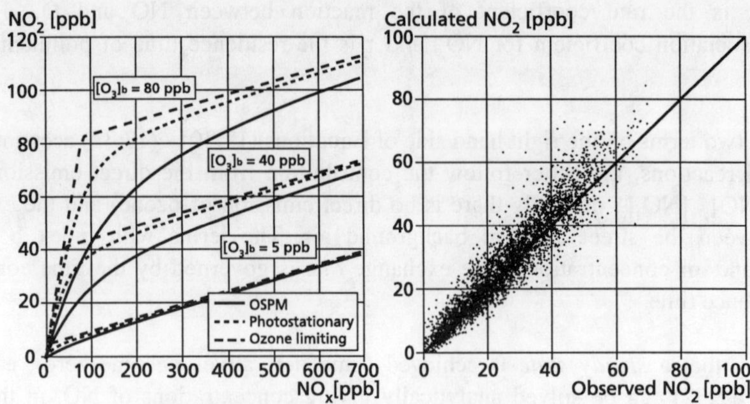

Figure 12.13 To the left, relationship between NO_2 and NO_x for three values of background ozone. Results are shown using Equation (12.33) and two other models (see the text for explanation). To the right, comparison between measured and modelled concentrations of NO_2 for the street Jagtvej.

For comparison, predictions from two other simple models are also shown in Figure 12.13 (right). The first implies that a photostationary state is achieved in the street. This is equivalent to the assumption that the residence time of pollutants in the street is infinite, or at least much larger than the chemical reaction time scale. This approach is used in the CPB model of Yamartino and Wiegand (1986). The second model is based on the assumption that production of NO_2 in the street is equal to the amount of available ozone. This means that not only the residence time is large enough, but also that the photodissociation of NO_2 can be neglected. The latest approach is used in the Nordic Computational Method for Car Exhaust Gases (Larssen 1992). Both the photostationary assumption and the ozone limiting method clearly overpredict the NO_2 concentrations. The difference between the OSPM approach and the two other models decreases, however, with increasing NO_x concentrations. The deviations are also expected to decrease with decreasing street ventilation (increasing residence time). Because such conditions are also connected with the highest concentrations, the simple ozone limiting method used by Larssen (1992) for prediction of the highest NO_2 concentrations is not unreasonable.

In Figure 12.13 (right) is shown a comparison between concentrations of NO_2 computed by OSPM and measurements from the monitoring station in Jagtvej. The agreement is very good, proving thus that the method based on the simplified chemistry involving NO - NO_2 O_3 reactions only and the Equation (12.33) performs well. Other chemical mechanisms may however be important in some extreme cases. Importance of NO oxidation by molecular oxygen is discussed in Chapter 8.

12.8 Outstanding problems and conclusions

Application of some few simple models for calculation of dispersion of traffic pollution in street canyons was discussed. It was demonstrated that even such simple and highly parameterized models, as e.g. OSPM, can handle a broad range of dispersion conditions and provide reliable results. The successful performance of OSPM is mainly due to parameterizations which were based on careful examination of available experimental data covering a broad range of conditions, but this is also the limitation of the model. Considering the "broad range of conditions" we have to emphasise that this never means - all conditions. Some of the problems that still lack a practical solution will be discussed in this section.

The most severe pollution episodes are usually associated with calm or very low wind speed conditions. Actually, differences in prevailing wind speed conditions seem to explain much of the differences in urban pollution levels (Vignati et al. 1996). Considering flow and dispersion regimes in street canyons at low wind speeds, we have already mentioned the important role of the traffic created turbulence but some thermal effects can also be important as well. We have here in mind the modification of flow regimes due to differential heating of building walls. This effect is most pronounced for east-west oriented canyons where, due to solar insolation, the temperature of a north wall can be several degrees warmer than the temperature of a south wall (on the Northern Hemisphere). Field observations (Nakamura, Oke 1988) and numerical simulations (Mestayer et al. 1995; Sini et al. 1996) have shown that such differential heating can substantially modify the wind flow in a canyon. If the heated wall is windward, this can even lead to reversing of the flow and thereby largely influence distribution of pollution in the street. Quantification of this effect in the form of relationships between solar radiation and wind flow is necessary in order to incorporate these phenomena into applied pollution models.

Traffic pollution models typically make use of wind flow data related to the roof level. Such data are rarely available and transformation of wind measurements undertaken at some few locations in the city or even outside the city are necessary. Taking into account that urban areas are strongly inhomogeneous, this task is far from trivial. Meso-scale meteorological models can be used in this case but very fine mesh resolution is required to resolve the building structures in the city. This is not possible with presently available models where urban areas are treated as bluff bodies and only the large topographical features are taken into account (Chapter 11).

Rotach (1995) has shown that the relationship between the roof level wind speed and the speed aloft depends on the atmospheric stability conditions. This can be especially important in the case of stable conditions when the canyon ventilation velocity might be reduced due to attenuation of the roof level speed and perhaps turbulence.

Local modification of wind flow and turbulence might also be due to some pronounced building formations nearby the measuring site. Based on wind tunnel modelling, Kennedy and Kent (1977) have demonstrated that a twofold decrease of CO concentrations observed at a street site in Sydney, Australia, could be explained by construction of a tower block that has largely influenced wind flow and thereby dispersion conditions at the measuring site.

Surroundings of the street canyon can in general have significant influence on flow and dispersion conditions in the canyon itself. Already the early experiments by Hoydysh et al. (1974) demonstrated that an upwind fetch of at least 8 to 10 street canyons is required before the wind flow can be considered stationary. Recent wind tunnel experiments by Meroney et al. (1995) have shown that ventilation of a canyon in an urban environment is less than ventilation of the same canyon but in an open country environment. The shape of roofs of surrounding buildings was also shown to influence the concentration distribution in street canyons (Rafailidis, Schatzmann 1995).

The question may arise, whether is it possible at all to manage such a variety of different conditions and parameters by mathematical models. Perhaps not, at least not by a single model. Do we need universal models? Are the simple models not adequate for most of the purposes? Perhaps yes, especially regarding routine applications.

More detailed models are however required when the problem is connected with interpretation of measurements. Measurements are, as rule, the basis for pollution surveillance programmes. Measurements provide the most direct information on pollution conditions (at least at the measuring sites) but as stand alone, they cannot be used to explain the relationships between sources and ambient pollution. Such relationships are necessary if conclusions based on the surveillance programmes have to be used to any thing else than just reporting the present conditions. Interpretation of measurements in terms of source-receptor relationships, meteorological conditions, the local street conditions e.t.c. can only be done using well performing dispersion models.

12.9 References

Albrecht, F. (1933) Untersuchungen der vertikalen Luftzirkulation in der Grossstadt, *Met. Zt.*, **50**, 93-98.

Berkowicz, R., Hertel, O., Sørensen, N.N., Michelsen, J.A. (1997) Modelling air pollution from traffic in urban areas, in: Perkins, R.J., Belcher, S.E. (editors) *Flow and Dispersion through groups of obstacles,* pp. 121-141, Clarendon Press, Oxford.

Bower, J.S., Broughton, G.F.J., Stedman, J.R., Williams, M.L. (1994) A winter NO_2 smog episode in the U.K., *Atmospheric Environment,* **28**, 461-475.

DePaul, F.T., Sheih, C.M. (1986) Measurements of wind velocities in a street canyon, *Atmospheric Environment,* **20**, 455-459.

Eerens, H.C., Sliggers, C.J., Van den Hout, K.D. (1993) The Dutch method to determine city air quality, *Atmospheric Environment,* **27B**, 389-399.

Hertel, O., Berkowicz, R. (1989a) Modelling pollution from traffic in a street canyon. Evaluation of data and model development, *DMU Luft A-129.*

Hertel, O., Berkowicz, R. (1989b) Modelling NO_2 concentrations in a street canyon, *DMU Luft A-131.*

Hertel, O., Berkowicz, R. (1989c) Operational Street Pollution Model (OSPM). Evaluation of the model on data from St. Olavs street in Oslo, *DMU Luft A-135.*

Hotchkiss, R.S., Harlow, F.H. (1973) Air pollution transport in street canyons, *EPA-R4-73-029.*

Hoydysh, W.G., Dabberdt, W.F. (1988) Kinematics and dispersion characteristics of flows in asymmetric street canyons, *Atmospheric Environment,* **22**, 2677-2689.

Johnson, W.B., Ludwig, F.L., Dabbert, W.F., Allen, R.J. (1973) An urban diffusion simulation model for carbon monoxide, *JAPCA* **23**, 490-498.

Johnson, G.T., Hunter, L.J. (1995) A numerical study of dispersion of passive scalars in city canyons, *Boundary-Layer Meteorology,* **75**, 235-262.

Kamenetsky, E., Vieru, N. (1995) Model of air flow and air pollution concentration in urban canyons, (Research Note), *Boundary-Layer Meteorology*, **73**, 203-206.

Kennedy, I.M., Kent, J.H. (1977) Wind tunnel modelling of carbon monoxide dispersal in city streets, *Atmospheric Environment*, **11**, 541-547.

Lamb, R.G., Hogo, H., Reid, L.E. (1979) A Lagrangian approach to modeling air pollution dispersion: development and testing in vicinity of a roadway, *EPA-600/4-79-023*.

Lanzani, G., Tamponi, M. (1995) A microscale Lagrangian particle model for the dispersion of primary pollutants in a street canyon. Sensitivity analysis and first validation trials, *Atmospheric Environment*, **23**, 3465-3475.

Larssen, S. (1992) Modell for bakgrunnsbidraget til NO_2-konsentrasjonen i gater (in Norwegian), *Project NILU 0-8839*, NILU, Lillestrøm, Norway.

Lee, I.Y., Park, H.M. (1994) Parameterization of the pollutant transport and dispersion in urban street canyons, *Atmospheric Environment*, **28**, 2343-2349.

Meroney, R.N., Rafailidis, R., Pavageau, M. (1995) Dispersion in idealized urban street canyons, *21st Int. Meeting on Air Pollution Modelling and its Applications*, Baltimore, 6-10 Nov. 1995, 317-324.

Mestayer, P.G., Anquetin, S. (1994) Climatology of cities, Diffusion and Transport of Pollutants, in: Rys, F.-S., Gyr, A. (editors), *Atmospheric Mesoscale Flow Fields,* Atmospheric Sciences Library, Kluver Academic Publishers.

Mestayer, P.G., Sini, J.F., Jobert, M. (1995) Simulation of the wall temperature influence on flows and dispersion within street canyons, *Third International Conference on Air Pollution, Porto Carras, Greece, 26-28 Sept. 1995.*

Moriguchi, Y., Uehara, K. (1993) Numerical and experimental simulation of vehicle exhaust gas dispersion for complex urban roadways and their surroundings, *Journal of Wind Engineering and Industrial Aerodynamics*, **46-47**, 689-695.

Nakamura, Y., Oke, T.R. (1988) Wind, temperature and stability conditions in an E-W oriented canyon, *Atmospheric Environment*, **22**, 2691-2700.

Oke, T.R. (1988) Street design and urban canopy layer climate, *Energy Bldng*, **11**, 103-113.

Palmgren, F., Berkowicz, R., Hertel, O., Vignati, E. (1996) Effects of reduction of NO_x on NO_2 levels in urban streets *Sci. Total Environ.*, **189/190**, 409-415.

Rafailidis, S., Schatzmann, M. (1995) Physical modelling of car exhaust dispersion in urban street canyons, *21st International Meeting on Air Pollution Modelling and its Applications*, Baltimore, 6-10 Nov. 1995, 170-171.

Rotach, M.W. (1995) Profiles of turbulence statistics in and above an urban street canyon, *Atmospheric Environment*, **29**, 1473-1486.

Sini, J.F., Anquetin, S., Mestayer, P.G. (1996) Pollutant dispersion and thermal effects in urban street canyons, *Atmospheric Environment*, **30**, 2659-2677.

Vignati, E., Berkowicz, R., Hertel, H. (1996) Comparison of air quality in streets of Copenhagen and Milan in view of the climatological conditions *Sci. Total Environ.*, **189/190**, 467-473.

Yamartino, R.J., Wiegand, G. (1986) Development and evaluation of simple models for flow, turbulence and pollutant concentration fields within an urban street canyon, *Atmospheric Environment*, **20**, 2137-2156.

Chapter 13

STOCHASTIC MODELS

OLF HERBARTH, UWE SCHLINK and MATTHIAS RICHTER
Department of Human Exposure Research and Epidemiology
UFZ - Centre for Environmental Research Leipzig-Halle Ltd.
P.O.Box 2, D-04301 Leipzig, Germany

13.1 Medical reasons for smog prediction

The aim of very short-term smog forecasts is to predict pollution levels which exceed medical threshold values and therefore may impair human health (Herbarth 1995).

Smog episodes are governed by a complex interaction between emitted pollutants and meteorological conditions. The concentration of pollutants in the air is thereby a function of emissions, meteorology, orography, etc. The overall system is characterised by numerous degrees of freedom. Dispersion modelling can be used to gain information about the human exposure to pollutants. This is then true for those conditions included in the calculation. In case of time-critical events (e.g. smog situations), however, the dispersion modelling has to be repeated for each sequential time step. This procedure is quite work-intense in situations that need a prompt response. Time is critical as far as smog episodes are concerned - as the key question is how will the concentration of pollutants develop on a short-term basis, i.e. over the next few hours.

Furthermore, dispersion modelling is always based on a three-dimensional grid. The calculated concentration of the pollutant is the average of the particular grid cell. However, localised point predictions are not possible with this method.

An alternative, more efficient approach may be, to start with locally observed air quality data and combine these with stochastic models. This approach would be suitable for:

- drawing conclusions about sources with emissions that vary over time;
- local short-term forecasting using temporal extrapolation.

13.2 Dose data

Studies in the field of environmental medicine show, that moving 3-h. and 24-h. averages of the measurements (referred to as M3 and M24) are especially suitable for this purpose. For example, in the case of winter smog, SO_2 is the indicator substance. According to these studies, a 24-h. average value below 0.6 mg SO_2/m^3 does not appear to increase the rate of human sickness significantly, and this value can therefore be used as a lower level for smog warning. According to dose-effect relations, situations in which M3 remains below 0.6 mg SO_2/m^3 (the "zero" case) provide no cause for concern whatsoever. Thus a useful prediction algorithm only occurs at the moment t_0, when M3 = 0.6 mg SO_2/m^3. On the basis of these threshold values prevailing in environmental medicine, we can distinguish between the following critical cases:

Pseudo occurrences, when it is expected that M24 will remain below 0.6 mg SO_2/m^3 over the next 24 hours.

Threshold breach, when it is expected that M24 will equal or exceed 0.6 mg SO_2/m^3 during the next 24 hours.

Therefore, the job of a prediction system is at the moment (t_0) when M3 reaches the threshold 0.6 mg SO_2/m^3 to forecast the likelihood of the immediate violation of this threshold.

The indicator components are either ozone and nitrogen oxides or sulphur dioxide, depending on the smog situation being monitored (summer or winter smog). Concentrations are continuously measured as half-hourly means by automatic stations (Figure 13.1). One aim of this extensive measuring system is to indicate smog situations. However, if smog situations are to be prevented, they must be predicted some time before they actually occur.

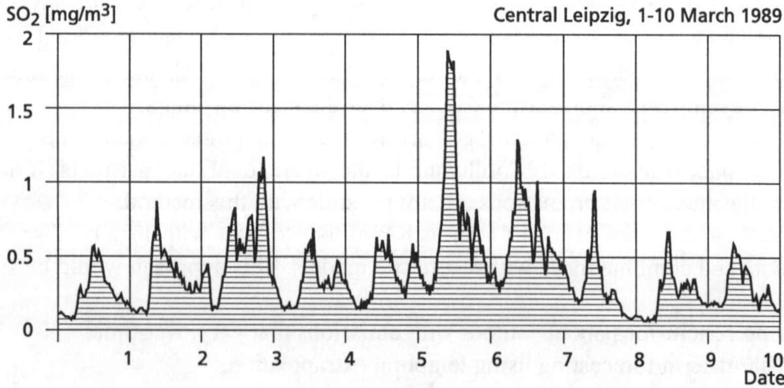

Figure 13.1 Example of a time-series of observed SO_2 concentration in Central Leipzig, 1-10 March 1988.

13.3 The Kalman filter algorithm

There are a number of algorithms for digital filtering that directly utilise observed values. In contrast to this the so-called recursive filters calculate the output via preliminary steps.

In case of a fixed smoothing strength a non-recursive filter shows restricted memory. However, recursive filters build on the completeness of the past input. For that reason, they are particularly suitable for extrapolation purposes (Hamming 1973). The Kalman-filter is a recursive filter (Kalman 1960). It is based on a state variable Y, the time development of which is described by a state equation. This equation already contains stochastic terms and, hence, is non-deterministic.

The following important assumption is made for the filter algorithm: the state variable Y does exist, but it is not directly observable. It can be measured only with superimposed noise (observation equation). This observation equation and the state equation is called the state space model. The task of the Kalman-filter is to recursively estimate the state variable from the observed variable. This situation may be demonstrated with a very simple state space model:

Observation equation: $z_t = y_t + \mu_t$, (13.1)

State equation (IRW model):

$$y_t = y_{t-1} + x_{t-1}$$
$$x_t = x_{t-1} + \eta_{t-1} \qquad (13.2)$$

The observed variable z_t results from the state variable y_t with additional white noise μ_t. The dynamic of y_t may be described by the integrated random walk (IRW) model formed by summing up a random walk series (x_t), with similar white noise, η_t.

The Kalman-filter then estimates y_t from the observed series z_t. Ng and Young (1990) presented the filtering algorithm in a recursive prediction-correction form and named it IRWSMOOTH. The smoothing parameter is the ratio var(μ)/var(η), the so-called noise-variance-ratio (NVR). IRWSMOOTH is a very good low-pass filter as its transfer-function is rapidly decreasing at the cut-off frequency (Ng, Young 1990).

The cut-off frequency f_{50} can be adjusted easily because of the simple relationship between NVR and f_{50} (Ng, Young 1990):

$$f_{50} = 0.158 \text{ NVR}^{0.25} \qquad (13.3)$$

The observation and state equations are then in a generalised vector form as follows:

$$z_t = H(t)\, x_t + \mu_t \qquad (13.4)$$

$$x_t = F(t-1)\, x_{t-1} + G(t-1)\, \eta_{t-1} \qquad (13.5)$$

with $x_t = (y_t \;\; x_t \;\; \ldots)^T$.

The Kalman filter algorithm is then based on the following successive steps (prediction, correction) which are calculated at first in the upward direction (t-1 → t) (Ho 1962; Young 1991):

Prediction:

$$x_{t|t-1} = F(t-1)\, x_{t-1} \tag{13.6}$$

$$P(t|t-1) = F(t-1)\, P(t-1)\, F^T(t-1) + G(t-1)\, Q(t-1)\, G^T(t-1) \tag{13.7}$$

Correction:

$$x_t = x_{t|t-1} + P(t|t-1)\, H^T(t)\, [1 + H(t)\, P(t|t-1)\, H^T(t)]^{-1} \{z_t - H(t)\, x_{t|t-1}\} \tag{13.8}$$

$$P(t) = P(t|t-1) - P(t|t-1)\, H^T(t)\, [1 + H(t)\, P(t|t-1)\, H^T(t)]^{-1}\, H(t)\, P(t|t-1) \tag{13.9}$$

The algorithm is applied recursively on the observations z_t which produces the conditional state $x_{t|t-1}$. Smoothing, however, requires an additional recursive downward filter (t+1 → t):

$$x_{t|N} = F^{-1}(t)\, [x_{t+1|N} + G(t)\, Q(t)\, G^T(t)\, L(t)] \tag{13.10}$$

$$L(t) = [I - P(t+1)\, H^T(t+1)\, H(t+1)]^T \cdot [F^T(t+1)\, L(t+1) - H^T(t+1)]\, [z_{t+1} - H(t)\, F(t)\, x_t] \tag{13.11}$$

where L(N) = 0.

13.4 Forecast model

Series of immissions have the characteristic of being autocorrelated (Herbarth 1982). This means that there is a conservation period during which the concentration deviates only slightly from the pollutant level reached. However, it also to a certain extent means that the concentration values recorded in the past contain information on the future development to be expected. The basic idea behind the forecasting algorithm consists of using this information in the form of time series parameters already realised. Such suitable parameters, termed "predictors", must be found. Evaluation of extensive measuring series and observation of an average pseudo occurrence and an average threshold breach suggest the use of the parameters M24(t_o) and Increase (M3) over the last 3 hours (M3A3) as predictors.

This is illustrated using the example of winter smog forecasting in Leipzig - a city with a population of about 500,000 located in the central German industrial region. As Figure 13.2 shows, the pseudo occurrences and the threshold breaches are differently distributed throughout the plane of the two predictors. If the actual situations are continuously incorporated, this system always adapts itself to the changed immissions conditions and delivers a forecast which is based on the latest knowledge of the smog situations monitored.

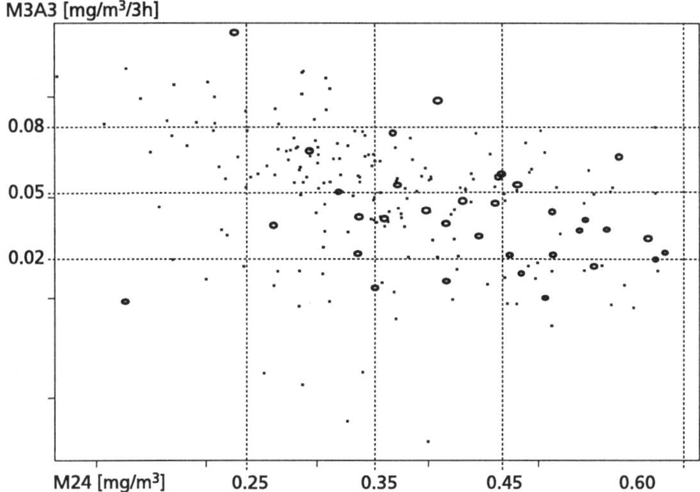

Figure 13.2 The 270 pseudo occurrences (dots) and threshold breaches (circles) in the plane spanned by the predictors M24 and M3A3.

The place of the two predictors can be divided into a series of boxes (Figure 13.2). A probability analysis of the cases in the individual boxes K can be used to obtain the likelihood of a breach T(B|k), assuming that the predictors adopt the values of the respective box. When using just one predictor (M24) the accuracy of this forecast algorithm is 81%, but rises to 84% if two predictors (M24 and M3A3) are used.

In order to further improve this success rate, these factual predictors can be augmented by another predictor - which in turn is also previously predicted. This predictor is the time difference Δt until the threshold value 0.6 mg/m³ is reached by M24. As this is of course not yet known at the time of forecasting (t_0), a time-series model must be drawn up for the course of the time series preceding t_0. Extrapolating this model will then deliver a forecast for the predictor Δt (Figure 13.3, next page).

The periodogram of the measuring series suggests the use of a *time-series model* additively composed of individual components (Schlink, Herbarth, Tetzlaff 1997). Each of these components is extracted from the measuring series by using a suitable Kalman filter (Section 13.3) and represents a fixed phase of the period spectrum. The measuring series is thus fragmented into a trend section, which only contains the deep frequencies (period > 1 day), two periodic series with periods of 24 and 12 hours, and a residual series.

Taking into account their typical characteristics, each of these components is then extrapolated and determined by summarising the course of the series forecast (Figure 13.4, next page), thus enabling Δt to be predicted by calculation.

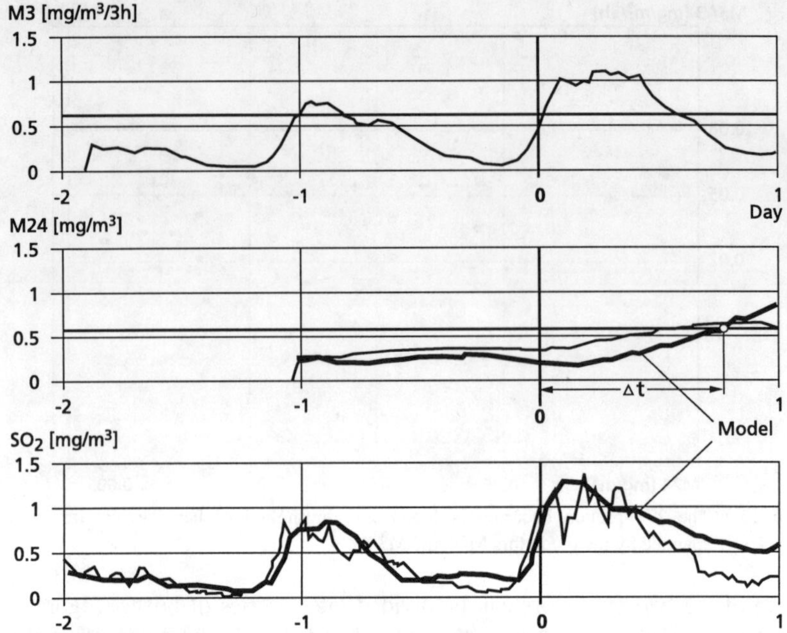

Figure 13.3 The forecast predictor Δt in an example situation; the modelled time-series is indicated by a bold line.

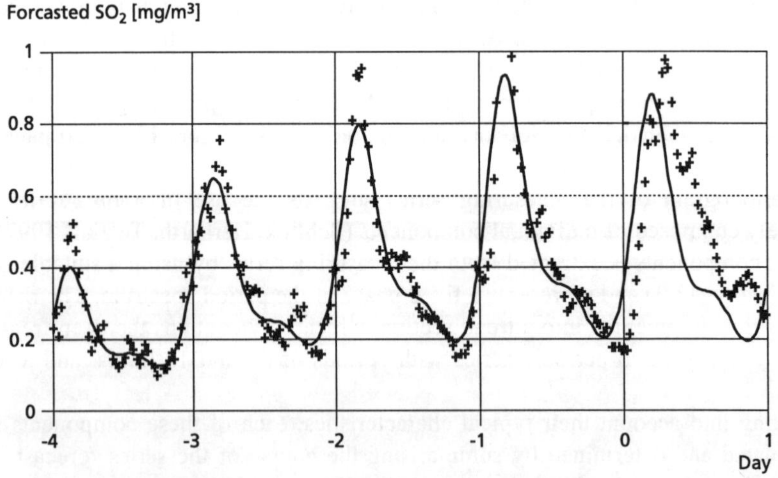

Figure 13.4 Forecast of the course of concentration for one day using the component model.

13.5 Conclusion

The stochastic technique outlined here enables the phenomena contained in the variability of air quality time-series to be used to design an online algorithm for very short-term smog warning. The accuracy achieved depends on the type and number of predictors used. The time horizon of very short-term forecasting is a few hours and thus provides sufficient warning for preventive steps to be taken.

The forecast algorithm is based on the time-series modelling of the air quality. It may be expected that the phenomena included in the algorithm such as daily seasonality and the description of smog events by help of the course of the trend do not depend on the specific measuring location, thus ensuring that the warning system has a certain universality. Adaptation to local conditions such as the orographic situation or the regional emissions structure is thus not required. Furthermore, the forecasting algorithm can also be used in a modified form to predict other pollutant components (Schlink, Herbarth, Richter 1996).

13.6 References

Hamming, R.W. (1973) *Numerical Methods for Scientists and Engineers*, McGraw-Hill.

Herbarth, O. (1982) Zeitreihenanalyse und Prognose lufthygienischer Konzentrationsreihen, Z. *gesamte Hyg.*, **28** 508-509.

Herbarth, O. (1995) Risk assessment of environmentally influenced airway diseases based on time-series analysis, *Environmental Health Perspectives*, **103**, 852-856.

Ho, Yu-Chi (1962) On the stochastic approximation method and optimal filtering theory, *Journal of Mathematical Analysis and Applications*, **6**, 152.

Kalman, R.E. (1960) A new approach to linear filtering and prediction problems, *Journal of Basic Engineering*, **82**, 35.

Ng, C.N., Young, P.C. (1990) Recursive estimation and forecasting of non-stationary time series, *Journal of Forecasting*, **9**, 173-204.

Schlink, U., Herbarth, O., Richter, M. (1996) Ozone data analyses to develop an immediate warning system, *Proceedings Air Pollution Conference '97*, 465-474.

Schlink, U., Herbarth, O., Tetzlaff, G. (1997) A Component Time Series Model for SO_2 data: Forecasting, Interpretation and Modification, *Atmospheric Environment*, **31**, 1285-1295.

Young, P.C., Ng, C.N., Lane, K., Parker, D. (1991) Recursive forecasting, smoothing and seasonal adjustment of non-stationary environmental data, *Journal of Forecasting*, **10**, 57-89.

Chapter 14

WIND TUNNEL EXPERIMENTS

MICHAEL SCHATZMANN and STYLIANOS RAFAILIDIS
University of Hamburg, Meteorological Institute
Bundesstrasse 55, D-20146 Hamburg, Germany

NIJS JAN DUIJM
Risø National Laboratory, System Analysis Department
P.O.Box 49, DK-4000 Roskilde, Denmark
Previously at:
TNO-Institute for Environment, Energy and Process Innovation
P.O.Box 342, NL-7300 AH Apeldoorn, The Netherlands

14.1 Application of wind-tunnel experiments

The intention of wind-tunnel experiments (or in general physical modelling, including water-tank experiments) is to reproduce the characteristics of the atmospheric boundary layer on a small geometric scale, typically between $1/100^{th}$ and $1/1000^{th}$ of the real world. In the previous chapters, numerical models were grouped into different classes depending on the scales they are applied to. Physical modelling is mainly applied to the scale of street canyons and individual buildings, in other words, to the obstacle resolving scale.

This scale is of particular importance for urban air pollution problems since most urban emissions of concern occur within or shortly above the canopy layer, i.e. the zone where the atmospheric flow is heavily disturbed by buildings and other obstructions, which is also the zone where the people live. Due to the combined effect of wind variability and the obstacles, the local concentrations differ substantially over short distances.

14.1.1 TYPES OF APPLICATIONS

One can distinguish two types of applications: The first type is aimed at *providing direct solutions for problems* which, due to the geometric complexity of the site, can not be solved otherwise, e.g. detailed studies on the flow and dispersion around buildings during the design phase. Wind-tunnel experiments can be used to find the optimum location of stacks and ventilation openings, avoiding re-entry of exhaust gases and impact on neighbouring buildings, pedestrian areas, etc. or to assess the effect of the building on the circulation of emissions from traffic. Often, the wind tunnel is also used to determine wind pressures on the building facades and the impact on the pedestrian wind climate, and this information can also be used to design the ventilation system.

Used in this way, i.e. as an engineering tool during the design phase, Computational Fluid Dynamics (CFD) or grid models are becoming an alternative to physical modelling. Still an important advantage of physical modelling is that, compared to numerical modelling, there is no need for closure assumptions with empirically fitted constants in order to describe turbulence.

The other type of applications is aimed at *collecting general information about flow and dispersion in generalised situations.* This information can be used to develop either specific dispersion models for typical problems, e.g. street-canyon models (Chapter 13) or more general grid models. Here, wind-tunnel experiments offer a wide potential in improving our understanding of flow and dispersion at the local, obstacle resolving scale:

- Systematic parameter-variation studies can be carried out by changing only one particular parameter over the whole range of interest, whereas all others are kept constant. Certain trends in the data can be made visible.
- Complete and quality-assured data sets can be generated which contain not only concentration data but also the underlying mean and turbulent velocities. All boundary conditions can be kept constant over sufficiently long time periods. Mass and momentum fluxes can be determined and the consistency of the data (conservation of mass etc.) can be checked.
- The development of new sub-grid scale turbulence closure schemes, as they are needed for high-resolution micro-scale models, can be supported by pertinent experiments carried out under carefully controlled conditions.
- Laboratory experiments can contribute substantially to the value of field data by repeating the field tests in a small-scale model of the site and its surroundings. Uncertainty in the field data can be quantified by repetition of the experiment with constant source and boundary conditions, and investigating the effect of averaging times. Gaps in the data due to the fact that certain ambient conditions were not met during the field tests can be closed.
- The wind tunnel can even help to investigate the effect of the usually insufficient spatial resolution numerical models provide. This can be done in two consecutive measurements within which subsequently to a detailed scale model, a model containing only rectangular blocks matched to the numerical grid is used.

These are only a few examples, which demonstrate that wind tunnels can play an important role in urban air pollution meteorology, and that combined approaches (physical modelling plus field experiments or numerical simulations) can lead to solutions unobtainable in any other way.

14.2 Properties of wind-tunnel boundary layers

The reliability of results from physical model studies depends strongly on the quality of the mean and turbulence characteristics of the simulated boundary layer. Some results of such characteristics measured in a wind tunnel boundary layer will be presented and compared with those of the real atmospheric boundary layer.

14.2.1 DESIGN OF WIND TUNNELS

Most wind tunnels, which are used to generate the small scale boundary layer, have a design similar to that sketched in Figure 14.1. They consist of an inlet nozzle with flow straighteners, a long flow establishment section, a test section and a fan driven by a speed controlled electric motor.

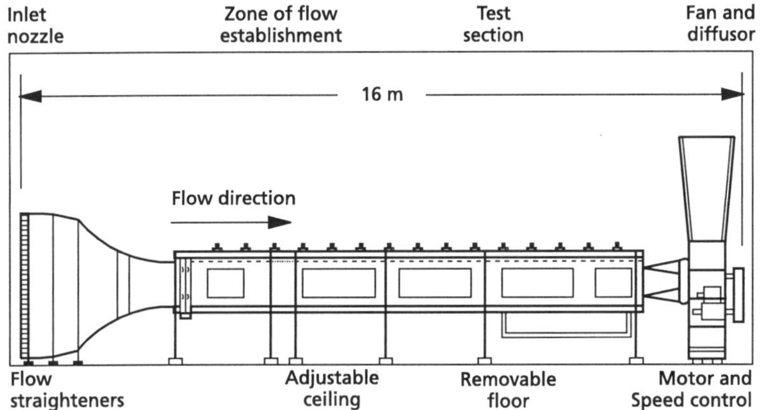

Figure 14.1 Neutrally stratified boundary-layer wind-tunnel of Hamburg University. Size of cross section: 1.5 m x 1.0 m.

Most tunnels work with suction in order to keep any disturbance of the flow inside the tunnel to a minimum. An adjustable ceiling is used to establish a zero-pressure-gradient boundary layer. The flow establishment section is necessary to create a boundary layer structure which is in equilibrium with the boundary conditions. The change of wind direction with height or density stratification can not be simulated in such a tunnel. However, there are other tunnel designs available which are specialised just on the simulation of stable and unstable boundary layers or even elevated inversions (e.g. Schatzmann et al. 1995).

Control over the boundary layer structure in the wind tunnel, i.e. the vertical profiles of mean velocity and turbulence can be achieved by use of specific combinations of vortex generators (e.g. Irwin 1981) and artificial roughness elements distributed over the bottom of the flow establishment section (Figure 14.2, next page).

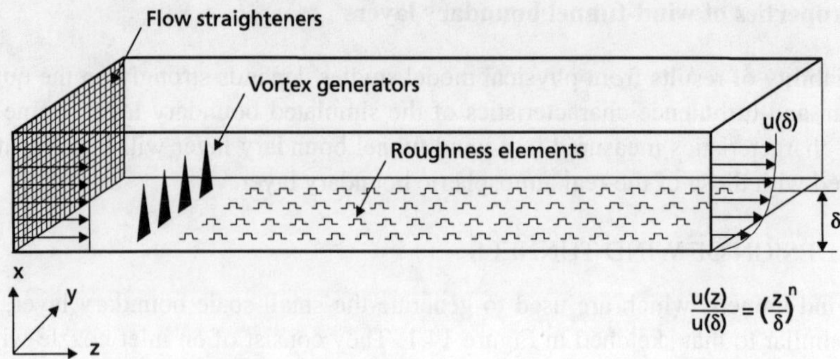

Figure 14.2 Generation of a neutrally-stratified boundary-layer wind profile by use of vortex generators and artificial roughness elements.

14.2.2 THE VERTICAL VELOCITY PROFILE

As in dispersion calculations for licensing purposes, the vertical velocity profile of the wind tunnel boundary layer is usually approximated by the power law:

$$\frac{u(z)}{u(\delta)} = \left(\frac{z}{\delta}\right)^n \tag{14.1}$$

with u(z) the velocity at height z, and u(δ) the velocity at boundary layer height δ. Alternatively use is made of a logarithmic representation:

$$\frac{u(z)}{u(H)} = \frac{\ln(z/z_0)}{\ln(H/z_0)} \tag{14.2}$$

with z_0 the roughness length of the ground surface, and H a characteristic height, e.g. the height of a main building. The logarithmic representation can be applied for problems within the surface layer (the lower 10-20% of the turbulent boundary layer). In that region the flow structure does not depend on the boundary layer height δ.

Figure 14.3 gives an example of a mean vertical velocity profile in double logarithmic presentation obtained in the tunnel. In this particular case, a boundary layer thickness of $\delta = 450$ mm and a profile exponent of $n = 0.207$ was obtained. Logarithmic velocity profiles are often presented on single logarithmic plots, where they appear as straight lines.

To obtain model/prototype similarity not only the mean velocities, but also the turbulent properties of the boundary layer have to be matched. Since the Reynolds number, that represents the ratio between inertial and viscous forces, will be some orders of magnitude smaller in the wind tunnel than in the field, some fundamental differences must be expected. Overrepresentation of viscous action in the model results mainly in two consequences: (1) the turbulent energy spectrum is cut off in the large wave number range and, (2) the internal structure of the boundary layer is somewhat different, e.g. the viscous sublayer will be thicker.

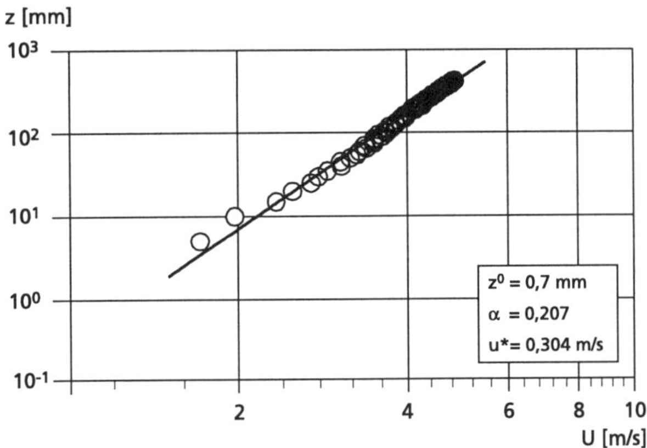

Figure 14.3 Vertical profile of wind velocity at the end of the zone of flow establishment.

The first point is of minor importance for dispersion experiments, since the mixing process is dominated by the large turbulent eddies. In the case of urban areas, the second point is also irrelevant because as long as the height of the roughness elements exceeds the thickness of the viscous sublayer, the roughness elements dominate the structure of the boundary layer. Consequently, the fair agreement generally found by comparing the turbulence characteristics of wind tunnels with those of real atmospheric boundary layers is not surprising.

Figure 14.4 (next page) shows this for the vertical distribution of the horizontal velocity fluctuation in comparison with field data (ESDU 1974). Fair agreement is also obtained as far as power spectra are concerned. The data taken by Leitl and Schatzmann (1998) compare well with the curves proposed by Kaimal (1972) and Simiu and Scanlan (1986) (Figure 14.5, next page). Quality assurance requires that any wind-tunnel study comprises documentation on the comparison between the wind-tunnel boundary layer and the real boundary layer, similar to Figures 14.3, 14.4 and 14.5.

14.3 Scaling techniques

Certain similarity requirements must be fulfilled in order to transfer results from small-scale wind tunnel experiments to full scale. These similarity laws are usually obtained by dimensional analysis, a method which makes use of the fact that physical equations must be dimensionally homogeneous and hence the parameters occurring therein can only appear in certain combinations. For a continuous release with the exhaust gases having the same density as the surrounding air, a non-dimensional functional relationship for pollutant concentration c at a given receptor point within the canopy layer is:

$$\frac{c}{c_0} = f\left(\frac{U_{ref}\,t}{H}, \frac{U_{ref}H}{\nu}, \frac{U_0}{U_{ref}}, \frac{A_0}{H^2}, \frac{z_0}{H}, \frac{\delta}{H}, \varphi, \frac{x}{H}, \frac{y}{H}, \frac{z}{H}, \frac{l_1}{H}, \ldots, \frac{l_n}{H}\right) \qquad (14.3)$$

with c_0 the pollutant concentration in the exhaust gases, t the time, H the height of a typical building, U_{ref} the reference velocity (e.g. at height H in the undisturbed boundary layer flow), U_0 the velocity of the exhaust gases, A_0 the cross sectional area of the exhaust (so that the pollutant emission $Q = c_0U_0A_0$), x, y, z the co-ordinates of the receptor point, ν the kinematic viscosity of the air, z_0 the roughness length, δ the boundary layer thickness (or mixing height), φ the wind direction and l_1 to l_n the numerous length scales which determine the complete geometry of the urban landscape in the vicinity of the source.

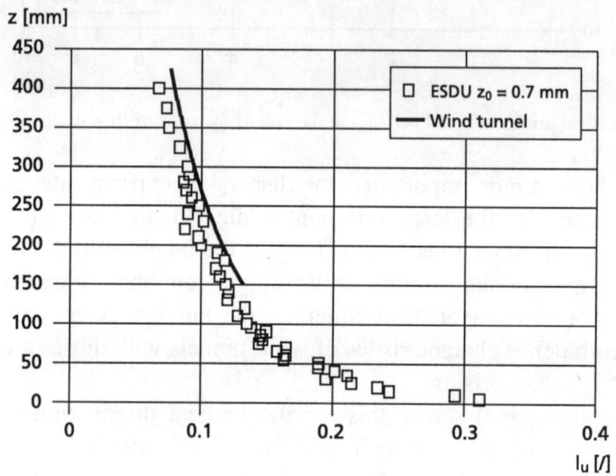

Figure 14.4 Vertical turbulence intensity profile in comparison with ESDU (1974) field data.

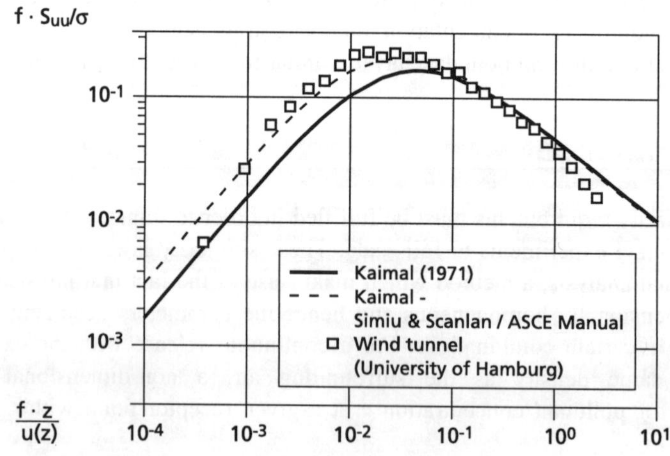

Figure 14.5 Power spectra measured in the wind tunnel in comparison with field spectra according to Kaimal et al. (1972) and Simiu and Scanlan (1986).

In a small scale model it is neither possible nor necessary to match all the similarity parameters listed in the equation to full scale values. The $U_{ref} \cdot t/H$-term disappears when we restrict our analysis to suitably defined time mean values. The Reynolds number $Re_H = U_{ref} \cdot H/\nu$ determines the flow around the obstacles. It is often significantly smaller in the wind tunnel than in the field, but the effect of Re_H on the concentration patterns is very small as long as it exceeds a critical value. This value can easily be found experimentally by increasing the reference velocity, while keeping all other non-dimensional parameters the same. The critical value is exceeded once the non-dimensional concentrations around the buildings do not change any more. The value of the critical Reynolds number depends on the building shape. It is typically around 10^4 for rectangular buildings, but for smooth cylinders it can be much more. The velocity ratio U_0/U_{ref} and the non-dimensional source area A_0/H^2 determine the size and effect of the exhaust jet or plume. If the momentum of the exhaust is negligible as, e.g., in the case of the simulation of traffic emissions in a street canyon, modelling requirements can be relaxed. These two ratios can be combined together with the non-dimensional concentration c/c_0 to obtain a single non-dimensional parameter $c \cdot U_{ref} \cdot H^2/Q$.

The parameters z_0/H and δ/H determine the approach flow well upwind from the built-up area, as discussed in the previous section. The boundary layer thickness δ/H will often not be matched since the wind tunnel boundary layer is seldom thick enough. As long as both in the model and in the real situation $\delta/H << 1$, i.e. when the problem is located in the surface layer, this has only marginal implications since the vortices shed from the buildings govern the dispersion pattern.

The remaining parameters in Equation 14.3 (page 265), the wind direction φ, the (reduced) co-ordinates of the receptor point and the length ratios can easily be matched in a geometrically scaled model.

We will not discuss the modelling implications of exhaust gases with densities different from the surrounding air. In that case, additional non-dimensional parameters enter Equation 14.3 that decrease the freedom of the wind-tunnel modeller, but satisfactory solutions are still possible (e.g. Snyder 1981).

Neither will we discuss the simulation of stable or unstable boundary layers. One may argue that thermal stratification is of little importance as long as the flow patterns and turbulence are dominated by the mechanical interaction between wind and buildings. It is therefore expected that thermal effects are less relevant in urban areas close to the surface than in rural areas or for dispersion from high stacks.

14.4 Transfer of small-scale results to full-scale applications

Provided all requirements expressed in Section 14.3 are met, Equation 14.3 states that in both model and full-scale identical non-dimensional concentrations will be found at receptor points defined by the reduced co-ordinates x/H, y/h and z/H. However, that statement holds only if all idealisations underlying Equation 14.3 are properly fulfilled.

There are some fundamental differences to be expected between results from field measurements, wind-tunnel experiments and numerical simulations (Schatzmann et al. 1997).

This is demonstrated in Figure 14.6 for the example of a small area source which continuously discharges a passive tracer into a street canyon. The evolution of concentration with time (in excess above background) is shown at the same receptor point and under identical ambient conditions. We assume that the objective is to determine a representative time mean concentration in all three cases.

14.4.1 FIELD EXPERIMENTS

High resolution field measurements provide usually highly intermittent signals, i.e. periods of near zero concentration are interspersed with non-zero fluctuating concentrations. It is to be expected that the intermittency of the signal depends largely on the turbulence structure within the canyon and the wind direction fluctuations. If the size of the assumed area source is small compared to the cross section of the street canyon, the concentration varies with time as shown in Figure 14.6 (top). Long averaging times are required in order to produce a meaningful time-mean-value. It is to be expected that the commonly used 10 or 30 min measurement cycles are not long enough. Longer averaging times (of more than one hour), however, are usually not feasible since the atmospheric boundary conditions change due to the diurnal cycle. Repetition of the same experiment for nominally identical conditions is, in practice, very hard to achieve. The conclusion is that large error bars should be attached to time-averaged concentrations determined in field situations as described.

14.4.2 PHYSICAL MODELLING

When the same dispersion problem is modelled in a wind tunnel or water channel, the concentration signal presented in Figure 14.6 (centre) is obtained. If all similarity parameters are matched properly, the time series will resemble that of the field test, but it will be somewhat less intermittent since the low frequency wind directional variations are not present in a ducted flow. Therefore, without further correction, the time mean concentrations based on physical modelling are usually higher than those obtained in the field.

An important advantage of wind tunnel measurements in comparison to field tests, however, is that the boundary conditions can be chosen to be appropriate to the problem being solved, and that numerous repetitions of the same case can be made in order to determine the inherent variability of the dispersing cloud characteristics.

14.4.3 NUMERICAL MODEL RESULTS

Finally, at the bottom of Figure 14.6 the evolution of concentration with time, as obtained from a common grid model, is displayed. We assume that the model considers turbulent fluctuations only within the turbulence parameterisation. Then, with constant boundary conditions, it delivers a stationary concentration value. In contrast to the point-by-point experimental data, this value represents not only a time-mean but also a space-mean concentration of the whole volume of the grid cell. If these fundamental differences are not properly taken into account, a comparison of results from the three different sources would resemble the proverbial comparison of apples and oranges.

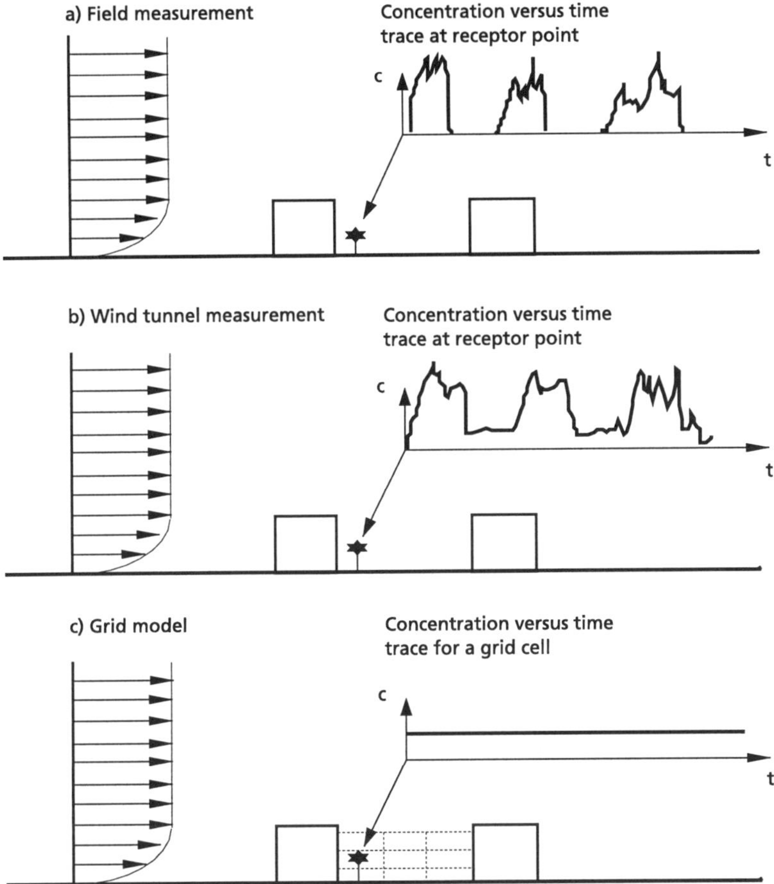

Figure 14.6 Comparison of concentration versus time traces typical for field measurements (top), wind tunnel measurements (centre) and numerical model results (bottom) (excess concentrations only) (Schatzmann et al. 1997).

14.5 Application examples

14.5.1 VENTILATION OF BUILDING EXHAUST GASES

One possible method to transfer wind-tunnel data to the full scale situation is demonstrated by means of measurements performed in the windtunnel of TNO, Apeldoorn, The Netherlands. The experiments addressed the problems associated to gas-fired heating units, exhausting at the facades of multi-storey apartment buildings with external balconies (Figure 14.7, next page). The exhaust gases from one apartment can easily cause problems on other balconies. The situation corresponds to a row of three similar apartment buildings in Zutphen, a small town in the eastern part of The Netherlands. The apartment building is about 12.5 m high (5 storeys), the exhaust is at the 4th, and the measurement point at the top level.

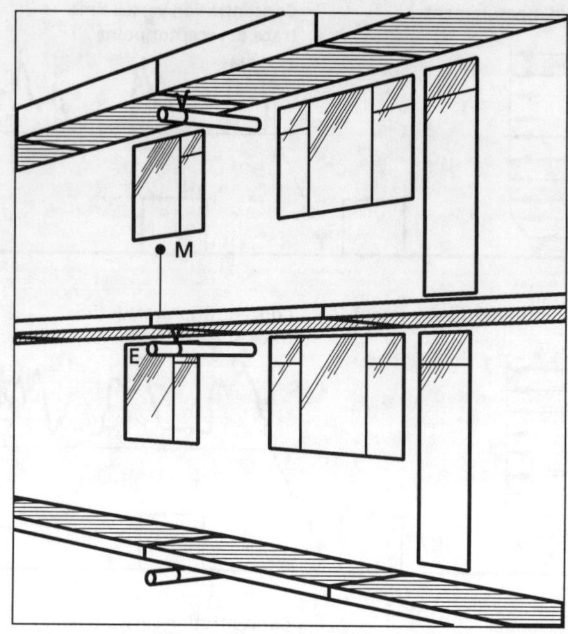

Figure 14.7 The upper two storeys of the apartment building near the measurement position. E is the exhaust pipe, M is the sample location.

The building has an L-shape (Figure 14.8), and the measurement position is about 20 m (about 1/4th of the length of that wing) from the inside corner of the building, facing north. This position was selected because a small set of full scale measurements was available. The exhaust pipe has an internal diameter of 80 mm and an exit velocity of 1.9 m/s (full scale). The model is at a 1 to 100 scale.

Wind-tunnel measurements were performed for different wind speeds and for every 5° wind direction. The results, with wind speeds expressed in full scale values, are shown in Figure 14.9. This Figure contains both the data as they come directly from the wind-tunnel measurements as well as the data after they have been transformed to a full scale averaging time of 1 hour. This transformation is based on a numerical superposition of wind direction variation on the measured wind tunnel data.

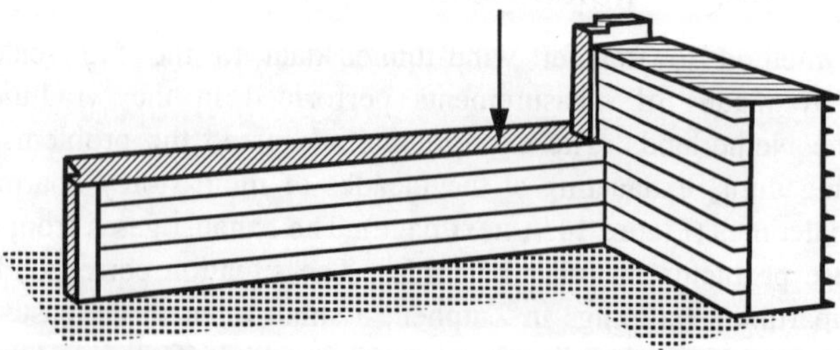

Figure 14.8 The apartment building presented in the example. The arrow points at the location where the emission takes place.

In other words, the 1-hour averaged concentration for e.g. 135° wind direction contains contributions from wind-tunnel data obtained between 110° and 160°. The lack of wind-direction variation in the wind tunnel compared to the full scale wind-direction variation in one hour is thus added artificially. The amount of wind-direction variation to be added to the wind-tunnel data is determined by comparing full scale measurements of plume width over simple, flat terrain with the wind-tunnel (Duijm 1994).

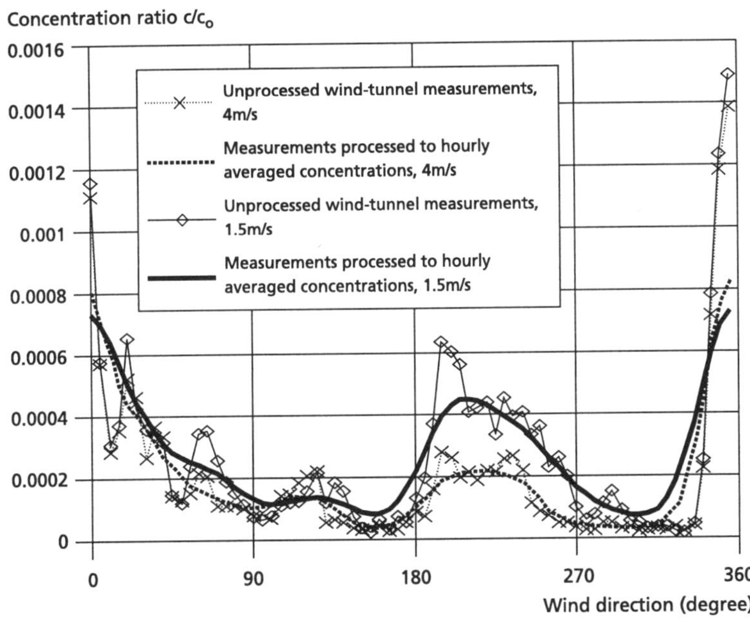

Figure 14.9 Example results of wind-tunnel measurements, showing the effect of time averaging. The curves present the variation of non-dimensional concentration with wind direction.

From Figure 14.9 some observations can me made. Due to the additional wind direction fluctuations, the hourly averaged concentrations show less variation and lower maximum concentrations, as expected. The unprocessed wind tunnel measurements show a lot of variation, which is not due to measurement uncertainties, they are reproduced at different wind speeds. This shows the unpredictability of the effect of small changes in boundary conditions like wind direction on the turbulent flow behaviour. The figure shows also a striking difference for northerly winds (hitting the building face directly) and southerly winds. In the first case, the concentrations do not depend on wind speed, i.e. the dilution due to the turbulence in the jet itself determines the concentration. In the second case, the exhaust and the measurement point are in the recirculation zone downwind of the building. In that case, the concentration is inversely proportional to the wind speed.

This example shows how the wind tunnel can provide information about very practical air pollution problems, and how the results can be transferred to the real situation.

14.5.2 URBAN BUILDING SPACING AND STREET POLLUTION DISPERSION

The second example shows how wind tunnel modelling may help study systematically the influence of the density of buildings in a town on the air quality inside the urban street canyons. Snyder (1994) has studied the effect of stratification on the flow and dispersion around simple cubical buildings and demonstrated that the behaviour resembles the neutral density case for densimetric Froude numbers of 2.5 and above. It has also shown that this critical threshold can be increased further for buildings with a larger cross-wind frontal area, as the typical urban building arrangement is. Therefore, reliable urban dispersion studies may be performed in the relatively simpler neutral-stratification boundary layer (BL) wind tunnels, rather than in convective BL wind tunnels or water tanks, especially when the near-field effects close to buildings are mainly of interest.

Building spacing is often considered significant for street air quality in towns. This effect has been studied systematically by Meroney et al. (1996), under a geometric scale of approximately 1:500, in the atmospheric BL tunnel shown in Figure 14.1 (page 263). For this study a fully developed BL was developed with a wind profile exponent n = 0.28, displacement height 2 mm and roughness length z_0 = 1 mm, determined from a logarithmic law fitting to the data from the lower 120 mm above the floor. The ratio between friction and free stream velocities, $U^*/U(\delta)$, equals 0.065, resulting to a roughness Reynolds number, U^*z_0/ν, greater than the 2.5 recommended by Snyder (1972). The urban model was two-dimensional and consisted of 60 x 60 mm bars to model flat-roofed buildings. The study was performed with 5 m/s free-stream wind, blowing normal to the street. Blockage of the tunnel test section did not exceed 6%. With this configuration the obstacle Reynolds number ReH exceeds the value of 3400 suggested by Hoydysh et al. (1974) and ensures independence from viscous effects. Tracer gas was released at street level through the line source shown in Figure 14.10. Two different line source positions were tested, once halfway between the buildings and secondly at a constant H/2 distance from the leeward wall of the upwind building.

The resulting mean concentrations C (ml/m^3) on the perimeter of the canyon were measured at the positions shown in Figure 14.11A (page 274) and are shown as non-dimensionalised concentrations $K = CUHL/Q_S$ for different street canyon aspect ratios and surrounding fetch conditions (Figure 14.11B-D). QS/L is the line source strength, in ml/(ms). For instance, in the open-terrain cases (Figure 14.11B and Figure 14.11C), the test street canyon is surrounded by flat terrain covered only by roughness elements (Figure 14.2, page 264). In the urban-case (Figure 14.11D), on the other hand, street canyons of a similar geometry form the upstream and downstream fetches surrounding the test section.

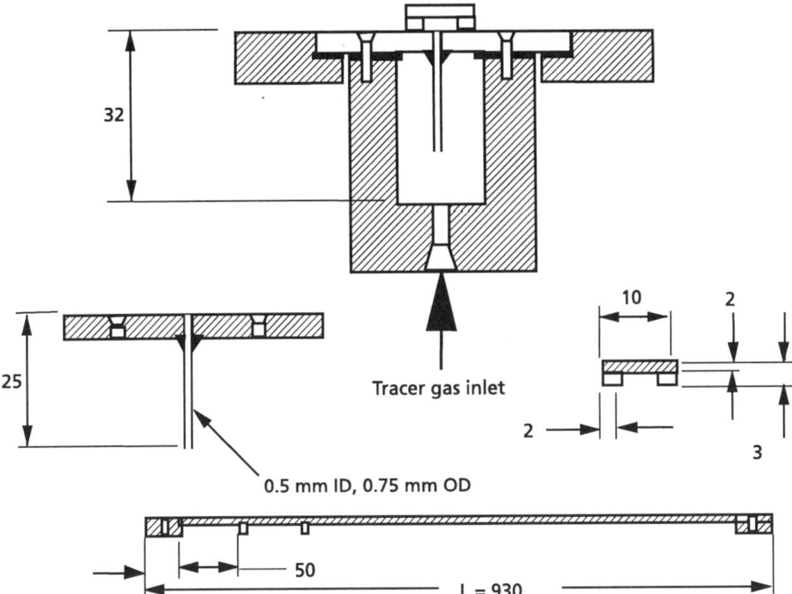

Figure 14.10 Line source design, using a total of some 300 discharge tubes per m (dimensions are shown throughout in mm). The capping bar running above the tubes, to remove discharge momentum, is shown at the bottom.

A clear feature distinguishing the urban fetch from the open country dispersion is the total absence of pollution on the upwind building roof. To explain this, recourse is made of laser sheet visualisation, the principle of which is shown in Figure 14.12 (page 275). Feeding, instead of tracer gas, paraffin smoke through the line source, enables the visualisations exemplified in Figure 14.13. These reveal that, in the open country, the vortex inside the street canyon is much more unstable than in surrounding urban roughness environment. This may be described as the "canyon breathing" discussed by Meroney et al. (1996). In the same case a recirculation bubble consistently forms above the upwind roof, into which pollution from the street is "sucked" from below. In the fully-developed urban internal BL above a homogeneous urban fetch, by contrast, skimming flow (Oke 1988) develops above the roofs and this recirculating bubble is totally absent. This leads to the total absence of pollution measured on the upstream roof (compare Figure 14.11D with Figure 14.11B or Figure 14.11C). This observation is feature of the flow in all canyon aspect ratios tested.

For all oncoming wind conditions and canyon aspect rations tested the canyon wall at the upwind building receives the highest pollution, which increases from the roof to the street. When the source is placed at a constant distance from the upstream building (Figure 14.11B) increasing the street canyon width has little effect on the concentrations measured on this upwind wall. Everywhere else in the canyon, widening of the canyon leads to lower pollution. A comparison between Figure 14.11C and Figure 14.11D at all measurement positions, shows that the urban canyons suffer worse pollution than the equivalent aspect ratios under open country oncoming wind conditions, even at quite wide building spacings. This is explained not only because, in the urban case, pollution is not "sucked" towards the absence of upwind roof, which in effect would increase the interface area between canyon and free-stream flow, but also the canyon vortex is much more stable in such canyons. Therefore, canyon "breathing" occurs much less frequently and street pollution is more effectively "trapped" inside the canyons.

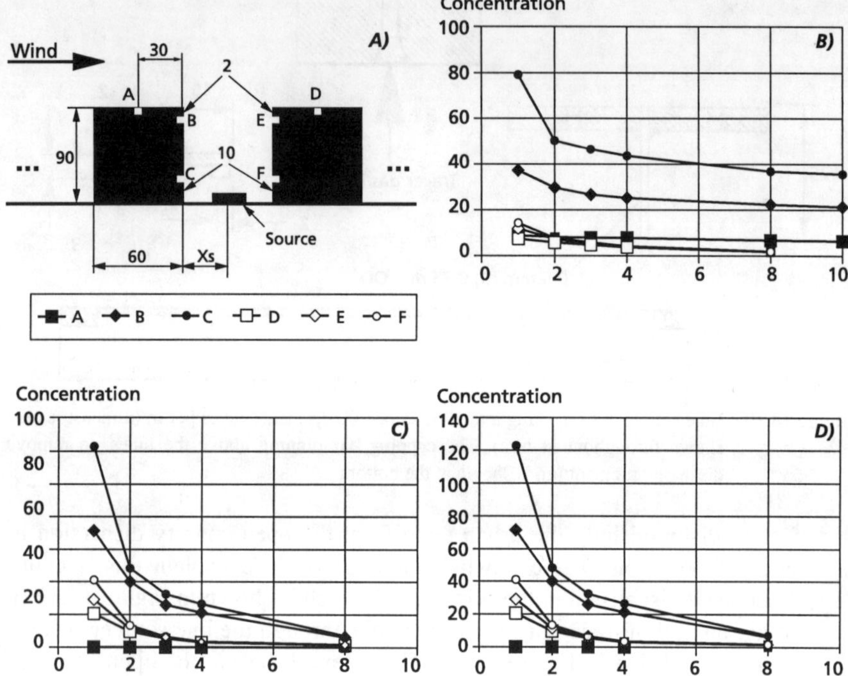

Figure 14.11 Concentration for street canyons with different ratios between streets and building heights (B/H).

A. Positions of the measuring points.
B. Open-terrain configurations; the line source is positioned in the test canyon at X_s = H/2 downwind of the leeward wall of the upstream building.
C. Open-terrain; the line sources is placed halfway between the two buildings.
D. Urban street canyons; the line source is halfway between the two buildings.

For a description of the mathematical modelling of air pollution in street canyons see Chapter 12. Compare especially Figure 14.13 with Figure 14.12.

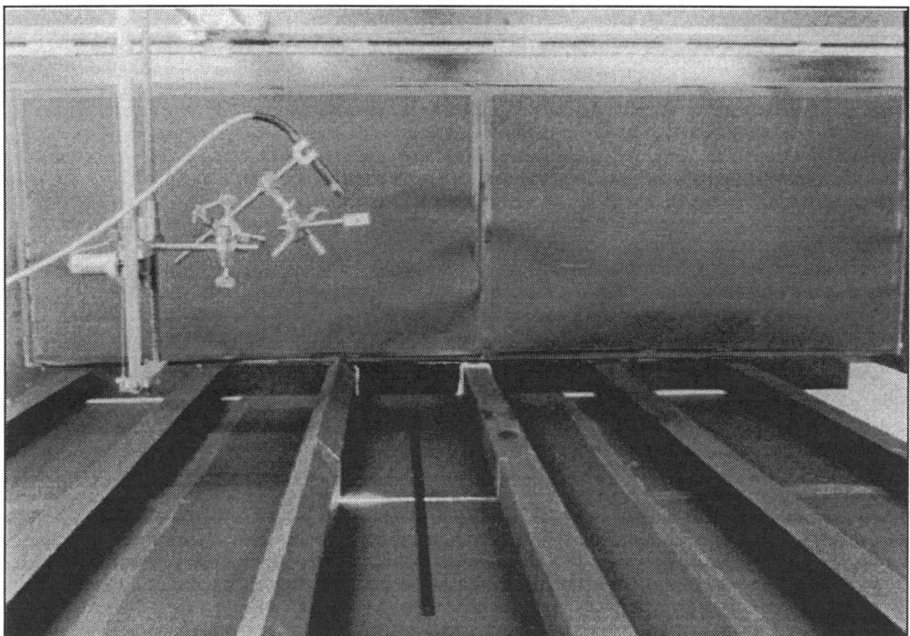

Figure 14.12 Laser sheet illumination arragement used to visualise the dispersion of smoke from the line source placed halfway between the buildings.

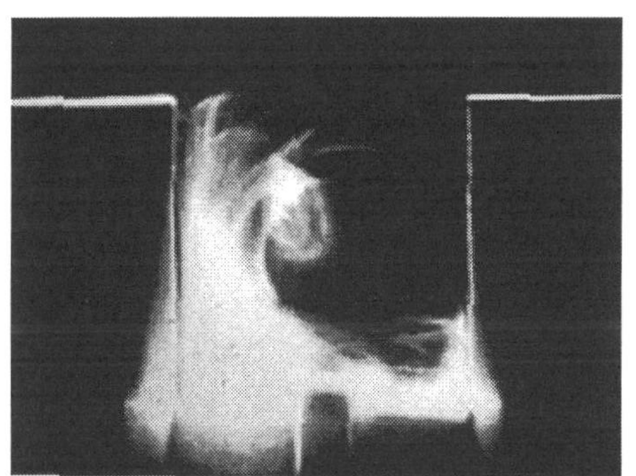

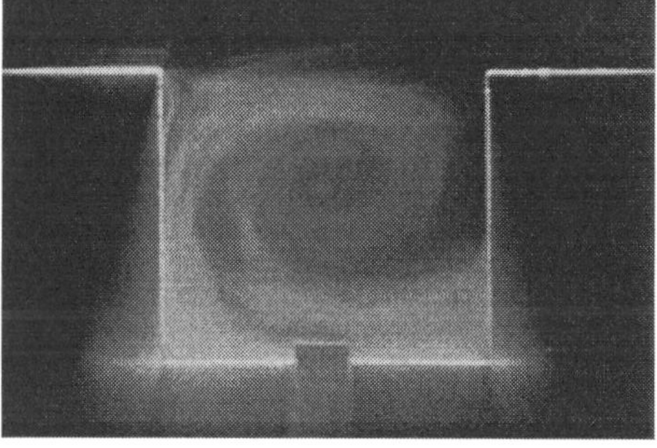

Figure 14.13 The typical air recirculations observed in identical square street canyons when placed amidst open terrain (left) or homogeneous urban fetch roughness (right). The wind blows from the left. The thickness of the light sheet is approximately 2 mm and the exposure time 4 s.

14.6 References

Duijm, N.J. (1994) Long-term air quality statistics derived from wind-tunnel investigations, *Atmospheric Environment*, **28**, 1915-1923.

ESDU (1974) *Characteristics of atmospheric turbulence near the ground, Part II: Single point data for strong winds (neutral atmosphere).* Engineering Science Data Unit, 251-259 Regent Street, London W1R 7AD, UK.

Hoydysh, W.G., Griffiths, R.A., Ogawa, Y. (1974) A Scale Model Study of the Dispersion of Pollution in Street Canyons, *67th Annual Meeting of the Air Poll. Control Assoc.*, Denver, CO, 9-13 June, 24p.

Irwin, H.P.A.H. (1981) The design of spires for wind simulation, *J. Wind Eng. Ind. Aerodyn.*, **7**, 361-366.

Kaimal, J.C., Wyngaard, J.C., Izumi, Y., Cote, P.R. (1972) *Spectral characteristics of surface layer turbulence, Quart. J. Roy. Met. Soc.*, **98**, 563-588.

Leitl, B., Schatzmann, M. (1998) Physikalische Modellierung von Geruchsausbreitungsvorgängen in Grenzschicht-Windkanälen, Berichtband, *VDI-Kolloquium "Gerüche in der Umwelt"*, Verein Deutscher Ingenieure, P.O. Box 101139, D-40002 Düsseldorf (in German, under preparation).

Meroney, R.N., Rafailidis, S., Pavagheau, M. (1996) Dispersion in Idealized Urban Street Canyons, in: Gryning, S.E., Schiermeier, F. (editors), *Air Pollutino Modelling and its Application XI*, pp. 451-458.

Schatzmann, M., Donat, J., Hendel, S., Krishnan, G. (1995) Design of a low-cost stratified boundary-layer wind tunnel, *J. Wind Eng. Ind. Aerodyn.*, **54/55**, 483-491.

Schatzmann, M., Rafailidis, S., Pavagheau, M. (1997) Some remarks on the validation of small-scale dispersion models with field and laboratory data. *J. Wind Eng. Ind. Aerodyn.* (in print).

Simiu, E., Scanlan, R.H. (1986) *Wind Effects on Structures; Part A: The Atmosphere*. J. Wiley and Sons Inc.

Snyder, W.H. (1972) Similarity Criteria for the Application of Fluid Models to the Study of Air Pollution Meteorology, *Bound.-Layer Meteorol.*, **3**, 113.

Snyder, W.H. (1981) Guideline for fluid modeling of atmospheric diffusion. Environmental Sciences Research Laboratory, U.S. Environmental Protection Agency, Research Triangle Park, NC 27711, *EPA Report No 600/8-81-009, NTIS Report No PB81-201410.*

Snyder, W.H. (1994) Some Observations of the Influence of Stratification on Diffusion in Building Wakes, in: Castro, I.P., Rockliff, N.J. (eidtors), *Stably Stratified Flows: Flow and Dispersion over Topography*, pp. 301-324, Clarendon Press, Oxford, Endland.

V

AIR QUALITY MEASUREMENTS

Air quality measurements are fundamental for monitoring of air pollution, studies of dispersion processes, atmospheric chemistry processes, exposure studies and development of air quality models. The techniques must be accurate, must be specific for the pollutants in question and must fit the purpose of the measurements with respect to sensitivity, measurement range, averaging time etc. In addition, a well defined measurement strategy should be followed. Only measurements with a well documented quality and strategy can be used for the above purposes. Numerous different types of measurement techniques have been developed. Some techniques are well established and have been used many years for air quality monitoring. They could be internationally standardised by standardisation organisations (ISO or CEN). Other techniques could be newly developed and only used for front line research for later development to standard methods. However, in any case the documentation is crucial for the applicability of the techniques and data.

Air quality *measurement techniques* (Chapter 15) can be based on two main principles: sampling followed by subsequent analysis and in situ measurements by automatic monitors. The sampling could take place by collection in liquids, impregnated surfaces, impregnated filters, filters, canisters, columns etc. The sampling techniques range from passive collection of pollutants to active collection by sucking air through the sampling system. The subsequent analyses could be standard chemical analyses or especially developed analytic techniques. Automatic methods range from electrochemical monitors over optical instruments to remote sensing methods based on light absorption spectrometry in open paths. The measurement techniques for particle measurements require special attention. Particle measurements must include the total amount of particles (number or mass) as well as the size fractions. Since no real reference material exist for particles, standard measurement methods must be defined.

The measurement equipment must be protected from temperature variations, weather influence, vibrations etc. in order to work properly. The instruments are therefore often installed in air conditioned houses/containers. The necessary equipment for calibration and maintenance should be available at the stations or in the laboratory. Commonly used methods are described and special attention is given to calibration and interferences.

In most European countries air *quality monitoring programmes* (Chapter 16) have been established since the 1950's. However, the monitoring objectives have not always been well defined. During the 1970's the design and optimisation of monitoring networks got higher priority (i.a. based on the work of WHO) and the development of *air quality management systems* was initiated. Today the design of networks, location and number of stations, accuracy and representativity are of central importance for the networks established in connection with the EU-directive on air quality. Air quality monitoring is not only a question of compliance with air quality limit values. It is also a central part of the air quality assessment and management process, which includes health/exposure studies, cost/benefit and cost/effective analyses, decisions on measures to be taken for reduction of the air pollution etc.

In order to have an effective monitoring programme the monitoring objectives must be clear. The programme should be able to give an adequate picture of the air pollution with a minimum use of resources. It must be taken into consideration to apply different types of measurement techniques and strategies and possibly supplemented by air quality modelling.

The number and location of monitoring sites must fulfil the legal requirements. Other considerations concern the scale of the air pollution problem, e.g. is the air pollution of local origin (e.g. traffic)? is there a significant regional contribution (long range transport of pollutants)? or is the large scale phenomenon important (e.g. ozone episodes)? The types of stations can be classified, and at present the most comprehensive classification schemes have been established under Exchange of Information Decision prepared by EC.

The pollutants to be included can be very comprehensive. In Chapter 16 lists and classifications are given in relation to population exposure, material exposure and ecosystem. Examples of modern air quality management systems are given.

Crucial for valid monitoring data are *quality assurance and quality control*, QA/QC (Chapter 17). *Quality Assurance* involves the management of the entire process, which includes all the planned and systematic activities which are needed to assure and demonstrate the predefined quality of data, to provide adequate confidence that an entity will fulfil the requirements for quality. *Quality Control* comprises the operational techniques and activities that are undertaken to fulfil the requirements for quality. Chapter 17 treats data quality objectives comprising precision, accuracy, representativity, data capture, time coverage, comparability and definitions. Realistic and necessary data quality objectives are suggested in relation to new EU-directives and networks, e.g. EUROAIRNET. A complete QA/QC plan is outlined including QA, QC and quality assessment.

Chapter 15

MEASURING TECHNIQUES

IVO ALLEGRINI
Instituto sull'Inquinamento Atmosferico, C.P. 10
I-00016 Monterotundo Scalo (Roma), Italy

15.1 Introduction

Pollutants directly emitted from sources (primary pollutants) or formed in the atmosphere as the result of chemical transformations (secondary pollutants) are known to contain compounds that are harmful to human health, materials and ecosystems. These species must be evaluated according to their spatial and temporal distribution in order to establish efficient control strategies. Such goal may be achieved through monitoring networks or campaigns based on well selected and developed measuring facilities. These include instruments for the sampling and for the direct measurements of atmospheric pollutants of interest in environmental studies. The criteria for selection of methods could be technical, scientific, practical, economical etc.

In some cases the measurements are parts of an international or national measurement or monitoring programme, which prescribes a certain measuring method, i.e. reference or standard method.

The measuring facilities may be fixed or mobile. A fixed station provides data on a continuous basis at a given location. If adequately selected, the location may describe the environmental situation in a more or less wide territory and be used to study the evolution of atmospheric pollutants over a long period of time. Conversely, mobile systems are very useful for the evaluation of pollutants over a limited amount of time, i.e. process studies. They are mainly used for the preliminary evaluation of pollution and for the understanding of particular processes of interest for the area or the phenomenon under investigation.

In a formal way, it can be said that measurements provide the following information about concentrations of pollutants:

$$C_{i,j,t} = f(i, j, t_1, t_2) \quad (15.1)$$

where:
c: Concentration of the pollutant
i: Type of pollutant to be considered
j: Location of the monitoring apparatuses
t_1: Start of integration time
t_2: End of integration time (For continuous monitor: $t_1 = t_2 = t$)

15.2 Station equipment

The basic design of a station intended for fixed monitoring networks or measuring campaigns is depicted in Figure 15.1. It consist typically of a container, which hosts the equipment for the sampling and measuring the pollutants of interest at the site and the equipment for data acquisition and transmission. The container consists e.g. of a sturdy steel reinforced, plastic or aluminium structure, which in most cases is fixed to the ground by means of a proper concrete platform. There are many available designs, sizes and shapes and most of them include racks to accommodate the instruments. Some containers are divided into two vans. The first intended for the instruments and a second, which accommodates the pumps for the analysers and for the sampler. This solution prevents heat, noise and vibrations building up into the main unit. The container must also include an air conditioning system to keep a constant temperature inside the container.

The sample intake of the monitoring stations includes often two intakes. One is used for the sampling of gaseous components and is made of inert material to avoid losses of reactive species such as ozone or nitrogen dioxide. It is heated to a few degrees above ambient temperature to avoid water condensation. In addition, in order to avoid chemical reactions (e.g. between ozone and nitrogen monoxide), the air flow is high. Usually the residence time of the gas in the sample intake is limited to a fraction of a second. A second inlet is serving instruments for the collection of particulate matter. This intake may include a size fractionating sampling head in order to ensure that particles enter the inlet cut off at a well defined aerodynamic diameter. Such a sampling head is mandatory for e.g. PM_{10} measurements.

Installations in a typical monitoring station include in addition to air conditioning, intrusion alarms, electric power distribution facilities and the telephone communication system in order to provide data transmission to the main computer. Other installations are related to gas supply e.g. for zero air generation or calibration of apparatus.

Sampling and analytical instruments are the main components in any monitoring station and will be described and discussed in the following.

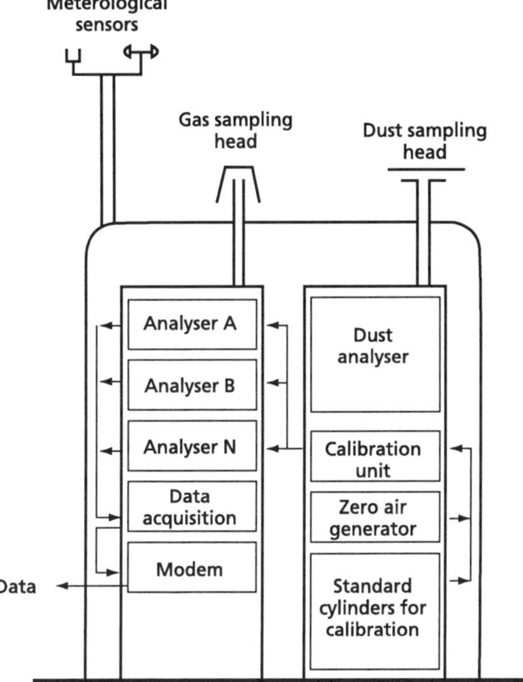

Figure 15.1 Schematics of a typical monitoring container.

15.3 Sampling and analysis

Sampling of atmospheric pollutants has been performed for many decades of the history of air pollution problems. A classical example is the measurement of the particulate lead concentration in air, which requires that a known volume of air is sucked through a suitable filter. After the collection the particles are dissolved in acids and lead is determined by atomic absorption spectrometry. This type of measurement, which includes sampling and successive analysis, are usually refered to as "off-line" measurement since relevant data are only available some time after the sampling. It is obvious that sampling and analysis is necessary, when the pollutant of interest cannot be determined by an automatic method, or when the available automatic method is not suitable at the concentration range of interest.

The *general principles of an air sampler* is depicted in Figure 15.2 (next page). Air is drawn through the sampling line and the accumulation device by means of a suitable pump. This is selected according to the total pressure drop developed in the sampling line and according to the nominal flow rate which is needed for the selected application. The interface between the atmosphere and the sampling line is provided with a sampling head made of a suitable tubing or an intake, e.g. PM_{10} or $PM_{2.5}$ or in the case of denuders, by the denuder itself. Temperature and pressure of the inlet air should be recorded because the final concentration of the pollutants of interest is expressed as unit mass of pollutant per unit volume of sampled air; thus a correction to standard temperature, T, and pressure, P, conditions must be made.

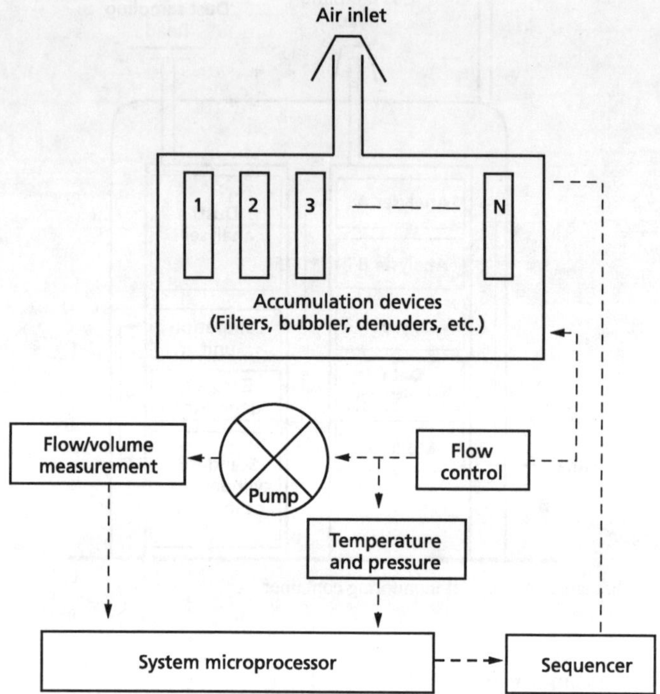

Figure 15.2 The general principles of a typical air pollution sampler.

A flow meter can be installed in the sample line for control of the air flow. After the pump, the volume of the sampled air is measured by a suitable gas meter. In modern samplers, the volume is recorded by electronic transducers for the flow rate and the total volume registration and correction for changes in atmospheric temperature and pressure. In addition, the sampling instrument may also include a device for automatic control of the sampling flow rate because the pressure drop along the sampling line may change with time. This is, for instance, the case of particulate collection of filters. The accumulated particles increases the pressure drop over the filter causing a decrease in the sampling flow rate. An active control keeps constant flow rate of the air in the line allowing the system to meet some operational specifications. For particles, the specification would be the nominal flow rate at which the sampling inlet operates a specified cut off size, e.g. PM_{10}.

The resulting sensitivity of the methods intended for sampling and analysis is controlled by the sensitivity of the analytical method, δQ (μg/ml), used after sampling and by the amount of material collected in the atmosphere. Determination of the duration of sampling will typically be determined by:

$$\delta C = \frac{\delta Q v t_{1,2}}{\Phi} \tag{15.2}$$

where Φ (m^3/min) is the sample flow rate, $t_{1,2}$ (min) is the sample duration of sampling, v (ml) is the volume of the solvent and δC ($\mu g/m^3$) is the wanted concentration sensitivity.

Since the sampling flow rate is usually controlled by the collecting device within a restricted range of values, the only way to increase the sensitivity for the measurement of a given pollutant, is to increase the collection time $t_{1,2}$. This is the reason why sampling based methods are also called "*accumulation methods*".

The above type of sampling systems and accumulation methods are frequently used for non traditional pollutants. They cover both organic and inorganic components in either gas-phase and particulate matter, thus covering any need that could arise in the management of atmospheric pollution, including the characterisation of species relevant for acid deposition and photochemical pollutants. The accumulation methods are also important for gathering information on the sources of atmospheric pollution by analysis of the distribution of individual species and components, e.g. receptor modelling.

Sampling of atmospheric pollutants may also be carried out *conditional.* This technique allows the automatic sampling of the pollutant of interest, when conditions important for the management of a certain air pollution episodes occur. For instance, conditional sampling may be carried out according to particular wind directions to intercept a polluting plume from a source, or according to the concentration values of another pollutant. An example is the accumulation sampling of benzene carried out when the carbon monoxide levels exceed a certain value in order to measure the peak values of benzene. In addition, conditional sampling may be carried out during alert situations caused by high values of criteria pollutants.

In order to perform conditional sampling, it is essential to handle samplers which have the capability of managing instructions from the main network computer or from the station microprocessors. Electronic actuators and transducers of physical parameters (e.g., temperature and pressure) allows for a very easy management of these tasks.

15.4 Sampling and analysis of particles

The most frequently used method for measurement of particles is collection on filters and subsequent analyses by traditional chemical or physical analyses. The filter can be produced of fibres (paper, glass, Teflon etc.) or membranes produced of polymers (nylon, esters of cellulose, polycarbonate etc.) For all types of filters a so-called pore size can be determined. A collection efficiency can be defined as the percentage of particles, which are collected at the filter. The collection efficiency of a certain filter depends on the aerodynamic particle size, the air velocity through the filter and possible chemical reactions on the filter. The collection efficiency is a result of several mechanisms leading to collection on the surface of the filter material, e.g. impaction, interception, diffusion, sedimentation and electrostatic forces. The impaction and interception are the dominating mechanisms for large particles and/or at high air velocities. The diffusion is dominating for small particles and/or low air velocities. The same mechanisms determine the transport of particles into the lungs. It is important to select filters and air velocities for high collection efficiency in the whole range of particles sizes. A high efficiency for large and small particles often leads to low efficiency for particles in the range 0.2-0.5 μm, as shown in Figure 15.3.

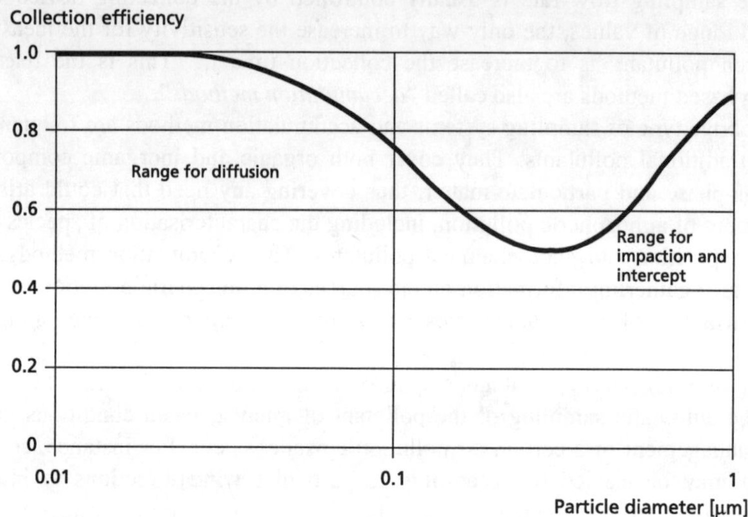

Figure 15.3 Collection efficiency of a typical membrane filter for different aerodynamic particle diameters at a given air velocity.

15.4.1 SIZE FRACTIONATION OF PARTICLES

Transport in the air and deposition on surfaces or in the human air-ways depends on the particle sizes. It is therefore crucial for the understanding of the processes and for determination of transport and deposition to determine the number or, the mass of particles for different particles sizes.

The most often used type is a filter sampler with a sampling head in the inlet, which cuts off the larger particles. Particles size fractions are defined as PM_{10} or $PM_{2.5}$ (particulate matter of particles smaller than 10 or 2.5 μm diameter respectively). TSP (total particulate matter) is often particulate matter with a more or less well defined upper cut off.

A better controlled size fractionation can be obtained by a so-called *impactor*. An example, a cascade impactor, is shown in Figure 15.4. The nozzles in the impactor become smaller and smaller resulting in larger and larger air velocities from stage to stage of the impactor, leading to smaller and smaller particles to be collected on the collection plates. The smallest particles are collected on a back-up filter.

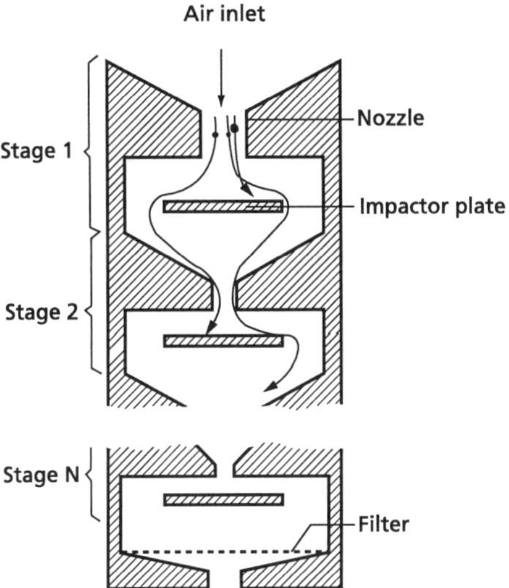

Figure 15.4 A general design of a cascade impactor.

15.4.2 ANALYSIS OF COLLECTED PARTICLES

A very simple way of analysis of particles, which has been used for many decades, is *determination of light reflection from the filter surface* after collection of particles. This method gives an non-specific measure of the particulates from incomplete combustion of coal, oil, petrol, wood or solid waste (soot or black smoke). It is still used as an indicator for man made pollution. A similar method is based on *changes in absorption of beta-radiation* before and after collection of particles on filters. A third method is based on measurement *changes in frequency of a vibration crystal* as a function of amount of particles deposited on the crystal surface. One of the advantages of these methods is that the measurements can be carried out automatically with short averaging time and the data can be send automatically to a central database.

The most frequently used method for determination of the particle mass is *increase in mass of the filter* after collection of air samples. After conditioning the filter is weighed before and after exposure. The conditioning must be carried out at a well defined temperature and relative humidity in order to take into account the absorption of water in the filter material and the particulates.

The filter samples can be analysed by traditional chemical methods for elements (heavy metals), inorganic ions (nutrients) or organic compounds (e.g. PAH). Frequently applied methods are wet chemical analyses, ion chromatography, x-ray fluorescence, PIXE, AAS and gas chromatography.

15.4.3 OTHER METHODS FOR PARTICLE MEASUREMENT

Numerous other methods have been developed for particle measurement. *Optical methods* are widely used. The methods are based on absorption or scattering of light. Light scattering depends on the refraction index and the size of the particles and by this it is possible to determine the particle size spectrum in details. The nephelometer is a simple instrument based on the total amount of scattered visible light and gives a measure of the visibility.

Other methods are *condensation nuclei counters, diffusion batteries, mobility analysers, cyclones, electrostatic or thermal precipitators* etc. These methods can be used for determination of size distribution of ultra fine particles, which are believed to play a major role for adverse human health effects. Figure 15.5 gives an overview different measurement methods.

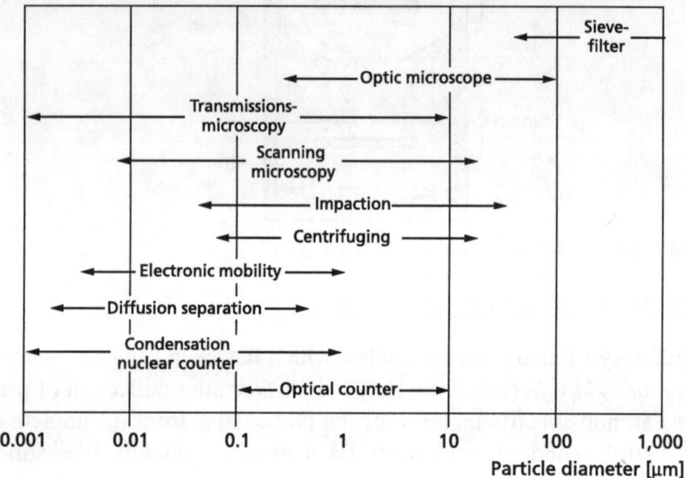

Figure 15.5 Summery of measurement methods suitable for different size fractions of particles.

15.5 Automatic methods

Automatic analysers are more or less complex instruments for the direct measurement of the concentration of a given polluting species. Figure 15.6 shows that the central component of an analyser is an electronic transducers which convert a concentration into an electrical signal. The signal may be electronically treated in order to communicate the relevant data to main network computer. An analyser provides an electric output V_{out} function of the concentration of the pollutant $C_{i,j,t}$ such that:

$$V_{out} = \phi\left[C_{i,j,t}, \alpha, \beta, \delta\right] \tag{15.3}$$

where the term ϕ is the transfer function, α is a term related to the presence of the interfering compounds and β is a term related to calibration factors; δ is a term related to other different types of deviations which can be e.g. thermal drifts and other electronic instabilities. Since the function ϕ is not only very complex, but in many cases also unknown, the automatic measurements of pollutant concentrations can be affected by large sources of error.

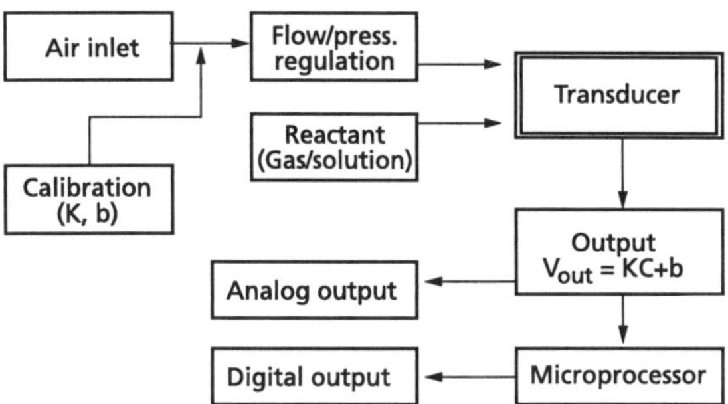

Figure 15.6 The central components of an automatic monitor.

Analysers are available for several atmospheric pollutants. However, they must be selected according to requirements which also depend upon the use of data gathered from monitoring stations. The most important requirements are the following:

- *Sensitivity*: The minimum increase in pollutant concentration which can be detected.
- *Minimum detectable amount*: The smallest concentration which can be reliably measured by the system.
- *Interference rejection*: The ability of the analyser to reject possible interference from other species present in the atmosphere.
- *Response time*: Stepping the input concentration from zero to a known value, it is the time needed to record a signal corresponding to a fixed percentage of the expected final output.
- *Drift*: At constant concentration input, instabilities and deviations from the expected output are observed. Usually a short term and a long term drift are measured and reported.
- *Integration time*: For many instruments, such as, for instance, those intended for particulate matter, the response is given over a period of time, so that the instrument does not provide a continuous output.

In addition to the above characteristics, other parameters may be defined taking into account the general performance of the analysers. The choice of an instrument is also affected by variables such as cost of the equipment, cost of maintenance, ease of operation, long term reliability, need for consumable etc. which must be taken into account if an optimised network has to be set-up.

It is not always possible to use the same instrument for *different monitoring tasks*. Consider for instance a typical pollutant as sulphur dioxide. When the measurements are addressed to the protection of public health, the response of the instruments should be fairly rapid (in the order of a minute or less, whereas the sensitivity should only be compatible with the alert concentrations. These are, in this case, in the order of several hundreds of $\mu g/m^3$, and thus a sensitivity of a few parts per billion is fully satisfactory. On the other hand if the monitoring network is a regional one, the expected concentrations are in the range between a few to a few tens of $\mu g/m^3$, and thus high sensitivities are required. In this case it would be reasonable to sacrifice, for instance, the integration time, by using accumulation methods, which require steps for sampling and analysis, but ensure very high sensitivities and very low minimum detectable concentrations.

15.6 Remote sensing

In the future a number of methods based on advanced open path spectrometry will be developed. It has been possible due to better transducers and advanced computer technology. Many component in the air can be measured specifically by remote sensing. By these methods it is not necessary to collect samples on filters, in containers or in measurements chambers, and many possible errors or artefacts can thus be avoided. Many of these methods are based on optical absorption spectrometry, e.g. Beer-Lambert's law:

$$I = I_0 e^{-\alpha NL} \tag{15.4}$$

where I_0 and I are the incoming light and the transmitted light respectively. N is the concentration of the absorbing molecules, L is the length of the light path, and α is the absorption cross section at the wave length in question.

Differential Optical Absorption Spectrometry (DOAS) (Platt, Perner 1980) is one of these methods. The principle of the method is outlined in Figures 15.7 a and b. By DOAS it is possible to measure concentrations of many air pollutants simultaneously. The method has been developed for e.g. SO_2, NO, NO_2, O_3, NH_3, CH_2O, HONO, C_6H_6, C_7H_8, $C_{10}H_{10}$, NO_3-radicals, HCl and CS_2.

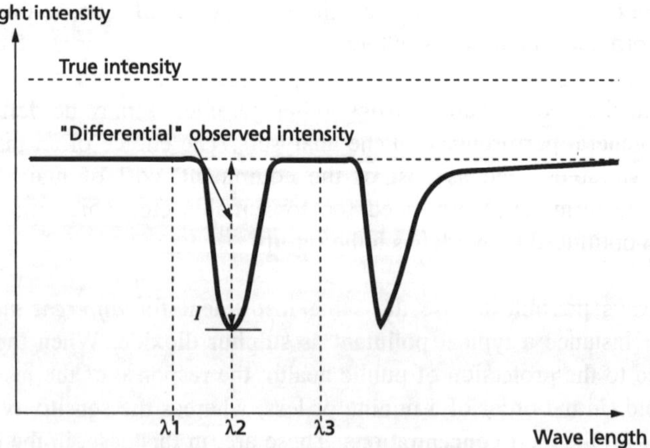

Figure 15.7a The fundamental principles of the DOAS technique.

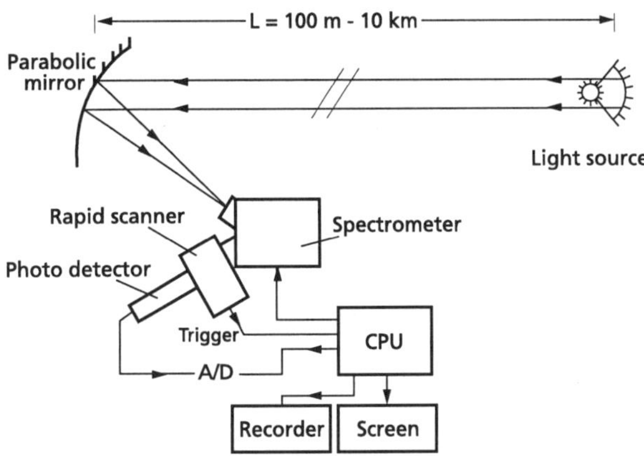

Figure 15.7b Schematics of a DOAS system. The path length in ambient air is normally 200-1000 m. Because the thermal turbulence causes large fluctuations of the broad banded absorption, the spectra must be recorded very fast, typically faster than 100 spectra per second.

Optical absorption in the infra red range has been used many years to measure the air concentrations, because many components have specific absorption peaks in this wavelength range. In the DOAS technique is used absorption in the UV and visible spectral range, where many important pollutant have absorption fine structures.

Fourier Transformed Infrared Spectrometry (FTIR) is a method to analyse absorption spectra. A Michelson interferometer with a fixed and a movable mirror and a beam splitter. The method gives a very selective and sensitive analysis of absorption spectra. In recent years many commercial field instruments have become available.

Emission measurements from cars will normally be made in the laboratory, (Chapter 5). However, a method has been developed for measurements on the road. The method is called *FEAT (Fuel Efficient Automobile Testing)*. It is based on infrared absorption in a light path across the street. The light absorption is measured by a fast technique, when the exhaust plumes from single cars pass the light beam. In practise, measurements must be carried out many time during the exhaust plume's passage. It is possible to measure the emission from every single car of. e.g. CO, hydrocarbons and NO. The advantage of the method is that it is possible measure many cars in a short time and that the measurements are carried out under real driving conditions on the road.

Remote sensing from satellites has already been used for measurements of the air pollutant on a large geographical scale. In the future, satellites may also be used for measurements in urban areas.

15.7 Calibration

With reference to Equation 15.2 the simplest case (which is very hard to reach in the practice) consists of a sensor characterised by $\alpha = \delta = 0$ and a linear response (V_{out}) function versus the concentration. In this case the function can be simplified to:

$$V_{out} = KC_{i,j,t} + b \qquad (15.5)$$

with K the calibration factor and b a residual signal from the instrument, when no pollutants are present (offset). As shown in Figure 15.8, the calibration procedure allow the determination of K and b. This is possible by using a calibrator, i.e. a device capable to provide artificial mixtures of the pollutants in pure synthetic air resulting in known and constant concentrations.

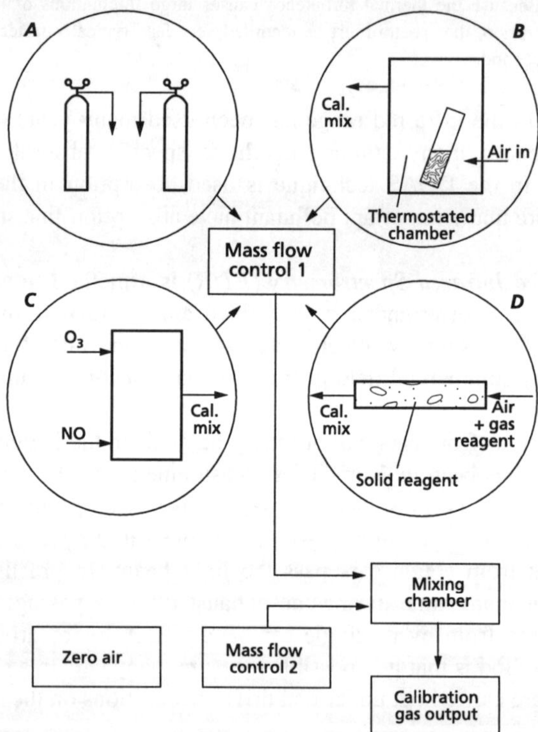

Figure 15.8 Calibration system for air pollution monitors. The calibration gases are gas mixtures (A), permeation tubes (B and D) or "gas phase titration" (C).

By input of the mixtures to the analyser and recording the output signal, K can be determined (calibration). The use of pure air will allow the determination of b (zeroing). If the response function is not linear or for more precise calibration, it is necessary to generate many different concentrations and more detailed recording of the output. The data allow a precise and accurate determination of the calibration curve even though the transfer function is not linear (multi-point calibration). Calibration and zeroing are fundamental operations in all types of measurement instruments. They can be carried out automatically every day and the quality control/quality assurance procedures are mainly addressed to the control of the calibration factors. Calibrating equipment is an important part of the station equipment. In several cases, the calibrators are incorporated into the analysers, and in other cases they are provided externally.

The present state of the art of calibration procedures offers several ways to prepare artificial mixtures of polluting gases. *Permeation tubes* has been extensively used. They were intended for species, which can be liquified under pressure, so that liquid phase is in equilibrium, at a rigorously constant temperature, with the gas phase in a small plastic container. Gas molecules are dissolving into the container walls and permeate through the walls at a constant rate. A suitable diluting flow of synthetic air provides a constant throughout of mixtures. The advantage of permeation tubes resides in the fact that the permeation rate may be characterised by weighing, thus they are standardised by means of primary mass measurement. Permeation through plastic walls is also used inversely by permeation into a pipe to generate standard mixtures of important pollutants, e.g. ammonia and hydrogen chloride. This is achieved by means of a plastic tubing immersed in a concentrated aqueous solution of pollutant. When air is delivered inside the tube, the output will consist of a gas solution of the pollutants. Even in this case, the temperature must be rigorously controlled since the permeation rate is strongly dependent upon the absolute temperature of the aqueous solution.

When possible, *standard mixtures* are generated by diluting primary or secondary concentrated mixtures of gases. Primary standard mixtures (cylinders) of gas pollutants in pure nitrogen are available through NIST (U.S. National Institute for Standard and Technology) in concentrations ranging from few to several tens of parts per millions. The concentration may be scaled down in the parts per billions range by diluting the primary standards. The dilution is achieved with high accuracy and precision by means of mass flow controllers, which are electronically controlled devices capable to supply well known amount of gas. By mixing the output of the primary standard cylinders with the diluting air, it is possible to obtain well characterised mixtures. In addition, by means of a proper modulation of the main flow and of the diluting flow, it is possible to achieve multi-point calibration. However, primary cylinders are rarely used in calibrating analysers in fixed stations. In fact, primary cylinders are more conveniently used to precisely calibrate transfer standard cylinders, which are used in field applications. Traceability of the standard mixtures, after dilution, to primary standards, is achieved by means of a rigorous calibration of the mass flow controllers.

Another method for the calibration of measuring apparatuses for atmospheric pollutant Z is based upon by the chemical reaction between a species X, for which a gaseous standard mixture can be generated, and another component Y: $X + Y \rightarrow Z$.

If it is known that the reaction is quantitative, then a standard mixture, Z is obtained. One example of this technique is the generation of standard mixtures of nitrous acid (HNO_2) in pure air, which can be conveniently prepared from a gaseous standard mixtures of hydrogen chloride reacting with a fluidised bed on sodium nitrate.

A technique involving chemical reactions is the so-called *GPT "Gas Phase Titration"*. In this case, the chemical reaction, which is taken into consideration, is the oxidation of nitrogen oxide to dioxide by means of ozone: $NO + O_3 \rightarrow NO_2$.

An NO air stream of known concentration is mixed with a known source of ozone having a concentration which is stochiometrically less than that of NO. The technique does not require the knowledge of the ozone concentration even though the ozone source output must remain constant though the calibration procedure. By increasing the ozone output from zero to high values, it is possible to convert increasing amounts of NO to generate mixtures of NO/NO_2 which may be conveniently used to calibrate, in a multi-point way, both channels of a chemiluminescence NO_X analyser. In addition, the GPT method allows the direct measurement of the converter efficiency.

15.8 Interference

The occurrence of interference in any instrument intended for the measurement of atmospheric pollutants is one of the most important problem for the selection of instruments and for data interpretation. Recalling Equation 15.2 and assuming that no deviations other than interfering species are present (δ=0), and that the instrument is well calibrated, we have:

$$V_{out} = \sum_i \phi_i \left[\alpha_i\right] \tag{15.6}$$

or:

$$V_{out} = \sum_k \left[\alpha_k C_k\right] \tag{15.7}$$

where C_k is the concentration of the interfering component, k, and α_k the term which indicates the response of the instrument to the interfering component. Ideally, if the species of interest is *i*: $\alpha_k \ll \alpha_i$. However, in practice, it is sufficient that $\alpha_k C_k \ll \alpha_i C_i$. Usually, the manufacturers specify the possible interference through an element of technical specification, which is well known as equivalent interference. This is the expected signal related to the species, is given by unit concentration of interfering, k. In order to show the implication of interference in the measurement of atmospheric pollutants, two cases will be described in details. Those related to the chemiluminescence analysers for nitrogen oxides - and those relevant for the measurement of dust concentration by beta gauge monitors.

15.8.1 EXAMPLE: CHEMILUMINESCENCE ANALYSERS

These analysers are based on the reduction of nitrogen containing compounds (NO_X) by means of a device known as a "converter" (It contains molybdenum or gold at high temperature) where they are converted into NO. NO is then oxidised to NO_2 by means of ozone resulting in a reaction where light is emitted (chemiluminescence):

$$NO_x \rightarrow NO \; ; \; NO + O_3 \rightarrow NO_2 + h\nu.$$

Therefore, the light emission is a measure of the total NO_X in the air stream. In a parallel channel, air is treated with ozone without any prior treatment, so the concentration of NO only is obtained. By subtraction of the output of the second channel from the output of the first channel, a measurement of NO_2 is obtained providing that the only NO_X species are NO and NO_2, i.e.:

$$[NO_x] = [NO] + [NO_2] \tag{15.8}$$

The resulting output from the two channels, assuming that no other deviations are present, may be expressed as:

$$V_A = K_A[NO] + b_A \qquad V_B = K_B[NO] + b_B \tag{15.9}$$

However, NO_X includes not only nitrogen monoxide and dioxide, and many other species can be reduced in the converter. The most important are nitric acid (HNO_3) and nitrous acid (HNO_2). Therefore:

$$V_b = K_B([NO] + [NO_2] + [HNO_2] + [HNO_3]) + b_B \tag{15.10}$$

Thus, if the difference between Channel A and B outputs is taken as the values for the concentration of NO_2, several requirements must be fulfilled:

- Both channels must be calibrated and offset. This means that a calibration with standard NO bottles or NO_2 permeation tubes are not sufficient. Only Gas Phase Titration is able to perform simultaneous calibration of both channels.
- The concentration of HNO_3 must be maintained as low as possible. This goal can be simply reached because HNO_3 is sticking over most surfaces, thus the inlet tubing are usually very efficient in removing this gas.
- The concentration of HNO_2 must be maintained very low or the conversion efficiency must be very low. This is accomplished by means of photolytic conversion.

Usually the concentrations of nitrous and nitric acid do not exceed 10 ppb. But this is about 10% of the established limit for nitrogen dioxide (200 $\mu g/m^3$), thus possible interferences must be taken into account.

15.8.2 AN EXAMPLE: BETA GAUGE MONITOR

The method for determining the mass concentration of particulate matter is straightforward, since it is based on the attenuation of beta radiation from a source when they interact with a filter. By measuring the beta transmission of a filter before and after sampling by placing it between a radioactive beta source and a detector, it is possible to evaluate the mass of particulate m by means of the following relationship:

$$I = I_0 \exp(-\mu m) \; m = K \log \frac{I_0}{I} \tag{15.11}$$

where K is a constant, I is the beta intensity after sampling, and I_0 is the beta intensity before sampling. This simple principle is affected by important errors, which may cause strong deviations from the ideal response. These errors are due to two main causes:

- Air absorption between the sources and the detector.
- Natural radioactivity associated to particulate matter.

Air absorption may cause error because the measurements before and after sampling are carried out in different periods of time. If a change in temperature or pressure occurs, a change in air density also occurs. This simulates a mass change which is not associated to particulate matter. Just to give an indication about this error, if a change in pressure of 0.3 % occurs and if 10 cm^3 of air are present between the source and the detector, a mass change of 45 μg occurs. If the sampling flow rate is in the order of 20 l/min and the sampling step is carried out for 2 hours, the resulting deviation is 18 $\mu g/m^3$. In many cases, this would not be acceptable. Errors associated to changes in air density may be corrected by measuring the actual temperature and pressure in the beta gauge.

Natural radioactivity is associated with the presence of radioisotopes such as those related to the Radon and Thoron progeny which are attached to suspended particulate matter. Natural radioactivity may cause severe interference problems since, especially in presence of volcanic rocks, may give large errors. In order to calculate the possible interference on the measurement of particulate mass, it is sufficient to recall that the limiting factor of the beta gauge precision is the uncertainties of the beta counts. This can be evaluated according to Poisson distribution where the Standard deviation is the square root of the mean value. Thus, assuming that natural radioactivity is present, Equation 15.11 becomes:

$$I = I_0 \exp(-\mu m) + I_{NR} \tag{15.12}$$

This errors can only be neglected if:

$$I_{NR} \leq \sqrt{I} \tag{15.13}$$

Very often this is not the case, therefore the amount of natural radioactivity should be corrected otherwise unrealistic results may occur. This is probably the most important drawback in the measurement of airborne particulate matter by means of beta gauge It is interesting to underpin that such errors are not due to chemical, but to physical reasons.

15.9 Types of instruments

Instruments also include samplers according to which pollutants are accumulated in proper devices and then analysed in the laboratory. A complete presentation of the techniques available for pollution monitoring is beyond the scope of this chapter, and only a general summary is given leaving to specialised papers and book the technical details. Table 15.1 shows techniques which are widely used for the measurement of individual components and the type of measuring approach.

Table 15.1 Techniques for air pollution monitoring.

Pollutant	Monitoring technique	Type
SO_2	Hydrogen peroxide acidimetric	Bubbler
	Colorimetric pararosaniline	Bubbler
	Ultra-Violet fluorescence[1] (ISO 1996)	Automatic
	Differential Optical Absorption Spectroscopy	Automatic
	Flame photometry	Automatic
	Diffusion denuders / Ion Chromatography	Denuder
NO_2	Chemiluminescence[1] (ISO 1985)	Automatic
	Diffusion Tubes (CEN 1998b)	Denuder
	Differential Optical Absorption Spectroscopy	Automatic
O_3	Chemiluminescence	Automatic
	Ultra-Violet photometry[1] (ISO 1998)	Automatic
	Differential Optical Absorption Spectroscopy	Automatic
CO	Infrared Absorption	Automatic
	Electrochemical cells	Automatic
Particulate matter	Hi-Vol. samplers	Sampler
	Smoke shade reflectance	Sampler
	Beta Gauge	Automatic
	Oscillating filter	Continuous
	Wide Range Aerosol Classifier (WRAC)[1] (CEN 1998a)	
PAN	Gas Chromatography/Electron Capture	Automatic
	Gas Chromatography/Chemiluminescence	Automatic
Benzene	Gas Chromatography	Automatic
	Absorption tubes	Sampler
	Differential Optical Absorption Spectroscopy	Automatic
VOCs	Gas Chromatography	Sampler
	Absorption tubes	Sampler
PAH	High Performance Liquid chromatography	Sampler
	Gas Chromatography / Mass spectrometry	Sampler
Metals	Filtration / X ray fluorescence	Sampler
	Filtration / atomic absorption spectrometry	Sampler
Acidity	Diffusion tubes / Ion chromatography	Denuder

[1]Reference methods for EU limit values.

Most of the instruments are based on spectrophotometric techniques, which allows good sensitivity and interference rejection. The instruments, as stated before, are designed in order to minimise interference and to maximise response time and sensitivity. In addition, since the market of such instruments is very active, reliability and low cost are essential prerogatives. Maintenance and costs of consumables must also be kept to a minimum. More information about the measurement methods can be found in the literature (Finlayson-Pitts, Pitts 1986).

15.10 Conclusions

In principle the problem of measuring atmospheric pollutants can be considered solved. The state of the art of modern instrumentation is such that reliable results are obtained at fixed monitoring stations. The achievement of environmental analytical chemistry is also able to provide sensitive and "up to date" methods for the analysis of several species of environmental interest, even at very low concentrations. However, what is important in the management of atmospheric pollution is the management of the problem and the possible way to solve it in cases where the limit values are exceeded. In such cases, there is the need to obtain not only data, but data of sufficiently good quality.

Even though this goal may be reached for several criteria pollutants, several measurements are currently performed with poor quality assurance and quality control. In other words, there is the diffuse opinion that data gathered by the instruments are inherently correct. This statement is not always true and pollution network managers should give to this point more attention.

15.11 References

CEN (1998a) prEN 12341. Air quality - Determination of PM_{10} fraction of suspended particulate matter - Reference method and field test procedure to demonstrate reference equivalence of measurement methods, Final draft, CEN, Brussels.

CEN (1998b) prEN xxx. Ambient air quality - Diffusive samplers for the determination of concentrations of gases and vapours - Requirements and test methods, Part 1 and part 2, Draft, CEN, Brussels.

ISO (1985) Ambient air - Determination of mass concentration of nitrogen oxides - Chemiluminescence method, ISO 7996, 9 pp., To be used as first draft of CEN standard, in preparation.

ISO (1996) Ambient air - Determination of sulphur dioxide. Ultraviolet fluorescence method, Draft international standard ISO/DIS 10496, 8 pp.

ISO (1998) Air quality - Determination of ozone in ambient air - Ultraviolet photometric method. Final draft ISO/FDIS 13964.

Finlayson-Pitts, B.J. and Pitts, J.N.Jr. (1986). Atmospheric chemistry. Fundamentals and experimental techniques, Wiley, New York, 1098 pp.

Platt, U., Perner, D. (1980) Direct measurements of atmospheric CH_2O, HNO_2, O_3, NO_2, and SO_2 by Differential Absorption in the near UV, in: Killinger, D.K., Moordian, A. (editors), 1983, *J. Geoph. Res.*, **85**, C12 7453-7458.

Chapter 16

MONITORING NETWORKS AND AIR QUALITY MANAGEMENT SYSTEMS

STEINAR LARSSEN
Norwegian Institute for Air Research
P.O.Box 100, N-2007 Kjeller, Norway

16.1 Introduction

Experience in the design and establishment of urban air quality monitoring networks has been gained since the 1950's, especially in Europe and North America. The considerable costs of running large monitoring programs lead many investigators, in the 1960's and early 1970's, to try to answer the question of how many sampling sites are necessary to estimate average and maximum urban air pollution concentrations (Munn 1981). No general guidelines have come out of this kind of work, although there is an obvious continued interest in the topic, for the same reason: the high costs of running large programs.

Earlier work that deal with the question of design and optimisation of urban monitoring networks include the following, chronologically: Rossano and Thielke outlined a systems approach to air quality surveillance program design (WHO 1976); The WHO treated air monitoring program design for urban and industrial areas in a special publication (WHO 1977) as a background for the Global Environmental Monitoring System (GEMS) program. Regarding number of stations, a general guide was given (Table 16.1): The question of optimisation of urban sampling networks was treated by Vucovich et al. (1978), and a design example for St. Louis was carried out using a combined dispersion, statistical and variation analysis model approach (Vukovich et al. 1978). Munn (1981) has produced a treatise on the design of air quality monitoring networks, which treats monitoring objectives, siting criteria and network design methods, including statistical and modelling methods. This treatise is still applicable as the most recent overview of available methods. Later, the European Commission has prepared a handbook for urban air improvement, which treats the monitoring network as part of the more complete assessment of urban air pollution (Butterwick et al. 1992). This is also the topic of a Guidebook for development of air quality management strategies prepared for the World Bank (Larssen et al. 1998a), namely: The role of the urban air quality monitoring network in the process of assessing urban air quality, and managing its improvement in a cost-efficient manner.

The question of network design and siting of stations is of as large interest today as it has been previously, due to cost as well as accuracy/representativity considerations. Even so, still no generally applicable quantitative guidelines have been worked out. At the same time, as EU-directives on air quality continue to put stricter requirements to the assessment of urban air quality, the requirements to monitoring also increase. Recently, the European Commission, in its proposed Council Directive for limit values for SO_2, NO_2, PM and lead (EC 1997a) has given criteria for the minimum number of stations in urban areas to check compliance (Table 16.2, page 306). Fortunately for the air quality managers of local as well as national designation, the tools for more complete and useful air quality assessments in urban areas, e.g. air quality models, are being improved. Also, the use of Air Quality Indicators (AQI), to simplify the monitoring needs to fewer compounds while keeping a sound background for evaluation of effects, are being promoted (Chapter 23). This will over time lessen the requirements to pure monitoring, which is now the method most frequently used to assess urban air quality.

This chapter deals with the role of monitoring as part of urban air quality assessment, and current practices of and requirements to urban monitoring in Europe.

16.2 Monitoring as part of the air quality assessment and management process

The air quality monitoring program is always a part of a broader *air quality assessment program* of the environmental authorities on the local, regional or national scale. The monitoring program is often the most important and developed part of the assessment program, and the part with the longest history. Other important parts of the assessment program are the emission inventory, the use of dispersion models and methods to estimate the exposure of the population (as a whole, or groups, or individuals) to air pollution. This assessment of exposure is again a basis for the abatement of air pollution based upon cost criteria (cost/benefit or cost/effectiveness analysis). The comparison of costs of control measures for air pollution reduction with the reduced costs of damage resulting from the air pollution abatement can result in abatement with least cost.

Figure 16.1 shows in principle the place of the monitoring system within this complete system of air quality assessment and cost assessment. This can also be termed the Air Quality Management System (AQMS). The activities that are necessary to develop an air quality management strategy are shown in Figure 16.2 (page 300).

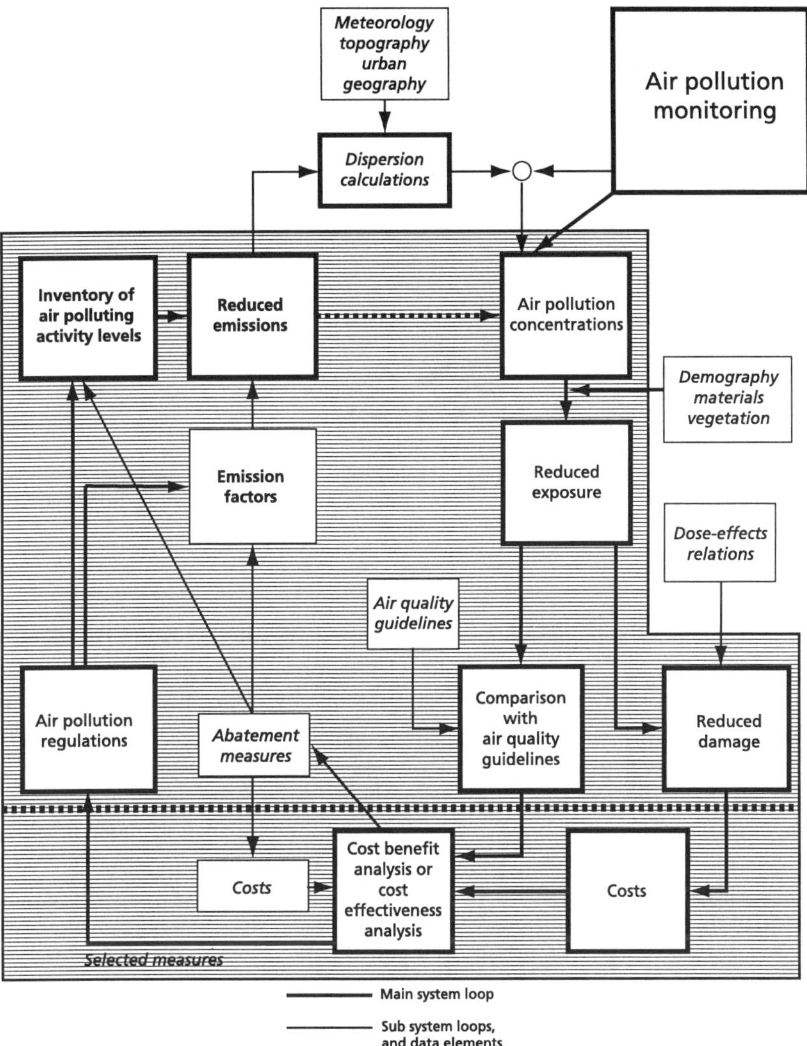

Figure 16.1 The place of the air quality monitoring program (AQMP) within the complete urban air quality management system (AQMS) (Larssen et al. 1998a).

16.3 Monitoring objectives

The design of an *urban air quality monitoring program* is decided by the objectives behind the need to monitor the air pollution. The monitoring objectives are again influenced by the extent of the broader air quality assessment program existing or being developed in the urban area in question. The general objective of an air quality monitoring program is to provide data for an adequate characterisation of the air

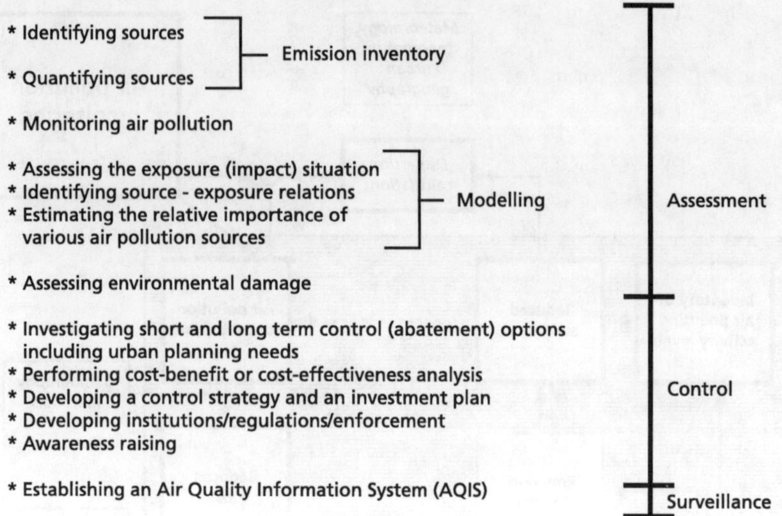

Figure 16.2 The elements in the development of an Air Quality Management Strategy (Larssen et al. 1998a).

pollution of an area, with minimum use of resources (time, costs). The data are to provide relevant information to governments, enterprises and the public for a range of purposes and applications connected to air quality assessment and air quality management. This objective is, however, far too general for use as a background for the network design in an analytical manner.

To design, establish and run an urban air quality monitoring program as a part of an assessment program involves making decisions about topics such as those listed below. The objectives of the monitoring program influence those decisions:

- The AQ assessment is to be based upon monitoring alone, or in combination with the use of dispersion modelling;
- Monitoring sites are to be located in a regular grid system or placed selectively at sites representative for defined micro-environments and exposure situations;
- Monitoring will be done at fixed sites or on moveable (mobile) platforms;
- Selection of compounds, methods;
- Data quality level.

There is a wide range of monitoring objectives in terms of the requirements they put on the actual design of the program. These objectives range e.g. from monitoring to indicate whether the air pollution levels in a city are of concern relative to possible damage to objects (the population, the ecosystems, materials), to e.g. monitoring as part of a long-term air quality management system to develop cost-effective abatement strategies.

Objectives of air quality monitoring include those listed on the opposite page. In a sequence of increasing requirements to the completeness and accuracy of the monitoring system they must:

- permit a preliminary assessment of air quality, to determine further needs for monitoring, e.g. to determine compliance with air quality limit values (AQLVs);
- facilitate a general description of air quality, and its development over time (trend);
- present on-line air quality information to the public, and to forecast near-future air quality;
- produce emissions/exposure relations;
- quantify damage to health, materials and vegetation, and exposure/effects relations;
- support development of cost-effective abatement strategies;
- support legislation (in terms of air quality directives);
- influence/inform/assess effectiveness of previous/future policy.

The current state of knowledge of air quality, its causes and monitoring is fairly advanced in many European cities. The current "state-of -the-art" objectives of urban monitoring programs in Europe are thus that the monitoring program shall permit:

- determination of the degree of compliance with AQLV,
- detection of trends with sufficient accuracy, and effects of on-going abatement programs,
- on-line presentation of present air quality,
- determination of source/exposure relationships,
- prediction of effects of planned abatement.

The ability to forecast short-term (near future) air quality might also be added to this list.

The last two objectives (as well as forecasting) are fulfilled only when the monitoring program is combined with programs for emission inventorying and dispersion modelling.

Current European air pollution legislation will to some degree require systems fulfilling most of these objectives in cities with air pollution levels approaching AQLVs. Simpler systems fulfilling only the first two objectives above can be used for preliminary assessments, but will need upgrading over time. More complete systems which also support near-future air pollution forecasting, damage assessment and air quality management based on cost-benefit analysis require integrated systems of (on-line) monitoring, emission data bases and operative dispersion models.

16.4 Network design topics, basic approach and number of stations

It is important to realise that the pollution concentration at any point is a sum of impacts or contributions originating from different sources on different scales. Depending upon the location of the point, the total concentration is a sum of

- a natural background concentration,
- a regional background concentration,
- an average urban (own) contribution in the area around the point,
- local impacts from nearby sources, such as streets, point sources (industry, heating plants), etc.

There are two major *approaches for designing* a monitoring system network (Butterwick et al. 1992):

- Sampling sites are located close to intersection points on a regular geometric grid superimposed on the city. This approach has been used e.g. in Germany. Care should be taken to adjust the site location where the grid intersection is close to a specific source. This method leads to a large number of sites, and provides the possibility of constructing iso-lines of air pollution concentrations based solely on measurements (with spatial resolution down towards the size of the grid).
- Sampling sites are located selectively at sites considered to be representative for more defined local environments, exposure situations or source activities, such as urban background, residential areas in various parts of the city, industrial or traffic sites, and also extra-urban background sites.

The latter approach is generally preferred as being more cost-efficient, and especially when dispersion modelling is used as part of the assessment program. It provides measured concentrations which indicate the exposure in residential areas or hot-spot locations, and provides data for comparison with, and possible adjustment of the dispersion model. With proper selection and location of measurement sites, it is possible to use the comparison between measured and model calculated concentrations actively to check emission data for specific sources (Grønskei, Walker 1995).

Another consideration in the basic approach to network design is *the scale of the air pollution problem*:

- The air pollution is of predominantly *local origin*. The network is then concentrated within the city, with only one (or very few) station(s) outside the urban area to monitor the regional background. This is applicable for primary compounds such as CO, lead, PAH and benzene when the city is not influenced significantly by other nearby large cities or major sources.
- There is a significant *regional contribution* to the problem. More emphasis must then be put on monitoring/modelling also the regional part. In most cities in Europe, this applies to compounds such as ozone, NO_2, PM (fine fraction, i.e. PM_{10} and smaller).
- Examples of *large-scale phenomena* are "winter smog episodes" of north-western Europe, and photochemical pollution episodes near the Mediterranean. Such episodes may be caused by the aggregated effect of emissions in a large area during prolonged anti-cyclonic periods in winter with poor dispersion (winter smog), or by the effect of large land-sea breeze domains where emissions in large city plumes are caught in a recirculation pattern, allowing time for photochemical reactions and build-up of ozone (photochemical episodes near the Mediterranean).

The investigation and control of such phenomena require monitoring and modelling on a scale much larger than the scale of the individual city affected.

An air quality monitoring network basically consists of two parts: *the air pollution monitoring part* and *the meteorology (dispersion parameter) monitoring part.*

The air pollution (AP) monitoring part of the network is treated in more detail below. The meteorological monitoring part, on the other hand, will not be covered in any depth in this chapter. It shall just be mentioned that meteorological data are needed for at least two reasons:

- For the interpretation of the temporal and spatial variation of the data from the AP monitoring, , there is an obvious need for meteorological data:
  - wind speed and direction;
  - parameters describing atmospheric turbulence and stability, such as temperature profiles (measurements at two or more heights), or direct turbulence measurements;
  - mixing height; and
  - ground air temperature

  Such data should be available on an hourly basis. The height of observations may vary, typically from 2 to 50 meters above ground. The number of measurement sites depends on the city's topographical complexity. Stations may be located near AP monitoring sites, near areas with large emissions, near areas with high population density (potentially large population exposure), or to monitor the flow in topographically dominating areas of the city (e.g. valleys). The sites must be chosen to represent the undisturbed meteorological conditions of the area they are to represent.

- The same comments apply for the provision of input data to dispersion modelling also here. For state-of-the-art urban-scale modelling, the meteorological data should provide hourly spatial fields of the meteorological/dispersion parameters, either by interpolation, or using a wind-field model. The calculation of dispersion parameters from the meteorological parameter measurements to be used in the dispersion models, usually require the use of a meteorological preprocessor.

The *objectives of the monitoring* has an impact on the network design and siting of stations:

- If the objective is to monitor compliance with air quality standards, those standards themselves may specify a particular sampling protocol and site location.
- If the objective is to evaluate risk to human health through population exposure estimates, the second design approach (on previous page) should be preferred.
- If the objective of the monitoring is to form an integral part of an Air Quality Management System, the monitoring network must:
  a) provide data for actual exposure to air pollutants in general residential and hot-spot locations;
  b) provide data usable for controlling and modifying dispersion models;
  c) monitor changes in air quality resulting from pollution abatement actions.

There is no well defined methodology for determining the *number of sites* necessary. One method for arriving at a rough indication of the minimum number of sampling stations needed as a function of population density for a typical air quality network was deviced by the USEPA already in 1971 (Munn 1981). According to this method, automatic continuous sampling equipment in general involve fewer stations than a network with integrating sampling instruments (24-h. average or more) (Figure 16.3). Later, WHO (1977) made a suggestion for the GEMS Air program regarding the number of stations for various compounds (Table 16.1).

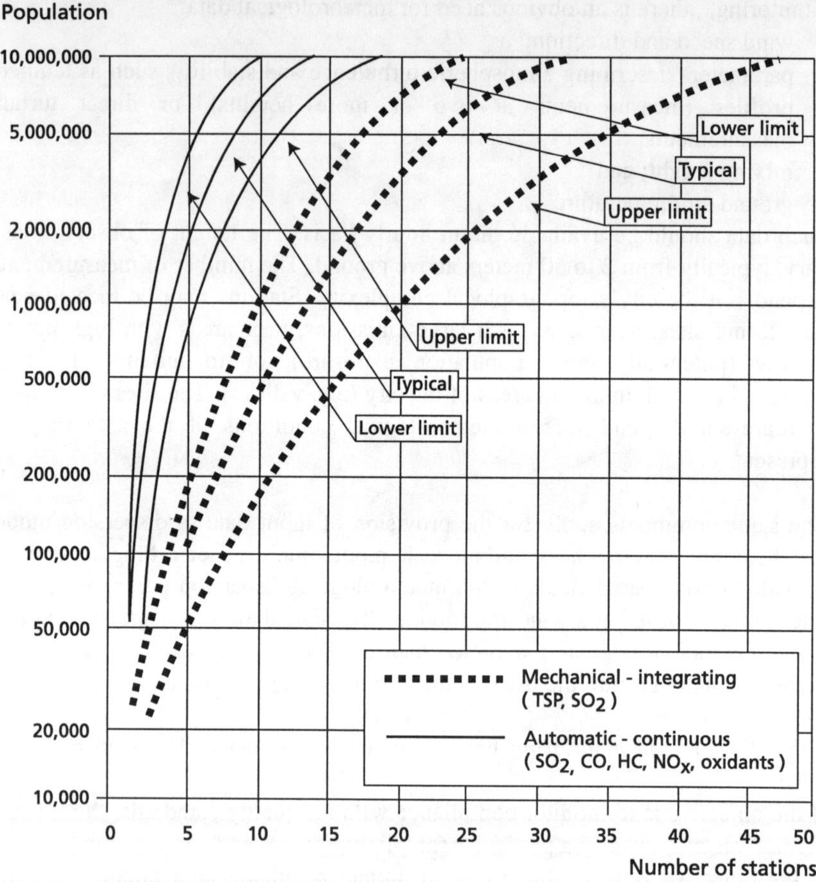

Figure 16.3 Minimum number of measurement stations needed for a typical air quality measurement network (based upon Munn, 1981).

These methods and recommendations are probably not commonly used in network design in Europe today. More often the actual design of the network is the result of a rather subjective "analytical" procedure, based upon knowledge of the area, its sources, and distribution of the population.

Table 16.1 Suggested average numbers of stations for air quality trend monitoring in urban areas of given population (WHO 1977).

Urban population (million)	Average number of stations per pollutant					
	Total susp. particulate matter	Sulphur dioxide	Nitrogen oxides	Oxidants	Carbon monoxide	Wind speed and direction
<1	2	2	1	1	1	1
1.0-4.0	5	5	2	2	2	2
4.0-8.0	8	8	4	3	4	2
>8.0	10	10	5	4	5	3

Modifying factors are as follows:

- In highly industrialised cities the number of stations for suspended particulate matter and sulphur dioxide should be increased.
- In areas where large amounts of heavy fuel are used the number of stations for sulphur dioxide should be increased.
- In areas where not much heavy fuel is used the number of stations for sulphur dioxide may be reduced.
- In regions with irregular terrain it may be necessary to increase the number of stations.
- In cities with extremely heavy traffic the number of stations for nitrogen oxides, oxidants and carbon monoxide may need to be doubled.
- In cities with a population of 4 million or more, with relatively low traffic, the number of stations for nitrogen oxides, oxidants and carbon monoxide may be reduced.

The latest recommendation of criteria for determining the number of sampling points have been given in the proposal by the European Commission for a directive relating to new limit values for SO_2, NO_2, particulate matter and lead (EC 1997a).

These criteria are set in order to determine:

- the *minimum number of sampling points* when *continuous monitoring methods* are used to assess compliance with the EU limit values for the protection of human health, and
- *alert thresholds in zones and agglomerations* where continuous measurement is the sole source of information.

The criteria are given in Table 16.2 (next page).

When selecting the actual number and location of monitoring sites in a city, the following steps should be investigated (Van Aalst et al. 1998):

1. A simplified map of suitable scale (depicting the whole urban area of interest), showing the locations of urban central districts, residential areas, areas of dense traffic, the main road system, industrial areas and large industrial plants and complexes.
2. A preliminary emissions inventory to assist in identifying the most polluted areas and locations.
3. The results of preliminary dispersion modelling to identify more closely areas warranting investigation.
4. A low-cost preliminary monitoring study. For example, diffusion tube samplers for NO_2 can be used at very low cost to provide spatially-detailed data on the

distribution of NO_2. This may be used to identify a smaller number of representative locations for fast-response continuous samplers.

5. The spatial scale of variability of the pollutant. This is very different for e.g. carbon monoxide (very high close to major roads) and peroxyacetyl nitrate (PAN) (high spatial uniformity).

Description of methods and techniques which can be used in such preliminary air quality assessment as a guide in network design in EU can be found in Van Aalst et al. (1998). One can also follow the more objective analytical procedures described by Munn (1981).

Table 16.2 Minimum number of sampling points for continuous measurement to assess compliance with limit values for the protection of human health and alert thresholds in zones and agglomerations where continuous measurement is the sole source of information (EC 1997a).

a. Diffuse sources

Population of agglomeration or zone	If concentrations exceed the upper assessment threshold	If maximum concentrations are between the upper and lower assessment thresholds	For SO_2, in agglomerations where maximum concentrations are below the lower assessment threshold
250,000	2	1	1
500,000	2	1	1
750,000	3	1	1
1,000,000	4	2	1
1,500,000	5	2	1
2,000,000	6	3	2
2,750,000	7	3	2
3,750,000	8	4	2
4,750,000	9	4	2
6,000,000	10	5	3
	For NO_2 and particulate matter: to include at least one urban background station and one traffic-orientated station		

b. Point sources

For the assessment of pollution in the vicinity of point sources, the number of sampling points for continuous measurement should be calculated taking into account emission densities, the likely distribution patterns of ambient air pollution and potential exposure of the population.

16.5 Classification of and information about monitoring sites

16.5.1 CLASSIFICATION SCHEME

It is well known that the concentration of air pollutants vary continuously in space and time, often with steep gradients especially in urban areas. For data from monitoring to be useful in the assessment of air quality, in addition to just stating the concentration in the measurement point, it is necessary to know the spatial representativity of the sampling point. This is necessary both to assess exposure based upon measurements, and to use data from measurement points to validate a calculated value from dispersion modelling. For this purpose, a classification scheme for the location of monitoring stations is useful. Such a scheme can be used both to classify existing stations, to the extent possible, and as a guide for establishing new stations.

The first parameter for station classification is its position relative to influence from (dominating) air pollution sources. From this originate the terms *Traffic* and *Industrial stations*, which are then dominated by those sources. A *City (Urban) Background station* is, however, located in such a way that no single source (industry, street or other) has a dominating influence on the pollution level at the station.

The second parameter for station classification is the type of zone where the station is located. This determines how the data from the station is to be used in the exposure assessment. Such zones may be *urban, suburban, residential, commercial etc.*

The classification scheme of the EC EoI (Reciprocal Exchange of Information) Decision (EC 1997b) is now used to classify stations in EU countries. In the EoI scheme, stations are classified according to the following criteria:

- Type of station (traffic, industrial, background)
- Type of zone (urban, suburban, rural)
- Characterisation of zone (e.g. residential, commercial, industrial, agricultural, and combinations of these).

Table 16.3 shows the classification scheme. Figure 16.4 (next page) gives details regarding the information about the stations requested in the EoI scheme.

Table 16.3 Exchange of Information (EoI) site classes.

Type of station	Type of zone	Characterisation of zone
Traffic (T)	Urban (U)	Residential (R)
Industrial (I)	Suburban (S)	Commercial (C)
Background (B)	Rural (R)	Industrial (I)
		Agricultural (A)
		Natural (N)
		Res/Com (RC)
		Com/Ind (CI)
		Ind/Res (IR)
		Res/Com/Ind (RCI)
		Agri/Natural (AN)

For traffic stations, EoI asks for additional information regarding:

- type of street (wide, narrow, canyon, highway or "other", e.g. cross roads, bus stop etc.),
- traffic amount (in 3 classes: <2,000, 2,000-10,000, >10,000 vehicles per day).

With the exception of the traffic volumes used to classify traffic oriented stations, there are no written criteria/definitions, quantitative or qualitative, on which to base the classification of stations. This may give rise to different interpretations.

I. INFORMATION CONCERNING NETWORKS

- ♦ Name
- ♦ Abbreviation
- ♦ Geographical coverage (local industry, town/city, urban area/ conurbation, county, region, entire country)
- ♦ Body responsible for network management
- ♦ Time reference basis (GMT, local)

II. INFORMATION CONCERNING STATIONS

1. General information
- – Name
- Reference number or code
- – Name of technical body responsible for the station
- – Type of station
- – Purpose of the station (local, national, EU dir., GEMS, OECD, EMEP ....)
- – Geographical coordinates
- – Altitude
- – NUTS level III
- – Pollutants measured
- – Meteorological parameters measured
- – Other relevant information: prevailing wind direction, ratio between distance from and height of closest obstacles,.......

2. Local environment/Landscape morphology
- – Type of zone
  - * urban
  - * suburban
  - * rural
- – Characterisation of zone
  - * residential,
  - * commercial,
  - * industrial,
  - * agricultural,
  - * natural.
- – Number of inhabitants of the zone.

3. Main sources of emission
- – public power, co-generation and district heating,
- – commercial, institutional and residential combustion,
- – industrial combustion,
- – production processes,
- – extraction and distribution of fossil fuels,
- – solvent use,
- – road transport,
- – other mobile sources and machinery (to be specified)
- – waste treatment and disposal,
- – agriculture,
- – nature.

4. Characterisation of traffic (only for traffic-orientated stations)
- – wide street
- – narrow street
- – canyon street
- – highway
- – others: crossroad, signal lights, parking, bus stop, taxi stop …

For each type, traffic in 3 classes: <2,000, 2,000-10,000, >10,000 vehicles a day.

III. INFORMATION CONCERNING MEASUREMENT TECHNIQUES

- – Equipment
  - * name,
  - * analytical principle,
- – Characteristics of sampling
  - * location of sampling point (facade of building, pavement, kerbside, courtyard),
  - * height of sampling point,
  - * length of sampling line,
  - * result-integrating time,
  - * sampling time,
- – Calibration
  - * type: automatic, manual, automatic and manual,
  - * method,
  - * frequency.

Figure 16.4 Information about stations requested by the EoI Decision (97/101/EC).

The background station class (as well as the T and I classes) has the subclasses urban, suburban or rural. Rural stations can be located fairly near or very far from large source areas, and this distance will to a large extent determine the pollution level. An additional classification of rural stations is beneficial, in order to be able to compare stations in different areas in Europe (Larssen et al. 1998b). The background station classes can then be described as follows:

- Urban and sub-urban background stations (B,U and B,S):
  Located within urban areas/agglomerations.
- Rural stations (B,R):
  - Near-city background stations:
    Located in rural/agricultural areas, with a distance of 3-10 km from built-up areas and other major sources.
  - Regional stations:
    Located in rural/agricultural areas, with a distance of 10-50 km from built-up areas and other major sources.
  - Remote stations:
    Located in rural/natural areas, with a minimum distance of 50 km to built-up areas and other major sources.

The reason for the subclass "near-city background stations" is to separate stations that are more influenced by nearby large city(s) from those (regional) stations which are influenced more by the ensemble of upwind sources (long-range transport), with no discernible influence from a single source area. Rural stations in areas with many closely located cities, such as in the Ruhr area and parts of The Netherlands, may be near-city background stations.

The *European Air Quality Monitoring Network (EUROAIRNET)* being established by the European Environment Agency (EEA) (Larssen et al. 1998b) is dependent upon a well defined station classification system in order that one of its main objectives be fulfilled: The ability to compare air pollution levels as measured at different stations in different networks and countries. The classification of stations under EUROAIRNET basically follows the EoI classification. However, to be able to use data from the EUROAIRNET to compare air pollution levels between cities or countries or different environments, some specific additional information is needed for some of the station classes, information which are not part of the EoI classification. Such additional information includes for instance:

For *traffic stations*:
- Traffic volume
- Traffic speed
- Distance from curb

For *background/rural stations*:
- Distance to nearest built-up areas and other major sources.

16.5.2 AREA OF REPRESENTATIVENESS

The area for which the concentration measured at the station is representative, can be called the *area of representativeness* of the station. A determination of the area of representativeness (quantitatively or qualitative evaluation) is of value when monitoring data are to be used to estimate exposure (of the population, materials or ecosystems), and also when used to validate dispersion models.

The area of representativeness varies with type of station: For a traffic hot-spot station it may be in the order of less than a few metres perpendicular to the street (while it may be representative for a rather long distance *along* the street), while for a regional station it might have a radius of tens of kilometres. It depends strongly upon the immediate environment of the stations, its morphology and sources and is not easily determined. It requires either extensive monitoring at several adjacent sites covering an area around the station, or rather detailed dispersion model calculations based upon detailed emission inventories, both for the area in question and the larger surrounding area.

In practice, such determinations are not often carried out. However, since the evaluation of the representative area is of large value when using monitoring data from a network to estimate exposure, the determination of station class should be accompanied by an evaluation of the station's area of representativeness. Typical ranges (radius of area) for the various station classes are listed in Table 16.4.

Table 16.4 Area of representativeness (radius of area) for various station classes. Range of values.

Station class	Radius of area
Traffic stations	1)
Industrial stations	10-100 m
Background stations:	
- Urban background stations	0.5-2 km
- Near-city background stations	1-5 km
- Regional stations	25-150 km
- Remote stations	200-500 km

1) A traffic station may be representative along a street, if the traffic and the street configuration is the same.

These indicative values should not be used directly, without an evaluation for each station. When evaluating the area of representativeness, account must be taken of:

- the emission variations in the immediate surroundings and possible localised influence of dominating sources further away,
- topographical features (both buildings and natural) influencing the dispersion and transport of the emissions,
- the nature of the monitor at the station, whether it measures in a point or along a line (such as the DOAS technique).

16.6 Selection of compounds, methods and averaging time

16.6.1 COMPOUNDS

There are several hundred specific air pollution compounds found in urban areas with a typical source mix (traffic, industry, fossil and biogenic fuel combustion, and refuse burning). Most of these are volatile and non-volatile organic compounds.

Compound specific EU directives exist and are being modified/developed for SO_2, NO_2, O_3, CO, benzene, lead and particulate matter (previously: SPM; in the future: PM_{10} and even finer particle fractions, such as $PM_{2.5}$, PM_1). The Framework Directive (EC 1996) lists those compounds, and in addition PAH (BaP) and the heavy metals As, Cd, Hg and Ni as compounds which should be taken under consideration. The EoI Decision lists a total of 37 compounds of air concentrations and deposition that countries should report, if available. The World Health Organization (WHO 1987) has described air quality criteria for 29 compounds (Table 16.5, next page).

All these compounds are of interest in an urban air quality monitoring program since they are all relevant in an evaluation of the effects of air pollution on humans, the ecosystem and/or materials. However, costs and practical implications necessitate a reduction of compounds to be measured to a prioritised list of indicator compounds. A selection can be based upon the general experience of air quality monitoring in European cities and on the list of compounds for which the EC has developed or are developing Limit/Guide/Target Values. Also the selection would depend upon the knowledge of the local air pollution situation, especially upon the emissions from industries located in or close to the city or residential area.

Table 16.6 (page 313) shows as an example the compounds and indicators that are proposed for the monitoring stations to be included in the European air quality monitoring network of the European Environmental Agency (Larssen et al. 1998b). Compounds and indicators are listed for 3 types of receptors: human population, ecosystems and materials.

The 3 levels of priority have been proposed to ensure that for the main criteria compounds, the EUROAIRNET will give a rather complete picture across Europe. For the local air quality assessment needs of any given city, the priorities may be different.

16.6.2 METHODS

For all compounds, either reference methods should be used, or else equivalent methods, i.e. methods that have been demonstrated to have a satisfactory correlation (in quantitative terms) with results from the reference method.

The European standards organisation CEN is presently working on harmonisation of measurement methods for the pollutants dealt with in the proposed new EU Daughter Directives (SO_2, NO_2, PM_{10}, Lead). It is anticipated that new standards will be available in time for the implementation of the Directives. The proposed new Air Quality Criteria of the EC (CEC 1997) refers to the existing reference methods for sulphur dioxide, nitrogen dioxide and lead to be carried forward and to a draft CEN standard for sampling PM_{10} to be adopted as a first step. The Air Quality Framework Directive (EC

1996) includes procedures for adapting measurement methods to technical progress when the new CEN standards are available for consideration. The same procedures will also enable criteria and techniques for other assessment methods to be adapted as necessary to technical progress.

Table 16.5 List of harmful substances in air considered by EU and by WHO.

Substances for which criteria have been considered by WHO (1987)	Substances selected by EC (1997b)	Substances for which Limit and/or Guide Values have been given by EC, or considered (in brackets)
Sulphur dioxide	Sulphur dioxide	Sulphur dioxide*
	Acid deposition	
Acid aerosols	Strong acidity	
Suspended particulates (total)	Suspended particulates (total)	Suspended particulates (total)
Suspended particulates (<10 μm)	Suspended particulates (<10 μm)	Suspended particulates (<10 μm)*
Black smoke	Black smoke	Black smoke
Ozone	Ozone	Ozone
Nitrogen dioxide	Nitrogen dioxide	Nitrogen dioxide*
Nitrogen oxides	Nitrogen oxides	
Carbon monoxide	Carbon monoxide	(Carbon monoxide)
Hydrogen sulphide	Hydrogen sulphide	
Lead	Lead	Lead *
Mercury	Mercury	(Mercury)
Cadmium	Cadmium	(Cadmium)
Nickel	Nickel	(Nickel)
Chromium	Chromium	
Manganese	Manganese	
Arsenic	Arsenic	(Arsenic)
Carbon disulphide	Carbon disulphide	
Benzene	Benzene	(Benzene)
Toluene	Toluene	
Styrene	Styrene	
Acrylonitrile	Acrylonitrile	
	1,3 Butadiene	
Formaldehyde	Formaldehyde	
Trichloroenthylene	Trichloroenthylene	
Tetrachloroethylene	Tetrachloroethylene	
Dichloromethane	Dichloromethane	
Benzo(a)pyrene	Benzo(a)pyrene	
Polyaromatic hydrocarbons	Polyaromatic hydrocarbons	(Polyaromatic hydrocarbons)
Vinyl chloride	Vinyl chloride	
	Volatile organic compounds (total non-methane)	
	Volatile organic compounds (total)	
	Peroxyacetyl nitrate	
	Ammoniac	
	Wet nitrogen deposition	
	Wet sulphur deposition	

*)New EU Target Values have been proposed by the EC (1997a).

Table 16.6 Selected compounds and indicators to be included in EUROAIRNET (Larssen et al. 1998b).

	Population exposure		Materials exposure		Ecosystems exposure	
	Average time	Medium/ compound	Average time	Medium/ compound	Average time	Medium/ compound
Priority 1		*Air:*		*Air:*		*Air:*
	1h (24h)[1]	SO_2, NO_2, NO_X, O_3	24h or longer	SO_2, O_3, NO_2, temp., relative humidity	1h 24h	O_3, SO_2, SO_4, NO_2
	1h or 24h	PM_{10}, $PM_{2.5}$	"	*Precipitation:* mm, pH	aa	NO_X
	24h or longer[2]	Pb	aa	*Materials*[3]*:* Weight loss, Steel panels	24h	*Precipitation:* SO_4, NO_3, NH_4, Ca, pH, (H+)
Priority 2	1h	CO	24h or longer	*Air:* HNO_3 (gas)	1h	*Air:* VOC, NO_X
	1h or 24h	SPM (or TSP), BS	"	*Precipitation:* Cl, SO_4, NO_3		
	24h or longer[2]	Benzene, PAH, Cd, As, Ni, Hg	"	*Soiling:* PM_{10}, SO_4		
			aa	*Materials*[3]*:* Weight loss, Zinc panels		
Priority 3	Other compounds		aa	*Materials*[3]*:* Weight loss, Copper panels, Damage to calcareous stone		

aa: annual average/exposure.

1) To be able to fully evaluate the measured levels relative to guidelines, these compounds should be measured as 1-hour averages.

2) For these compounds, mainly long term average concentrations are of interest for the assessment of effects. However, measurement methods often take much shorter samples (e.g. 24-h. or weekly samples), and shorter samples are also needed in order to explain variations in terms of source contribution etc.

3) Measurements of weight loss of standardised panels of material, measured according to standard procedures (Swedish Corrosions Institute 1989).

Priority 1, Steel

Steel is the most frequently used reference material for characterisation of the corrosivity of the environment through out the world. Several ISO standards use this material since the corrosivity of steel is highly reproducible if the same production badge is used for the exposure.

Priority 2, Zinc

Zinc is used as reference material in standards in the same way as steel. Zinc tends to give slightly different results compared to steel mainly because zinc give larger spread in the exposure results.

Priority 3, Copper and calcareous stone

These two materials are to a less extent used as reference materials. However, they are important materials for our cultural heritage. Copper has a slow corrosion rate and may need longer exposure time than one year. Calcareous stone will differ in quality from stone quarry to stone quarry, and different countries are recommended to select their own reference material for stone among the most frequently used calcareous stone types in their country.

16.7 AQMS software systems

For today's environment authorities/managers there is a strong need for software tools to be able to efficiently perform their main task: To secure, through planning and abatement decisions, a continued acceptable or improved air quality, or development towards compliance with directives, standards or guidelines. The most effective tools will be those that contain modules for all the items indicated in the AQMS loop in Figure 16.1 (page 299), coupled into an integrated, operative tool, with a user-friendly interface.

The key feature of the modern environmental information and management system is the integrated approach that enables the user in an efficient way not only to access data quickly, but also to use the data directly in the assessment and in the planning of actions. The demands to the integrated system to enable monitoring, forecasting and warning of pollution situations has been - and will be - increasing in the future.

Today's environmental information systems combine the latest sensor and monitor technologies with data transfer, data base developments, quality assurance, statistical and numerical models and advanced computer platforms for processing, distributing and presenting data and model results. Geographical Information Systems (GIS) are important tools, particularly for the presentation of data. Figure 16.5 visualises the elements of a modern environmental surveillance, planning and information system tool for AQMS.

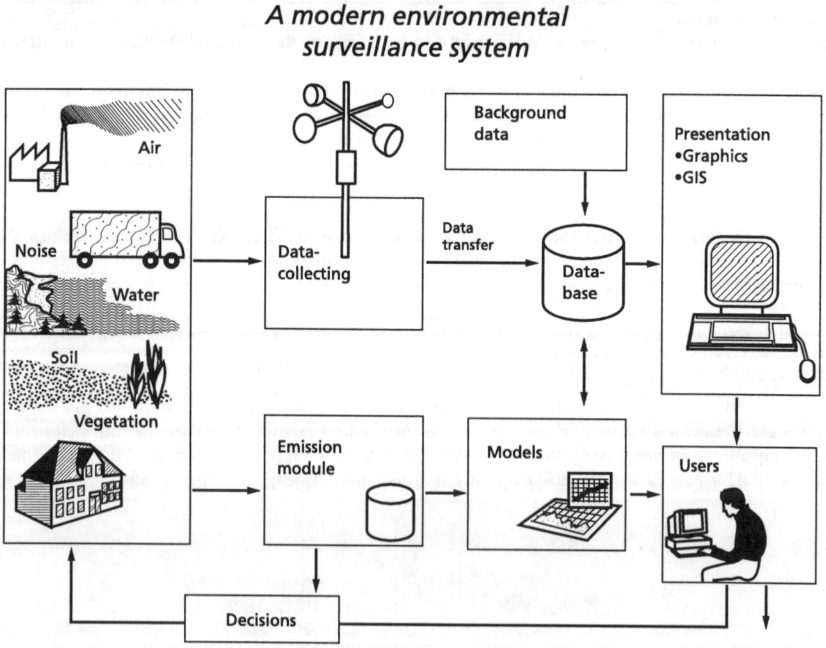

Figure 16.5 A principal structure of a modern environmental surveillance and information system.

The system should include:

- Data collectors; sensors and monitors.
- Data transfer systems and quality assurance/control procedures.
- Data bases including emission and discharge modules.
- Statistical and numerical models (included air pollution dispersion models and meteorological forecast procedures).
- User friendly graphical presentation systems including GIS.
- A decision support system.
- Data distribution systems and communication networks for dissemination of results to external users.

Several systems are currently being developed and have been demonstrated in selected areas in Europe. One of the important topics of the European Commission DG XIII Telecommunications, Information Market and Exploitation of Research, Telematics Application Programme (1994-1998) deals with this subject. Several major urban areas in Europe will thus be involved in the establishment and demonstration of such systems.

The SATURN subproject (Moussiopoulos 1998) of the EUROTRAC II program also deals with the development and documentation of integrated urban air quality management tools. Within SATURN, several research groups are developing/demonstrating integrated AQMS and decision support software tools, some of them available for purchase of licence: the ADMS system of Cambridge ERC, UK (Singles et al. 1998); the ENSIS/AirQUIS system of NILU, Norway (Bøhler 1995); the POEM system of Milan/Brescia (Catenacci et al. 1998); the OPANA system of universities in Madrid (San Jose et al. 1998).

Some of the key considerations in the development and efficient use of such systems are:

- "Speed of the model": Time needed for e.g. on-line calculation of the pollution concentration field, or in a number of receptor points, for a given (the actual, hourly) situation.
- Ease of use, e.g. by non-expert authorities.
- Domain of system, e.g.
  - short term/forecasting,
  - long term averages,
  - abatement strategy/planning tool.
- Operability regarding e.g. abatement scenarios.
- Spatial/time resolution.

The need for AQMS software tools which satisfy increasing demands to operability, accuracy and detail will be a driving force for developers, not only regarding the integrated tools, but also for the further development of the contents of each of the modules of the system, e.g. emission data bases, emission factors, dispersion models, dose response relationships.

16.8 References

Aalst, R. van, Edwards, L., Pulles, T., De Seager, E., Tombrou, M., Tønnesen, D. (1998) *Guidance report on preliminary assessment under EC air quality directives* (in preparation).

Butterwick, L., Harrison, R., Merritt, Q. (1992) *Handbook for urban air improvement 1991*, Brussels, Commission of the European Communities.

Bøhler, T. (1995) Environmental surveillance and information system, in: Power, H., Moussiopoulos, N., Brebbia, C.A. (editors), *Air pollution theory and simulation,* pp. 43-49, Southampton, Computational Mechanics Publications, (Air pollution III, vol. 1).

Catenacci, G., Maurizio, R., Volta, M., Finzi, G. (1998) *A Model for Emission Scenario Processing in Northern Italy. A contribution to subproject SATURN*. Presented at EUROTRAC-2 Symposium, 23-27 March 1998, Garmisch Partenkirchen, Germany (Executive summary).

EC (1996) Directive 96/62/EC on ambient air quality assessment and management (1996a), *Official Journal*, **L 296**, 55.

EC (1997a) Proposal for a Council directive relating to limit values for sulphur dioxide, oxides of nitrogen, particulate matter and lead in ambient air (presented by the Commission). Luxembourg, Office for Official Publications of the European Communities, *(COM(97) 500).*

EC (1997b) Council Decision 97/101/EC establishing a reciprocal exchange of information and data from networks and individual stations measuring ambient air pollution within the Member States (1997), *Official Journal*, **L 35**, 14.

Grønskei, K.E., Walker, S.E. (1995) A stochastic receptor model based on source oriented structure functions, in: Power, H., Moussiopoulos, N., Brebbia, C.A. (editors), *Air pollution theory and simulation*, pp. 43-49, Southampton, Computational Mechanics Publications, (Air pollution III, vol. 1).

Larssen, S. et al. (1998a) in: Shah, J., Nagpal, T., Brandon, C.J. (editors), *Urban Air Quality Management in Asia, Guidebook*, The International Bank for Reconstruction and Development (the World Bank), Washington D.C.

Larssen, S., Sluyter, R., Helmis, C. (1998b) Criteria for EUROAIRNET. The EEA air quality monitoring and information network. Draft 15 July 1998. Kjeller, European topic centre on air quality (RIVM, NILU, NOA, DNMI).

Moussiopoulos, N. (1998) *Urban Air Pollution: Current Knowledge and Future Research. A contribution to subproject SATURN*, Presented at EUROTRAC-2 Symposium, 23-27 March 1998, Garmisch Partenkirchen, Germany (Executive summary).

Munn, R.E. (1981) *The design of air quality monitoring networks*, London, MacMillan.

San José, R., Rodrigues, M.A., Toribio, R., González, R.M. (1998) *Integration of Operational Environmental Forecasting Models in Metropolitan Areas: OPANA, A contribution to subproject SATURN*. Presented at EUROTRAC-2 Symposium, 23-27 March 1998, Garmisch Partenkirchen, Germany (Executive summary).

Singles, R.J., Carruthers, D.J. (1998) *ADMS-Urban Air Quality Management System: Some Examples of Application and Validation in the UK, A contribution to subproject SATURN*. Presented at EUROTRAC-2 Symposium, 23-27 March 1998, Garmisch Partenkirchen, Germany (Executive summary).

The Swedish Corrosion Institute (1989) *Description of test sites, Stockholm (UN/ECE International Co-operative Programme on Effects on Materials, including Historic and Cultural Monuments*, report No. 2).

Vukovich, F.M., Bach, W.D., Clayton, C.A. (1978) Optimum meteorological and air pollution sampling network selection in cities. Vol. 1: Theory and design for St. Louis. Las Vegas, NV, U.S. Environmental Protection Agency, *EPA-600/4-78-030*.

WHO (1976) in: Suess, M.J., Craxford, S.R. (editors), *Manual on urban air quality management*, Copenhagen, World Health Organization (WHO Regional Publications, European Series, 1).

WHO (1977) *Air Monitoring Programme Design for Urban and Industrial Areas*, (WHO Offset Publication No. 33), World Health Organization, Geneve.

WHO (1987) *Air quality guidelines for Europe*. Copenhagen, World Health Organization (WHO Regional Publications, European Series, 23).

Chapter 17

QUALITY ASSURANCE AND QUALITY CONTROL

STEINAR LARSSEN
Norwegian Institute for Air Research
P.O.Box 100, N-2007 Kjeller, Norway

COSTAS HELMIS
Department of Applied Physics, University of Athens
Panepistimioupolis, GR-015784 Athens, Greece

17.1 Introduction

In air quality measurement systems, the *Quality system*, comprising Quality Assurance/Quality Control (QA/QC) is concerned with all the activities that assure that a measurement meets defined standards of quality.

The quality terms relevant for QA/QC procedures and criteria can be defined as follows (ISO 8402 1994):

- Quality is the totality of characteristics of an entity that bear on its ability to satisfy stated or implied needs.
- Quality Assurance involves the management of the entire process, including all the planned and systematic activities, which are needed to assure and demonstrate the predefined quality of data, to provide adequate confidence that an entity will fulfil the requirements for quality.
- Quality Control comprises the operational techniques and activities that are undertaken to fulfil the requirements for quality.

The Quality Assurance activities cover all the pre-measurement phases, ranging from definition of data quality objectives to equipment and site selection and personnel training, whereas the Quality Control operational functions cover directly measurement-related activities such as routine checks, calibration and data handling. An extended review of the above mentioned procedures is given by Lalas and De Saeger (1997).

The principal objectives of QA/QC are to:
- identify errors;
- quantify errors;
- reduce errors.

These errors can be of two types: procedural and technical.

Procedural errors result from unclear and ineffective management, inadequately training staff, improper planning, lack of adequate QA, lack of data tracking and other problems related to how the work gets done.

Technical errors are directly related to the methods and technologies used. They may result from mathematical errors, use of incorrect data, use of incorrect methodology and assumptions, etc. In general, avoiding or minimising procedural errors will reduce the likelihood of technical errors.

The first step in designing and implementing a QA/QC plan is to define the monitoring objectives. Then *Data Quality Objectives* (DQO) must be established to ensure that the data collected are sufficient and of adequate quality for their intended uses derived from the monitoring objectives.

The data quality objectives may be defined in terms of the following parameters, which are indicators of data quality (see Figure 17.1 for definitions):
- Precision.
- Accuracy and/or trueness.
- Representativeness.
- Data capture.
- Time coverage.

There should be *consistency between the monitoring objectives and the DQOs.* This can also be formulated more stringently: The monitoring objectives must be the basis for specifying the DQOs, such that fulfilling the DQOs will ensure that the monitoring objectives will be fulfilled. For example, if a monitoring objective is to detect a time trend with a certain accuracy over a certain period, this will effectively determine the necessary accuracy of the measurements.

Having defined the data quality objectives, it is necessary to establish a QA/QC plan. The QA/QC plan is a technical document that shall specify all the QA/QC activities required to achieve the DQOs. It should also describe how the data will be assessed for precision, accuracy, representativeness and completeness. Finally, it should describe the mechanisms to be used when corrective actions are necessary.

The quality system of a monitoring network must explicitly define the responsibility and authority for each of the activities contributing to data quality, and the coordination between them (Schaug 1998). Each network should have a designated Quality Assurance Manager, responsible for implementing the Quality System, and for activities improving the quality.

Precision: The closeness of agreement between independent test results obtained under stipulated test conditions.

Precision depends only on the distribution of random errors and does not relate to the true value or the specified value.

The measure of precision is usually expressed in terms of imprecision and computed as a standard deviation of the test results. Less precision is reflected by a large standard deviation.

"Independent test results" means results obtained in a matter not influenced by a previous result on the same or similar test object. Quantitative measures of precision depend critically on the stipulated conditions.

Repeatability and reproducibility conditions are particular sets of extreme conditions (ISO 5725-1, 1994).

Repeatability: Precision under repeatability conditions (ISO 5725-1, 1994).

Repeatability conditions: Conditions where independent test results are obtained with the same method on identical test items in the same laboratory, by the same operator, using the same equipment within short intervals of time (ISO 5725-1, 1994).

Reproducibility: Precision under reproducibility conditions (ISO 5725-1, 1994).

Reproducibility conditions: Conditions where independent test results are obtained with the same method on identical test items in different laboratories, with different operators, using different equipment (ISO 5725-1, 1994).

Accuracy: The closeness of agreement between a (one) test result and an accepted reference value.

The term accuracy, when applied to a set of test results, involves a combination of random components and a common systematic error or bias component (ISO 5725-1, 1994).

Trueness: The closeness of agreement between the average value obtained from a large series of test results and an accepted reference value.

The measure of trueness is usually expressed in terms of bias.

It was referred to as "accuracy of the mean" which is not recommended (ISO 5725-1, 1994).

Representativeness: This parameter expresses the degree to which the air pollution measurement data are adequately representative, both of the *location* in which monitoring is taking place, and of the *time period* to be covered. The location (spatial) part can be quantified by the *area of representativeness*: the area in which the concentration does not differ from the concentration measured at the station by more than a specified amount. The temporal part is covered by the data capture and time coverage indicators below.

Data capture: The percentage of measurements made which are judged to be valid measurements.

Time coverage: The percentage of time covered by the operational time of the measuring device.

Comparability: This is a qualitative parameter expressing the confidence with which one set of air pollution measurement data can be compared with another. Data representative of air pollution levels of a location should be possible to compare with measurement data of another similar location. It should be noticed that data of known precision and accuracy and with a high degree of representativeness and completeness can be compared with confidence, for stations of the same location class.

Figure 17.1 Definition of data quality indicators.

17.2 Data Quality Objectives (DQO)

Table 17.1 provides examples of the DQOs for some monitoring networks/programmes. DQOs are given there for accuracy, precision and data time coverage. The main monitoring objectives behind the DQOs are also specified in the table. Regarding the details of the DQOs for spatial representativeness, the reader is referred to the references listed below.

The EU Regulatory Monitoring is required by the European Commission to show compliance/detect non-compliance with its air quality directives (EC 1997a). For the EUROAIRNET network of the EEA, DQOs have been proposed e.g. to enable comparison of air quality at different stations and areas (urban, etc.) across Europe (Larssen et al. 1998). The DQOs of the EMEP monitoring programme are set mainly to provide data for control of the models for dispersion and deposition on the regional scale (Schaug 1998). The DQOs of the WMO - Global Atmospheric Watch (GAW) monitoring programme have been set in order that present trends be detected over 5 years or less (WMO 1992).

Table 17.1 show that DQOs for accuracy and precision may vary dependent upon the monitoring objective. It may also be that the DQOs do not purely reflect the requirement set by the monitoring objective, but are set with a view also to what is possible to achieve with the typical monitoring techniques and practices used today.

Table 17.1 Data Quality Objectives of some monitoring programmes.

Monitoring programme/ Monitoring objective	Compounds	Accuracy	Precision	Data time coverage
EU Regulatory Monitoring [1)]				
Detect non-compliance with directives	SO_2, NO_2	15% [2)]		90% annual
	PM, Pb	25% [2)]		"
EUROAIRNET mon. network [3)]				
Compare air quality across Europe	Gases	10%	2 ppb	90% annual
	Particles	10%		"
EMEP				
Provide basis for control of models		15-25% [4)]		90% annual
WMO-GAW	Examples:			
Detect trends over short term (5 years)	O_3	15% or 3 ppb	10% or 1 ppb	80% monthly
	NO_2	20% or 50 ppt	10% or 25 ppt	"
	$PM_{2.5}$	0.05 + 5% M	10%	90% monthly

1) Minimum DQOs. Final approval of the directive, (EC 1997a) is pending (as of July, 1998).

2) Combined accuracy and precision.

3) Proposed DQOs (as of July, 1998).

4) Total "uncertainty" (combined accuracy and precision) for sampling and analysis combined; dependent upon compound.

17.3 The complete QA/QC plan

The elements included in a complete QA/QC plan are:

17.3.1 QUALITY ASSURANCE

- Setting monitoring objectives and DQOs, as mentioned above.
- Procedures for site selection and air quality monitoring network design shall be described.
  Tools for the selection of the stations' positions are, among others, indicative air quality measurements, emission sources inventories, and application of air pollution modelling.
- A main feature of the air quality monitoring network, national or local, should be the central institution or laboratory that is responsible for the implementation of the QA/QC plan.
  Requirements to the central institution or laboratory include:
  - It should manage the overall activities of the QA/QC plan on local or national basis.
  - It must be the organisation that validates in a final stage and reports directly or indirectly the air pollution measurement data.
  - It should possess the appropriate facilities and equipment, including primary calibration standards, for the implementation of QA/QC plan, mainly for the Quality Assessment procedures.
  - The staff must be qualified with special training or have considerable practical experience.
- Instrumentation should be selected according to justifiable criteria, and must completely fulfil the requirements of a reliable QA/QC plan. The instrumentation may include:
  - Measuring devices (automatic, semiautomatic, manual).
  - Calibration instrumentation and standards such as :
    - Primary standards (for the central laboratory), that are defined as the substances or a mixture of substances whose specific properties with respect to the purpose of the measurement are known. These properties are determined by measuring base quantities or by deriving quantities from these measurements.
    - Secondary standards (field usable) whose value is based upon comparison with a primary standard.
  - Measurement data management and processing equipment.
  - Infrastructure equipment e.g. sampling lines, station shelters, HVAC systems etc.
- The central laboratory must have:
  - The capability to prepare primary standards (e.g. static dilution, permeation tubes) or acquire them.
  - Equipment suitable for the implementation of the below mentioned Quality Assessment procedures.

- Adequate education and training of personnel are prerequisites for reliable measurement capability. Thus, a documented personnel training program must be included.
- A detailed QA/QC manual, including, with proper justification, all the procedures relevant to the QA/QC plan. All QA/QC operations should be performed in the way they are described in the manual.

17.3.2 QUALITY CONTROL

The quality control includes preparation of protocols and implementation of procedures such as:

- Site operation and equipment maintenance. This includes routine (and non-routine) site visits.
- Calibration. This is the most important operation in the measurement process. Calibration is the process of establishing the relationship between the output of a measurement process and an known input (e.g. primary and secondary standards). A calibration plan must be developed and implemented for all measuring equipment.
- Data validation procedures. These should include the procedures of the EoI Decision (EC 1997b), taking into consideration calibration and technical problems, off-scale measurements and unusually high variations. Finally, doubtful or potentially erroneous measurements should be detected using either historical data or existing relationships with other pollutants. The validation results should accompany the data set, as a separate list. All data should be marked either as not yet validated (code T), validated (code V), or erroneous/doubtful (code N).
- Completeness: the EoI specifies the following: one hour average time series should consist of at least 75% of valid data, while 24 hour average time series require at least 50% valid data and no more that 25% of successive rejected data (code N), (EC 1997b). For the mean and the median value, more than 50% of the data should be valid, while for the 98% and 99,9%, more than 75% valid data points are required. Also the ratio between the valid data for the two seasons of the year shall not be greater than 2. (Winter is from January to March and from October to December and summer from April to September). The network operator may well set stricter requirements than these.

  Reporting and documentation should e.g. comply with the EoI Decision (EC 1997b).

For a QA/QC plan to be characterised as complete, it should contain, as a minimum, the above mentioned QA and QC procedures.

17.3.3 QUALITY ASSESSMENT

Quality Assessment techniques are usually performed independent of and in addition to normal QC checks. Quality Assessment procedures may be:

- Ring test; a process taking place at a competent laboratory and consists of consecutive measurements carried out by the circulation of one or more reference material samples.

- Inter-calibration of networks; a technique addressing the need to directly inter-compare the measurement procedures used in different networks. A nominated laboratory, capable to provide standard materials (e.g. standard gases), other required apparatus, and personnel to perform the fieldwork, carries out the inter-calibration.
- Round robin tests; usually the nominated laboratory mails a set of calibration standard gas cylinders to the participating stations or networks in turn.
- Audits. There are two types of audit procedures:
  - System audit is the "in situ" inspection of the measurement system taking into account all the measurement elements such as sample collection and analysis, data processing etc. It is a qualitative appraisal of quality. The auditor inspects the QA/QC plan and the relevant documentation of the station and fills up the appropriate checking list.
  - Performance audit is a quantitative appraisal of quality that is carried out by standard material of the auditor. The standard should be measured by the instrumentation of each station, and an assessment of the instrument performance (e.g. precision, accuracy, response time) is drawn up.

17.4 The minimum QA/QC plan

The assurance of data quality is very important in ensuring full usefulness of the monitoring data. Quality assurance procedures are costly, however, and network managers will tend to evaluate which extent of QA/QC procedures are necessary for their use of the data. The QA/QC procedures should not be reduced below a minimum. A minimum, documented QA/QC plan comprises the following quality procedures:
- DQOs are set on a minimum basis regarding:
  - Accuracy and precision.
  - Data capture.
  - Time coverage.

Regarding accuracy, precision, data capture, and time coverage, both values and a detailed description of the method used for the estimation must be given. Those values should comply, whenever possible, with EU legislation. According to the draft of the proposed Council Directive relating to limit values for SO_2, NO_X, PM and Pb (EC 1997a) the minimum DQO values for SO_2, NO_2, PM and Pb measurements are those presented in Table 17.2.

Table 17.2 Accuracy and completeness minimum values for air quality measurements, according to EU Directives (EC 1997a).

	SO_2, NO_2	PM, Pb
Accuracy and precision[1)]	15%	25%
Data capture, annual basis	90%	90%
Time coverage, annual basis	100%	100%

[1)] Of individual measurements.

Further the quality procedure comprises:

- A reporting organisation, collecting the data, performing a subjective quality check on the data, and finally collect, report and archive the output of the QA/QC.
- Site selection according to justifiable criteria.
- Site description (position on maps of relevant scale, local sources, immediate environment) available for all sites.
- The measuring methods in the form of either reference or reference equivalent according to EU legislation and/or internationally accepted standards. Both reference and equivalent methods shall be those officially documented by an approved testing institute or laboratory.

  A reference method is defined as a measurement method for the complete determination of a specific air pollution compound which can be handled by many users and which is based on well-founded experience over many years. Also, the method has to be successfully tested experimentally, by an officially approved testing institute or laboratory (Lahman 1992).

  An equivalent method is defined as a measurement method experimentally proven to be equivalent to a reference method, by an officially approved testing institute or laboratory, using an officially approved suitability test based on documented technical requirements (Lahman 1992).

- A documented calibration program along with an instrument performance checking program. This should include at least:
  - For automatic methods and for each measuring device:
    Regular zero-span checks, multi-point calibrations, and precision-checks.
  - For manual methods and for each measuring device:
    Flow and leak checks, routine calibration procedures in the laboratory, precision-checks with collocated identical samplers, and determination of the method's accuracy on a regular basis.

  The frequency of the above mentioned procedures should be set according to the network manager's experience and in any case ensure, in an unambiguous way, that the DQOs are met. All calibration activities shall be logged and reported.
- Data validation procedures complying with the EoI Decision (EC 1997b).

Proper equipment and procedures to carry out the minimum QA/QC plan includes:

- Certified reference material approved-traceable to official primary and secondary standards (e.g. calibration gases, permeation tubes), mostly for the implementation of the inter-calibration procedure.
- Operational (secondary or transfer) calibration standards for routine procedures, e.g. calibration, precision check, zero-span check.

These standards must be validated against certified standards, traceable back to official primary standards.

- Proper technical equipment.
- Sampling and analysis Standard Operating Procedures (SOPs).

17.5 References

EC (1997a) Proposal for a Council Directive relating to Limit Values for SO_2, NO_2, particulate matter and lead in ambient air, Brussels (COM(97)500 final, 97/0266 (SYN)), *Official Journal*, **C009**, 6.

EC (1997b) Council Decision 97/101/EC establishing a reciprocal exchange of information and data from networks and individual stations measuring ambient air pollution within the Member States (1997), *Official Journal*, **L 35**, 14.

ISO (1994) *Quality management and quality assurance*, Vocabulary, Geneve, International Organization for Standardization (ISO 8402).

Lahman, E. (1992) *Determination and evaluation of ambient air quality, Manual of ambient air quality control in Germany*, Berlin, Federal Minister for the Environment, Nature Conservation and Nuclear Safety.

Lalas, D.P., De Saeger, E. (1997) *A common minimum quality assurance program, European Topic Centre for Air Quality of the Environment Agency*, RIVM-NILU-NOA-DNMI, Subproject MA 2-6 (Final draft).

Larssen, S., Sluyter, R., Helmis, C. (1998) *Criteria for EUROAIRNET, The EEA air quality monitoring and information network*, Draft 15 July 1998, Kjeller, European topic centre on air quality (RIVM, NILU, NOA, DNMI).

Schaug, J. (1998) The status of quality assurance within EMEP; recommendations for further action. Chemistry Forum '98, Warsaw University of Technology, Warsaw, April 1998 *(NILU F 5/98)*.

WMO (1992) Report of the WMO Meeting of Experts on the Quality Assurance Plan for the Global Atmospheric Watch. Garmish-Partenkirchen *(WMO Report No. 80)*.

VI

IMPACTS OF URBAN AIR POLLUTION

As it appears from the previous chapters, the character of urban air pollution in the industrialised world has changed in the recent decades. Generally the levels of sulphur dioxide and black soot from domestic heating have been reduced. On the other hand the increasing traffic have led to emissions of nitrogen oxides and various organic compounds. In spite of the introduction of catalytic converters, these problems have not yet been fully solved. The same applies in particular to small particles in the exhaust from diesel engines. These changes are reflected in the impacts of the urban air pollution.

It is well established that *human health and well-being* are affected by urban air pollution, which has both acute and chronic impacts (Chapter 18). Although sulphur dioxide and particulate matter are now less important in themselves, they are still relevant as indicator components. More important however, are various organic compounds and small particles. In most cases the impacts are demonstrated in statistical analyses, but the exact mechanisms are poorly understood.

Degradation of materials by air pollution (Chapter 19) has been observed for centuries - mostly in the form of attacks on stone and iron by acidifying sulphur compounds. With the reduced levels of sulphur dioxide in recent years a multi-pollutant situation has arisen in which also other compounds and their reaction products are important. Simultaneously synergistic effects appear.

Another aspect is the change in materials. New building materials are designed to be more resistant and more resistant materials are applied in the restoration of our cultural heritage. On the other hand new applications - especially within the increasing use of miniaturised electronics - must be taken into account.

While concerns about air pollution in cities are focused on its impacts on human health and, to a lesser extent, damage to materials and cultural monuments, its effects on *the natural environment* (Chapter 20) should not be forgotten. There are three reasons for this. Firstly, it has a utilitarian value. For example, food is grown by many city-dwellers for their own consumption in gardens and allotments, while peri-urban and urban fringe areas often are a focus for high-value horticultural crops because of their proximity to markets or airports. Secondly, it has an aesthetic value, improving the quality of life through its visual appearance. Thirdly, sensitive elements of the natural environment may provide useful bio-indicators of spatial and temporal trends in air quality. In some cases, there may be multiple reasons for interest: for example, planting trees on a city street may have a value in providing shade to buildings, reducing the need for air conditioning in summer; may intercept traffic-generated pollutants; may improve the visual appearance for residents and visitors; and may provide a substrate for sensitive epiphytic lichens which can be used as bio- indicators.

In the western world the London smog mainly caused by soot (smog = smoke + fog) with a visibility down to a few meters is a thing of the past. Still however, significant *reductions of visibility* (Chapter 21) can be observed in European cities, but mostly as an aesthetic loss.

Chapter 18

HEALTH IMPACTS

OLF HERBARTH
Department of Human Exposure Research and Epidemiology
UFZ Centre for Environmental Research Leipzig Halle
Permoserstraße 15, D-04318 Leipzig, Germany
and
University of Leipzig, Faculty of Medicine
Environmental Hygiene and Epidemiology
Liebigstraße 24, D-04103 Leipzig, Germany

18.1 Introduction

Many people live in urban areas and their numbers are continuously increasing worldwide. These people accept - if necessary - increases in their exposure to numerous hazardous influences, such as noise, air pollution, etc. It is tolerated as urban living conditions more often than not are associated with more amenities, generally a better standard of living, greater cultural diversity as well as a greater variety for the consumers and often enough also greater job opportunities.

Among the inescapable exposures, air pollution and noise are the most conspicuous. And yet, the majority of the people accept the decrease in the quality of life associated, for instance, with air pollution, however, only insofar as one's health is not affected. A direct effect on one's own health is thereby excluded. It is just not realised or may simply be shut out. However, subjective threats may still influence the personal evaluation of the urban-living-associated exposures, which ultimately may be expressed in a changed state of health. An exception in this subjective appraisal are extreme emission events. At that moment, the threat to one's health becomes very real. Extreme events, therefore, contribute greatly to an increase in the concern between environment and health. This is especially relevant with regard to air pollution associated morbidity and mortality. In this context such events as the "London Smog" or the "Los Angeles Smog" episodes are to be mentioned.

In 1969, the World Health Organization (WHO) introduced a long-term programme to monitor and control environmental pollution, with air pollution playing a major role. The European Community has initiated numerous activities as well to investigate air pollution and its effects on human health.

In the following chapter only air pollution and its effects on human health are dealt with and limited to the urban environment. The focus is on problems related to questions such as:

- Which pollutants or pollutant compounds are associated with increased risks to human health?
- Which disorders (syndromes, symptoms) are primarily influenced by exposure to this pollution; will it be possible to associate, not yet detected pollutant compounds with specific emitting sources?
- Is it possible to set exposure threshold levels or determine the risks involved?

The following presentation does not claim to cover the subject to its full extent. It only tries to depict exemplary which exposure, if and when a burden, is associated with adverse health or well being effects.

This presentation will not so much be concerned with the substances, but rather deal with source-dependent complexes taking the special circumstances of urban areas into account. Even if numerous complex mixtures contribute to adverse health effects, many can be attributed on a first inspection to a few dominating emitting sources. Furthermore, methods are discussed which lend themselves to determine health effects. These methods include ways which will allow, at least from a theoretical point of view, to derive at threshold values. "Urban stress" and "chemical risk", cited in the Dobris assessment (EEA 1995), as two of about 12 dominating environmental problems are dealt with as well. The final declaration of the 2nd European Conference on Environment and Health (Helsinki, June 20-27, 1994; WHO ECEH 1995) identified problems, which among others, demand preventive actions with regard to:

- Urban health
- Ambient outdoor and indoor air pollution

Basically, the dominating sources of exposure within an urban area are:

- Heating and energy production
- Traffic emissions
- Exposures emitted from sources indoors in which people tend to spend much of their time (interior furnishings, construction materials, their utilization and human activities).

Then there are industrial facilities, their location within the urban area are also important as point source of the pollution. In general, however, this would be a specific problem of the city under investigation. As this presentation is concerned with ambient exposures, emphasis will only be on the two exposure complexes, heating and energy production as well as traffic. Industrial facilities contribute to the ambient exposure burden, but is also a case of occupational exposure indoor frequently.

18.2 Methods for exposure and effect research

Two main areas of questions:

1) those concerning the possible association between air pollution and health effects:
- Does an association exist between air pollution and adverse health effects?
- Which disorders are affected the most?
- Which specific substance or complex of components (also indicator components) plays a more dominant role with respect to the illness under investigation?
- Can a susceptible subgroup especially at risk be identified within the urban population under investigation?

and

2) questions concerning the pathomechanism:
- Which specific reaction caused by the interaction of the air pollution components in the organism leads to disturbances of well-being?
- Are specific markers (indicators) detectable which reflect the exposure?
- Can prepathologic states be identified, indicative of a shift towards physiological processes with possible adverse effects already before the manifestation of or at time of outbreak of the illness.

Epidemiology is aimed at assessment of air pollution effects in humans. Epidemiology can be used to determine an association with a specific probability, but does not establish a cause-effect relationship with an absolute level of significance. Epidemiology is concerned with the distribution of disease, physiologic factors, socially mediated illnesses and with factors affecting their distribution. Moreover, when dealing with levels of concentrations, generally below the toxic level, only epidemiologic methods will lead to any indication of increased frequencies of illness events. All factors discussed here in this context, which have shown an effect on a disease, are associated with air pollution. In this respect air pollution is the independent parameter and well-being or health problems the dependent variable, in addition to other associated confounding variables. These include those concerning one's life style, individual behaviours (smoking, hobbies, etc.), occupational exposure, etc., and last, but not least, the polluted indoor air, already mentioned.

Answers to the second set of questions infer to more clinically oriented research. However, epidemiologic investigations do play an important role here too, as the concentrations dealt with in these studies are such that observations of individual cases will rarely lead to any cause-effect determination, i.e., a statistically significant effect.

Independent of these arguments, the two question areas are related. Epidemiology today is not only applicable in relation to clinical events. A combination of clinical and epidemiologic research may lead to determination of indicators which could be useful as markers to characterize prepathologic states. Individual-oriented environmental epidemiologic studies applying methods of environmental medicine are gaining in importance. Traditional epidemiology, originally population-oriented, is now more and more complemented by an individuum-oriented element. A significant reason for this development in epidemiology is the constant search for factors which affect the development of disease. The aim is to find a link between a specific health-indicator and a specific environmental factor. The more complicated the association, the more important it becomes to examine a multitude of possible influencing or associated factors. Considering, the present financial and personnel situation, studies will only be feasible when the numbers of participating probands are reduced, the assessment of the effect of the influencing factors more clinically oriented and medical interventions offered.

One prerequisite for such environmental epidemiologic research is a detailed knowledge of the exposure situation. In the past, effects associated with sulphur dioxide and particulate matter were mainly under discussion, but at present volatile organic compounds and dust content matter are becoming increasingly relevant. This is also true for exposure investigations carried out close to the persons affected. Indicator components were monitored more globally (spatially and temporally), representative for a specific area, today, the attempt is made to obtain a more accurate and specific profile of exposure the study population is exposed to. Only a detailed knowledge of both, the exposure and the effects, will help to come to any conclusion concerning the risks involved (Figure 18.1).

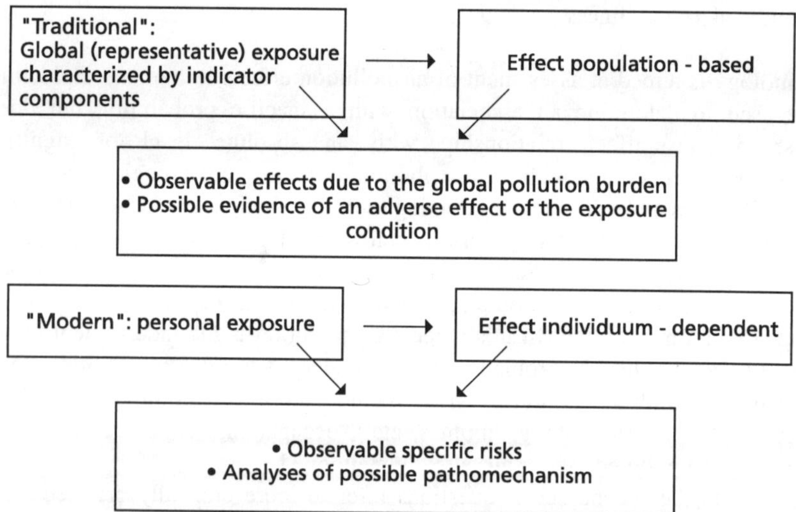

Figure 18.1 "Traditional" and "modern" pathways in epidemiology to determine associations between exposure and effect.

Exposure conditions have changed greatly over the last few years with respect to ill health relevant components. For instance, the level of concentrations of the primary, combustion-derived components, like suphur dioxide and particulate matter (fly ash) has decreased substantially. Yet, the level of nitrogen dioxide and aerosols has increased comparatively. Such changes can bring about changes in morbidities, which may become apparent in intervention studies and, thus, allow to draw conclusions as to the risks involved. This opens the possibility for specific preventative actions.

18.3 Urban air pollution effects

In general, it can be said that trends of effects associated with air pollution components and component mixtures have been observed. However, frequently the evidence is not clear-cut and most study results are not comparable. It is not that exposure effects are undefinable and arbitrary but rather lies in the differences of the study designs applied. One underlying reason for the development of standardized study designs, measurement techniques and methods of data analyses (COST 613) was to make studies comparable.

At present, the importance of volatile organic hydrocarbons and substances bound to dust particles as well as substances contained in the dust particles is increasing, yet, indicator components, such as SO_2 and particulate matter, will remain very relevant. This is reflected in the priority listing of WHO (WHO ECEH 1994), which was developed because of the necessity of determining threshold limit values (Table 18.1).

Table 18.1 Substances and mixtures of substances considered for inclusion in the second edition of guidelines.

Priority I	Priority II	Priority III
Carbon monoxide	Nickel	Vanadium
Nitrogen oxides (nitrogen dioxide)	Platinum	Carbon disulphide
Ozone/photooxidants	Asbestos/fibres	Hydrogen sulphide
Sulphur dioxide	Styrene	Acrylonitrile
Particulate matter [1]	1,3-butadiene	1,2-dichloroethane
Sulphur dioxide/particulate matter	Fluoride	Vinylchloride
Lead	Dioxins/polychlorinated biphenyls	Gallium
Cadmium	Total volatile organic compounds/(indoors)	Rhodium (new technology metals)
Manganese	(Alternative fuels) [2]	
Mercury	Dichlormethane	
Arsenic		
Hexavalent chromium		
Formaldehyde		
Radon		
Toluene		
Benzene		
Trichloroethylene		
Environmental tobacco smoke		

[1] Including polycyclic aromatic hydrocarbons.
[2] Including alcohols, ethers and aldehydes.

These components will allow inferences to be drawn with regard to an observable emitting source or, at least, to part of a group of emitting sources. This may especially be important in relation to any regulatory and intervention measures to reduce emission levels for health reasons. Even personal prevention relies on a specific, with indicator component associated, spatially and temporally varying pollution burden.

Generally, one has to differentiate between acute and chronic effects. In as much if recurrent acute effects will finally lead to a chronic condition has not been elucidated as yet. Causes for acutely high pollution levels tend to be smog situations associated with meteorological conditions. Other temporally limited but periodically occurring emission episodes of high concentrations from specific sources (for instance, the heat-up period of coal-fired residential heating systems, rush-hour traffic, etc.) or other extreme pollution situations due to accidents may also lead to possible adverse health effects.

Besides those substances typical of winter conditions, with the exception of particulate matter that remains relevant throughout the year, ozone and its precursoer substances are very important during the summer. Here too, associations to lung function and respiratory symptoms (specially cough) have been implicated (WHO 1995).

A number of investigations, carried out within the framework of European-sponsored research, had the aim to determine air pollution-associated health risks in order to determine preventative measures, should they be required. Analyses of morbidity in cities with differing source-dependent air pollution profiles, i.e., the types of emission and resulting burdens, allow an estimate of the risk due to the dominating component-associated exposure. A variety of investigations address the risk associated with exposure to typical urban ambient air and besides that of such dominant factors as smoking.

The focus of this research to this day was on urban relevant factors concerning airway diseases, allergies and especially asthma. It should be pointed out that other parameters associated with other adverse health effects have been investigated mainly in relation to specific events.

Most of the known information concerns airway disorders. Respiratory diseases, accounting for 9.3 to 12.3 percent of all mortality cases in the 1970's, has been reduced to 5.9 to 7.3 percent in the late 80th. However, these illnesses continue to play an important role. Among them bronchitis, asthma and possibly emphysema are the most prevalent ones. Recent studies (Wjst 1996; Lundbeck et al. 1993; Burney et al. 1990) show that asthma prevalences are increasing. The role air pollution plays in this effect is not clear as yet. However, independent of this, evidence is accumulating that adverse respiratory health effects are associated with chronic as well as acute exposure under present air pollution standards set for European cities (Brunekreef et al. 1995; Herbarth 1995; Herbarth et al. 1996). The spectrum of effects ranges from temporary, reversible impairment of lung function to permanent dysfunctions and clinical forms of respiratory diseases. Populations at risk such as children, the elderly and individuals of ill health are especially susceptible. Considering the main pollution components, the following is evident: pollution containing suspended particles and SO_2 has been found to lead to a decrease on average of 4 to 7 percent in the ventilatory function of 3 to 9 million people in the most polluted cities in Europe. It accounts for 0.2 to 0.4 percent of exacerbations of chronic respiratory illnesses (WHO Regional Publication 68, 1996). Particulate matter appears to be a main factor in this relationship.

Moreover, increases in sensitizations and allergic disorders have been observed world-wide. This is assumed to support the relationship with asthma. However, the role of indoor air pollution should not be neglected.

18.4 Effect-associated parameters

18.4.1 DOSAGE

Differentiation has to be made between the concentration measured, the actual personal exposure and the dose. Only upon contact with a noxious substance in the ambient air can an incorporation into an organism take place. Exposure is one such event. In contrast to the concentration, that merely specifies the amount of the noxious material per m^3 air, exposure means that an interaction between a human being (or other living creature) and its environment has to occur. Exposure can be specific or total. Total exposure to a specific noxious substance means the total amount taken in by the organism with all its possible routes of entry. One example would be heavy metals, which human beings can take in inhalatively with the inhalable particulate matter in the ambient air as well as orally with the food and water.

Exposure per se, however, does not signify that an adverse effect has to occur. The first important point is the kind of specific exposure or exposure mixture. Crucial is, for this purpose, that the exposure takes place under conditions, which facilitate (1) the entry of the noxious substance into the organism and (2) the reaching of the target organ. The dosis at the target tissue is dependent on the interaction of exposure and the physiological or biochemical activity states of the organism. This is important for all components in the air mixture, particularly with respect to children. Children have relative to their body weight a higher air turn-over rate than adults. Moreover, they are generally much more active physically and are, because of their body height, more exposed to certain pollutants. Thus, based on the interplay of these factors, a higher exposure and a higher dose relative to their body weight should be expected at the target organ.

18.4.2 EXPOSURE DURATION

Exposure effects can be divided into effects associated with subacute, acute, subchronic and chronic exposure, whereby the duration of exposure and the concentration may be interchangeable to a certain degree (Figure 18.2, next page). The effect-inducing dose is then either characterized by a low concentration over a longer period of time or by a high concentration over a short exposure duration, however, the effects may not necessarily be the same. Nevertheless, both factors, the concentration and the duration of exposure, are responsible for the effective dose at the target tissue (organ), i.e., the amount of the noxious substance relative to the body weight.

18.4.3 ORGAN PRETAINING

Based on these arguments, inhalation of air-borne substances are of great importance. This portal of entry has a few peculiarities. The overall surface area of the lung at normal inspiration is approximately 60 to 80 m^2. It is the largest, primary internal contact area with the environment. Moreover, most inhaled noxious substances reach their target organ directly, with no interference of the intermediate filtering function of the liver. The effect may be compared to an intravenous injection.

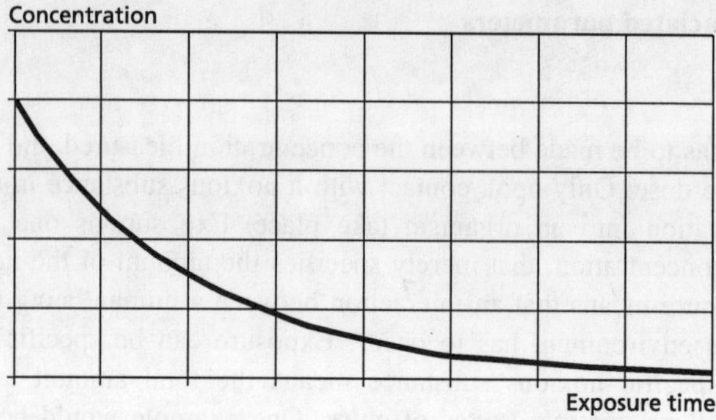

Figure 18.2 Schematic representation of a dose-response-relationship.

In general, the physiological conditions and anatomical structure of the respiratory system determine the extend of resorption and retention as well as the interdependent effects with the pollutants in the air mixture at the site of contact. Here the water-solubility plays a special role. The less water-soluble a substance is, the deeper it may penetrate down the respiratory tract. In this case, synergistic effects of the various noxious substances are important, which are not at all very well understood at this time. One example would be the SO_2-associated adverse effects on lung clearance and the possibly linked increase in residence time of particle (or dust)-like components at the target organ.

Figure 18.3 depicts the main structural components of the respiratory system. Starting with the diameter of the air-venting system, several different clearing mechanisms guard against intruding noxious substances. They basically fall into the two existing body defence mechanisms - immunologically and non-immunologically controlled reactions. The first defence are non-immunologic reactions, especially as far as the elimination of the noxious substances is concerned. The three basic mechanisms are filtration, mucociliary and phagocytic elimination (Table 18.2).

Table 18.2 Particle-size dependent clearance.

Exposure	Clearance mechanism
Particles > 10 µm	Filtration
Particles 2-10 µm	Mucociliary clearance
Particles 0,01 - 1 (2) µm	Phagocytic clearance

18.5 Effects of specific air pollution complexes in urban air pollution

The main aspect of the following presentation centres around the question, whether the typical air pollution burden in urban areas, attributable to dominant emitting sources, can induce specific changes in health status, trigger the occurrence of or lead to exacerbations of diseases.

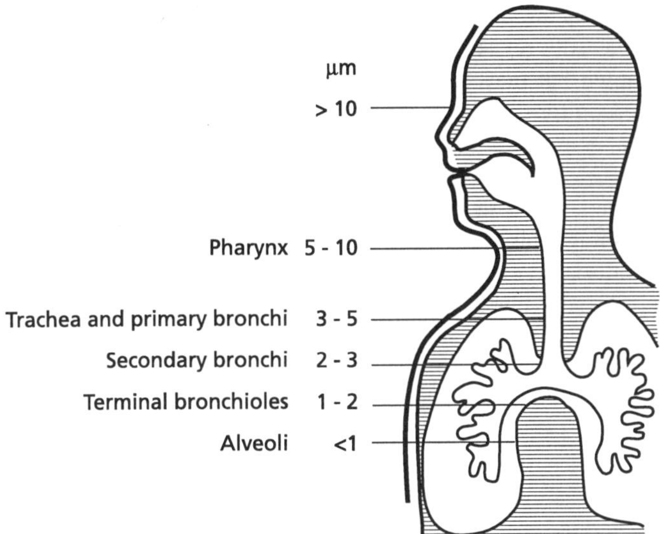

Figure 18.3 Basic structure of the human respiratory system. The various parts can be reached by particles of the indicated size.

The inter-city comparisons described above are frequently influenced by additional independent and possibly confounding variables, which are regionally or city specific. Among those are climatic and socio-economic factors as well as lifestyle-associated influences which may be region-specific. These factors have to be statistically controlled for and eliminated as part of the assessment of the relationship between differential air pollution levels and their associated risk. Intra-city investigations avoid such disadvantages as the number of confounding factors are reduced. One disadvantage is that the sample sizes of the investigated populations are generally small as only those populations should be included which are relatively locality-bound, i.e., less mobile. They are generally exposed to specific air pollutants and the exposure should differ significantly from that of another city area. Aim of such studies is to investigate the effects of specific city-area-typical exposures.

Intra-city comparisons are seldom used as the statistical requirement is considerable and the study design has to be very well co-ordinated. Herbarth and colleagues (1996) investigated the relationship between intra-city area exposure and specific disease patterns. The results indicated that areas with predominantly domestic-heating-attributed emissions (indicator component SO_2) influenced bronchitis-type illnesses, whereas areas with predominantly traffic-attributed emissions (indicator component benzene) affected allergic and asthmatic disorders. The study population were pre-school children, a less mobile population group. Their range of mobility is restricted to their area of residence which allows for a more accurate classification of their exposure. On the average, observed risk increases ranged between 1.5 to 2 times. The risk increased to a factor of three for children parentally predisposed to asthma and exposed to traffic-attributed emissions in their area of residence.

Indicator components are only representative of the dominant emission profile but are not necessarily causally related to the effect. On the other hand it may be possible that the same (unknown but may be causal) components coming from other sources than traffic are the reason for the effects. Similar results, specially with regard to asthma and specific pollution complexes as shown by indicator components, have been reported in a recent study carried out in Paris (Medina 1996). The author reports on increased hospitalizations dependent on traffic-attributed primary and secondary components of typical summersmog episodes. Dose-dependent effects were observed for asthma, ischemic heart disease, exacerbations of respiratory illnesses and the daily mean concentrations of SO_2, black smoke, NO_x and O_3. The association was especially evident between respiratory illnesses and exposure to the described indicator components.

The International Study of Asthma and Allergy in Childhood (Isaac 1993) confirms these results as well. Data from the collaborative centres confirm a positive association between the prevalence of wheezing and allergic rhinitis with the density of traffic. The associated risks increased between 1.2 and 2 times (Weiland et al. 1994).

18.5.1 IRRITANTS

Recent studies also point to an association of "indicator"-irritative gaseous substances and allergies. It is assumed that irritant gases such as sulphur dioxide and nitrogen oxides and possibly particulate diesel exhaust gases induce inflammation of the respiratory tract. This may facilitate the penetrance of allergens into the body thereby increasing the probability of an allergic response.

Besides the described effects of the indicator components for the two main emitting sources (domestic heating or combustion-derived energy production and traffic), other substances of air pollution have become increasingly the focus of recent investigations:

- volatile organic compounds,
- particle-bound organic compounds and
- dust, whereby the size-dependent chemical complex of the dust particles as well as their morphology are included in any assessment.

Untoward health effects are reported mainly in regard to particulate matter (PM) and specifically to PM_{10}. Summarizing, it has been estimated that smog situations or other short-term pollution episodes account for 7-10% of all lower respiratory illnesses in children. This proportion reaches 21% in the most polluted cities. It has been observed, that the number of asthmatics who experience exacerbations of symptoms doubles when the concentration of respirable particles (PM_{10}) remains at 100 $\mu g/m^3$ over a three day period. Hospital admissions for respiratory conditions increase by 20%. If the daily concentration of total suspended (TSP) frequently exceeds 150 $\mu g/m^3$, 3-7% of new cases of obstructive airway diseases can be expected (WHO 1996).

It should be pointed out that PM and SO_2 were highly correlated. Nevertheless, the relevance of industry as well as traffic as sources of PM is gaining in importance. This is the reason that epidemiologic research has focused more and more to the particular adverse health effects of PM. These effects range from mild to severe and include changes in pulmonary function, respiratory symptoms and daily mortality rates. The studies indicate that adverse effects already do occur at very low concentrations of the particles, which could mean that the effect may be due to its size. Based on this information, it seems almost necessary to differentiate between PM_{10} and $PM_{2.5}$ and/or other fractions of PM.

18.6 Risk analyses and threshold limit values

Based on the analyses of the temporal distribution of noxious substances and morbidity rates among populations, a possible association can be inferred from the pollution burden and the exposure. The same conclusions can be drawn from the spatially differentiated exposure-morbidity analyses. Combining the two allows more data to be included in the analyses of associations. The advantage is that a wide spectrum of concentrations is included which may lead to derive at a dose-effect association. Spatial comparisons alone will not allow such deductions, as, generally, only an average burden of the region of investigation is included, leaving two sole reference points with respect to the concentration. Applying this analysis in a large-scaled study, Herbarth (1995) determined a dose-effect relationship for respiratory illnesses and the indicator component SO_2 (Figure 18.4).

Figure 18.4 depicts the dose-effect relationship between respiratory illnesses among adults and children and the indicator component SO_2. The areas below the curves mark the area of "no effect" for either adult or children respectively. Should the concentration, dependent on the duration of exposure, fall into this range, an increase in the morbidity above the one to occur just by-chance is not likely. Does the concentration, however, increase above the indicated curves, an increased morbidity in respiratory illnesses can be expected taking the susceptibility/sensitivity of the population (children, adults, etc.) into account.

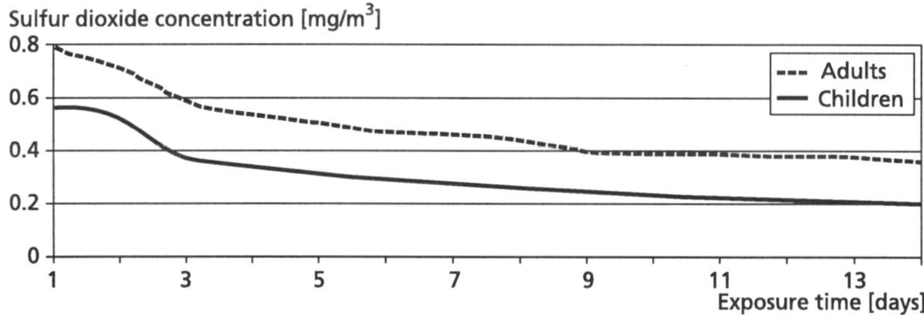

Figure 18.4 Dose-effect relationship for respiratory illnesses and indicator component SO_2.

It should be stressed that indicator components are indeed indicators. These components are indicative of a specific pollution burden attributable to a specific urban emission complex. It does not indicate a cause-effect relationship between the indicator component and the disease under investigation. The effect may possibly be causally related to any other component of the emission complex. While the curve lines in

Figure 18.4 mark the boundary beyond which an increase in the morbidity is likely to occur, Table 18.3 specifies the threshold, below which no adverse effects will occur. This means, no adverse physiological reactions should be detectable by present-day methods. Dose-effect relationships will allow to determine various threshold limit values. The WHO (1995) has proposed threshold limit values for several indicator components (Table 18.3).

Table 18.3 Threshold limit values for selected indicator components.

Component	Threshold limit value	Exposure time
Ozone	120 µg/m³	8 h
Nitrogen dioxide	200 µg/m³	1 h
	40-50 µg/m³	A year
Sulphur dioxide	500 µg/m³	10 min
	125 µg/m³	24 h
	50 µg/m³	A year
Carbon monoxide	100 mg/m³	15 min
	60 mg/m³	30 min
	30 mg/m³	1 h
	10 mg/m³	8 h

A summary of these guidelines is included in the WHO publication (1995). The presentation includes an exposure evaluation, followed by the health risk assessment and recommendation guidelines for these "classical" air pollutants.

18.7 Summary

Despite the implementation of many emission-reducing measures, the so-called classical pollutant substances are still among the most health-relevant noxious substances of urban areas. The main ones are sulphur dioxide, carbon monoxide, nitrogen oxides and particulate matter, all corresponding to the two dominant sources of pollution complexes in urban air, energy production (domestic heating) and motor vehicle traffic. Increasingly, however, the volatile organic compounds, substances adsorbed on or contained in particulate matter are gaining in importance.

Undoubtedly, airway diseases are the disorders most influenced by these urban air components. Allergies and allergy-associated diseases (such as asthma) are becoming more and more a concern. Evidence is also increasing that bronchitis-like illnesses are associated with a high probability with "winter"-typical exposures, indicated by sulphur dioxide. Allergic disorders, including asthma, on the other hand are associated rather with exposure to emissions typical of motor vehicle traffic and indoor pollution.

Within this context, it should be pointed out, that other than the above mentioned emission complexes lead to respective exposures. The one particularly noteworthy is indoor pollution caused by materials used indoors, the furnishings as well as be due to the interaction with functional human activities.

We are just at the beginning of getting answers to the questions regarding the range of response patterns to mixture of substance complexes and further effective collaborative scientific actions will be required.

18.8 References

Asher, M.J. et al. (1995) International study of asthma and allergies in childhood (ISAAC): rationale and methods, *Eur. Resp. J.*, **8**, 483-491.

Brunekreef, B. et al. (1995) Epidemiologic studies on short-term effects of low levels of major ambient air pollution components, *Environmental Health Perspectives*, **103**, 3-13.

Burney, P.G.J. et al. (1990) Has the prevalence of asthma increased in children? Evidence from the National Study of Health and Growth 1973-1986, *British Medical J,*. **300**, 1306-1310.

EEA (1995) European Environment Assessment, *Europe's environment: the Dobris assessment*, Luxemburg, Office for Official Publications of the EC.

Herbarth, O. et al. (1996) An epidemiologic study of the effects of traffic- and domestic heating-attributed emissions on respiratory illnesses and allergies among children, *Epidemiology*, **7**, 586.

Herbarth, O. (1995) Risk Assessment of Environmentally Influenced Airway Diseases Based on Time-Series Analysis, *Environmental Health Persp.*, **103**, 852-856.

ISAAC-study (1993) *Manual*, Auckland (NZ) and Münster (FRG).

Lundbeck, B. et al. (1993) The prevalence of asthma and respiratory symptoms is still increasing. In: Lundback, B. Asthma, chronic bronchitis and respiratory symptoms: prevalence and important determinants, The obstructive lung disease in northern Sweden, Study I. Umea, *Umea University, Medical Dissertation No. 387.*

Medina, S. et al. (1996) Urban air pollution is still a public problem in Paris, *World Health Forum*, **17**, 187-193.

Weiland, S.K. et al. (1994) Self-Reported Wheezing and Allergic Rhinitis in Children and Traffic Density on Street of Residence, *AEP 4 3 243-247.*

WHO ECEH (1995) WHO European Centre for Environment and Health, *Concern for Europe's tomorrow*, Stuttgart, Wissenschaftliche Verlagsgesellschaft mbH.

WHO ECEH (1994) WHO European Centre for Environment and Health, Update and Revision of the Air Quality Guidelines for Europe, *EUR/HFA target 21.*

WHO (1996) *Regional Publications, European Series, No. 68, EEA Environmental Monograph, No. 2*: Environment and Health 1, Overview and main European issues.

WHO (1995) Regional Office for Europe: Update and revision of the air quality guidelines for Europe, *EUR/HFA target 21.*

Wjst, M. (1996) Epidemiology of asthma in childhood - an international comparison, *Allergologie*, **19**, 234-243.

Chapter 19

MATERIALS DAMAGE

JOHAN TIDBLAD and VLADIMIR KUCERA
Swedish Corrosion Institute
Roslagsv. 101, Bldg. 25, SE-104 05 Stockholm, Sweden

19.1 Introduction

The detrimental effect of air pollutants emitted by burning fossil fuels has been known for a very long time, one of the first mentioned was the adverse effect of smoke from coal burning which was acknowledged several centuries ago in England. Sulphur compounds emitted to the atmosphere were suspected to be the cause of acidification already in the end of the 19th century. Systematic laboratory exposures in the 1930's demonstrated the corrosive effect of SO_2 on metals. This was later also proved by field exposures and SO_2 was for a long time considered to be the main corrosive pollutant. Over the last decade successful attempts to reduce SO_2 emission have been executed in Europe, resulting in decreasing corrosion rates for many materials. SO_2 can no longer be regarded as the only important corrosion stimulator, a new "multipollutant" situation has arisen in which SO_2, NO_2, O_3, and their reaction products are the main actors. In the following treatment the main parameters and their general effects on materials degradation are presented. After that, specific considerations for different groups of materials as well as the impact on cultural heritage are outlined. The chapter concludes with a description of economic evaluation and the concept of acceptable levels and mapping.

19.2 Description of the degradation process

The end product of the degradation of a material is well known. With time, the materials are converted to their natural state, i.e., the original compounds from which they were manufactured. However, the time scale for this process is of utmost practical importance and can vary significantly depending on material and environment.

Figure 19.1 shows a view of the different regimes important in materials degradation. The moisture layer is formed by physical adsorption of water from the atmosphere. It is often a prerequisite for the degradation process, i.e., atmospheric corrosion can be regarded as a discontinuous process occurring only when the surface is sufficiently wet. The water provides a way for the pollutants in the atmosphere to transform into species aggressive to the material and the water amount can vary from a few nm in thickness (1 nm = 10^{-9} m) to a water layer in the range 100 μm thick in the case of rainfall. The damaged material, i.e., the corrosion products, separates the base material from the aggressive solution and can be either protective or non-protective. Furthermore, the nature of the damaged zone can either be a reaction product of the base material, incorporating compounds from the atmosphere as in the case of metal corrosion, or a depletion of the compounds making up the base material, as in the development of a leached layer in glass corrosion.

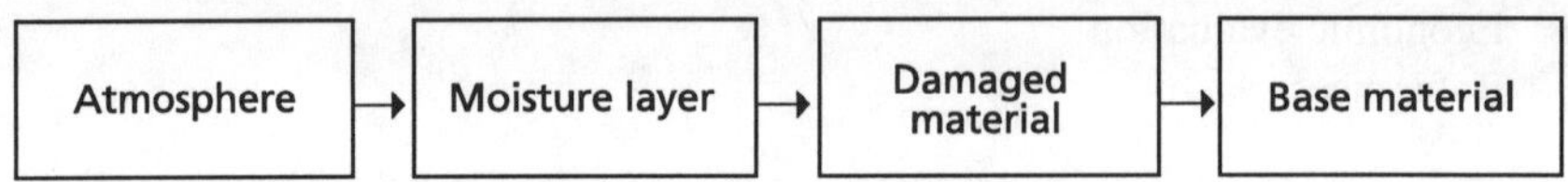

Figure 19.1 Schematic view of the atmospheric deterioration process.

Many parameters can influence the damage to materials, which is an interplay between chemical, physical and biological parameters. It would lead too far to describe all these processes in great detail and this treatment will focus on the aspects specific to the urban situation, i.e., man-made pollutants and their interplay with natural factors. It is, however, important to recognize that corrosion is a process that occurs even in the absence of pollutants and the key question is to which extent urban conditions affect and accelerate the "natural" or background corrosion of materials.

The most important pollutants acting as corrosive agents are sulphur and nitrogen compounds including secondary pollutants, and particulates. It has clearly been demonstrated that pollutants enhance the natural corrosion process for several metallic and non-metallic materials even if it may be difficult to quantify the contribution of pollutants in the weathering process.

Depending on the way pollutants are transported from the atmosphere to the corroding surface, two types of deposition processes are recognised in atmospheric corrosion, dry deposition and wet deposition. Wet deposition refers to precipitation whereas dry deposition refers to the remaining processes, including gas-phase deposition and particle deposition.

A dose-response function is a mathematical function that expresses the corrosion as a function of the environment for a particular material. For an unsheltered material the dose-response relation may include terms for dry deposition as well as wet deposition. Dose-response functions for different materials are included in Section 19.3. A compilation of available dose-response functions for building materials was done recently (Butlin et al. 1994). The term damage function is not synonymous to dose-response relation. A damage function is a function in which the service life or life time is included, i.e., dose-response functions quantify the effect on a material whereas damage functions quantify the effect on a material used in a specified application.

19.2.1 CLIMATE

Climate has a significant role in degradation of materials, hence the term "weathering". Wind and water can cause erosion and temperature fluctuations result in freezing-thawing cycles. The volume expansion of water turning to ice results in significant stress to any porous material. Sun radiation can also be a direct factor, especially for the degradation of polymers.

Temperature and relative humidity are important since they are the main factors that determine the thickness of the moisture layer in absence of rainfall. On the urban scale a closer view on mesoclimatic parameters of temperature and humidity shows that the temperature in the centre may be 2-3 degrees higher than in the outskirts and that a gradient can also be found in relative humidity that reaches higher values in the cold areas on the periphery (Chapter 7). This effect partially compensates the higher SO_2 deposition in the centre of an urban area resulting in a more even level of corrosivity than what would otherwise be the case (Kucera, Fitz 1995).

The time of wetness (TOW) is a commonly used concept for inorganic materials, in particular for metals, that refers to the time when corrosion occurs, i.e., when the moisture layer is present. This concept is useful when describing the degradation process and when classifying different climatic regions from a corrosion point of view but TOW is difficult to calculate. Since the surface always contains some small amount of water, a thickness has to be defined below which corrosion of practical importance is not considered to occur. The most common practical approach is to calculate the time when the temperature is above 0°C and the relative humidity is above 80%. The TOW value resulting from this procedure is a crude estimate since the true time of wetness depends on many other parameters, such as type, mass, orientation, and degree of sheltering of the metal, hygroscopic properties of corrosion products and deposited particles, type and level of pollutants, air velocity, and solar flux (Leygraf 1995). Figure 19.2 (next page) shows the effect of climate on sandstone degradation using the concept of time of wetness and the dose-response relation given later (Section 19.3).

19.2.2 EFFECTS OF DRY DEPOSITION

Dry deposition refers to the process by which particles and gases are transferred from the atmosphere to the material surface. The deposition velocity, i.e., the ratio of surface flux to concentration for a particular gas, depends not only on the conditions in the atmosphere but also on the thickness of the moisture layer, the reactivity of the material and the corrosion products.

Indoor corrosion is in many aspects similar to corrosion outdoors in sheltered positions in that dry deposition is the main transport process of pollutants. However, there are many differences as well. Compared to outdoor environments there are more corrosion stimulators present even if concentrations of individual pollutants often are lower. Therefore, the complexity of the situation outlined below is even more important indoors. Another aspect of indoor corrosion is that the materials stored indoors often are more sensitive to attack. In museums, even a low corrosion rate has serious consequences to the valuable artefacts.

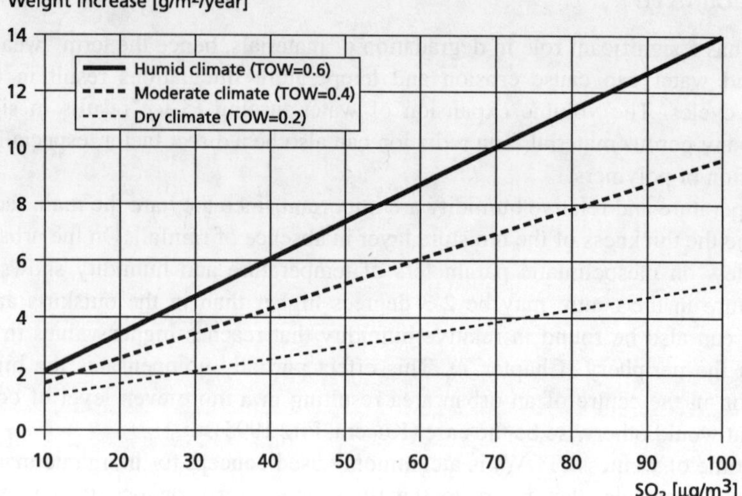

Figure 19.2 Weight increase of sandstone calculated from a dose-response relation (Equation 19.10, page 353) showing the effect of different climates on the deterioration rate.

The new multipollutant situation

The general difficulties associated with field exposure programs are related to time variations and the corresponding biases introduced. Short term time variations result in correlations between pollutants and long term variation makes it difficult to obtain reliable steady-state corrosion rates. The climatic variations and man-related activities together with the complex chemistry result in diurnal and seasonal variations for most of the atmospheric species. As a result, correlations of pollutants are not uncommon and have a large effect on evaluation of field exposure programs. It is often very hard, or impossible, to discriminate the individual effects of each pollutant. Another difficulty can be that the pollutant responsible for attack is not measured. Large exposure programs often extend over several years. One crucial parameter is to decide if and when a steady-state corrosion rate has been reached. The most extensively examined, and probably most important pollutant in atmospheric corrosion - sulphur dioxide - has gone through dramatic changes during the last decades. In most developed countries, the SO_2 concentration is gradually decreasing and significant changes can sometimes be seen from year to year. As a consequence, the kinetic behaviour is difficult to deduce. Sometimes it may even be impossible to decide if corrosion is linear or decreasing with time. This is the reason why it is so valuable to have comparative data from laboratory exposures. In carefully planned and controlled laboratory experiments, it is possible to systematically explore and quantify effects which are correlated in the field.

In the field of atmospheric corrosion, SO_2 is the single most investigated gaseous pollutant. Its detrimental effect on materials in general is since long an indisputable fact. With the reduced levels, SO_2 is no longer the only main pollutant and many other need to be considered. A new, multipollutant situation has arisen in which NO_2, O_3, and their reaction products are important but it should be stressed that SO_2 is still one of the main actors in the new situation, although not as dominant as in the past.

Gaseous pollutants including synergistic effects

Sulphur dioxide is one of the main contributors to the degradation of materials. SO_2 is dissolved in the moisture layer forming sulphite and, after oxidation, sulphate:

$$SO_2(g) + H_2O \rightarrow H^+ + HSO_3^- \rightarrow \ldots \rightarrow 2H^+ + SO_4^{2-} \quad (19.1)$$

This process results in an acidification of the moisture layer which enhances the corrosion process. The SO_2 deposition rate depends mostly on the material, it is often higher for sensitive materials, and varies between 0.01 and 2 cm/s.

SO_2 can also be oxidized in the atmosphere and thereby contributes to the acidity of wet deposition. The droplets of sulphuric acid (H_2SO_4) can be neutralized by ammonia (NH_3) forming particles of ammonium sulphate (NH_4HSO_4). Sulphate can be deposited in the form of wet deposition and the particles can be deposited in the form of dry deposition according to the reaction sequence:

$$SO_2 \rightarrow H_2SO_4 \rightarrow NH_4HSO_4 \rightarrow (NH_4)_2SO_4 \quad (19.2)$$

Sulphate is also frequently present in corrosion products.

The role of nitrogen oxides in the degradation of materials has not yet been fully elucidated. Nitrogen is mainly emitted as NO formed in the combustion process and then further oxidized to NO_2 and HNO_3 by photochemical reactions and possibly neutralized by NH_3 forming particulate nitrates (NH_4NO_3):

$$NO \rightarrow NO_2 \rightarrow HNO_3 \rightarrow NH_4NO_3 \quad (19.3)$$

Nitrates are not frequently found in the corrosion products. For unsheltered surfaces this can be explained by the high solubility of many nitrate precipitates. For materials protected from rain, the low nitrate content may depend on the low deposition velocity for NO_2 or that it is possible for nitrogen to leave the moisture layer in the form of $HNO_2(g)$. Even if NO_2 in itself is much less detrimental to most materials than SO_2 it can still be important. O_3 and also HNO_3 are aggressive secondary pollutants to NO_2. Furthermore, the combination of NO_2 and SO_2 results in a much higher corrosion rate compared to the effect from the individual pollutants, i.e., there exists a synergistic effect.

Ozone (O_3) is a principal pollutant, in the past mainly associated with the degradation of natural rubber but most organic materials containing carbon double bonds are sensitive to its effect, such as painted surfaces, polymers and textiles. It is, however, a general oxidant and thus, for inorganic materials O_3 has a synergistic effect in combination with SO_2. The effects of ozone on the degradation of organic materials are further described in Section 19.3.

HNO_3 is a secondary pollutant formed by the oxidation of NO_2. It is a strong acid with a high deposition velocity that is relatively independent of the relative humidity which increases its importance for dry and warm climates. The effects of this pollutant has so far been the subject of only a few studies but it is potentially harmful for many materials.

In laboratory exposures synergistic effects are often detected (Tidblad, Kucera 1996). The term "synergistic effect" means that the corrosion attack for a material exposed to a mixture of gases is greater than the sum from individual exposures. In practice, the term almost exclusively refers to the combination of SO_2 with a second pollutant, often NO_2 or sometimes O_3. The effect of the second pollutant has been attributed to the oxidation of S(IV) to S(VI), i.e., sulphite to sulphate. A qualitative difference exists between the role of NO_2 and O_3 in this respect. The oxidation of sulphur dioxide by O_3 is stoichiometric,

$$SO_2 + O_3 + H_2O \rightarrow SO_4^{2-} + 2H^+ + O_2 \quad (19.4)$$

whereas the oxidation by NO_2, at least on zinc, appears to be catalytic, i.e., NO_2 is not consumed. Figure 19.3 illustrates the synergistic effects for zinc (Svensson, Johansson 1993), as obtained from laboratory exposures. Illustrative as the figure is, however, one should be careful to extrapolate results from laboratory exposures to field conditions. Even though synergistic effects frequently are detected in the laboratory they remain to be proven in the field due to the correlation of gaseous pollutant concentrations. This is particularly true for O_3 and NO_2, which are often negatively correlated in the field, if annual averages are used. Since both NO_2 and O_3 can have similar effects to various materials a statistical analysis of field exposure data will always have difficulties in separating effects from these pollutants.

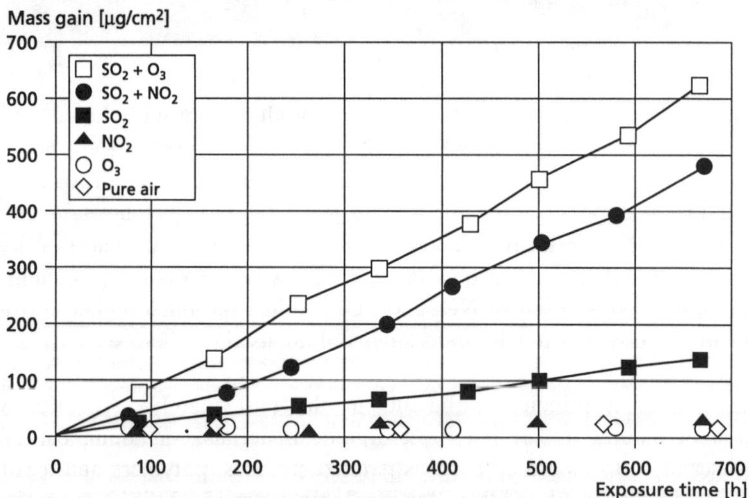

Figure 19.3 Mass gain of zinc samples exposed in air at 95% RH; influence of SO_2 (225 ppb), NO_2 (400 ppb) and O_3 (400 ppb), (Svensson, Johansson 1993).

19.2.3 PARTICULATES AND SOILING

(This section has been compiled in cooperation with H. Horvarth, author of Chapter 9).

Particles normally enhance, but may in some cases reduce the *deterioration rates of materials*, e.g., basic particles deposited on a surface may neutralise the effect of other pollutants, such as SO_2. The deposition of particles is characterised by the deposition velocity, similar to gas deposition. Particles deposit due to sedimentation, impaction, and diffusion depending on the size of the particles. Sedimentation is important for particles larger than a few micrometers and mainly occurs at horizontal surfaces.

Particles in this range have a high deposition velocity (typically 1 mm/s). Since the life time of these particles in the atmosphere is short, they are found on surfaces near the source. Particles smaller than a few micrometers, i.e., submicron particles, have a much lower deposition velocity (typically 0.05 mm/s) thus less particles will reach the surface. Since diffusion is the main factor, the deposition takes place at any surface. Furthermore, submicron particles contains soot, therefore although a less effective deposition occurs, the effect is well visible by the dirty appearance of the surface.

Particles containing NH_4NO_3, $(NH_4)_2SO_4$ and NH_4HSO_4 play an important role in atmospheric corrosion and this is related to their ability to increase the time of wetness. They are hygroscopic and, in the most simple case, the particle starts to absorb water when the relative humidity exceeds a critical level, determined by the equilibrium properties of a saturated solution of the salt. In practice, however, the particles deposited on the surface are a mixture of different compounds and this results in a difficulty to define a critical relative humidity. The ionic content of the particles prolongs the time of wetness by other mechanisms as well, for instance, the freezing point of the moisture layer is affected and can be significantly below 0°C.

In addition to the prolonging of the time of wetness, ionic particles enhance the corrosion by providing corrosion stimulators. Consider the fate of an ammonium sulphate particle, $(NH_4)_2SO_4$. When the relative humidity is sufficiently high, a saturated solution of ammonium sulphate forms on the surface. Ammonia, NH_3, leaves the moisture layer irreversibly in its gaseous form, increasing the acidity and the corrosivity of the moisture layer.

Soiling is another effect of particulate matter. It is an optical effect, a darkening of the surface that can be measured as a change in light reflectance, and is generally related to deposition of airborne particulate matter. Dirty buildings are a normal appearance in all larger towns. Old churches, e.g. the dome at Cologne are almost black, although they are built of stones with fairly bright colours. This disagreeable appearance of otherwise beautiful buildings is caused by the atmospheric pollution. But this is not only restricted to old buildings, greenhouses need to be cleaned, or solar cells have less output due to soiling. The alteration of the visual appearance may be unacceptable even if the base material is virtually unaffected and the costs related to cleaning may be substantial. Furthermore, the substances constituting the soiling matter, i.e., carbon particles, may indirectly take part in the degradation process by acting as catalysts for various chemical reactions, particularly for the conversion of SO_2 and NO_x into sulphuric acid and nitric acids (Newby et al. 1991).

19.2.4 EFFECTS OF WET DEPOSITION

Similar to particles, the effect of wet deposition can be either detrimental or beneficial, depending on the conditions. Figure 19.4 shows a simplified picture of the flow of matter involved in the degradation process. For sheltered conditions, wet deposition and run-off are excluded and the corrosion, i.e., the destruction of the base material, is identical to the formation of the corrosion products. The wet deposition has two effects on the corrosion process. On one hand, it transports chemically active compounds present in the rain to the surface, thereby increasing the corrosivity of the moisture layer. On the other hand, it washes away chemically active compounds previously deposited on the surface, with the opposite effect. Thus, for a specific material and environment choosing a sheltered exposure condition rather than unsheltered may, or may not, increase the corrosion rate.

Another important aspect of the corrosion process in unsheltered exposure conditions may be derived from Figure 19.4. The run-off may contain significant amount of the base material. This is important to consider from an ecological point of view. If a sustainable city development is to be achieved it is important to quantify this source of heavy metals, keeping in mind that it not necessarily equals the corrosion rate as part of the metal may be bonded in low soluble corrosion products.

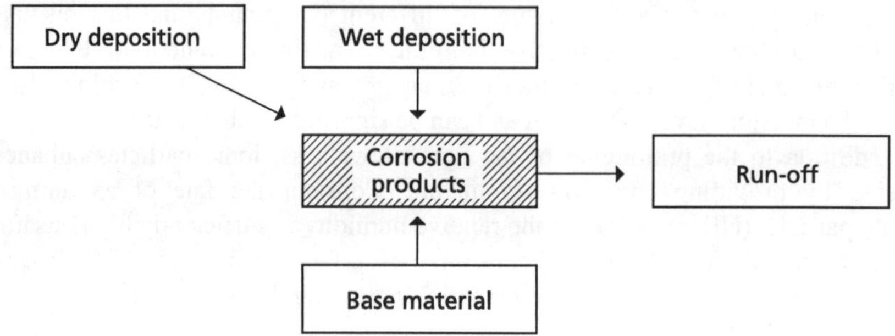

Figure 19.4 Flow of materials in the degradation of unsheltered materials with developed corrosion products.

19.3 Effects on different materials

Even if the description of the degradation process outlined above has a general validity the degradation mechanism is specifically dependent on the material. The specific effect of a pollutant or mixture of pollutants depends mostly on the material. Inorganic materials, such as metals and stone materials, are generally more sensitive to SO_2 (Arroyave, Morcillo 1995) while organic materials, such as painted surfaces, polymers and textiles, are more sensitive to O_3 (Lee et al. 1994). However, synergistic effects of different pollutant combinations, such as SO_2/NO_2 and SO_2/O_3, are often present and the magnitude of these effect vary considerably, depending on material. In the following the degradation process for inorganic and organic materials are briefly described, followed by examples of specific applications including cultural heritage.

19.3.1 INORGANIC MATERIALS

The degradation of inorganic materials is closely related to mineral chemistry. This is obvious for stones. Metals are produced from minerals and in the degradation process the metal strives to return to its natural state. Metal corrosion products are therefore similar in structure, if not identical, to naturally occurring minerals.

Metals

Leygraf recently reviewed the atmospheric corrosion of metals (Leygraf 1995), which is the basis for this summary and which the interested reader should consult.

The first water monolayer forms at approximately 20% relative humidity and is quite immobile. Subsequent water layers become increasingly mobile and after the third layer the properties of bulk water is reached. The relative humidity when this occurs varies from 50% to 90% depending on the metal and this interval coincides with that of the "critical relative humidity" above which corrosion rates start to be significant. The most important pollutants for metals are those that are acidifying, either directly or indirectly. Direct acidifiers include SO_2, NO_2 and carbonyl compounds like HCHO and CH_3CHO forming organic acids. Indirect acidifying pollutants are those that promote the oxidation of S(IV) to S(VI). The release of hydrogen ions approximately doubles since HSO_3^- is the stable form of S(IV) while SO_4^{2-} is the stable form of S(VI).

Atmospheric corrosion of metals is of electrochemical nature, i.e., the process can be separated in anodic and cathodic parts

$$Me \rightarrow Me^{n+} + ne^- \quad \text{(anode reaction)} \qquad (19.5)$$

$$O_2 + 2H_2O + 4e^- \rightarrow 2OH^- \quad \text{(cathode reaction)} \qquad (19.6)$$

In reality, however, the use of electrochemical measurements in atmospheric corrosion is scarce, due to the difficulties of reproducing and measuring within the moisture layer. The initially formed corrosion products are oxides, hydroxides and even oxyhydroxides. During prolonged exposure, however, a variety of corrosion products may form including ligands originating from the atmosphere, i.e., sulphate, nitrate and chloride.

Atmospheric corrosion outdoors include different periods, an induction period, a period for establishing stationary conditions and, if constant thickness, composition and atomic structure of corrosion products are reached, a stationary period. The two initial periods decrease in length in more aggressive environments. The degradation process may be considered as discontinuous, only occurring when the moisture layer is present.

Indoors, on the other hand, the aqueous layer is often governed by relatively constant humidity conditions. This means that there are practically no wet-dry cycles indoors and that the influence of indoor humidity can hardly be described by introducing a time of wetness factor. Concentrations of corrosion stimulators are in general lower indoors compared to outdoors, except NH_3 and HCHO which have indoor sources. Dry deposition velocities are also lower due to lower air velocities. As could be expected, the corrosion rates are generally lower indoors. On the other hand, corrosion rates are harder to predict since a variety of corrosion stimulators can be present. Furthermore, the microclimate can vary substantially indoors.

Within a UN-ECE exposure program (Kucera et al. 1995) dose-response functions for a number of materials have been derived including among others weathering steel, zinc, aluminium, copper, bronze, limestone, sandstone, alkyd/steel and nickel. As an illustration, the dose-response functions for unsheltered zinc and sheltered nickel after 4 years of exposure are:

$$ML_{Zn} = 14.5 + 0.043\ TOW[SO_2][O_3] + 80\ Rain[H^+] \tag{19.7}$$

$$WI_{Ni} = 5.5 + 15.8[SO_2] \tag{19.8}$$

where ML_{Zn} is the mass loss of unsheltered zinc in g/m^2, WI_{Ni} is the weight increase of sheltered nickel in $\mu g/cm^2$, TOW is the time of wetness in parts of year, "[]" denotes gaseous concentrations in $\mu g/m^3$, Rain is the average amount of precipitation in m/year and $[H^+]$ is the hydrogen ion concentration in precipitation in mg/l. The term $TOW[SO_2][O_3]$ corresponds to the dry deposition and the term $Rain[H^+]$ to the wet deposition.

Stone materials

Stone in architecture, their properties and durability is described in the highly recommended book by Winkler (1994), which has recently been revised and extended and upon which this summary is based.

Stone is characterised by a wide range of compositions, textures and structures. Due to their heterogeneous nature, the durability of stones is quite variable even for a constant environment. When characterising a particular piece of stone a number of important features are used, i.e., porosity, water sorption ability, bulk specific gravity, hardness, strength, elasticity and thermal properties. Porosity is one of the most important and is defined as the ratio of pore space to total volume. The porosity can vary from 0.5% as for granites and marble up to 20% to 25% for the most porous limestones and sandstones.

In contrast to metals, stone deterioration is not of an electrochemical nature. Otherwise, the weathering agents for stone are similar to those for metals, sulphate originating from sulphur dioxide being the most important, and the process of weathering is directed from the original minerals towards those of greater regional stability. The presence of oxygen leads to oxidation and water absorption leads to hydration or dissolution. Nitrogen oxides can be converted to nitric acid which is corrosive to stones, though less damaging than sulphuric acid. The chemical process of weathering is aided by mechanical deterioration and freezing-thawing cycles. Humid conditions makes it possible for moisture to enter into the narrow stone capillaries.

$CaCO_3$ is the major component in natural calcareous stones such as limestone and marble, which is the more purely crystallised form. Sandstones also contain magnesium carbonate, $MgCO_3$, in varying proportions. Calcareous stones have the highest mass loss rates of the common building stones. The natural heterogeneity of most rocks makes the modelling of weathering rates very difficult. The concept of time of wetness used for metals is in principle also valid for stones, but the porosity and hygroscopic properties of the natural stone makes it even more difficult to estimate its true value.

Dissolution is the dissociation of a mineral in a solvent and the solvent motion in acid rain is a very important factor in the dissolution process. The movement keeps the solvent in contact with the stone undersaturated and thus increases the dissolution. Gypsum or $CaSO_4{\cdot}2H_2O$ is a secondary mineral formed from calcite in the presence of sulphate in the atmospheric environment. Calcite has a solubility of 40 to 85 ppm in 20°C freshwater while gypsum has a solubility of 2,400 ppm. Sulphates can attack carbonate rocks in a dual way, either by direct dissolution due to the acidic conditions or by the conversion to the more soluble sulphites and sulphates. An important degradation mechanism in SO_2 polluted environment is therefore conversion to gypsum followed by dissolution. Also, gypsum has a larger volume than calcite which results in tension and possible fractures. In sheltered position the gypsum is not dissolved and instead forms a surface layer on the stone which may eventually crack and peel off. The future of stone corrosion will depend on successful implementation of air pollutant abatement strategies. With these actions, stone may once again deserve its name, "rock of ages".

The UN-ECE dose-response functions after 4 years of exposure for unsheltered limestone and sheltered sandstone are (Kucera et al. 1995)

$$ML_{\text{Limestone}} = 34.4 + 5.96TOW[SO_2] + 388Rain[H^+] \tag{19.9}$$

$$WI_{\text{Sandstone}} = 2.84 + 0.88TOW[SO_2] \tag{19.10}$$

where ML and WI are mass loss and weight increase, respectively in g/m^2, TOW is the time of wetness in parts of year, $[SO_2]$ is the concentration in $\mu g/m^3$, Rain is the rainfall in m/year and $[H^+]$ is the concentration in rain in mg/l. The term $TOW[SO_2]$ corresponds to the dry deposition and the term $Rain[H^+]$ to the wet deposition.

19.3.2 ORGANIC MATERIALS

While degradation of inorganic materials mostly are associated with SO_2 and NO_2, the degradation of organic materials have traditionally been associated only with O_3 in addition to the most important natural factors, temperature and solar radiation (Tidblad, Kucera 1996). The major effects of O_3 have been noted for organic polymers possessing double bonds in their structure. Two damage mechanisms may occur. Chain-scissoring results in a reduction in average molecular weight and loss of tensile strength. Cross-linking of polymers increases the rigidity, reduces the elasticity and brittleness may result. O_3 is a principal pollutant most associated with the degradation of rubber but most organic material, such as painted surfaces, polymers and textiles, are sensitive to O_3. The exposure of O_3 leads to fading and embrittlement of paints, cracking of rubber and fading of dyes in textiles or reduction of textile strength (Lee et al. 1994; Lewry 1991). However, its effect can be difficult to distinguish from direct sunlight damage. The materials affected could be subdivided in thin organic layers like paints and coatings, bulk organic items like gaskets, sealants or PVC windows and rubber materials like car window wipers or rubber roofing materials. O_3 at the concentrations found in the indoor atmosphere of many museums poses a fading hazard to the pigments used in works of art. The pigments have different sensitivities depending on the base material (Grosjean et al. 1993).

Interaction of NO_2 in the solid phase with macromolecules containing carbon double bonds can be one of the most important ways of initiating the free-radical reactions of polymer ageing in an industrial atmosphere. The formation of a nitroxyl radical (R_2-NO·) in reaction of NO_2 with carbon double bonds of the polymers occurs as a free radical cell process. Reaction of raw rubbers with NO_2 is accompanied by a reduction of the concentration of C-H bonds, and formation of nitro (R-NO_2) and nitrate (R-O-NO_2) groups.

19.3.3 SPECIFIC APPLICATIONS

Although different materials and their characteristics have been described specific considerations may be necessary depending on the application. Here, two such will be discussed, cultural heritage and electronics.

Cultural heritage

Pollutants affect also historic and cultural monuments which create an important part of our cultural heritage. Monuments, which have withstood the effects of the environment for hundreds or even thousands of years, suffered in the past decades from rapid deterioration like the Acropolis in the heavy polluted atmosphere of Athens.

In principle, there are two ways of preserving our built environment, to improve the conditions of the ambient environment and to develop better materials. For cultural heritage the latter choice is not an option. Instead, the development of conservation procedures needs to be persued, an important topic that will not be reviewed in this context. For maintenance planning of a specific object, the microclimate is of utmost importance as has been demonstrated at the Royal Palace in Stockholm and in the Prague Castle (Rendahl, Kucera 1993).

Figure 19.5 demonstrates the big difference in the level of pollutants in the two cities. Measures taken in the last two decades in Stockholm, comprising a reduction of the sulphur content in oil and diesel fuel and an extensive expansion of the district heating system, has lowered the SO_2 concentration to rural levels. In contrast to Stockholm the corrosion rates are much higher in the polluted atmosphere in Prague.

A wide range of materials are used in our cultural heritage. The principal construction materials of architectural monuments are stone, brick and plaster. Medieval stained glass has attracted attention and in France alone approximately 80,000 m^2 are at risk. In terms of sheer numbers of artefacts no class of cultural property approaches that of library and archive materials. Textile deterioration include the fading of dyes and the degradation of weighted silk (Baer 1991).

Electronics

There are several reasons why corrosion of electronics are becoming increasingly important. First, there is a fast growing use of computerised equipment which is often

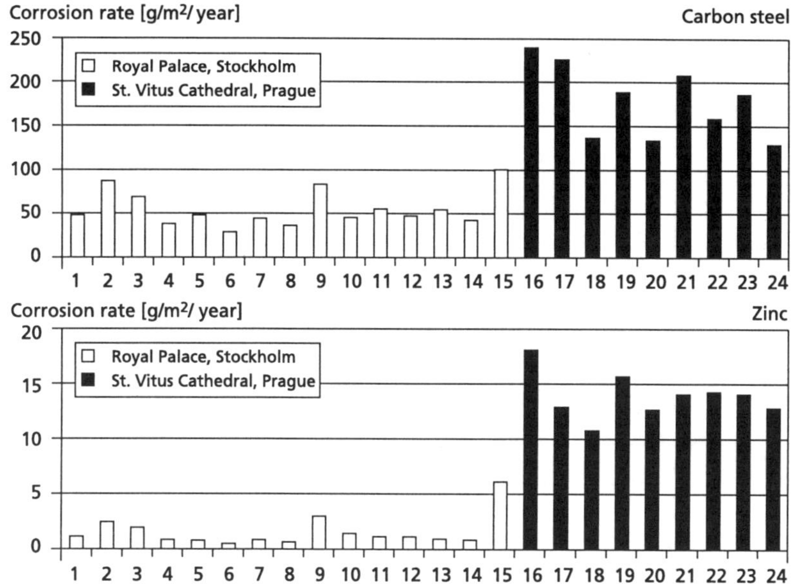

Figure 19.5 Corrosion rate of carbon steel and zinc after 1 year of exposure; site 1-15 Royal Palace, Stockholm, 16-24 St Vitus Cathedral, Prague. The SO_2 concentration was on average 5 $\mu g/m^3$ in Stockholm while it was almost 10 times higher in Prague (Rendahl et al. 1996).

located also in polluted urban and industrial atmospheres or in traffic environments. Second, the miniaturisation trend with lower signal voltages and higher packaging density makes the components more and more sensitive to a particular pollution situation (Henriksen et al. 1991).

In addition to corrosion, the degradation of electronics can be ascribed to a variety of electrical, mechanical and chemical mechanisms. Failures due to corrosion can be caused by increased contact resistance in electromechanical connectors, which is most common, overheated components, loss of electromagnetic shielding, attacks to critical signal paths and short circuits. The design of an electronic system should therefore be analysed at various levels, from system level to component level.

Nickel is a material that is particularly sensitive to sulphur dioxide. The dose-response relation has already been given and Figure 19.6 (next page) shows the corrosion attack on this electric contact material as a function of SO_2 concentration, highlighting some cities in Europe.

19.4 Simultaneous trends in corrosion rate and pollution

A way to quantify the effect of decreasing SO_2 levels on corrosion rate is to evaluate absolute corrosion after a definite short time, instead of corrosion rate. Repeated exposures at regular intervals enable comparative studies of trends in corrosion and environment. This, so-called "trend analysis", is often performed as one-year exposures

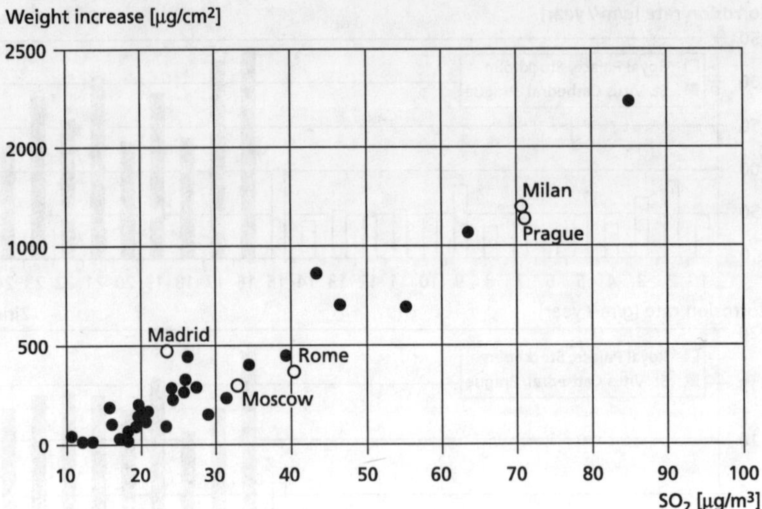

Figure 19.6 Weight increase of nickel exposed in a sheltering box after 4 years of exposure in UN-ECE Exposure programme vs. SO_2 concentration of the ambient air.

but has its own drawbacks. If corrosion products develop slowly, the one-year corrosion attack may not be representative of the long-term behaviour. Thus, statistical trend analyses can only, without major sources of error, be performed on a homogenous group of sites having similar corrosion products.

The trend analysis within the UN-ECE Exposure programme constitutes an important part of the short/medium-term aim of the program (Tidblad, Kucera 1996). The analysis is performed as 1-year exposures of carbon steel and zinc and can be divided in trends of environment and trends of corrosivity. Figure 19.7 shows the decreasing trend in SO_2, NO_2, acidity, as well as corrosivity. All values are compared to the initial value, corresponding to the 1987/88 period (100%). Of the environmental parameters measured SO_2, NO_2 and H^+ exhibit a decreasing trend, with SO_2 having the strongest and NO_2 the weakest. Both carbon steel and zinc show strong negative trends. The decrease in corrosivity occurred first in Scandinavia, between 1987-88 and 1989-90 and later in Western and Central Europe, between 1989-90 and 1992-93. A statistical analysis shows that SO_2 is the largest contributing factor for the decreasing trend. The contribution due to decreasing H^+ in precipitation is on average much smaller than that of dry deposition.

19.5 Economic evaluation

The economic impact of acidifying pollutants is mainly concentrated to urban areas where high densities of population and of materials coincide with high pollution levels. Atmospheric corrosion is thus in most areas a local effect caused mainly by a country's own emission and only to a lesser extent affected by long distance transport of pollutants.

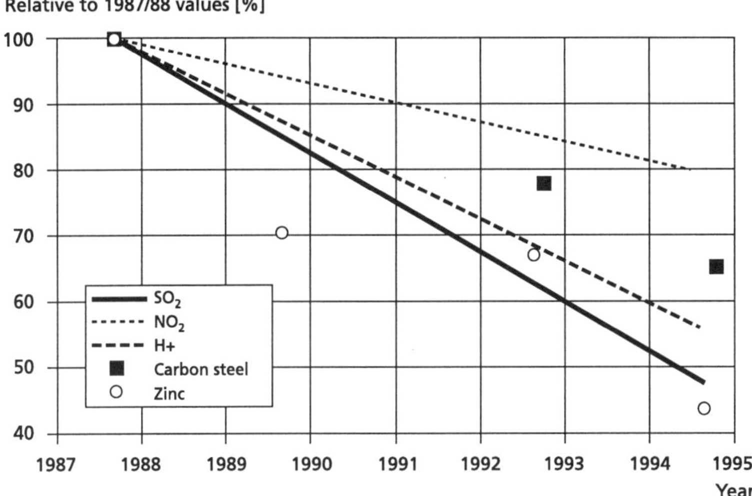

Figure 19.7 Trends of SO_2, NO_2, acidity (H^+), and corrosion of unsheltered unalloyed carbon steel. All values are expressed relative to the initial (1987/88) value.

The degradation to materials results in significant costs for the community. It is thus of great interest for policy decisions, aiming at reductions of the costs, to map areas with increased risk and to calculate the corrosion costs at different pollution scenarios.

19.5.1 MAPPING OF CORROSION AND ACCEPTABLE POLLUTION LEVELS

The basic guidelines for mapping critical levels and loads within the framework of the UN-ECE Convention on Long-Range Transboundary Air Pollution (LRTAP) are described in the Mapping Manual (Gregor et al. 1996).

A critical load is defined as the highest deposition of a compound that will not cause chemical changes leading to long-term harmful effects. Atmospheric corrosion and deterioration of materials is a cumulative, irreversible process which proceeds even in the absence of pollutants. Thus, the critical load/level approach used for ecosystems has to be modified in relation to degradation of materials as even the lowest concentration of pollutants causes an increase in the deterioration rate. This leads to the concepts of acceptable corrosion rates and pollution levels.

The acceptable corrosion rate (K_a) is defined as the corrosion which is considered "acceptable" based on technical and economic considerations for the application of the material. In reality this may not be a practical approach. Therefore it is recommended that the acceptable corrosion rate should be related in a simple way to corrosion rates in areas with "background" pollution (K_b). It has proven to be useful to define a level of acceptance, n, which is dimensionless

$$n = K_a/K_b = K_a/K_{10\%} \qquad (19.11)$$

It has been decided that the background corrosion rate should be calculated as the lower 10-percentile from the UN-ECE Exposure programme data set ($K_{10\%}$) after four years of exposure. The $K_{10\%}$ values are shown in Table 19.1. At present it is recommended that values of n between 1.2 and 2 are used. When possible, however, n should be calculated from Equation 19.11 (previous page) by deciding an acceptable corrosion rate, K_a, and then using the $K_{10\%}$ values given in Table 19.1.

Table 19.1 Background corrosion rates for different materials based on $K_{10\%}$ (10-percentile corrosion rates) from the UN-ECE Exposure programme.

Material	$K_{10\%}$ (g/m², 4 year)
Weathering steel	204
Zinc	17.1
Aluminium	0.67
Copper	14.0
Bronze	11.6
Limestone	44.0
Sandstone	40.0

For a fixed level of acceptance, n, it is now possible to calculate acceptable pollution levels, assuming that dose-response functions are available. For example, consider the case of unsheltered limestone, Equation 19.9 (page 353). The acceptable corrosion rate, K_a, is estimated by the dose-response functions:

$$K_a = 44n = 34.4 + 5.96TOW[SO_2] + 388Rain[H^+] \qquad (19.12)$$

The equation implicitly defines an acceptable pollution situation but this situation can be reached by different combinations of the pollutants. The acceptable SO_2 level, $[SO_2]_a$, can be calculated as

$$[SO_2]_a = (44n - 34.4 - 388Rain[H^+])/5.96TOW \qquad (19.13)$$

which means that for sites in dry, warm regions with low wet deposition (Rain[H^+]) and short time of wetness (TOW) comparatively higher SO_2 levels are "acceptable".

The ultimate goal of the mapping activities in the field of materials is the calculation of costs of damage caused by air pollutants to materials. However, for practical reasons two levels of mapping can be used. Level 1 mapping encompasses corrosion rate, acceptable corrosion rate, acceptable pollution levels and also individual degradation parameters. Level 2 mapping includes data on the stock of materials at risk and the damage costs associated with corrosion of these materials. Figure 19.8 shows an example of level 1 mapping for zinc in the Oslo area (Haagenrud et al. 1996).

Calculation of cost on technical materials and cultural heritage

A model has been developed that comprises the determination of amounts, geographical distributions and corrosion status of base materials and surface finishing at inspections of randomly selected buildings in different pollution strata (Kucera et al. 1993).

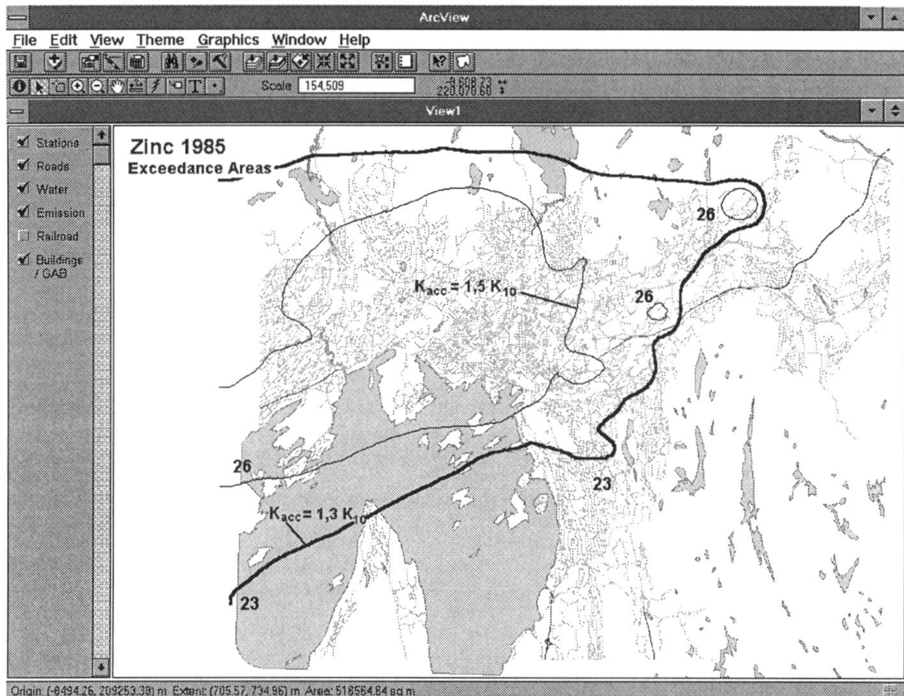

Figure 19.8 Mapping of exceedance areas for acceptable zinc corrosion = 1.3 resp. 1.5*K10%, i.e., 30% resp. 50% above background corrosion in the Oslo area, year 1985, exhibited in the Arc View AIRQUIS CORROSION (Haagenrud et al. 1996).

Figure 19.9 shows a schematic description of the different steps included in the model. First a division into pollution strata is made. Based on levels of pollutants, mainly SO_2, the area is subdivided into smaller areas (strata), each considered to have a similar pollution situation. In each of these strata, a materials inspection and inspection of physical damage are made on a random selection of buildings of different categories. This information is then used to obtain the stock of material at risk in each pollution strata.

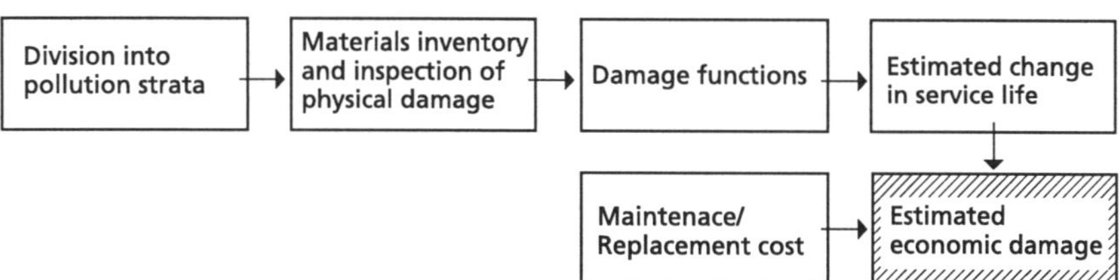

Figure 19.9 General approach for assessing cost of corrosion damage.

The damage functions are obtained from the dose-response functions described previously and these are obtained from field exposure programmes and using the total available knowledge about materials degradation. The estimated change in service life can be obtained from the damage functions or more favourably from observations of physical damage at the inspections of the selected buildings and/or from different sources of well documented practical maintenance intervals or guidelines for maintenance of buildings. Maintenance and replacement costs are the direct costs associated with the maintenance/replacement procedures, for instance labour costs. The estimated economic damage can then be calculated by the equation:

$$K_{add} = KS[\ 1/L_p - 1/L_c\ ] \tag{19.14}$$

where K_{add} is the additional cost for maintenance/replacement, K is the cost of repair/maintenance, S is the surface area of material, L_p is the maintenance interval in polluted areas and L_c is the maintenance interval in clean areas.

A recent study using the described model has shown that the *total cost savings in Europe* from reduced damage to buildings, after implementation of the second sulphur protocol, are in total over 9,000 millions US$/year (ApSimon, Cowell 1997). The damage included reduced lifetime of galvanized steel structures like roofs and fencing, more frequent maintenance painting on steel structures and wooden constructions and damage caused to rendering facades and to roofs covered by bitumen felt.

At a UN-ECE Workshop on Economic Evaluation of Air Pollution Abatement and Damage to Buildings and Cultural Heritage held 1996 in Stockholm an economic analysis of direct O_3 costs was presented for the UK region. An extrapolation from earlier analyses in USA gave an annual cost of £170-345 million. A more detailed analysis of costs in the UK gave £25-63 million for antiozonant protection to manufacturers and £25-189 million to consumers. Annual costs associated with the effect of O_3 on the yearly cost of repainting, using dose-response functions available in the literature, were estimated at £0-60 million for a change from 15 to 20 ppb O_3 and at £0-182 million for a change from 15 to 30 ppb O_3 (Holland et al. 1997).

Historic and cultural monuments are a special class of objects and it can be discussed if the loss of an original object of art can be expressed in monetary terms. Nevertheless, the artefacts are of great value to society and, therefore, economists have developed the *contingent valuation method (CVM)* to measure the benefit that people derive from an object such as a cultural monument. The technique relies on direct questioning of the population, confronting them with choices regarding a building's future. Statistical techniques can then be applied to measure the benefit that people derive from the object. This will include the value for both those who use the building and those who appreciate it, but do not use it.

It should be emphasised that all studies probably give very conservative estimates of corrosion costs since the cost for damage to the cultural heritage and cost for corrosion damage indoors are usually not included due to lack of data. Despite that, the studies of damage to materials show that the savings may balance a considerable part of the total abatement costs.

19.6 References

ApSimon, H.M., Cowell, D. (1997) Proc. UN-ECE Workshop on Economic Evaluation of Air Pollution Abatement and Damage to Buildings including Cultural Heritage, pp. 211-216, *Swedish Environmental Protection Agency, Report No. 4761.*

Arroyave, C., Morcillo, M. (1995) The effect of nitrogen oxides in atmospheric corrosion of metals. *Corros. Sci.*, **37**, 293-305.

Baer, N.S., Sabbioni, C., Sors, A.I. (eidtors) (1991) *Science, Technology and European Cultural Heritage*, Butterworth-Heinemann Ltd., Oxford.

Butlin, R., Yates, T., Murray, M., Paul, V., Medhurst, J., Gameson, T. (1994) Effects of pollutants on buildings. Building Research Establishment, Garston, Watford, *DOE Report No: DoE/HMIP/RR/94/030.*

Gregor, H-D., Werner, B., Spranger, T. (editors) (1996) *Manual on methodologies for Mapping Critical Loads/Levels and geographical areas where they are exceeded*, UN-ECE Convention on Long-Range Transboundary Air Pollution Umweltbundesamt, Berlin.

Grosjean, D., Grosjean, E., Williams, E.L. II (1993) Fading of artists' colorants by a mixture of photochemical oxidants. *Atmospheric Environment*, **27A**, 765-772.

Henriksen, J., Heinonen, R., Imrell, T., Leygraf, C., Sjögren, L. (1991) Corrosion of Electronics, A handbook based on experiences from a Nordic research project. *Bulletin No 102, Swedish Corrosion Institute*, Stockholm, Sweden.

Holland, M., Watkiss, P., Cupit, M., Lee, D. (1997) *Economic Evaluation of Air Pollution Damage to Materials*, in: Proceedings of the UN-ECE Workshop on Economic Evaluation of Air Pollution Abatement and Damage to Buildings including Cultural Heritage, p.p. 52-64, Swedish Environmental Protection Agency, Stockholm, Sweden.

Haagenrud, S.E., Henriksen, J.F., Skancke, T. (1996) in: Sjöström, C. (editor), E & FN Spon, *Proceeding 7th Int. Conf. Dur. Build. Mater. Compon.*, Chapman & Hall, London, UK, Vol. I, p. 209.

Kucera, V., Fitz, S. (1995) Direct and indirect air pollution effects on materials, including cultural monuments. *Water, Air, and Soil Pollution*, **85**, 153.

Kucera, V., Henriksen, J., Knotkova, D., Sjöström, C. (1993) Model for calculations of corrosion cost caused by air pollution and its application in three cities, *10th Eur. Corros. Congr., Barcelona, Paper No. 084.*

Kucera, V., Tidblad, J., Henriksen, J., Bartonova, A., Mikhailov, A.A. (1995) *International cooperative programme on effects on materials, including historic and cultural monuments, Report No. 18, 1995, Statistical analysis of 4 year materials exposure and acceptable deterioration levels*, Swedish Corrosion Institute, Stockholm, Sweden.

Lanting, R.W. (1984) Materials damage by photochemical oxidants., Grennfelt, P. (editor), *Ozone, Proc. Int. Workshop*, IVL, Göteborg, pp. 44-59.

Lee, D.S., Holland, M., Falla, N. (1994) An assessment of the potential damage to materials in the U.K. from tropospheric ozone, *AEA/CS/18358021/001/FINAL REPORT*, AEA Technology, Oxfordshire, UK.

Lewry, A. (1991) The effect of Ozone on Organic Materials, *BRE Report N72/91*. BRE, Watford, UK.

Leygraf, C. (1995) Atmospheric corrosion., in: Marcus, P., Oudar, J. (editors), *Corrosion Mechanisms in Theory and Practice*, Marcel Dekker, New York, pp. 421-455.

Newby, P.T., Mansfield, T.A., Hamilton, R.S. (1991) *Sci. Total. Env.*, **100**, 347.

Rendahl, B., Kucera, V., Vlckova, J., Knotkova, D., Sjöström, C., Norberg, P. (1996) in: Sjöström C. (editor), *Proceeding 7th Int. Conf. Dur. Build. Mater. Compon.*, Chapman & Hall, London, UK, Vol. I, p. 219.

Svensson, J.-E., Johansson, L.-G. (1993) A laboratory study of the effect of ozone, nitrogen dioxide and sulfur dioxide on the atmospheric corrosion of zinc, *J. Electrochem. Soc.*, **140**, 2210.

Tidblad, J., Kucera, V. (1996) The role of NO_x and O_3 in the corrosion and degradation of materials, *Report No. 1996:6E, Swedish Corrosion Institute*, Stockholm, Sweden.

Winkler, E.M. (1994) *Stone in architecture, properties, durability*. 3rd completely revised and extended edition, Springer-Verlag, Berlin.

[illegible] and [illegible] [illegible]

Chapter 20

IMPACTS ON URBAN VEGETATION AND ECOSYSTEMS

MIKE ASHMORE
Department of Environmental Science, University of Bradford
West Yorkshire BD7 1DP, United Kingdom

20.1 Introduction

20.1.1 AIMS OF CHAPTER

This chapter aims to provide a brief overview of the significance of air pollution for vegetation and biodiversity in the urban environment, illustrated by a limited number of examples from European cities. Since it is not intended to be a detailed in-depth analysis, but merely to indicate some of the key areas of relevance, and the evidence of their significance, references to a number of books and review articles have been made which provide more detailed information.

20.1.2 MECHANISMS OF AIR POLLUTANT IMPACTS

The impacts of air pollution on vegetation and animals in European cities has been known for centuries; thus the diarist John Evelyn wrote a pamphlet on air pollution in London in 1661, in which he noted: "It is the horrid smoake which kills our Bees and Flowers abroad, suffering nothing in our gardens to bud, display themselves or ripe; so as our Anemonies and many other choycest Flowers will by no industry be made to blow in London or the Precincts of it...". A number of air pollutants are known to be toxic to vegetation (phytotoxic), including sulphur dioxide, nitrogen dioxide, ozone, particulates, and metals. Carbon monoxide is not phytotoxic at the levels found in urban areas, while there is some uncertainty about the effects of nitric oxide. Animal species may be directly affected by air pollutants, or may be indirectly affected by changes in vegetation on which they feed, and by bioaccumulation of toxic elements, such as metals and organic compounds, in the food chain.

The effects of pollutants on individual living organisms generally show a characteristic relationship with the pollutant concentration, or dose, in which four regions are delineated:

1. At high concentrations, organisms may be killed outright.
2. At lower concentrations, the growth rate may be altered or the pattern of development of the organism may change.
3. At even lower concentrations, subtle physiological, morphological, biochemical or behavioural effects may occur, which may influence responses to other stresses, and alter the interactions between species within a community.
4. Finally, there may be positive effects of certain pollutants at low concentrations; for example, both sulphur and nitrogen are essential plant nutrients, while copper and zinc are essential micronutrients in animal diets.

While effects are often related to environmental concentrations, the absorbed dose, or bioavailable concentration, is often a better indicator of the potential for environmental impacts. For example, when stomata are closed at night, or under drought conditions, less air pollution can penetrate to the internal leaf tissue, although pollutants may still be deposited on the plant surface. Similarly, for metals, such as lead, which are tightly bound to the soil matrix, only a relatively small fraction of the total soil concentration is available for uptake by plants or micro-organisms.

A key feature of the urban atmosphere is the presence of mixtures of pollutants, with different pollutant mixtures being present at different locations. For example, the effects on vegetation of pollutant combinations, such as SO_2 and NO_2, may be much greater than the effects of the individual gases, meaning that relying on experiments using single pollutant gases may lead to underestimates of the effects of ambient air quality in urban areas (Mansfield, McCune 1988). Furthermore, when considering the ecological effects of pollutants on communities, it is important to consider a range of other factors, including competition, climate and soil type (Ashmore 1997).

20.2 Impacts on urban agriculture

20.2.1 SIGNIFICANCE OF URBAN AGRICULTURE

An estimated global population of 800 million people are engaged in urban agriculture world-wide, mainly in Asia, Africa and Latin America (UNDP 1996). In some European countries, such as the UK and Germany, most of these are community gardeners, who work in their own garden or in small communal plots to grow vegetables and fruit for their own consumption. More intensive commercial production, often of high-value horticultural crops, tends to be found in the peri-urban areas. However, in some areas, such as the more densely-populated regions of The Netherlands, more intensive agricultural activity is incorporated into urban planning (UNDP 1996).

20.2.2 TRANSECT STUDIES

Some of the earliest studies to investigate the impacts of urban air pollution used the technique of growing plants along a transect of increasing pollution levels. For example, Cohen and Ruston (1925) carried out a series of studies in the years before Word War 1 in which different crops were grown along a transect into the city of Leeds. The deposition of sulphur and particles was 10-20 times higher at the most polluted site, and large reductions in the growth and yield of a range of crops was found at the most polluted urban sites. Although there were undoubtedly other factors varying systematically along this gradient, such as temperature and light, there is little doubt that air pollution was the dominant factor in the observed effects. More recent studies in UK cities, using multivariate statistical analysis, have provided evidence of continuing impacts of urban air pollution (e.g. Ashmore et al. 1988) on vegetation, although on a less dramatic scale.

20.2.3 CHAMBER STUDIES

Another technique which has been used to assess the impacts of air quality is to grow plants in chambers ventilated with ambient air or air filtered to remove pollutants from the atmosphere. One of the earliest such studies was carried out in Manchester in the 1950's by Bleasdale (1973), and showed reductions in ryegrass yield in unfiltered air of 17-33%, which was attributed to SO_2. However, other more recent studies in UK cities have shown more variable results, with both stimulations and reductions in growth in unfiltered air being reported (e.g. Fowler et al. 1988). One explanation of the growth stimulations in some of these later studies is that relatively small concentrations of both SO_2 and NO_2 can have beneficial effects on plant growth. In more southerly cities, where the concentrations of ozone are higher, substantial effects of filtration on yield may be found. For example, Schenone et al. (1993) reported a 20-30% yield increase in bean yield as a result of filtering air at a site in suburban Milan.

20.2.4 EFFECTS ON PESTS AND PATHOGENS

There are a number of ways in which air pollution may alter the severity of insect pests and plant diseases (Bell et al. 1993); these include direct effects on the organism, effects on the leaf surface, changes in numbers of pests and predators, and changes in leaf chemistry. There are a number of reports of increased insect pest outbreaks alongside busy roads, which would imply a potential for effects in urban areas, while a transect study into London has shown that aphid growth rates were greater on plants which had previously been grown closer to central London, where SO_2 and NO_2 concentrations are higher (Figure 20.1). Proof of the potential effects of urban air quality was provided by Dohmen et al. (1984), who compared the growth of an aphid pest of beans in central London in chambers ventilated with ambient air, or air filtered to remove air pollutants, and found an increased insect growth rate in unfiltered air sufficient to cause a 60% increase in aphid numbers within 10 days.

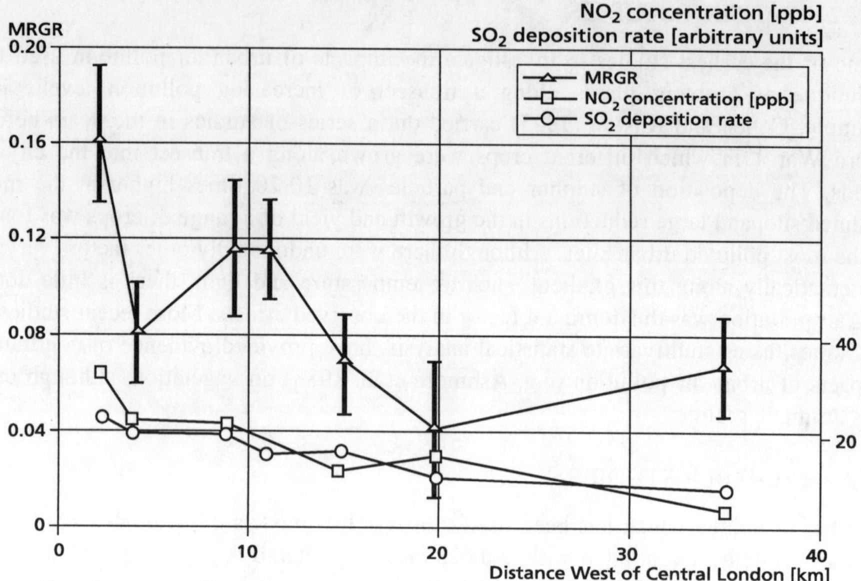

Figure 20.1 Four-day mean relative growth rate (MRGR) (with 95% confidence limits) of the aphid *Metapolophium dirhodum* on barley grown for 6 weeks along an air pollution gradient westwards from Central London (Bell et al. 1993).

Urban air may also have effects on fungal and viral diseases of plants. Certain fungal diseases, such as tar spot of sycamore and black spot of roses (Saunders 1980) are known to be sensitive to SO_2, and used to be absent from many urban areas. However, some workers have attributed the absence of tar spot on urban trees to sweeping of leaves in autumn removing a reservoir of infection for the following year.

20.3 Effects on urban trees

20.3.1 SIGNIFICANCE OF URBAN TREES

Trees have an important role to play in improving the appearance and amenity value of the urban environment. In addition, they may ameliorate the microclimate by providing shade and through transpirational cooling. Trees also provide an effective sink for many pollutants, because of their large size, high surface area to volume ratio of foliage, and their surface roughness.

20.3.2 IMPACTS ON URBAN TREES

Many of the points in Section 20.2 apply equally well to urban trees - however, the size and longevity of trees means that experimental studies of the effects of atmospheric pollution are difficult to perform. This longevity also means that the absence of mature specimens of particular tree species in cities reflects the pollution levels or planting policies many years ago; for example, a survey of an area which included a number of towns and cities in northern Britain in the mid-70's showed a high correlation between the concentration of SO_2 and the presence of Scots pine trees. For sulphur dioxide, in particular, there is strong evidence that the effects of the pollutant are exacerbated by low winter temperatures, and this means that the performance of evergreen species which are known to be sensitive to this pollutant, such as Scots pine and Norway spruce, may be poor in cities of central and northern Europe.

20.3.3 DEPOSITION TO URBAN TREES

Attempts have been made to model the contribution of tree planting on air pollutant concentrations over urban areas. Most evidence suggests that, over the scale of a city, urban tree planting would have only a small effect on deposition; however, the planting of trees adjacent to point sources, or alongside roads, could be more beneficial, and their effect in reducing particulate concentrations would be greater than the effects on gaseous pollutants. There is some evidence for reduced dispersion of particulates from roads with tree or shrub barriers (e.g. Tian, Lepp 1977). Species such as spruces are in theory more effective, because of their high leaf area index, and the presence of foliage through the winter. However, such species also experience greater pollution loads, and spruces are among the most sensitive species to sulphur dioxide.

20.4 Impacts on biodiversity

20.4.1 EFFECTS ON LICHENS AND BRYOPHYTES

The reduced biodiversity, and in extreme cases absence, of lichens in urban areas, as a result of elevated sulphur dioxide concentrations, is a well-established phenomenon in cities throughout Europe. The term "lichen desert" has been used to describe the centre of cities where trees lack any fruticose or foliose lichens. Sulphur dioxide concentrations have fallen dramatically over recent decades in western European cities, but the reinvasion of lichens in response to this decline has been variable and patchy (Gilbert 1993); for example, in Paris, reinvasion of Jardin de Luxembourg only recommenced in the late-80's, and many species recorded there in the middle of the last century have not yet reappeared. Hawksworth and McManus (1989) reported that, while some sensitive species were now found close to central London, others have hardly reinvaded at all. There is evidence that other factors, such as the acidity of the tree bark which responds very slowly to falling pollutant levels, are also important (Bates et al. 1990).

20.4.2 AERIAL POLLUTION AND ANIMAL POPULATIONS

There are a large number of studies showing that proximity to specific pollutant sources, such as smelters or roads, increases tissue concentrations of toxic compounds, and affects the population size, of a range of invertebrates and vertebrates. However, very few studies have compared general urban populations, although higher tissue concentrations of lead and other metals have been reported in urban populations of a small number of species, including pigeons, starlings and squirrels (Figure 20.2). There is also little evidence of a direct effect of air pollutants on the size of urban populations. Although studies suggest that insectivorous birds such as the house martin are reduced in frequency in urban areas, this may be due to lower insect numbers. For example, there is evidence of direct effects of peak SO_2 concentrations increasing mortality of a mite species in Brussels (Andre et al. 1982).

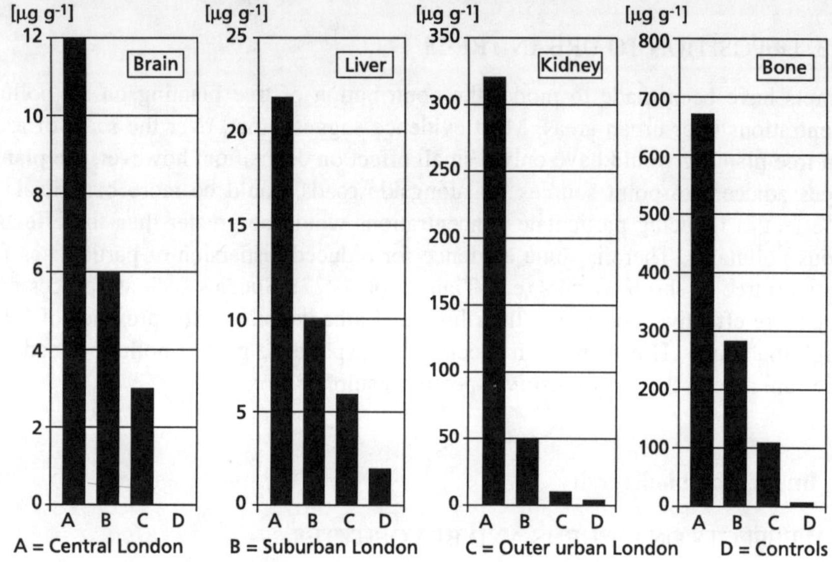

Figure 20.2 Mean tissue lead concentrations (µg/g dry weight) in feral pigeons *Columba livia* from urban and control sites. Data from Hutton, Goodman (1980).

One problem in interpreting any change in species numbers is that air pollution is just one of a wide range of factors associated with the urban environment which can influence its biodiversity, and there may be complex interactions between them. For example, Numata (1982) reported that an important evergreen tree species in Tokyo is subject to unusually heavy insect outbreaks in each spring which is gradually reducing vitality; while the change in plant chemistry due to SO_2 and NO_2 discussed above may be important, it was also suggested that the disappearance of insectivorous birds, which itself may be related to reduced overall forest cover, noise, disturbance and pollution, is also an important factor.

20.4.3 EVOLUTION OF TOLERANCE

Pollution, like many other environmental factors, can act as a selective pressure leading to the evolution of different phenotypes or ecotypes. The most well-known examples of evolution lie in the studies of industrial melanism in moths and ladybirds; for example, darker forms of the moth have a selective advantage in areas with smoke-blackened trees with a limited lichen cover, because of the reduced predation. Improvements in air quality have led to a reversal of the evolutionary trend, with a decreased proportion of darker forms in the population (Askew et al. 1971).

Studies of vegetation in urban areas have also demonstrated the potential for the evolution of tolerance. Thus studies of grass species in London, Manchester and Liverpool in the 1970's demonstrated that the tolerance to SO_2 was higher in more polluted areas, and that evolution of tolerance could evolve within a few years of sowing a fresh grass sward (Bell et al. 1991). However, these more tolerant genotypes also appeared to be at a selective disadvantage in the absence of SO_2, and there is evidence from more recent studies that the tolerance is rapidly lost from populations as ambient SO_2 concentrations fall.

20.5 Bioindication

20.5.1 VALUE OF BIOINDICATION

The response of biological organisms to air pollution also offers opportunities for monitoring of the urban environment. Biological monitoring can offer advantages over standard physico-chemical methods, especially in providing detailed spatial coverage at low cost; such monitoring should be viewed as complementary to, rather than a substitute for, other methods. Several books provide a detailed description of biological monitoring methods (e.g. Burton 1986; Arndt et al. 1987), but essentially three major types of response are used:

- The absence of sensitive species in situ.
- Characteristic symptoms of foliar injury, or physiological responses, either using species in situ, or using selected sensitive species grown under standard conditions at different sites around the city.
- Use of plants or animals as bioaccumulators, for example of metals or organic compounds, either by surveys of existing flora or fauna, or the use of standard transplants or cultures.

20.5.2 EXAMPLES OF BIOINDICATION

Methods of using *lichen biodiversity* to indicate levels of SO_2 have been developed in a number of countries (e.g. Hawksworth, Rose 1970; Herzig et al. 1987). Lichens have also proved useful for mapping metal deposition, which they accumulate effectively. Figure 20.3 (next page) shows one example, a recent study of the Trieste area, in which both responses (1) and (3) are used. Lichen diversity is used to map an Index of Atmospheric Purity (IAP), which coincides with the urban areas; in contrast the map for chromium, based on analysis of a single lichen species, shows the highest levels to be associated with a point source outside the urban areas (Castello et al. 1995).

While lichens are an example of the use of vegetation in situ, other studies rely on *standard bioassays*. For example, standard grass cultures, grown in containers at heights of 1-2m to avoid contamination by soil splash, have been used in a number of European countries to map the inputs of metals or organics; the grasses are harvested and analysed after exposure for a fixed period of time. Similarly, standard cultures of a sensitive tobacco cultivar (Bel-W3), exposed for 1-4 weeks, have been used to map the distribution of phytotoxic ozone in and around a number of European cities, in this case using the extent of typical symptoms of visible injury to leaves.

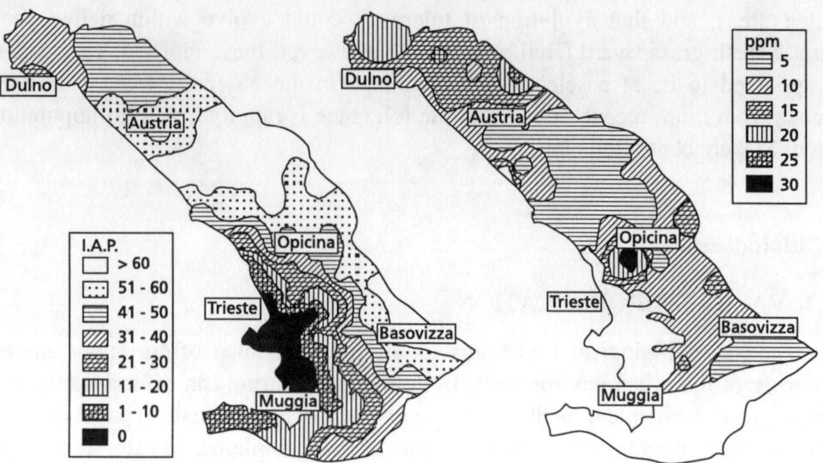

Figure 20.3 Left: Air quality map, based on lichen diversity, of the Province of Trieste. Numbers refer to the Index of Atmospheric Purity (IAP, see text). The black area delimits the lichen desert. Right: Deposition pattern of Cr in the survey area, as indicated by the metal content of the lichen *Parmelia caparata*.

20.6 Air quality standards

While air quality management in urban areas is dominated by the need to protect human health, it is important that the need to protect the wider environment is also recognised. Critical levels, defined as the concentration above which adverse effects may occur on sensitive receptors, according to current knowledge, have been established by UN-ECE for use in assessment of the policy significance of transboundary air pollution (Ashmore, Wilson 1993), and could also be applied in the urban environment. They also form the basis for new ecotoxicological air quality guidelines which have recently been adopted by WHO. The values are summarised in Table 20.1, which shows that the critical levels for SO_2 and ozone distinguish between different types of vegetation.

The air quality guidelines for ozone use the concept of the accumulated dose above a threshold concentration of 40 ppb (hence AOT40), rather than an average concentration, as this basis of expression is better related to plant response (Fuhrer et al. 1997). Each hourly value is subtracted from 40 ppb, and all the positive differences then summed; hence the guidelines are equivalent to seasonal 8h mean concentrations of about 45 ppb for crops and 47 ppb for trees, if the value of 40 ppb is exceeded on every day.

An important difference from the equivalent standards for human health is that for nitrogen oxides; the value in Table 20.1 includes both NO and NO_2, rather than just NO_2 as for human health. Given the high NO concentrations in urban areas, this makes the criteria more stringent than those based on NO_2 alone. Similarly, the guidelines for ozone and SO_2, in the case of lichens, may be more stringent than those based on effects on human health.

Table 20.1 Summary of air quality guidelines for effects on vegetation.

Pollutant	Guideline value	Type of species
SO_2	30 µg m^{-3} (annual mean)	Agricultural crops
	20 µg m^{-3} (annual mean)	Trees and semi-natural vegetation
	15 µg m^{-3} (annual mean)	Trees and semi-natural vegetation in harsh environments
	10 µg m^{-3} (annual mean)	Lichens
NO_x	30 µg m^{-3} (annual mean) (sum of volume mixing ratios of NO and NO_2, converted to mass units as NO_2)	All vegetation
O_3	3000 ppb.h (AOT40) (during daylight hours over a three-month growing season)	Agricultural crops and herbaceous semi-natural vegetation
	10000 ppb.h (AOT40) (during daylight hours over a six-month growing season)	Trees

20.7 Conclusions

This chapter briefly summarises some of the ways in which air pollution can interact with flora and fauna in the urban environment, and presents some evidence of these effects in European cities. There have been rapid changes in air quality in many European cities over the past twenty-five years, and more changes can be expected as new regulations to control emissions take effect. However, we have insufficient knowledge to predict the effects of these changes on urban ecosystems with any certainty, and it is possible that responses may occur quite slowly. Urban air quality management should give more attention to changes in flora and fauna, which are important factors in the quality of life of urban residents. Biological responses may provide a valuable method both of determining spatial variation in air quality and identifying changes in local air quality; furthermore, their more active use offers the potential to promote community awareness of indicators of local air quality.

20.8 References

Andre, H.M., Bolly, C., Lebrun, P. (1982) Monitoring and mapping air pollution through an animal indicator: a quick new method, *J appl. Ecol.*, **19**, 107-111.

Arndt, U., Nobel, W., Schweizer, B. (1987) *Bioindikatoren*. Eugen Ulmer GmbH & Co., Stuttgart.

Ashmore, M.R. (1997) Pollution, in: Crawley, M.J. (editor) *Plant Ecology*, pp. 568-581, Blackwell Science, Oxford.

Ashmore, M.R., Wilson, R.B. (editors) (1993) *Critical Levels of Air Pollutants in Europe*, Department of the Environment, London.

Ashmore, M.R., Bell, J.N.B., Mimmack, A. (1988) Crop growth along a gradient of ambient air pollution, *Environ. Pollut.*, **53**, 99-121.

Askew, R.R., Cook, L.M., Bishop, J.A. (1971) Atmospheric pollution and melanic moths in Manchester and its environs, *J. Appl. Ecol.*, **8**, 247-256.

Bates, J.W., Bell, J.N.B., Farmer, A.M. (1990) Epiphyte recolonisation of oaks along a gradient of air pollution in south-east England, 1979-1990, *Environ. Pollut.*, **68**, 81-99.

Bell, J.N.B., Ashmore, M.R., Wilson, G.B. (1991) Ecological genetics and chemical modifications of the atmosphere, in: Taylor, G.E.Jr., Pitelka, L.F., Clegg, M.T. (editors), *Ecological Genetics and Air Pollution*, pp. 33-59, Springer-Verlag, New York.

Bell, J.N.B., McNeill, S., Houlden, G., Brown, V.C., Mansfield, P.J. (1993) Atmospheric change: effects on plant pests and diseases, *Parasitology*, **106**, 811-824.

Bleasdale, J.K.A. (1973) Effects of coal smoke pollution gases on the growth of ryegrass (*Lolium perenne* L.), *Environ. Pollut.*, **5**, 275-285.

Burton, M.A.S. (1986) Biological Monitoring of Environmental Contaminants (Plants). Report no. 32, Monitoring and Assessment Research Centre, London.

Castello, M., Nimis, P.L., Cebulec, E., Mosca, R. (1995) Air quality assessment by lichens as bioindicators of SO_2 and bioaccumulators of heavy metals in the province of Trieste (NE Italy), in: Lorenzini, G., Soldatini, G.F. (editors), *Responses of Plants to Air Pollution*, pp. 233-243, Agricoltura Mediterranea Special Volume.

Cohen, J.B., Ruston, A.G. (1925) *Smoke. A study of town air*, Edward Arnold, London.

Dohmen, G.P., McNeill, S., Bell, J.N.B. (1984) Air pollution increases *Aphis fabae* pest potential, *Nature*, **307**, 52-53.

Fowler, D., Cape, J.N., Leith, I.D., Paterson, I.S., Kinnaird, J.W., Nicholson, I.A. (1988) Effects of air filtration at small SO_2 and NO_2 concentrations on the yield of barley. *Environ. Pollut.*, **53**, 135-149.

Fuhrer, J., Skarby, L., Ashmore, M.R. (1997) Critical levels for ozone effects on vegetation in Europe. *Environ. Pollut.*, **97**, 91-106.

Gilbert, O.L. (1992) Lichen reinvasion with declining air pollution, in: Bates, J.W., Farmer, A.M. (editors), *Bryophytes and lichens in a changing environment*, pp. 159-177, Oxford Science Publications, Oxford.

Hawksworth, D.L., McManus, P.M. (1989) Lichen recolonisation of London under conditions of rapidly falling sulphur dioxide levels, and the concept of zone skipping, *Bot. J. Linn. Soc.*, **109**, 99-109.

Hawksworth, D.L., Rose, L. (1970) Qualitative scale for estimating sulphur dioxide air pollution in England and Wales using epiphytic lichens, *Nature*, **227**, 145-148.

Herzig, R., Liebendorfer, L., Urech, M. (1987) Flechten als Bioindiikatoren der Luftverschmutzung in der Schweiz: Methoden-Evaluation und Eichung mit wichtigen Luftschadstoffen, *VDI Berichte*, **609**, 619-639.

Hutton, M., Goodman, G.T. (1980) Metal contamination of feral pigeons, *Columba livia*, from the London area: Part 1 - Tissue accumulation of lead, cadmium and zinc, *Environ. Pollut.*, **22**, 207-217.

Mansfield, T.A., McCune, D.C. (1988) Problems of crop loss assessment when there is exposure to two or more gaseous pollutants, in: Heck, W.W., Taylor, O.C., Tingey, D.T. (editors), *Assessment of Crop Loss from Air Pollutants*, pp. 317-344, Elsevier, London.

Numata, M. (1982) Changes in ecosystem structure and function in Tokyo, in: Bornkamm, R., Lee, J.A., Seaward, M.R.D. (editors), *Urban Ecology*, pp. 139-147, Blackwell, Oxford.

Saunders, P.J.W. (1970) Air pollution in relation to lichens and fungi, *Lichenologist*, **4**, 337-349.

Schenone, G., Botteschi, G., Fumagalli, I., Montinaro, F. (1992) Effects of ambient air pollution in open-top chambers on bean (*Phaseolus vulgaris* L.). I. Effects on growth and yield, *New Phytol.*, **122**, 689-697.

Tian, T.K., Lepp, N.W. (1977) Roadside vegetation: an efficient barrier to the lateral spread of atmospheric lead? *Arboricultural Journal*, **3**, 79-85.

UNDP (1996) *Urban agriculture: food, jobs and cities*. United Nations Development Programme, New York.

Chapter 21

REDUCTION OF VISIBILITY

HELMUTH HORVATH
Institute of Experimental Physics, University of Vienna
Boltzmanngasse 5, A-1090 Vienna, Austria

21.1 Introduction

Poor visibility is a side effect of atmospheric pollution. It is easily recognizable with the unaided eye by the reduced number of distant buildings, trees, hills and airplanes that can be seen. During such conditions the sky usually has a fairly uniform whitish or brownish color, clouds are difficult to recognize, and the colors of distant objects are faint. At night a halo can be seen around distant light sources, the sky is brighter than in clean environments, making the observation of stars difficult.

All these effects can be caused by the attenuation, scattering, and absorption of light by the polluted atmosphere. The different constituents of the polluted air have unequal importance for the optical effects of the atmosphere. The mass of the different constituents of a typical European urban atmosphere, the extinction coefficient and their contribution to the attenuation of light is given in Table 21.1. The attenuation of light is clearly dominated by the contribution from particles although being a trace component representing 30 ppb of the mass. Each of the other components, even the air molecules (sum of N_2 and O_2) is considered to contribute less than 10%.

Table 21.1 Contribution of various pollutants to the optical attenuation coefficient at a wavelength of 550 nm.

Substance	Air	Particles	NO_2	SO_2	O_3	Unit
Concentration	1300	$40 \cdot 10^{-6}$	$50 \cdot 10^{-6}$	$50 \cdot 10^{-6}$	$200 \cdot 10^{-6}$	g/m^3
Extinction coefficient	11.62	160	8.6	$10 \cdot 10^{-6}$	1.0	Mm^{-1}
Contribution to extinction (reduction of visibility)	6.4	88.3	4.7	< 0.061	0.6	%

21.2 Aerosol optical relations

The transmittance τ of light through an aerosol extending a distance x is the ratio of the transmitted to the incident flux, and is given by Lambert Beer law $\tau = \phi/\phi_o = \exp(-\sigma_e \cdot x)$. With σ_e the extinction coefficient (for nomenclature and definitions see e.g. Horvath 1994) or attenuation coefficient of the aerosol, which is given in units of reciprocal length, e.g. m^{-1}, km^{-1} or Mm^{-1}. The extinction coefficient can be considered as the relative reduction in light flux per unit length, thus $\sigma_e = -d\phi/(\phi \cdot dx)$. The attenuation of the light flux is possible in two ways: (1) The particles scatter a light flux $d\phi_s$ of the incident flux ϕ which is diverted to direction different from the one of the incident light beam. No light is lost since, the photons are scattered eleasically. We can write this as $-d\phi_s = \phi \cdot \sigma_s \cdot dx$, with σ_s the scattering coefficient. (2) By absorption the photons transfer their energy to the particle, increasing its internal energy, and eventually the photon energy is converted to heat. This fraction of absorbed flux can be written as $-d\phi_a = \phi \cdot \sigma_a \cdot dx$ with σ_a the absorption coefficient. Obviously $d\phi = d\phi_s + d\phi_a$ thus $\sigma_e = \sigma_a + \sigma_s$.

Due to scattering of light by the particles/molecules a layer of air appears bright when illuminated e.g. by the sun. When illuminated by irradiance E (Figure 21.1) the radiant intensity scattered by the volume element is given by $dI = E \cdot \gamma(\phi) \cdot dV$, with $\gamma(\phi)$ the volume scattering function of the aerosol. The radiance of this volume element is obtained by dividing by the surface dA perpendicular to the direction of observation thus $(dL)_s = d^2I/dA = E \cdot \gamma(\phi) \cdot dx$. The radiance is the flux emitted per surface area and solid angle. In all our considerations below we will assume surface elements and solid angles as infinitesimal. Thus the light propagating to an observer or instrument can be considered as parallel. Therefore the same attenuation laws apply as for a parallel flux of light, i.e. when passing a distance dx through a layer with extinction coefficient σ_e the reduction in radiance due to attenuation is $(dL)_e = -L \cdot \sigma_e \cdot dx$.

21.2.1 APPARENT RADIANCE OF A DISTANT OBJECT

An observer (or an instrument) looks at an object at distance R having an inherent radiance L(0) being the radiance of the object at distance zero (Figure 21.1). We consider the light coming from the object which is contained in a cone having its apex at the observer and the base at a small part of the object. The light reflected or emitted by the object passes through the atmosphere and this light is attenuated by the aerosol. At the same time light is scattered by the particles in between the object and the observer, adding diffuse light to the light coming from the object. A lamina of thickness dx shall be illuminated by irradiance E, e.g. by sunlight. The light scattered by the particles within the volume element causes an increase in radiance given above by $(dL)_s = E \cdot \gamma(\phi) \cdot dx$. At the same time light passing through the volume element is attenuated, causing a loss in radiance $(dL)_e = -L \cdot \sigma_e \cdot dx$. The sum of the two expressions is the net change in radiance, yielding $dL = -L(x) \cdot \sigma_e \cdot dx + E \cdot \gamma(\phi) \cdot dx$ or:

$$\frac{dL}{dx} = -L \cdot \sigma_e + E \cdot \gamma(\phi) \tag{21.1}$$

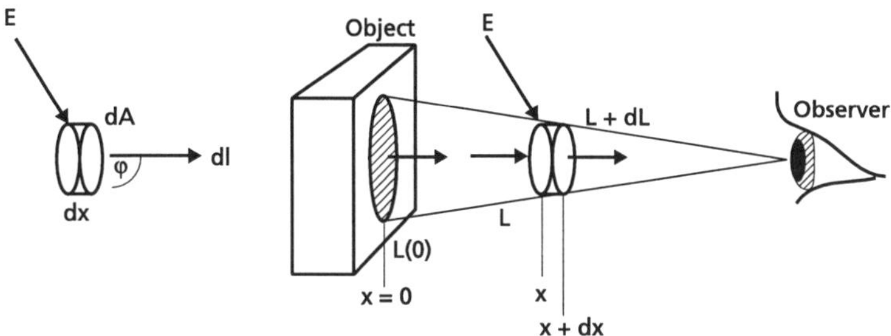

Figure 21.1 Left part of the figure: Radiance of a distant volume element.
Right part of the figure: Derivation of radiative transfer equation.

This is a crude formulation of the equation for radiative transfer. It considers illumination from only one direction. In reality the volume element is not only illuminated by light coming from one direction but also multiply scattered light, reflections from clouds and ground are important. Thus for a more complete formulation $E \cdot \gamma(\phi)$ must be replaced by an integral expression, containing the irradiance produced by the illumination form different directions and the corresponding volume scattering function. Furthermore it must be considered that σ_e, E and $\gamma(\phi)$ is a function of x. Integrating Equation 21.1 for a homogeneous illumination and aerosol, and considering that the radiance at distance zero equals the inherent radiance L(0) one obtains:

$$L(x) = L(0) \cdot \exp\left(-\sigma_e \cdot x\right) + \frac{E \cdot \gamma(\phi)}{\sigma_e}\left[1 - \exp(-\sigma_e \cdot x)\right] \qquad (21.2)$$

If σ_e, E, and $\gamma(\phi)$ are functions of x, a solution is also possible, (Horvath 1971), the full radiative transfer equation can only be solved exact for very special cases such as isotropic scattering, but good approximations have been derived (e.g. Chadrasekhar 1960).

The right hand side of Equation 21.2 consists of two parts: The first is the radiance of the object attenuated by the atmosphere decreases exponentially with distance. The second term is the radiance caused by the scattering of sunlight and other radiation by aerosols. It increases with distance towards an equilibrium value $E \cdot \gamma(\phi)/\sigma_e$. For horizontal vision it is the radiance of the horizon L_h, since looking at the horizon can be considered as looking at an object at infinite distance, thus $L_h = L(\infty) = E \cdot \gamma(\phi)/\sigma_e$. The apparent radiance of an object at distance x therfore is:

$$\mathrm{L(x) = L(0) \cdot \exp\left(-\sigma_e \cdot x\right) + L_h\left[1 - \exp\left(-\sigma_e \cdot x\right)\right]} \qquad (21.3)$$

With increasing distance the object more and more resembles the horizon, both in radiance and color. Recognition of an object depends on the apparent radiance, but even more on how well the object can be discerned from its background. Relevant for this is the contrast C which is the relative difference between the radiance L_o of the object and the radiance L_b of the background, thus $C = (L_o - L_b)/L_b$: The contrast is negative, if the object is darker than the background and it is positive, if it is brighter. Using (2a) the contrast of the object at distance x against the horizon as background is obtained as:

$$C(x) = \frac{L(0) - L_h}{L_h} \exp(-\sigma_e \cdot x) = C(0) \cdot \exp(-\sigma_e \cdot x), \tag{21.4}$$

with $C(0) = (L(0)-L_h)/L_h$ the contrast of the object against the horizon at distance zero or the inherent contrast. Equation 21.4, found by Koschmieder (1925), gives a relation similar to the exponential Lambert Beer law for the attenuation of light.

The apparent contrast of an object decreases exponentially with distance, and will eventually reach the contrast threshold ε, where the eye cannot recognize a difference between the object and its surroundings, therefore the object is invisible for the observer. The distance, where the contrast reaches the threshold, is called the visibility V and is obtained from (21.4) as:

$$V = -\frac{1}{\sigma_e}\left(\ln \varepsilon - \ln\left|C(0)\right|\right) \tag{21.5}$$

The contrast threshold ε is a property of the eye and must be determined experimentally. As with all biologic measurements, considerable scatter occurs, which requires many measurements. The most complete study has been performed by Blackwell (1946), who made more than 220,000 individual measurements. Later measurements have confirmed these data (Gorraiz et al. 1985). For a wide range of illumination levels and sizes of the object the contrast threshold is determined to 0.007, both for negative and positive contrasts.

21.2.2 STANDARD VISIBILITY

Except for glare and twilight the threshold is ε = 0.007 for objects of the usually occurring size. Therefore an observer looking very careful at a just visible object, having a contrast of 0.007 will see it in 50% of the cases. For everyday's life it is difficult to define the visibility as the distance at which an object can be seen in 50% of the cases, just consider a pilot finding the runway only every second time. Experiments, where human observers have been asked to identify a just visible object in the landscape, revealed that a contrast threshold between ε = 0.02 and ε = 0.05 is a good value. The value ε = 0.02 is the traditional value used by Koschmieder (1925) and the visibility of a black object seen against the horizon as background (C(0) = -1) is given by $V_s = -\ln 0.02/\sigma_e = 3.91/\sigma_e$. This value is called the standard visibility, Koschmieder visibility or visual range. For nonblack objects having an inherent contrast C(0) or other contrast thresholds, the visibility is obtained by

$$V = \frac{V_s}{3.91} \ln\left[\left|\frac{C(0)}{\varepsilon}\right|\right] \tag{21.5}$$

The radiance of objects usually seen in an urban environment changes with amount of sunlight and daylight hitting the surface of the object. Especially if the object is in its own shade, the radiance can be quite low.

A few selected data are shown in Figure 21.2 (left part) together with the visibility of these objects in relation to the standard visibility (right part). In its own shade the contrast is usually between -0.9 and -0.5, which means that the visibility is between 0.97 and 0.82 of the standard visibility. The only exception is for bright surfaces such as concrete or plaster, which may be brighter than the horizon, i.e. it has a positive contrast. A contrast very close to the contrast of an ideal black object, is found for black pine, and a ridge covered with black pine is almost a perfect black object. From this it may be concluded, that the standard visibility, although a hypothetical value, is a good measure for the real visibility, which is between 3 and 20% lower than the standard visibility.

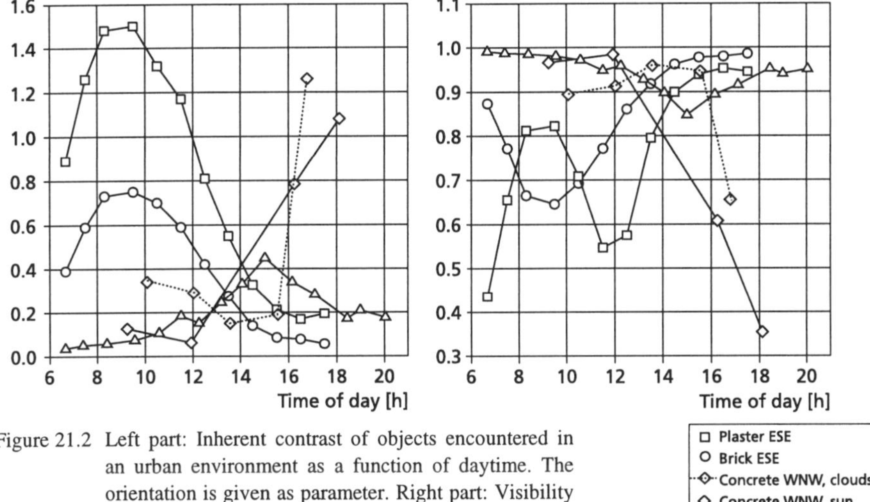

Figure 21.2 Left part: Inherent contrast of objects encountered in an urban environment as a function of daytime. The orientation is given as parameter. Right part: Visibility of the various objects relative to the standard visibility.

21.3 Properties of the aerosol

21.3.1 EXTINCTION COEFFICIENT

The visibility is inversely proportional to the extinction coefficient of the aerosols, and it is dominated by the interaction of the aerosols with light. The optical properties of the aerosol particles depend on size, shape and refractive index of the particles and the wavelength of light. The extinction coefficient of N spherical particles of radius r per m^3 can be calculated as $\sigma_e = N \cdot r^2 \cdot \pi \cdot Q_e$. The extinction efficiency factor Q_e, a pure number, is the ratio of the light attenuated by the particles and the light incident on the cross section of the particles. The value of Q_e can be obtained by e.g. the Mie-computer codes available in the literature (e.g. Reist, Wilson 1989; Bohren, Huffman 1984).

For particles having a size distribution n(r), the extinction coefficient is obtained by

$$\sigma_e = \pi \cdot \int_{r\,min}^{r\,max} n(r) \cdot Q_e \cdot r^2 \cdot dr \tag{21.6}$$

For a typical European aerosol the function $n(r) \cdot Q_e \cdot r^2$ whithin the integral is plotted in Figure 21.3. The most important region of particles is between 0.1 and 2 μm diameter. Both at the upper and the lower limit, the particles are less effective in attenuating light. Thus the amount of attenuated light by a given mass of particles depends highly on the particles size. It is useful to define the mass extinction coefficient or specific extinction cross section σ_e/M which is $3 \cdot Q_e/4 \cdot \rho$ for monodisperse particles and

$$3 \cdot \int_{r\,min}^{r\,max} n(r) \cdot Q_e \cdot r^2 \cdot dr \,/\, 4 \cdot \int_{r\,min}^{r\,max} n(r)\, \rho \cdot r^3 \cdot dr \tag{21.7}$$

for particles having a number size distribution n(r). Some values are shown in Figure 21.3 for monodispersions and are listed in Table 21.2 for size distributions.

Table 21.2 Specific extinction (attenuation) cross sections for particles of given size distribution. The volume size distribution is lognormal with a geometric standard deviation of 1.7 and a geometric volume mean diameter as given in the table. The specific cross section is given in units of m^2/g..

Refractive Index	Density [kg/m³]	Mean diameter 0.3 μm	0.5 μm	0.8 μm	5.0 μm	Substance
1.53	1769	3.4	4.2	3.5	0.43	Ammonium sulfate
1.59	2930	2.4	2.8	2.2	0.26	Limestone
1.56-0.47·i	1000	10.2	7.9	5.4	0.76	Soot aggregates
1.5	950	5.8	7.3	3.8	0.51	Organic material
1.5-0.1·i	1600	4.3	4.7	3.8	0.51	"Urban aerosol"

It is evident, that the specific cross section depends on the mean diameter of the particles. Also the density plays a role, denser particles (compare e.g. organics and ammonium sulfate) have a smaller specific cross section since the interaction with light is the same but the particles have more mass. The most efficient interaction with light is for particles with a diameter of 0.5 μm, for strongly light absorbing soot it is at smaller diameters. The particles in the size range of the accumulation mode produce the major contribution to visibility reduction. The coarse mode particles are only one tenth as effective. Thus the specific cross section of an urban aerosol strongly depends on the mass contribution of the coarse mode. The typical European urban aerosol has 70% of the mass in the accumulation mode with d_v = 0.5 μm and 30% of the mass in the coarse mode with d_v = 5.0 μm: For such a typical European urban aerosol the specific cross section is 3.5 m^2/g, if the coarse mode contributes 70% of the mass, the value would drop to 1.8 m^2/g. An experimentally determined specific extinction cross section thus will strongly depend on the cutoff of the sampling device. A PM 2.5 will mainly give the extinction cross section of the accumulation mode.

Combining the specific extinction cross section value $\sigma_e/M = 3.5\ m^2/g$ for typical European urban aerosol with $V_s = 3.9/\sigma_e$ we obtain $V_s \cdot M = 1.1\ m^2/g$. This means: A prism with a base area of 1 m^2 and a height equal to the standard visibility, contains 1.1 g of aerosol particles, independent of the visibility.

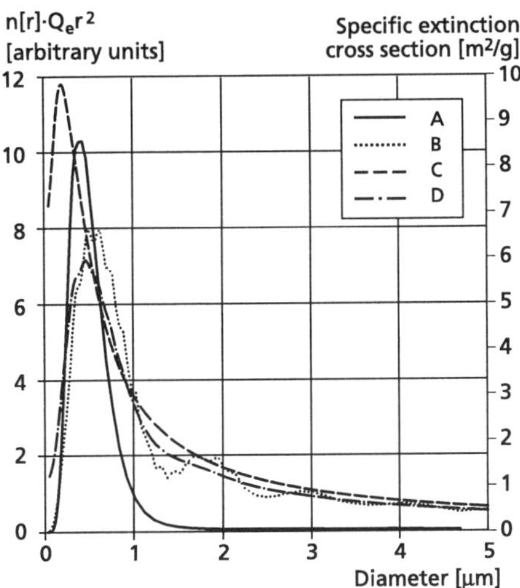

Figure 21.3 Size dependent extinction of aerosols. (A) Function which has to be integrated between the smallest and largest diameter in order to obtain the extinction coefficient for a typical European aerosol. (B) specific cross section (mass extinction coefficient) for particles with a refractive index of m = 1.5 and ρ = 1500 kg/m_3, representative e.g. for sulfates, (C) for m = 1.5 - 0.5·i and ρ = 1250 kg/m_3, a typical value for pure soot, (D) m = 1.5 - 0.1·i and ρ = 1500 kg/m_3, a typical value for an urban aerosol, which consists of internally mixed particles of absorbing and non absorbing substances.

21.3.2 LIGHT ABSORPTION

In an urban environment combustion processes such as heating of boilers or internal combustion engines emit soot, and produce NO_2, both are light absorbing. The gas NO_2 has negligible light scattering and only absorbs light. This means that the extinction coefficient is increased due to presence of NO_2 in the atmosphere but the light scattering coefficient and the volume scattering function remains unchanged.

Black carbon (soot) also scatters light; the effect is small for particles below 50 nm and approximately equal to the light scattering of soot for sizes in the range of 0.5 µm. Thus the soot produces an aerosol where part of the attenuation is due to absorption. Since the visibility is determined by the extinction coefficient, Equation (21.5) is unchanged. But the radiance of the horizon, $L_h = E \cdot \gamma(\phi)\ \sigma_e$ (cf. Equation (21.2)) is less for an absorbing aerosol in comparison to an aerosol exhibiting no light absorption. Multiple scattering enhances the effect, since the photons travel a longer distance due to multiple deflections and more light is absorbed. The "dirty" appearance of soot containing air is thus a consequence of the light absorption by the aerosol.

21.3.3 WAVELENGTH DEPENDENCE

The light extinction coefficient of air is high for blue light and low for red light. The λ^{-4} dependence of the Rayleigh scattering is well known. Since the attenuation of light is only due to scattering, high extinction in blue means also more scattered blue light than red light, thus a higher value of the volume scattering function for blue light. For the air molecules the wavelength dependent scattering function and the extinction coefficient are proportional.

For aerosols there is also a wavelength dependence of the extinction coefficient, which is weaker (see the examples in Figure 21.4). The steepest $\sigma_e(\lambda)$ curve is for the pure Rayleigh scattering, with increasing size of the particles, the wavelength dependence is less distinct, whereas for particles larger than a few micrometers the curves are flat. In addition the wavelength dependent light absorption of NO_2 is given. It is high for blue light and negligible for red light. The volume scattering function (not shown) is roughly proportional to the extinction coefficient, i.e. $\gamma(\phi)/\sigma_e$ shows little wavelength dependence.

The radiance of an infinite layer of aerosol is given by $E\cdot\gamma(\phi)/\sigma_e$. For non absorbing aerosols $\gamma(\phi)/\sigma_e$ exhibits little dependence on wavelength, the light spectrum of the radiance of the horizon will be the same as that of the illumination, thus white. This can be observed even in very clean air with unobstructed horizontal view; the horizon is white. If we look at the sky it appears blue in clean air. This is also evident, when applying Equation (21.2). The sky's radiance can be understood as the radiance of a black object (the space) at a distance x which is the path through the atmosphere, yielding $L(x) = E\cdot\gamma(\phi)/\sigma_e\cdot[1\text{-exp }\sigma_e\cdot x)]$.

For a non absorbing aerosol $\gamma(\phi)/\sigma_e$ has no wavelength dependence, thus the expression in parenthesis, $[1\text{-exp }(-\sigma_e(\lambda)\cdot x)]$, determines the the wavelength dependence of the light seen when looking at the sky.

For pure air this expression is larger for $\lambda = 400$ nm than for 800 nm (also shown in Figure 21.4) and thus the sky appears blue. The same effects occurs, if a dark object is observed through aerosol particles with diameters < 0.15 μm, the wavlength dependence of which is shown in Figure 21.4. This may occur for oil smoke, e.g. emitted by not well maintained engines or two stroke motors. On the other hand, for a typical urban aerosol (70% of the mass in the accumulation mode of $d_v = 0.5$ μm and 30% coarse mode mass with $d_v = 5$ μm) the wavelength dependence almost vanishes, (Figure 21.4) and the sky appears more or less uniformly white and for an aerosol containing carbon, it may even have a dull gray color.

For an aerosol containing nitrogen dioxide, $\gamma(\phi)$ remains unchanged, but σ_e is increased by the absorption coefficient of NO_2. Since the light absorption by NO_2 is larger at 400 nm, the radiance of the aerosol layer is less in blue than in the red (also shown in Figure 21.4). Less blue light means a yellow-red color, and if the layer is dark it appears brown, this can be frequently observed in towns.

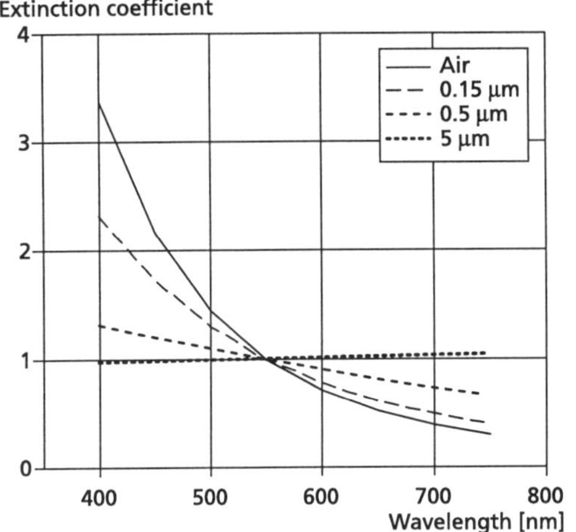

Figure 21.4

Wavelength dependence of the extinction coefficient for several aerosols and the absorption coefficient of NO_2. The curves are normalized to a value of 1 at 550 nm. For the aerosol a lognormal volume size distribution with σ_g = 1.7 is used, the mean diameter is given as parameter.

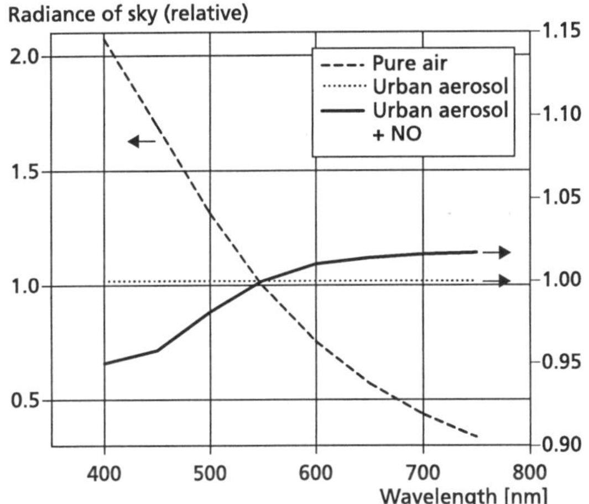

Wavelength dependence of the light scattered in a layer of air when looking 45° upwards. Dashed line: pure air. Dotted line: air containing urban particles with an extinction coefficient of 0.5 km^{-1}. Solid line: Same aerosol containing 100 $\mu g/m^3$ NO_2. The scale height, i.e. the distance the layer extends vertically was assumed 8 km for air and 2 km for the aerosol.

21.3.4 MEASUREMENTS OF VISIBILITY

Visibility is determined both by the aerosol and the human observer's ability to recognize differences in radiances. The traditional method to determine visibility requires a point from which objects, preferably in a horizontal plane can be seen against the horizon as background. On a day with excellent visibility a list of objects and their distance to the point of observation is determined. An observer performing a visibility measurement records the visible objects he can see, and the visibility is the distance where the farthest visible object is located.

When well done, the method is astonishingly reproducible. This method may have difficulties, if the terrain is not flat or the atmosphere is inhomogeneous, and the path of sight thus passes through layers of aerosol with different extinction coefficients. In that case the visibility is determined by the average of the extinction coefficient between the observer and the object, and when looking upwards and it usually is larger than for horizontal view. Visibility has been measured at many locations and records with decades of data can be used to investigate the history of air pollution (e.g. Trier, Firinguetti 1994).

If the extinction coefficient of the aerosol is known, the visibility can be determined by Equation 21.5. The extinction coefficient can be measured by determining the transmission of a parallel beam of light through the atmosphere. A transmissometer consists of a light source at one location and a detector a few tens of meters away or source and detector at the same location together with a reflector. At airports, visibility transmissometers with base lengths of 30 and 150 m are in use. The ratio of the signal for the aerosol under consideration and the signal with pure air between light source and detector is the transmission of the aerosol.

The determination of the transmission for pure air is difficult, usually the transmission on a very clean day is taken as reference. For good visibility (e.g. 40 km) the transmission through a 100 m layer is small, in our example $\tau = 0.99$, which requires very stable light source and detector. A folded path in a White cell is another possibility (Poβ, Metzig 1987). With a telephotometer (e.g. Horvath 1981) the contrast of distant targets can be determined and using (2) and (4) the visibility can be determined. Another way is to measure the scattering coefficient and convert it to visibility e.g. by the AEG/FFM scattered light recorder (AEG 1961) or the integrating nephelometer (Beutell, Brewer 1946). This method neglects light absorption.

21.4 Visibility in European cities

Visibility is governed to more than 80% by the aerosol and is thus highly variable since all the factors influencing the aerosol also influence the visibility: The source strength, the wind speed and the mixing height: more or stronger sources increase the particle concentration and thus decrease the visibility. Low wind speed and low mixing height cause less dilution of the aerosol from the sources, which lead to low visibility. But even with the same aerosol present, the visibility can vary by two additional factors which influence the visibility: illumination and humidity.

21.4.1 ILLUMINATION AND VISIBILITY

The radiance of an object depends on the angle of illumination (Figure 21.2). An object illuminated by the sun can be brighter than the horizon, and if its radiance is at least twice the radiance of the horizon, it will be more visible than a black object. This can e.g. be the case for a snow covered, sunlit mountain. However, dark objects illuminated by the sun have a radiance between 0 and $2 \cdot L_h$, and thus less contrast than bright object and therefore be less far visible. If the objects are illuminated from behind, the situation reverses. Dark objects in their shade, have a contrast close to -1 and are thus further visible than a bright object, with less contrast to the horizon.

21.4.2 HUMIDITY AND VISIBILITY

Frequently low visibility is observed in the morning, better visibility in the middle of the day and again low visibility in the evening. This is due to increased source strength in the morning and evening caused partly by traffic. However the change of particle size due to uptake of humidity must be considered: Many particles consist of water soluble substances such as ammonium sulfate and nitrates. These substances form a solution where the concentration depends on the humidity. If there is only little air movement, the insolation causes the temperature to increase during the daytime and the humidity to drop, which makes the particles shrink is size.

During the day the height of the mixing layer increases which also dilutes the aerosol and increases the visibility. A variation of visibility is thus caused by several factors. In Figure 21.5 the variation of the standard visibility is shown during one day, where considerable change in humidity occurred. A considerable decrease in visibility is possible by uptake of water without changing the mass concentration of the (dry) aerosol: e.g. for the urban aerosol in Mainz an increase in extinction coefficient by a factor of 2.2 was observed by Hänel (1976) when the humidity increased from 50% to 90%. For the maritime aerosol the increase is a factor 4.6.

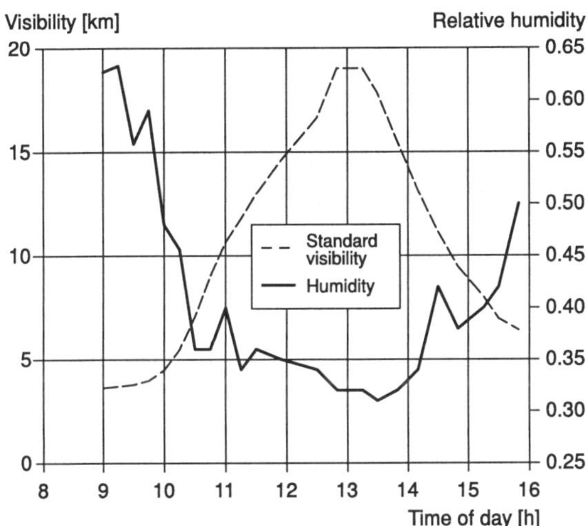

Figure 21.5 Example for the variation of visibility in Vienna on a day with stagnant air.

21.4.3 RANGE OF VISIBILITY IN EUROPEAN CITIES

An example, representative for an average European city is given in Table 21.1. The extinction coefficient of this aerosol is 181 Mm^{-1}, giving a standard visibility of 22 km. If a contrast threshold of $\varepsilon = 0.05$ is considered for seeing an object well, the visibility is 17 km. Since the pollution may be easily one third or three times the values assumed in Table 21.1 (page 373), and increased humidity may double the particle size, a range of visibility between 2 km and 60 km is easily possible.

21.4.4 OUTLOOK

Good visibility is pleasant and a part of the quality of life; everybody enjoys the few days with crystal clear air, which we may encounter in Europe, when fresh air masses rapidly move over the continent and do not pick up much pollution. Good visual quality in an area influences the decision to spend the vacation there and good view in a city will make the place more attractive and cause higher real estate value. Standard visibility is a good measure for the optical quality of the atmosphere, but a more complete picture is obtained if contrasts of details of the object, color changes and dicoloration of objects are treated. Visibility at night is very complex, since distributed light sources cause a general illumination level which, when driving on a street in light fog may change from one meter to the next. In aviation e.g. the runway visual range RVR is computed from the general illumination level, the extinction coefficient and the brightness of the lights boarding the runway. Mostly the aerosol particles are the reason for the bad visual quality of the atmosphere, but there are also exceptions: For a colourful sunset the presence of particles is essential.

21.5 References

AEG (1961) Allgemeine Electricitäts-Gesellschaft, Hamburg, *Brochure STR-V22-56-MS04*.

Blackwell, H.R. (1946) Contrast thresholds of the human eye, *J. of the Optical Society of America*, **36**, 624-643.

Bohren, C.F., Huffman, D.R. (1983) *Absorption and scattering of light by small particles*, Wiley Interscience, New York.

Beutell, R.G., Brewer, A.W. (1949) Instruments for the measurement of visual range, *J. Sci. Instr.*, **26**, 357-359.

Chandrasekhar, S. (1960) *Radiative transfer*, Dover Publications, New York.

Gorraiz, J., Habenreich, T., Horvath, H. (1985) Einfluβ kleiner Farbdifferenzen auf die Kontrastschwelle des menschlichen Auges bei verschiedenen Lichtarten, *Die Farbe*, **32/33**, 355-379.

Hänel, G. (1976) The properties of atmospheric aerosol particles as a function of the relative humidity at thermodynamic equilibrium with the surrounding moist air, *Advances in Geophysics*, **19**, 73-188, Academic Press, New York, USA.

Horvath, H. (1971) On the applicability of the Koschmieder visibility formula, *Atmospheric Environment*, **5**, 175-84.

Horvath, H. (1981) The University of Vienna Telephotometer, *Atmospheric Environment*, **15**, 2537-2546.

Horvath, H. (1994) Remarks and suggestions on normenclature and symbols in atmospheric optics, *Atmospheric Environment*, **28**, 757-759.

Koschmieder, H. (1925) Theorie der horizontalen Sichtweite, *Beiträge zur Physik der Atmosphäre*, **12**, 33-53 and 171-181.

Poβ, G., Metzig, G. (1987) Extinction coefficient measurements of ambient aerosol particles using a "White cell", *J. Aerosol Sci.*, **18**, 895-898.

Reist, P.C., Wilson, W. (1989) An interactive model for predicting scattering functions of polydisperse aerosols. Paper presented at the *International Conference on Aerosols and Background Pollutions*, Galway Ireland, June 13-15, 1989.

Trier, A., Firingetti, L. (1994) A time series investigation of visibility in an urban atmosphere -I, *Atmospheric Environment*, **28**, 991-996.

VII

POLICIES, LEGISLATION - AND A FINAL OVERVIEW

The impacts of urban air pollution depend not only on the nature of the sources, but also upon a series of non-technical factors, which can form the basis of an abatement policy. For the dominant source of pollution - traffic - especially two types of tools are important (Chapter 22): *The planning of the city structure* can be used to reduce the transport demand and promote the use of public means of transport. The *economic measures*, in the form of various taxes and fees, can discourage the use of private cars, either in general or in specific areas.

Both technically and politically it is important to demonstrate the development in urban air pollution levels by monitering of the air quality. On the other hand the pollution comprises hundreds of different harmful - or potentially harmful - compounds, which can not all be monitored individually. Therefore a few *indicator compounds* can be selected. They must not necessarily be important in themselves, but their levels and behaviour should reflect larger groups of compounds. It has also been attempted to construct *indices,* which e.g. show the weighted impacts of several compounds or the exposure of population groups. Chapter 23 describes the various possibilities and gives examples of applications.

The legal basis for regulation of air pollution is a series of *limit values* (Chapter 24). In Europe they are set down in EU directives, which are subsequently ratified in the individual member states. The primary aim has been to protect human health, but for SO_2 and NO_2 also the prevention of damage to the environment has been recognised.

The World Health Organisation (WHO) has adopted *guidelines* for a larger number of compounds. Although not legally binding they are used as background information and guidance to national and international authorities. They are also used as the starting point for developing new EU air quality limit values.

In Chapter 25 the air pollution in different European areas and major cities are compared both with respect to emissions and resulting air quality. The results of Chapter 25 are shortly summarised and put in perspective in the final Chapter 26.

Chapter 22

POLICIES TO REDUCE URBAN AIR POLLUTION

SUSANNE KRAWACK
The Danish Transport Council
Chr. IX gade 7,4., DK-1111 Copenhagen, Denmark

22.1 Introduction

There is no single technical or economic measure which will reduce traffic air pollution simply and efficiently. Furthermore, there is no simple strategy which will be efficient in all urban areas. This is, of course, due to the complexity of the problem of traffic air pollution as described in the previous chapters of this book. Consequently the problem of urban air pollution is treated very differently. In some cities like Athens, air pollution is an essential part of traffic planning problems, whereas in several other urban areas such problems as congestion and accessibility are given much higher priority.

In this chapter various strategies to reduce urban air pollution will be presented. The chapter will focus on types of measures that local or regional authorities can use to reduce the negative impact of air pollution from traffic. The previous chapters described technical measures such as improved vehicle technology and improved fuel quality and together with their impact on air pollution. These kinds of measures will not be dealt with here.

There are a number of reasons why air pollution problems are treated in different ways. Urban air pollution consists of various different substances, depending on the type of traffic in the particular city. More important, however, is the political priority given to various aspects of the air pollution problem. In some areas reducing the impact of local air pollution is highly prioritised and particular priority is given to reducing particle emission from diesel vehicles. In other areas there is more emphasis on reducing regional air pollution and emissions of NO_x and VOC are given highest priority.

The acceptance of the air pollution problem can vary significantly from city to city. In areas where concentrations of hazardous substances from time to time reach levels which can be harmful to human health, quite naturally, there is greater political awareness than in areas where concentrations generally are lower. The concentrations vary, of course, with variation in traffic density but also because of differences in climate - specially wind situations, as described in Chapters 3, 11, 12 and 25.

22.1.1 RELATIONS TO OTHER ENVIRONMENTAL IMPACTS

Air pollution is only one among several environmental and health related problems which are often addressed in a common action plan. Other problems like noise, traffic safety and severance from traffic often have, considerable weight in an integrated plan. These problems are treated together with the need for accessibility, reduction of congestion and other general traffic problems of urban areas.

In several cases measures to reduce one environmental factor will not lead to reduction of other environmental factors. To concentrate traffic on a few main roads will reduce the number of persons disturbed by traffic noise because double the amount of traffic only leads to a minor increase in the noise level. On the other hand to gather large traffic volumes can have the effect that air pollution limits are exceeded in these streets, so a measure that is adequate to reduce the impact of air pollution on a health may prove to be in conflict with the problems of noise pollution.

22.1.2 THE POTENTIAL TO REDUCE THE NEED FOR TRANSPORT

Strategies to address air pollution will generally stress the potential to:
– reduce the transport demand,
– change the mode choice towards less polluting transport means and
– regulate the traffic in order to reduce air pollution from traffic.

When reducing the need for transport, the overall transport mileage will fall and most of the environmental problems will also decrease. The transport demand is not, however, a well-defined volume for the single person, but relates to both how the urban structure provides services in local areas and to the cost of transport.

A certain volume of transport can be carried out in several different ways and pollution depends heavily on the choice of means of transport. This is, of course, only correct from an average point of view. The marginal decision is of less importance as the bus or the train will run whether a specific person chooses to use it or not. With a fixed modal split, car and bus traffic in an urban area can lead to different levels of air pollution. Speed and frequency of hard accelerations has an influence on emissions and thus traffic management can influence the level of air pollution in the urban area.

Since there is not a fixed volume of transport that defines the transport demand for a specific citizen, the financial position of the individual and socio-economic considerations in general have a great influence on travel behaviour. Furthermore, the general mode choice is important as well as distances to important services. Travel surveys carried out in several European countries show that both the total mileage travelled and the mileage travelled in private cars is clearly related to household income. On the other hand, the urban structure and the general density of the city also has a specific influence on the demand for transport, i.e. how the services which the citizen need are located in relation to dwelling areas.

In the following sections the impact of two important factors is discussed: urban structure and the cost of transport.

22.2 Land use planning

There is a clear relation between the *density of cities and mileage* as shown in Figure 22.1.

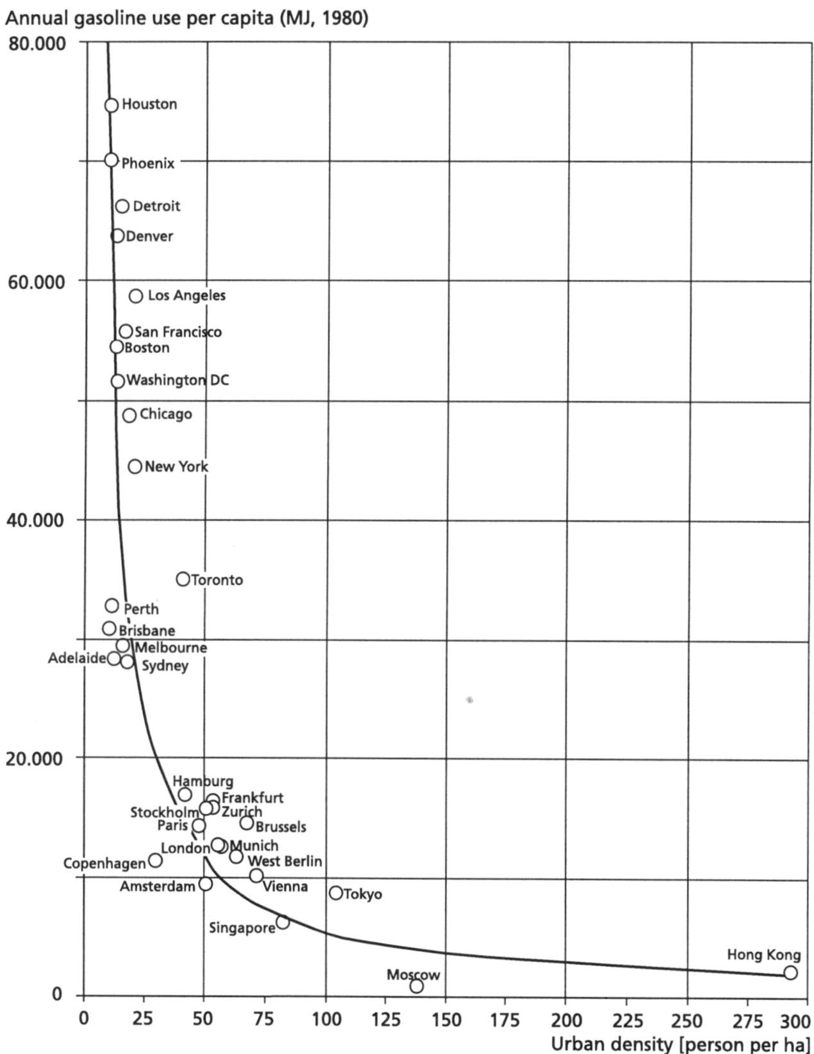

Figure 22.1 Gasoline use per capita versus urban density in 1980 (Newman, Kenworthy 1991).

In cities like Hong Kong and Tokyo with very high population density, fuel consumption for transport is significantly lower than in typical American cities where the population density is much lower. Accepting this relation it is, however, not clear whether it can be turned into an active measure to reduce the demand for transport. Behind the relation between population density and energy consumption for transport lie several types of explanations which so far have not been quantified. The simple explanation is, of course, that in dense urban areas distances to shopping, work and other necessary activities are shorter leading to a lower demand for transport. There is, however, a tendency for low income groups to live in the most densely populated urban areas and as car ownership and mileage are low in these groups, income might explain just as much as urban density. Another explanation is the role of public transport. Public transport systems are generally better in very dense urban areas than, for instance, in American cities with low density.

A Scandinavian study of the impact of urban structures on fuel consumption suggests, however, that around 40% of the energy used for transport in a household can be explained by land use planning variables, whereas 60% can be explained by other factors such as income, car ownership, family type, gender, age etc.

From *travel surveys* it is known that a great part of the population use their financial surplus to choose among activities at several levels. People choose a working place not primarily because of its location, but because of wages and the content of the work. Similarly, in Western Europe it is common to bypass the closest shops to go to large shopping centres, and when planning leisure activities people do not choose the nearest sports club, but select the type of sport or other activity of their wish. This indicates that even if the urban structure provided opportunities to reduce transport, it is not sure that the citizen will choose the nearest activity.

However, an urban structure, where distances to shopping facilities, working places and other activities are short, will provide a necessary background for other types of measures, which are more efficient for reducing the transport in urban areas.

Specific measures to increase density in existing urban areas are, however, limited. Some measures which have been used include developing a framework for land use planning, defining minimum densities of future residential areas, offices and other types of areas. Also restrictions for localizing large shopping centres can influence the mileage done by customers.

22.2.1 PLANNING OF NEW URBAN AREAS

When developing new, large urban areas, the traffic structure can be seen as the backbone of the town. If this backbone is a public transport line like a commuter train or a highly frequent bus service and not a road network, the precondition for a less car dependent urban area is established. Another important factor to promote public transport is to limit the urban area to a ribbon of the width of an acceptable walking distance to public transport (less than 1 km.). Figure 22.2 presents types of urban development which support the use of public transport.

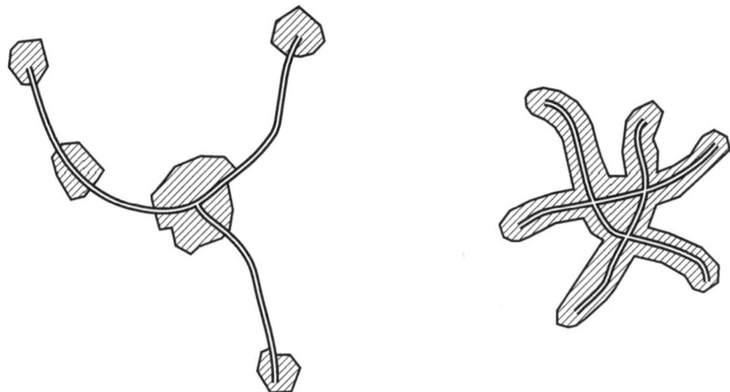

Figure 22.2 Public transport based urban development.

The land use plan in Figure 22.3 shows urban development for a *very dispersed area with low density*. This type of plan, typical of the urban development of the sixties, is rather difficult to facilitate with public transport, and as distances are rather long to any type of service, the population of such kinds of areas become very dependent on their private cars. Development of such areas will limit the potential for a future sustainable transport policy for the area.

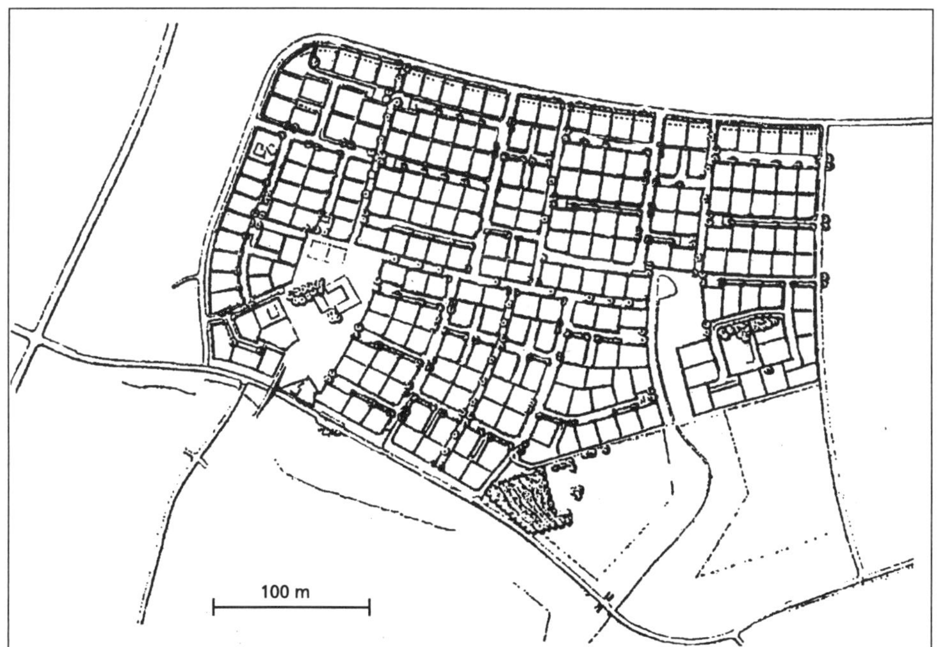

Figure 22.3 A dispersed and car-based urban development in Århus, Denmark (Gaardmand 1993).

The development of a new urban area west of Copenhagen (Figure 22.4, next page), on the other hand, is an example of an urban development plan which could add to a *strategy to reduce air pollution from transport.* Urban density increases close to the railway station to provide short walking distances for a great number of people. Furthermore, public facilities such as a shopping centre and kindergartens are located near the station to make it the natural centre of the town.

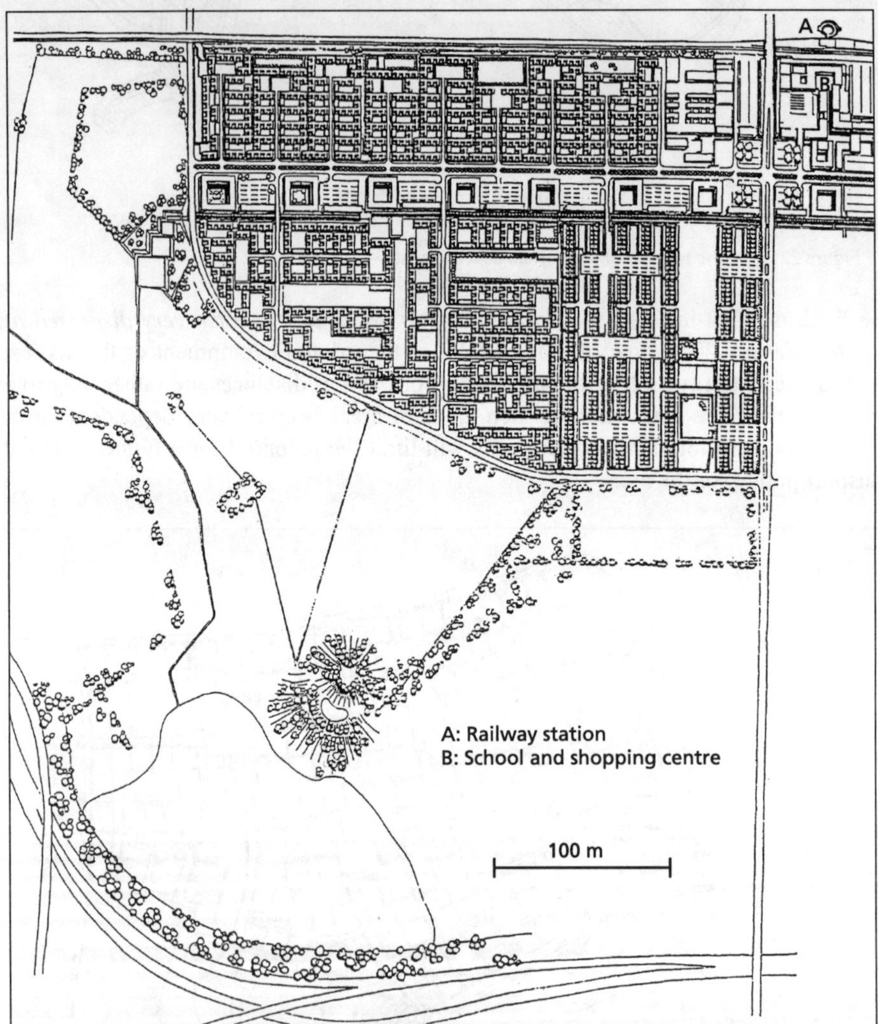

Figure 22.4 Albertslund west of Copenhagen is a typical example of a public transport based urban development (Gaardmand 1993).

It is, however, difficult to give quantitative indications of how great a difference in mileage alternative development schemes will lead to. A fair comparison can hardly be made as income, educational at level or other factors will vary substantially between urban areas and such differences can greatly influence travel patterns.

As mentioned earlier, the impact of urban structure alone might only have a limited influence on the travel pattern and the mileage in an area, but it can be a necessary background for introducing other, more efficient measures - to reduce urban air pollution from traffic.

22.3 Economic measures

Pricing transport is an efficient measure to reduce the mileage driven and, consequently, air pollution. As a general rule an average elasticity on mileage is – 0.2 - – 0.4, which means that an increase of the cost of transport by 10% will reduce the overall mileage by around 2-4%. This measure is not frequently part of a strategy to reduce air pollution. However, although most countries impose various kinds of taxation on transport in most cases taxation has been introduced as a fiscal measure and not for the purpose of obtaining a certain transport or air pollution situation.

Economic measures are, however, of vital part of the political debate and are an integrated part of EU transport policy. In the green book: "Towards Fair and Efficient Pricing in Transport" (COM 1992) the Commission argues for the polluter pay principle. There are two different principles in the discussion of how the polluter should pay for his pollution in a fair manner. One is to integrate the external cost in the transport price; the other is to set taxation to meet environmental targets.

22.3.1 INTERNALIZING EXTERNAL COSTS

The *external costs of transport* are costs related to a journey, but not met by the driver. The driver of a car will cover the cost of the vehicle, fuel and usually taxation, which to a certain extent covers the cost of the infrastructure. These may be called internal costs. In general, however, the driver will not cover the full cost of a possible traffic accident or the air pollution which is produced during the trip. These costs are covered by the taxpayer or to some extent not covered at all. This applies, for instance, to the nuisance caused by a new road crossing a nature resort. It can, however, be argued that the users of the nature resort and, in the end, society cover the costs in the form of a loss of natural value.

Discussions of how to value such factors are currently taking place. From a theoretical point of view, several methods lead to an estimate of the market price on the external effect.

One method used is based on the *willingness to pay*. This can for instance be done by a survey of how much people using a nature resort are willing to pay as an entrance fee in order to preserve the area. Another method is *avoidance cost*, which estimates the expenses to avoid the external effect. An example of this method is to use the cost of introducing new and cleaner technology such as the catalytic converter to estimate the cost of reducing NO_x emissions.

Even though in principle these methods can be used to estimate the same market price of a certain external effect, there are substantial differences in the estimates of these costs. This is, of course, due to the fact that not even all the external effects can be described in detail and evaluating the market price of almost unknown effects leads to great uncertainties. More research in this field will, however, gradually improve the knowledge of the external costs.

The theory of internalizing the external costs is that if the individual driver were to cover his full transport cost, the market would lead to an optimal balance between the demand for transport and the external effects of transport. In other words, this theory defines an optimal level of pollution.

22.3.2 PRICING TO MEET TARGET

Another theory of pricing often takes the view that the true cost of transport, including external costs, cannot be defined or not defined by consensus for the next decades. Another and more pragmatic way of defining the price for transport is to use the pricing instrument to meet the political targets for the transport and environment sector.

In this situation it is hardly necessary to estimate the cost of transport. A strategy of "*learning by doing*" can be used. If a city authority wants to reduce NO_x emission by 20% and want to use pricing of transport, the price can be adjusted until the target is met. This method has not so far been used within the transport sector, even though it might seem very simple.

The problem behind this approach is defining the targets with enough scientific background that the policy makers and the population support them and are ready to introduce the necessary changes in behaviour to meet the targets. It would be necessary to provide policy makers and the population with knowledge of what will happen to nature, eco-systems and human health if the measures are not met.

Another problem related to implementation of this strategy is to define a fair relation in taxation of different means of transport. In other words the system provided must relate taxation to the actual emission level of the different kinds of vehicles. If, for instance, a lorry emits 8 g/km of NO_x and a private car emits 0.5 g/km of NO_x, the taxation of the lorry should be 16 times higher than for the private car per km. This seems very simple, but in practice it is very complicated to provide a system of taxes which is fair for all kinds of vehicles and all kinds of pollution.

22.3.3 TYPES OF TAXATION

As mentioned earlier, taxation has mainly been introduced for fiscal purposes. It has, however, an important impact on air pollution in various way, which depend on the way the taxes are imposed. Taxation can influence the choice of vehicle and fuel type, the mileage of the single vehicle and technological development.

There are a number of different ways of using taxes in the transport sector:
- Purchase tax
- Annual taxation of car ownership
- Fuel taxes
- Toll roads
- Road pricing

A purchase tax is generally introduced as a kind of luxury tax and is imposed as a percentage of the price of the car. In Europe, purchase taxes vary from nothing in Sweden and Germany to 180% of the price of a new car in Denmark. There are three main effects of a high purchase tax: First, the number of cars is generally lower per 1000 inhabitants than in countries with a similar economy. For example, there are 400-480 cars per 1000 inhabitants in Germany and Sweden, whereas car ownership in Norway and Denmark is 320-390 per 1000 capita. Second, the average size of the cars is smaller as only a very small part of the population can afford luxury cars. Third, the purchase tax provide substantial revenue, which is of importance for the national governments.

The annual tax is generally introduced as a means of providing revenue. The annual tax exists in nearly all European countries but with big differences in levels. Annual taxes are generally higher in countries with no purchase tax. In several countries the annual tax is related to the weight of the vehicle, which is easy to administer. In some countries, for example Sweden and Denmark, the annual tax is related to the energy consumption of the car in order to provide an incentive for having less energy-consuming cars. If the total taxation is constant, the effect of increased annual tax and a reduced purchase tax is to increase the number of cars and reduce mileage. All in all, model calculations from both the Netherlands and Denmark show that the overall effect of such a change in taxation is a small reduction in energy consumption.

The level of *fuel taxes* vary a great deal in Europe. EU regulations impose a minimum level of fuel tax of ECU 245 per 1000 l of petrol. The effect of an EU minimum tax on fuel has reduced the differences in fuel prices in Europe. The effect has primarily been to raise the fuel tax in countries with low taxation whereas there has hardly been any reduction in taxation in countries with high taxes.

Fuel taxes have several effects. Higher fuel taxation leads to reduced mileage as every km becomes more expensive. Another important effect is that increased fuel tax will change the car fleet in the direction of more energy-efficient types of vehicles. In the long run this will add to the development of a market for less energy-consuming vehicles and will influence technological development. Finally, increased fuel tax will lead to purchase of fewer cars, as the price of driving one km will increase. The total effect of fuel tax on energy consumption and emissions is much higher than any other form of taxation.

Table 22.1, based on an international survey of elasticities, shows a relation between fuel tax and fuel consumption, and between fuel tax and total mileage:

Table 22.1 Elasticity between fuel price and fuel consumption, and total mileage (Godwin 1992).

	Short-term effect	Long-term effect
Fuel consumption	–0.2 - –0.3	–0.7 - –0.8
Total mileage	–0.16	–0.33

Two main points should be noted. The effect on total mileage is smaller than on fuel consumption, which is due to a shift towards more energy-efficient cars. The effect is much higher in the long run than on the short term, since people in the longer term can reduce their transport demand by moving or changing activities.

Finally, it should be mentioned that one important obstacle to using increased fuel prices as a means of reducing air pollution from transport is the strong belief that increasing mobility in a society leads to economic growth. Another important reason is to avoid cross-border trade in fuel, which is a result of significant differences in fuel prices on either side of a border. Of course, loss of tax-revenue is the primary reason for preventing cross-border sales of fuel.

Toll roads as a mean of pricing the infrastructure has been used in several areas in Europe. The idea is that every car that passes a given point must pay to continue the journey. The pay stations can be on highways or they can form a ring around a town in order to tax all traffic in the central area. The payment can be manual, but electronic payment with various types of smart cards is becoming more common. Infrastructure taxes are normally used to gather revenue to finance roads or other infrastructure investments. In France and Italy highways are toll roads and the taxes paid are used to finance the same roads. In Norway toll roads have been used in several towns in order to provide revenue for building new infrastructure to improve the mobility of the citizen. In these cases the price is set in order to optimize revenue and not to change transport behaviour. So far no toll systems have been implemented with the main purpose of reducing urban pollution or transport in general. Several plans for tolls have been developed in Stockholm, Amsterdam and other cities, but when it come to the final political discussion they have been rejected. It is, however, expected that a toll system in Randstadt in The Netherlands will be implemented in 2001. The purpose of this system is not mainly to collect revenue, but rather to influence behaviour. Severe congestion problems are the background for this decision. So time-saving and not the impact of air pollution would seem to be high on the political agenda.

Used as a behavioural instrument toll roads have the advantage that taxation only covers the area where transport reduction is of specific interest, either because of environmental or congestion problems. The disadvantage of the system is the impact around the pay stations. A local area can be cut in two by line of pay stations, and it does not seem fair that people in this area must pay for a lot of short trips, which is not the case for people living inside the ring.

Road pricing is a concept that represents a further development of the toll system. The idea is area taxation. Each car will be provided with information as to which taxation zone it is in by satellite or by coils in the road network. The number of km travelled in the different zones defines the payment. This system has not been implemented anywhere and the guess is that at least 5-10 years of both technical and political development will be necessary before it can be implemented.

The system is based on a chip in each car that gathers the information on the distance travelled in each tax zone. The system can operate a very detailed division in space, time and types of vehicles and will be able to provide fair taxation in relation to the environment. Where the surroundings are very sensitive, the tax is high as well as in periods during the day where high pollution levels can be reached. On the other hand, it will be cheaper to drive a private car in the countryside where air pollution levels are low and where the impact of air pollution on human health is not as high.

One of the political discussions about this system is whether it will involve the registration of all movements of all people. Eventually this depends on the specific design of the road pricing system.

22.4 Transport modes and air pollution

Emissions of air pollution from motorized transport of one person for one km depend on the mode of transport, the way the vehicle is driven and, most important, on the number of persons in the vehicle. Comparisons of air pollution emissions from different transport modes have been carried out, both in a scientific context and in political discussions, to promote one transport mode or vehicle type in relation to others.

The specific technology of the vehicles, the type of fuel used in the vehicles, the number of persons in the vehicles and the chosen route travelled have a great influence on the result of such a comparison. A comparison between petrol cars and diesel buses carried out in Copenhagen with the actual number of persons in the vehicle is shown in Table 22.2.

Table 22.2 Emissions from individual and public traffic. Unit: g/person km

	CO	VOC	NO_x	Particles
Petrol car	9 - 15	0.4 - 0.8	0.5 - 0.7	0.01
Diesel bus	0.1 - 0.3	0.2 - 0.7	0.4 - 1.5	0.01 - 0.05

The intervals in the table above cover different type of urban roads and times of day. In peak hours there are more people in the buses but in these periods the driving pattern is less smooth, which leads to higher emissions. Furthermore, the number of passengers is lower in suburban areas compared to the central urban areas.

In general the emissions of CO and VOC are highest from private cars. Particle emissions are highest from diesel buses and only relatively less from cars. NO_x emissions are significantly higher from the private car in central areas where buses carry more than 25 passengers and driving patterns are uneven. In suburban areas the picture is reversed and NO_x emission from the bus is higher as the number of passengers is less than 15 in the bus and the driving is more even.

As a rule of thumb, the most environmentally friendly method of transport is vehicles running at maximum capacity. This means that transport of 4-5 persons will be carried out with the least emissions in one passenger car. Where 2-3 full passenger cars in a row are necessary to fulfil the transport demand, a bus will be an environmentally better solution. Where buses drive in a line to cover the demand for transport, a railway of some kind will represent the most environmentally friendly transport solution.

A road in a central urban area has a certain share of the different transport modes. An example from Copenhagen has the mix of vehicles and persons during an average day as shown in Table 22.3.

Table 22.3 Mix of vehicles and persons at Bredgade in Copenhagen.

	Passenger cars	Buses	Total
Number of vehicles	11,410	233	11,643
Number of persons	15,289	6,757	22,046

To illustrate the impact of *shifting the transport mode* in order to reduce air pollution, a number of model calculations are presented (Table 22.4). The baseline is the actual traffic situation, which is rather congested during peak hours. In one alternative all passengers shift to buses. In this scenario there will be no congestion, and driving patterns will be totally even. In the other extreme scenario, all persons drive passenger cars. In this situation there will be congestion during most of the day and driving will be dominated by "stop and go". A third scenario is to maintain the number of passenger cars at a level which provides a smooth driving pattern and leaves the rest of the persons to buses.

Table 22.4 Emissions from alternative mode scenarios.

	CO	HC	NO_x	PM
Only buses	1	3	64	156
Only passenger cars	177	176	116	56
Smooth driving	66	58	90	110
Base line	100	100	100	100

The scenarios show that even if there were adequate measures to change the mode choice in urban transport, it is not easy to define the optimum strategy from an air pollution point of view. The policy measures to shift the transport mode towards a higher proportion of public transport are not simple. The most efficient measures are to localize transport intensive activities near stations or other areas well served with public transport. Another efficient measure is restrictions for parking of private cars or pricing of parking areas.

The impact of *localization of major workplaces* in areas with good public transport is significant according to an analysis of 52 enterprises in the Copenhagen area (Figure 22.5). The assessment covers modal split in commuter traffic in various offices under the Ministry of the Environment and Energy, in the financial sector, and consultants. The transport mode used in commuter traffic seems to relate directly to where the offices are located. The type of employees is similar at these institutions concerning education and income.

At offices in the centre of Copenhagen there are severe parking restrictions, which makes it very expensive to use private cars. On the other hand there is a frequent public transport system. At offices situated in suburban areas, there is generally free parking space available for employees, but less frequent public transport.

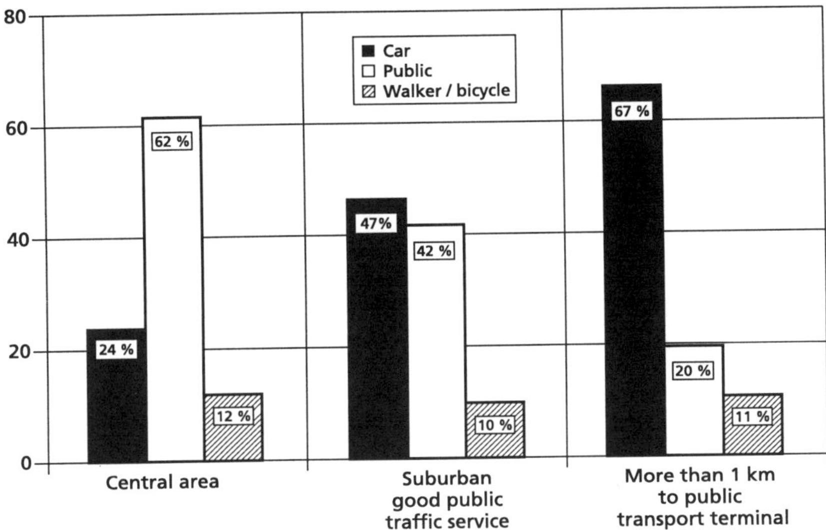

Figure 22.5 Mileage and mode choice in commuter traffic in the Copenhagen area (Hartoft 1997).

The analysis shows that both commuting distances and the mode of transport change significantly according to the location of the workplace. Similar results are found when major workplaces move from one type of area to another.

Parking restrictions either by pricing or time limits are - if enforced - a very efficient means of reducing commuting in private cars. The impact is, however, that the available parking space will be used for shopping, meetings or other activities. This means that one parking space can create several car trips in the urban area, whereas a commuter occupies the parking space with only two trips a day. Most urban authorities want, however, to reserve the parking places for shopping customers for other reasons. Parking restrictions can prevent the highest air pollution concentrations in peak hours.

To shift from motorized transport to *non-motorized traffic* (walking and bicycles) is, of course, an efficient means of reducing air pollution from traffic. This can be accomplished by giving the non-motorized traffic better conditions such as bicycle lanes and pavements, good and secure parking facilities for bicycles etc. A rather important factor for the volume of non-motorized traffic seems, however, to be a question of lifestyle more than a traffic planning issue only.

22.5 Traffic management and emissions

With a fixed volume of traffic and a fixed mode choice, there are still options to reduce air pollution from the traffic. The measures are speed limits and creating a framework for even traffic flow.

Emissions from vehicles depend on speed and how smooth the traffic flow is. If a vehicle is driving at a constant speed the emissions of CO and VOC are highest at low speeds whereas emissions of NO_x and energy consumption are highest at higher speeds.

A high proportion of accelerations increase the emissions, so "stop and go" driving, typical of congested urban areas, will result in much higher emissions than a steady traffic flow. When driving in actual traffic flows the average speed is, however, related to a certain emission level. This indicates that the share of accelerations and stops is at the same level at a certain average speed. The average speed can therefore be used as an indicator of emission levels. Figure 22.6 shows the relation between average speed and emissions of different substances for a *passenger car.*

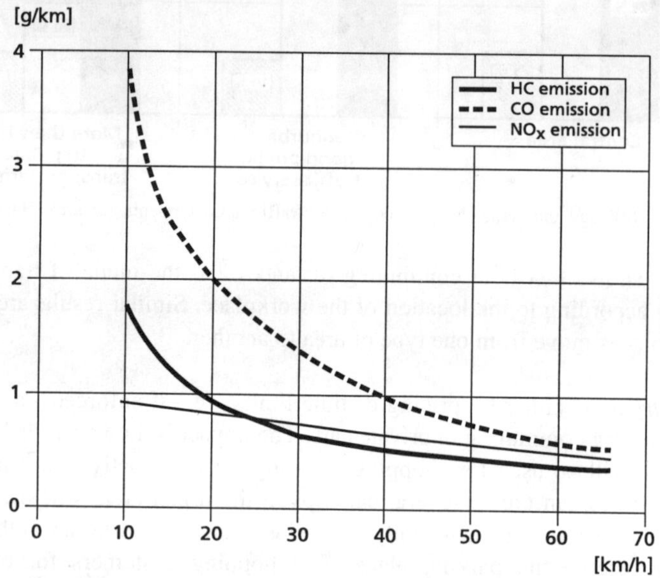

Figure 22.6 Average speed and emissions for a typical Danish passengercar (Ministry of the Environment 1991).

Increasing average speed from 20 km per hour to 40 km per hour will reduce emissions of VOC and CO by 40-45% and emissions of NO_x by 30%. In order to reduce emissions from a given transport volume, it will be efficient to avoid stops and accelerations. This means green waves, when traffic signals are regulated to meet steady flowing traffic with a green light. This is, of course, not possible to provide for all traffic flows in an urban area but can be provided for the main flows of traffic.

Similar arrangements can be made for *public transport.* By providing bus lanes in congested areas the travel speed of the buses will increase, which will lead to both reduced emissions from buses and to improved public transport service in relation to individual transport. Bus lanes will reduce emissions of NO_x by 5-15%. Providing bus lanes will, of course, reduce the road space for the individual traffic, and if that leads to a congested traffic situation the overall picture can be much higher emissions from the traffic.

The general approach to traffic management in order to reduce emissions ought to be a balance between the volume of traffic and the capacity of the road network. This can be obtained by planning the urban structure and using the economic measures as toll roads or road pricing. A much simpler way is to regulate the access to the urban area by traffic signals.

22.6 References

Book, K., Eskilsson, L. (1997) Transport, city structure and development control, *KFB rapport: 9, 1997* (in Swedish), Swedish Transport & Communications Research Board.

COM (95) 691 *Towards Fair and Efficient Pricing in Transport*, EU-Commission, Brussel.

Goodwin, P.B. (1992) A Review of New Demand Elasticities with Special Reference to Short and Long Run Effects of Price Change, *Journal of Transport Economics and Policy*, Vol XXVI No 2.

Gaardmand, A. (1993) *Land use planning in Denmark 1938 - 1992* (in Danish), Arkitektens forlag, Copenhagen.

Hartoft Nielsen, P. (1997) Localization, transport means and urban structure (in Danish), *Byplan* 6/97.

Kågeson, P. (1993) *Getting the prices right: a European scheme for making transport pay its true costs*, European Federation for Transport and Environment, Brussels.

Ministry of Environment (1991) *Air pollution from individual and public transport* (in Danish).

National Research Council (1994) *Curbing Gridlock peak period Fees to relieve traffic congestion*, Vol. 182, National Academy Press Washington DC.

Newman, P.W.G., Kenworthy, J.R. (1991) *Cities and automobile dependence: A sourcebook*, Gower Publishing Group, Hampshire, U.K.

Næss, P. (1995) Urban form and energy use for transport, A nordic experience *NTH 1995: 20*, Norwegian Technical Highschool.

OECD/CEMT 1995 *Urban Travel and Sustainable Development*, OECD, Paris.

Chapter 23

AIR QUALITY INDICATORS

PETER WIEDERKEHR and SEUNG-JOON YOON
OECD Environment Directorate
Paris, France

23.1 Introduction

To date, more than 2800 different anthropogenic air pollutants have been identified in ambient air samples; more than ninety per cent are organic compounds (including organo-metals), and less than 10% are inorganic pollutants. sixty per cent of the pollutants identified are present in gaseous or aerosol form (Graedel et al. 1986). Fuel combustion sources, especially motor vehicles, release about some 500 different components. However, ambient air levels have been determined for only a small fraction of the known pollutants and health and ecological effects data are available for only some 200 pollutants (WHO/Euro 1987, 1996). This has led to attempts to define air pollution indicators or indices for air quality, where a relatively small set of parameters, representing a large number of air pollutants, reflect relevant health and ecological impacts, and are related to major emission sources.

National and international monitoring and evaluation programmes have been established to track pollution in urban areas (city centres traffic exposed sites) (WHO/UNEP 1992), in rural and remote areas to evaluate large scale problems (e.g. acid rain, tropospheric ozone, greenhouse gases) (EMEP 1997). Also, monitoring and surveillance programmes were set up for toxic trace pollutants at local, regional and global scale (UNEP/GEUS 1991). Consequently, indicators have been defined and used for different purposes and fields of application. Dependent upon the problem that is being addressed, e.g., tracking levels and trends for public reporting, monitoring of compliance with air quality standards assessing progress of air pollution control programmes.

Given the large amount of monitoring data currently available in many countries, the principal purpose of developing indicators is to condense and simplify this information to make it suitable for public reporting and decision makers. Such indicators are also needed for trend analyses and comparisons between levels measured in different cities, metropolitan areas and different countries around the world. Indicators should also allow the assessment of air quality in relation to effects on human health, vegetation, animals and materials.

Air quality indicators are only a sub-set of a broader set of environmental indicators that comprise different environmental media (air, water, land), and impacts on man and the environment distinguishing between "pressure", "state" and "response" indicators for national and international reporting (OECD 1994). The indicators described here address only *air pollution problems in urban areas*; and therefore, by and large exclude regional and global problems.

23.2 Definition and objectives

Measurements of selected air pollutants in European cities (Chapter 25) have been carried out systematically since the 1950's focusing on SO_2, and soot (black smoke). In the 1960's, suspended particulate matter, lead and CO measurements were performed due to the growing concern over the increasing emissions from the traffic sector. In the 1970's and 1980's, these monitoring programmes were expanded in terms of the scope (i.e. more monitoring sites), the number of pollutants (e.g. SO_2, CO, NO_2, heavy metals, urban ozone), and the frequency of monitoring (continuous measurements).

In recent years, greater attention has been paid to the need for monitoring photochemical oxidants, notably O_3, and their VOC precursors. Although the instrumentation is now well developed (Chapter 15), relatively few countries monitor O_3 routinely as an indicator of photochemical pollution. Reliable monitoring instrumentation for VOCs has only recently been developed, and consequently urban VOC data are scarce. Furthermore, greater attention is being paid to characterising the nature of suspended particulate matter (SPM), where the heterogeneous nature means that total amount has only limited value for assessing health effects. Monitoring objectives in some countries have therefore narrowed to determine specific size fractions of particulate matter, e.g. PM_{10} (inhalable particles) and $PM_{2.5}$ (respirable particles), and chemical characteristics (e.g. heavy metals, such as mercury, cadmium and lead; organic matter including so called hazardous air pollutants (HAP), such as PAHs). Typical ranges of pollution levels are shown in Table 23.1.

Table 23.1 Typical levels of air pollutants in urban areas. Unit ppb.

Pollutants		***HAP's***	
Carbon monoxide	1800	Methyl chloroform	3
Sulphur dioxide	15	Benzene	4
Nitrogen dioxide	30	Tetrachloroethylene	0.6
Ozone	25	1.3-Butadiene	0.5
		Carbon tetrachloride	0.15
Metals		PAH: Benzo(a)pyrene	0.005
Lead	0.5		
Mercury	0.001		

Source: OECD 1995.

23.2.1 PURPOSE

The purpose of developing indicators is to condense a large number of monitoring data into a much smaller set of numbers that can be communicated to - and comprehended by - the public and the decision makers. In particular, the use of air quality indicators at local and national level covers the following major categories:

- Information of the public, e.g. day-by-day status of air quality, including warnings of sensitive population groups during episodes with high air pollution levels.
- Control of the conformity of recording of air pollution data to permit comparison between different locations.
- Establishment of trends over longer periods of time to permit, e.g. the control of the effectiveness of abatement measures.
- At the international level establishment, of relationships between pollution indicators and macro-economic indicators and parameters in order to provide a framework for comparing air quality between different countries.

Pollutant specific air quality indicators should provide:

- Links to health and ecological effects - both acute and chronic.
- Representativeness of major pollutants released by important economic sectors or activities.
- Presentation of the variability and characteristics of specific sites and locations in urban areas (city centres, traffic exposed sites, residential areas).

Usually, these criteria are met by a variety of air quality indicators that can be proposed for different purposes and areas of application. The indicators to be selected have to be linked to measuring methods and to existing and - if possible - future monitoring systems. These criteria are of importance in the development of control programmes and for international harmonisation of monitoring and surveillance programmes.

23.2.2 DEFINITIONS

As a simple indicator can be used the concentration level of a *single pollutant,* presented as an annual mean, one or more peak statistics or exceedences of specific WHO guideline values (Chapter 22). Major air pollutants are thus often used as *indicators* for the numerous air pollutants present in ambient air. These include usually six pollutants: sulphur dioxide (SO_2), total suspended particles (TSP), nitrogen dioxide (NO_2), carbon monoxide (CO), lead (Pb) and ozone (O_3), all of which are known to cause health effects at currently observed levels.

Combination of several pollutant indicators (or parameters) to generate a normalised number can result in an *air quality index.* The purpose is to relate daily air pollution information for multiple pollutants to short-term air quality standards or WHO guidelines, etc. and turn the data from these multiple pollutants into a single index number. Air quality indices have successfully been used in the United States, Canada, Mexico, France, the United Kingdom and other countries to report air pollution information to the general public. They provide a normalised number, such as 100, when a standard value is read, a descriptor word such as "good", "moderate," "unhealthy", etc. and cautionary language when appropriate for sensitive members of the population. Like air quality indicators, they can be used for presenting air pollution trends. A possible construction is:

$$\text{Air Quality Index} = \sum_{e=1}^{n} (\text{concentration})_i \,/\, (\text{guideline value})_i$$

where i is the number of individual pollutants measured.

The principal disadvantage is that an air quality index can mask significant trends in individual pollutants. For example, most indices in metropolitan areas in the United States are dominated by the pollutants O_3 and CO. When the index presents the trend of the maximum air quality index value, an increasing trend of NO_2 may go unnoticed.

A third type of indicator has been developed for the numerous trace pollutants, also referred to as *hazardous air pollutants* (HAPs). The term HAPs is being used for a very large number of chemical species, including many heavy metals and individual volatile organic compounds which are released from a large variety of sources and which have different effects. Sources may include both mobile or diffuse sources and strong point sources. Some hazardous air pollutants have a long lifetime and can have a cumulative effect on the natural environment. They can also accumulate in the human body and can be assessed by measuring their concentration in body fluids and tissues. Transformations in the atmosphere can convert chemical species to more toxic forms that can be deposited and occur in areas at long distances from their sources. In order to evaluate their combined impact on health, carcinogens can be represented by the cumulative risk indicator estimating the additive cancer risk of the total exposure to these pollutants (Section 23.5).

If ambient air quality data are not available, *surrogate indicators* such as emission inventories or economic activities related to emissions (e.g. fuel consumption or energy production) may substitute them.

23.3 Identification and selection of AQI

The most important requirement for the selection of air quality indicators is a consideration of potential effects of the various air pollutants on:
- Health and well-being of humans.
- Flora and fauna.
- Materials (buildings and cultural monuments).

To define the criteria for the development of an appropriate system of air quality indicators, a variety of factors that determine the air pollution load has to be taken into account:
- Air pollution conditions dependent upon emission source types (Chapters 4-6).

- Meteorological and topographical conditions, scale of the problem and season (Chapters 7-9).
- Data availability, measuring methods and available air pollution monitoring programmes (Chapters 15-17).
- Air pollution effects on human health, flora, fauna and building materials (Chapters 18-21).

The scale of air pollution levels and impacts vary considerably from microscale problems in street canyons to global air pollution problems. Typical scales described by different areas are given in Table 23.2.

Table 23.2 Air pollution scales, air pollution levels and typical effects.

Area	Size (km)	Typical Concentration Range ($\mu g/m^3$)	Typical effects of concern
Street canyon	0.01	100 - 1000	Acute health effects
City blocks	0.1-1	20 - 500	Health, Discomfort
Local	1-10	10 - 100	Health, Materials
Regional	~ 100	5 - 50	Acidification, Vegetation
Continental	~ 1 000	1 - 10	Acidification, Forest damage
Global	~10 000	< 1	Climate change, ozone depletion

Source: WHO/OECD 1991.

The time scale is of great importance. Short-term acute toxicity, represented by very high concentrations over short periods of time, acts differently from long periods of exposure. Short-term high concentrations are often linked with accidental releases or weather conditions leading to air pollution episodes. Different pollutants have to be considered on different scales in time and space.

23.3.1 CRITERIA FOR MAJOR AIR POLLUTION INDICATORS

Table 23.3 on the next page summarises the principal criteria used for developing air quality indicators related to health and ecological effects. The relevant measuring time period associated with the effects and the WHO guideline values are also listed.

23.3.2 DATA AVAILABILITY, ANALYSIS AND PRESENTATION

Standardised statistical analyses should be performed to assess air quality trends, changes in emissions or impact from specified sources. The severity of the air pollution problem or the air quality should be specified relative to air quality guidelines or formal levels of classification, e.g. good, moderate, unhealthy, hazardous (Chapter 17).

The number of hours and days, or percentage of time when the air pollution concentrations have exceeded air quality standards should be presented. This will also need minimum requirements of the database completeness. Long-term averages (annual or seasonal) should be presented relative to air quality standards. In the Norwegian surveillance programme the winter average values of SO_2 and NO_2 are presented on maps in per cent of the national air quality guideline values.

Table 23.3 Criteria and relevant averaging time for major air pollutants.

Pollutant	Acute health effects	Chronic health effects	Ecological effects	WHO Air Quality Guideline
CO	max. 1-h. average max. 8-h. running average	No chronic effects	n.d.	30 mg/m³ 10 mg/m³
Pb	No acute health effects	Annual average	Cumulative annual deposition	0.5 μg/m³
NO_2	max. 1-h. max. 24-h.	 Annual average	 Annual average	200 μg/m³ (150 μg/m³) 40 μg/m³
O_3	(max. 1-h.) max. 8-h. running average	 max. 8-h. running average	 Average (growing season) 5d - 6m	(150 - 200 μg/m³) 120 μg/m³ (60 μg/m³) 0.5 - 10 ppm·h
PM	max. 24-h.	 Annual average	 n.d.	effect-response
SO_2	max. 1-h. max. 24-h.	 Annual average	 max. 24-h. Annual average	---- 125 μg/m³ 50 μg/m³

Source: OECD/ENV 1998.

23.4 Major air pollutants

In Table 23.4 pollutant-specific indicators based on single parameters are presented as candidates for air quality indicators in urban areas:

- An average air quality statistic (annual or seasonal mean, as appropriate);
- A peak statistic (expressed as an upper percentile); and
- An exceedence statistic (expressed as the number of exceedences of the relevant health related short-term WHO guideline or national standard.

Table 23.4 Pollutant-specific Indicators for Major Air Pollutants (continuous monitoring of pollutants)

Species	Sampling time and statistics
SO_2	Annual average (related to WHO A.Q.G.) peak statistic: max. 1-h. or daily mean exceedence of guidelines/standards
TSP, PM_{10}; black smoke	Annual average peak statistic: max 24-h.
CO	Peak statistic: max. 8-h. average exceedence of 8-h. guidelines/standards
NO_2	Peak statistic: max. 1-h. mean, max. 24-h. average exceedence of 1-h. or daily guidelines/standard annual average to protect ecosystems/vegetation
O_3	Peak statistic: max. 8-h. mean exceedence of 8-h. guidelines/standards critical levels for vegetation (ecotoxic effects)

With respect to the *exeedence statistic*, an adjustment must be made for missing data. In the pollutant-specific summary of recommended air quality indicators which follows, it is assumed that the gaseous pollutants are sampled continuously, while some particulate sampling (PM_{10}, TSP) is done intermittently, such as once every 6 days. Only those WHO guidelines with an averaging time of one hour or greater for gaseous pollutants are recommended for indicator comparisons, because with the exception of CO, WHO guidelines principally deal with one-hour or longer averages.

In using the recommended ambient air quality indicators given above, each should meet an annual data *completeness criterion.* In general, gaseous monitors are operated continuously, producing a measurement every hour for a possible total of 8,760 hourly measurements a year or less in the case of a campaign during a specified season. At least 50% of scheduled measurements reasonably distributed in time should be carried out. For a site to be included in a 10-year trend analysis, it might be desirable to require an historical data completeness in at least 8 of the 10 years.

Examples of such reporting on air quality trends and frequency of occurrence of concentrations exceeding air quality guidelines or standards are given in Figures 23.1-23.3 presenting data from the Norwegian air quality surveillance programme (Hagen 1994, Sivertsen 1994).

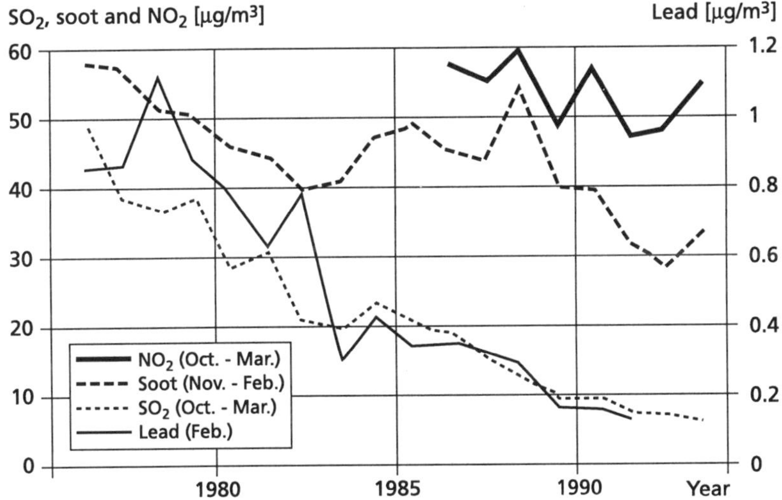

Figure 23.1 Trends of six-month winter average concentrations from 8 selected urban areas for soot, NO_2, SO_2 and lead in Norway; 1976-1993.

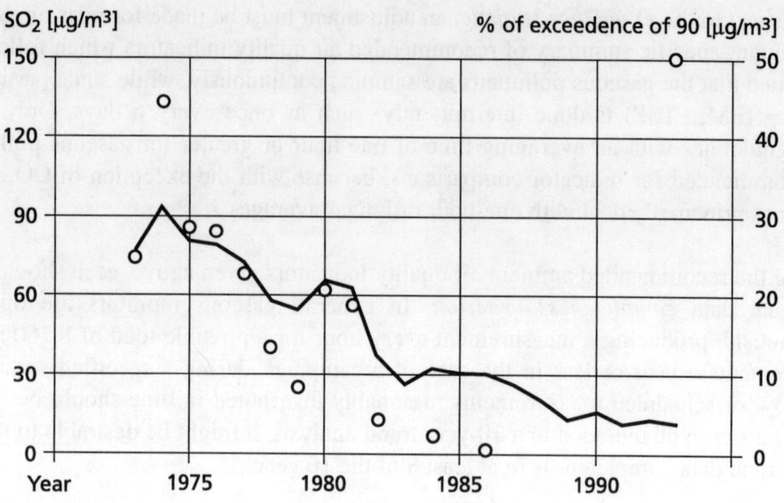

Figure 23.2 Trends and frequency of exceedence of air quality guidelines for SO_2 in Oslo (1973-1993).

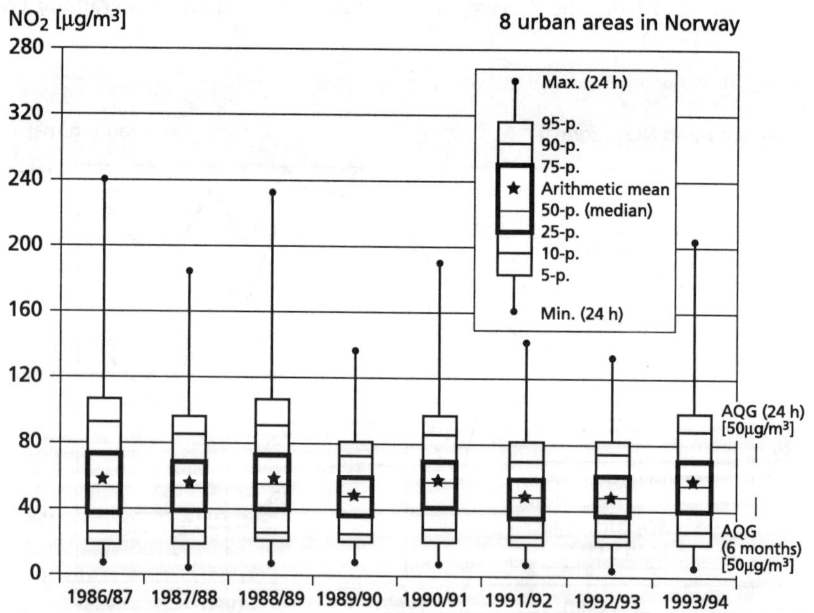

Figure 23.3 Trends of NO_2 concentrations in eight urban areas in Norway; boxplot statistic presentations 1986-1994.

23.4.1 EXPOSURE ESTIMATES FOR URBAN AREAS

In urban areas, where air quality monitoring programmes are linked to air pollution modelling for assessment purposes, it might be possible to establish one single *urban air quality indicator (UAQI)* for each air pollutant. This UAQI will be representative for the situation for the whole urban area that can directly be used for intercomparisons between different areas and different countries. It will also serve as a simple factor for trend analyses, to assess improvements of impact to the population.

To obtain these UAQI numbers, the monitoring programmes have to be supported by air quality dispersion models. Emission inventories and population distributions have to be established. The monitoring programme can be used in statistical optimisation analyses to improve the model results. Distributions of air pollution concentrations for each relevant pollutant are used to estimate exposure. The UAQI can then be presented as either a) the Number of people exposed to the specific air pollutants exceeding a guideline value (based upon health effects) or the number of hours or days that people are exposed to concentrations above AQG values; or b) the area covered by concentrations exceeding a certain value known to have an effect on vegetation or materials.

An *exposure index* can be defined as the number of people living in areas with air pollution levels exceeding the air quality guideline values (AQG) values. Table 23.5 provides an example from Norway of the number of people living in areas exceeding Norwegian AQG values in 1992. The figure includes a forecast for the year 2005. For SO_2 it can be added that the estimates recently performed based on new measurement data already indicate a decreased exposure index from 29,000 in 1992 to 13,000 persons in 1993/94.

Table 23.5 Number of people living in areas where air quality guidelines are being exceeded. (Norwegian Air Quality Surveillance Programme) (Sivertsen 1994).

	Concentration ($\mu g/m^3$)	Average time	Urban scale (1000 persons) 1992	2005
NO_2	50	6 m.	660	390
	75	24 h.	210	42
PM_{10}	70	24 h.	700	340
SO_2	90	24 h.	29	29

23.4.2 DEVELOPMENT OF ADVANCED AIR QUALITY INDICATORS FOR INTERNATIONAL REPORTING

In order to improve current reporting procedures at international level the OECD launched a project on advanced air quality indicators involving designated experts from member countries and representatives from the World Health Organization (WHO) and United Nations Environment Programme (OECD/ENV 1998). The indicators have been designed so that both the existing status and trends in air quality can be identified and are sufficiently sensitive to identify significant changes in the emissions of the countries or regions. The indicators can also be used together with health and ecological criteria to assess likely effects, and with sectorial economic, population or exposure indicators to develop relationships.

The indicators are based upon *individual pollutant concentrations from multiple sites*. The concentration averaging periods adopted are those used by the WHO in their Air Quality Guidelines for Europe and include those to evaluate both acute and chronic health effects and ecological effects (WHO/Euro 1987; 1996). Indicators have been initially developed for six traditional (criteria) pollutants for time averaging periods representative of those likely to cause acute and chronic health effects and/or ecological effects. The indicators were selected to primarily reflect health concerns, but also serve as indicators for ecological effects.

The list of criteria pollutants has also been expanded to include a preliminary list of indicators for hazardous air pollutants (HAPs) for which advanced air quality indicators need to be developed in the future. For this preliminary review the HAPs indicators have not been assessed. However, an attempt has been made to present a method for estimating cumulative cancer incidences from a number of hazardous air pollutants.

To manage and rationalise the large volume of data available from each site, only the *maximum annual value* of each indicator (e.g. 8-h. or 24-h. value) was utilised for each site for acute effect indicators, and *annual average value* for chronic health and/or ecological effect indicators. The application of both average and extreme value statistics from each site enables large data sets to be rationalised whilst maintaining data representativeness for a broad range of values. Furthermore, extreme value statistics are more sensitive than composite average values in identifying changes in air quality, and are of more relevance in evaluating acute health effects than composite average values of measurements from multiple sites. The use of the maximum and average values measured for each pollutant indicator at each site per year therefore more comprehensively meets the objectives of the indicators.

Furthermore, each monitoring site was classified as one of three broad categories: *Urban Traffic/Commercial, Urban Residential, and Rural.* The two urban categories together should be representative of the air quality situation where the majority of people live, while rural sites would serve as a reference of the large scale pollution levels observed.

Status and trends in pollution concentrations have been determined and evaluated for three OECD regions with air quality statistics from the following countries:

- North America (with data from USA);
- Western Europe (with data from Austria, Denmark, Germany, France, Netherlands, Sweden, Switzerland and UK);
- Pacific (with data only from Japan).

To present and evaluate trends in pollutant concentrations, *boxplot diagrams* were produced for each combination of region, site category and pollutant indicators. Boxplot diagrams are simple graphic presentations of frequency distribution functions, represented by selected percentile values (Figure 23.4). To construct the boxplot, data for each pollutant indicator were grouped into those from the same site category, region and year. The data from all the monitoring sites in each group were then ranked and the percentile values and the composite average value calculated.

To enable the comparison of air quality in large cities of OECD Member countries, bar charts of urban peak statistics were constructed. The bar charts show the range of concentrations in 1993 measured by monitors within a city for selected indicators (e.g. annual max. 8-h. ozone). The bar charts show the highest, composite average and lowest values of peak concentrations of each defined indicator as segments of the bar.

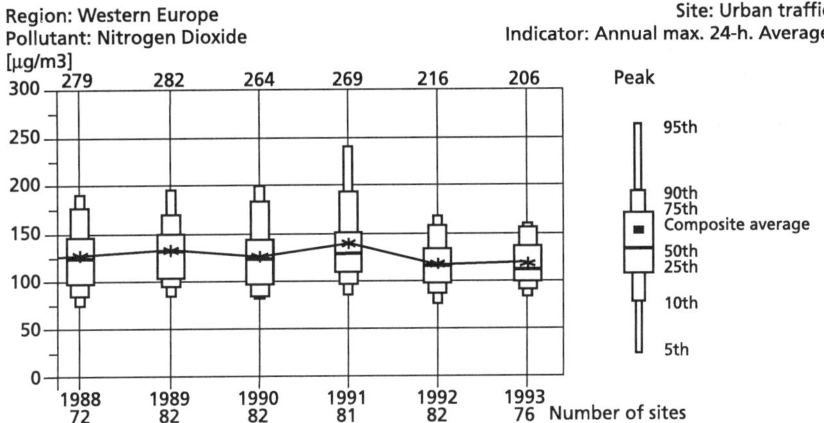

Figure 23.4 Example of boxplot diagram for air quality trend. Source: OECD/ENV 1998.

The cities included within the bar charts are those with populations greater than one million. For countries with less than 3 cities larger than one million of residents, at least the three largest cities of those countries are presented. The peak statistic bar charts presented here contain data from 35 cities for the Western European region, 28 from North America and 16 cities of the OECD Pacific region. An example bar chart is shown in Figure 23.5 (next page). Urban peak statistics bar charts have been produced in OECD's review for selected acute health effect indicators and annual mean lead concentration.

23.5 Indicators for hazardous air pollutants

The type of hazardous airborne substances vary widely and include chemical agents, physical agents like radionuclides, ionising and non-ionising radiation, biological agents (micro-organisms) and fibres such as asbestos. These additional substances are present in the atmosphere in much smaller concentrations than traditional pollutants, but are nevertheless toxic or hazardous.

In the context of this review the focus is on *chemical agents*, although recognising the important effects that other hazardous substances can have upon health. Hazardous air pollutants (HAPs) were defined for the purposes of indicator development as:

Gaseous, aerosol or particulate contaminants present in the ambient air in trace amounts with characteristics (toxicity, persistence) so as to be a hazard to human health, plant or animal life.

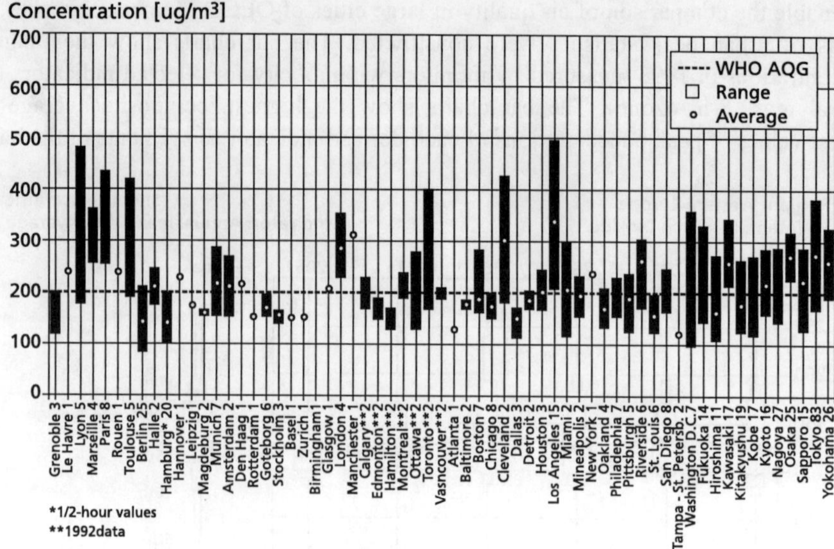

Figure 23.5 Example bar chart for urban peak statistics of nitrogen dioxide (averages and ranges of 1-hour values. Source: OECD/ENV 1998.

A preliminary set of representative indicators for urban HAPs has been proposed within the OECD project: 15 components have been selected among heavy metals and volatile organic compounds, as well as a surrogate HAPs indicator of black smoke (Table 23.6). These substances were selected using the following criteria:

- Significant potential for outdoor exposure;
- Significant health or ecotoxic effects; and
- Monitoring in an adequate number in OECD member countries and sufficient number of sites so that adequate representative data be available to derive the indicator.

For hazardous air pollutants of all kinds it is most important to establish and follow the time trend exposures based on whatever indicator is available for a given pollutant. Trends should preferably be based on concentration measurements in which case they can be related to a "No-effect-level", unit risk of cancer or other guideline values. If trends are based on regional or national estimates of releases such comparisons with guideline values cannot usually be made.

If concentrations for HAPs are not available, trend assessments can be based on annual estimates of production, utilisation or emission related statistics. Very few instances exist where HAPs are monitored in the environment on a consistent basis, with heavy metals measured from particulate samples as one exception. There may, in the future, be some monitoring of species that are widespread such as benzene.

Many toxic trace pollutants or hazardous air pollutants are released from point sources such as refineries, chemical manufacturing and waste disposal facilities. HAPs indicators describing and assessing such exposures are derived from local measurements usually in combination with source strength assessment and dispersion modelling.

Table 23.6 Preliminary List of Urban HAPs Indicators.

Pollutant	Period of measurement	Major source categories
Metals in suspended particulate matter (ng/m^3)		
Cadmium	Annual average	Process emissions (metal
Mercury		production, batteries, electrical
Lead		apparatus), waste incineration,
Vanadium		fuel combustion from stationary
Nickel		and mobile sources
Volatile Organic Compounds (μg/m^3)		
Ethene	Annual average	Representative for
Propene		photochemically active pollutants
n-butene		from fuel use
1,3-butadiene	Annual average	Motor vehicles
Formaldehyde		Petroleum refining
Benzene		
Toluene	Maximum of daily (24-hour) average	Motor vehicles
Benzo(a)pyrene (ng/m^3)	Annual average	Particulate matter from fuel (diesel, gasoline)use and small combustion sources
Trichloroethylene	Maximum of daily (24-hour) average	Solvent use (degreasing, surface cleaning, inks and adhesive)
Dioxins (TEQ/m^3)	Annual average	waste incineration, incomplete combustion of organo-chlorines in fuels and products
Black smoke or soot (μg/m^3)	Annual average	Surrogate HAPs indicator

Source: OECD/ENV 1998.

23.5.1 HEALTH RISK INDICATORS

Hazardous air pollutants (HAPs) that are carcinogens usually receive special attention. In the case of exposure to multiple carcinogens the impact on an exposed population can be estimated from toxicological data using extrapolation from animal-derived unit risk factors to human populations. These estimates must, however, be considered with caution in view of the large uncertainties in the risk assessments.

Many of the hazardous air pollutants are known or suspected to be carcinogenic, and for selected substances unit risk factors have been developed by several national and international institutions. A unit risk factor represents the probability of developing cancer, if an individual is exposed continuously to a concentration of 1 μg/m^3 of a given hazardous air pollutant during a lifetime of 70 years. This is usually determined by epidemiological and animal studies. Combination of unit risk factors with ambient HAPs data and exposed population could yield estimation of health-risk related information such as maximum individual lifetime risks and national cancer incidence.

The magnitude of the cancer risk is presented in terms of lifetime individual risk and annual cancer incidences. Dividing the lifetime incidence by 70 years yields an annual cancer incidence (Huut 1992). Table 23.7 shows a number of unit risk values for selected hazardous air pollutants and estimated cancer incidences in the United States.

Table 23.7 National cancer incidences for selected pollutants in the United States in the 1980's.

Pollutant	Unit risk factor (per µg/m³)	Population of areas with data (Millions)	Incidence per million	National cancer incidences
Metals				
Arsenic	4.3E-03	91	0.28	68
Beryllium	2.4E-03	91	0.01	1
Cadmium	1.8E-03	91	0.04	10
Chromium+6	1.2E-02	89	0.47	113
Nickel	4.8E-04	91	-	-
VOCs				
Benzene	8.3E-06	63	0.75	181
Benzo(a)Pyrene	1.7E-03	47	0.02	4
1,3-Butadiene	2.8E-04	3	1.02	244
Carbon tetrachloride	1.5E-05	41	0.15	36
Chloroform	2.3E-05	35	0.48	115
Ethylene dibromide	2.2E-04	32	0.28	68
Ethylene dichloride	2.6E-05	26	0.19	45
Formaldehyde	1.3E-05	31	0.52	124
Methylene chloride	4.7E-07	32	0.02	5
PIC	4.2E-01	47	3.63	872
Styrene	5.7E-07	6	0.01	2
Tetrachloroethylene	5.8E-07	53	0.03	6
Trichloroethylene	1.7E-06	44	0.03	7
Vinyl chloride	4.1E-06	7	0.05	13
Vinylidene chloride	5.0E-05	29	0.04	10

Source: Huut 1992; USEPA 1990.

To illustrate the estimated cumulative risk for a number of HAPs, Figure 23.6 shows the combined national cancer incidences for the USA and the city of Melbourne (Australia). The national cancer incidences in the USA are determined by twenty hazardous pollutants which include five heavy metals, B(a)P, 13 VOCs, and products of incomplete combustion (PIC). For the Australian case, eight substances have been incorporated such as PIC, butadiene, benzene, formaldehyde, B(a)P, acetaldehyde, and asbestos.

For the USA, eight pollutants account for over 90% of the total risk which include PIC, butadiene, benzene, formaldehyde, chloroform, cr+6, ethylene dibromide, and arsenic. The figure shows that total cancer incidences in the USA have decreased about 50% between 1970 and 1986. More than 95% of the cancer risks in Melbourne are related to three substances: PIC, butadiene and benzene, among them almost 70% of the cancer incidences is caused by PIC which is mainly composed of diesel and gasoline particles. PIC exposure will remain a major health issue even in 2005; therefore much more efforts will be required to control transport-related pollution.

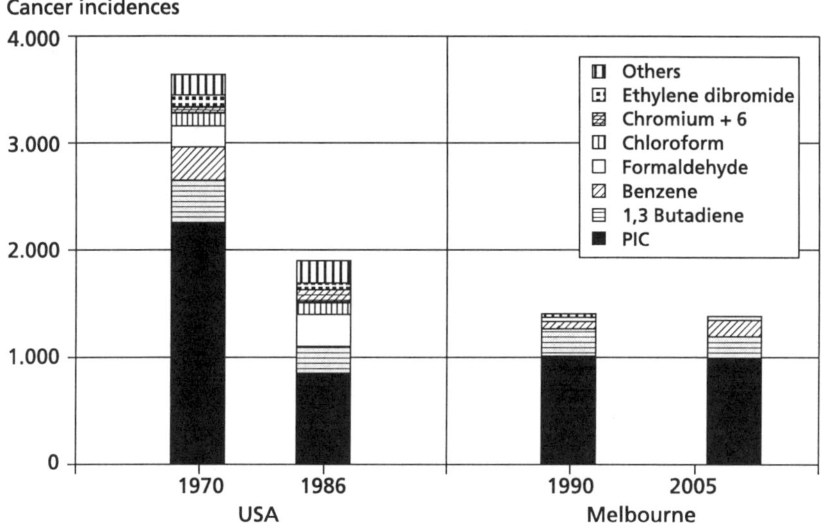

Figure 23.6 Illustrative examples of estimated annual cancer incidences. Source: USEPD 1990; Hearn (1995).

The health risk indicators proposed here have advantages as well as limitations. One of major advantages is that health risk indicators can make complex information more readily understandable by general public as well as decision makers. Combination of unit risk factors and ambient HAPs measurements may provide useful information for establishing priority setting for air toxic pollutants controls. These indicators can also be used to assess overall trends and effectiveness of environmental policies.

23.6 Conclusions

A large set of air quality indicators can be proposed and has been used in various countries for different purposes and areas of application. Single air quality indices in which several pollutants are being merged into one value have also been used in many countries, although they have some shortcomings for the users, notably that variations and trends of an individual component may be significant regarding impacts, but could be masked through the aggregation of values. Indicators for hazardous air pollutants need to be further developed and will largely depend on the availability of reliable monitoring data. The indicators to be selected for future use have to be linked to measurement methods (sensors and analysers) and to future monitoring systems, e.g. DOAS (Chapter 15).

The final choice of air quality indicators will depend on a variety of factors (e.g. type of pollution, extension of the problem, information needs), and the intended use will finally determine the selection of indicators, the compounds and averaging time. To monitor the evolution of emission trends and changes, it would be appropriate to develop a specific set of indicators. To compare the situation in different areas, it is necessary to develop a set of indicators based on a common priority list of pollutants that are being monitored in all areas.

The establishment and use of AQIs will always be dependent upon the representativeness of the monitoring sites and the quality of data. In some cases it might be appropriate or even necessary to supplement the monitoring programmes with data from pollution dispersion models.

23.7 References

EMEP (1997) *Transboundary Air Pollution in Europe,* Co-operative Programme for Monitoring and Evaluation of the Long-Range Transmission of Air Pollutants in Europe, EMEP/MSC-W Report 1/97, Norwegian Meteorological Institute, Oslo, Norway.

Graedel, T.E. et al. (1986) *Atmospheric Chemical Compounds*, Academic Press, London, United Kingdom.

Hagen, L.O. (1994) Rutineovervåking av luftforurensning (Air Quality Monitoring in Norway), April 1993 - Mars 1994, Norwegian Institute for Air Research *(NILU) OR 46/94*, Kjeller, Norway.

Hearn, D. (1995) Health Risks due to Motor Vehicle Emissions in Melbourne, *Clean Air* **29**, *2.*

Hunt, W. F. (1992) *Examination of Alternative Air Quality Indicators*, United States Environmental Protection Agency, Research Triangle Park, NC, USA.

OECD (1994) *Environmental Indicators - OECD Core Set*, Organisation of Economic Co-operation and Development, Environment Directorate, Paris, France.

OECD (1995) *Control of Hazardous Air Pollutants in OECD Countries*, Organisation for Economic Co-operation and Development, Paris, France, 1995.

OECD/ENV (1998) *Advanced Air Quality Indicators and Reporting*, OECD, Paris, France.

Sivertsen, B. (1994) The Use of Air Quality Indicators in Norway, Norwegian Institute for Air Research *(NILU) TR 19/94*, Kjeller, Norway.

UNEP/GEMS (1991) United Nations Environment Program, Global Environmental Monitoring System (GEMS), *Urban Air Pollution*, UNEP/GEMS Environmental Library, No.4, Nairobi, Kenya.

USEPA (1990) *Cancer Risk from Outdoor Exposure to Air Toxics*, US Environmental Protection Agency, Washington, DC, USA.

WHO/Euro (1987) World Health Organization, Regional Office for Europe, *Air Quality Guidelines for Europe,* Copenhagen, Denmark.

WHO/Euro (1996) World Health Organization, European Centre for Environment and Health, *Revised Air Quality Guidelines for Europe,* Bilthoven, The Netherlands.

WHO/OECD (1991) Air Quality Guidelines in the European Region, Report on a third workshop, *EUR/ICP/CEH 079/A*, WHO Regional Office for Europe, Copenhagen, Denmark.

WHO/UNEP (1992) World Health Organization - United Nations Environment Programme, *Urban Air Pollution in Megacities of the World,* WHO-UNEP, Blackwell, Oxford, UK.

Chapter 24

LIMIT VALUES

LYNNE EDWARDS
European Commission, Directorate- General XI
Environment Nuclear safety and Civil Protection, TRMF 01/51
Rue de la Loi, B-1049 Bruxelles, Belgium

24.1 Directives

The first EC Directive setting ambient air quality limit values was adopted by Council in 1980. Council Directive 80/779/EEC (EEC 1980) on Air Quality Limit Values and Guide Values for Sulphur Dioxide and Suspended Particulates and its amending Directive 89/427/EEC (EEC 1989) were adopted to protect human health and the environment against adverse effects from SO_2 and Suspended Particulates.

For this purpose, the Directive lays down limit values for SO_2 and Suspended Particulates which are mandatory throughout the territory of Member States. These limit values are linked - that is, permitted concentrations of SO_2 depend on the simultaneous concentration of particulate matter. The Directive also sets long term guide values. Member States are required to measure SO_2 and particulate matter, to ensure that the limit values are met, and to inform the Commission of any breaches of the limit value(s) and to undertake any necessary abatement measures. The Directive was followed by Directives setting air quality limit values for lead and nitrogen dioxide (EEC 1982) (EEC 1995). Table 24.1 shows the main air quality limit values in force in the European Union at the end of 1997. (There are additional limit values for sulphur dioxide and for particulate matter for different time periods and depending in part on the method used to measure particles).

24.1.1 PROBLEMS AND GOALS

The three Directives have many common features. Their *primary aim* was to protect human health, though it was specifically recognised in the case of SO_2 and NO_2 that meeting the limit values would also reduce damage to the environment. Each Directive requires Member States to take steps to ensure that limit values are not exceeded after a certain date, with decisions about what those steps should be left to Member States. Member States must measure concentrations, and report any breaches of the limit values to the European Commission.

Table 24.1 EC Air Quality Limit Values in force in 1997.

Pollutant	Directive	Parameter	Limit Value, µg/m^3
Sulphur dioxide	80/779/EEC	98 percentile of all daily mean values taken throughout the year.	250 (if particles < 150)
		98 percentile of all daily mean values taken throughout the year.	350 (if particles >= 150)
Particulate matter (measured as Black Smoke)	80/779/EEC	98 percentile of all daily mean values taken throughout the year.	250
		median of daily mean values throughout the year	80
Lead	82/884/EEC	annual mean	2
Nitrogen dioxide	85/203/EEC	98 percentile of all daily mean values taken throughout the year.	200

The Directives played a major part in reducing concentrations of the four pollutants which they deal with. There are now relatively few hot spots in the European Union where any of them is exceeded. However, although the Directives were based on the best scientific evidence available at the time, and in particular the work of the World Health Organisation, there has been a good deal of further research on the effects of air pollution on both human health and the environment since they were agreed. In addition, implementation of the existing Directives revealed a number of problems:

- The limit values were to be met in a relatively short time following adoption of the Directives. The Directives included more ambitious long-term objectives (guide values) for SO_2 and NO_2. These were intended to provide a higher standard of protection for human health and the environment over the longer term. They are not mandatory and have not been widely adopted as operational targets by Member States.
- There were large differences in air quality monitoring and assessment strategies amongst and within Member States.
- The methods used for monitoring air quality differed to some extent across the Union and results were not necessarily comparable.
- The quality of the measurements made by Member States varied greatly due to differences in calibration and quality assurance procedures.
- The Directives required Member States to report only exceedances of air quality limit values to the Commission. The resulting database on air quality across the Union is sparse. Better information is needed in order for the effectiveness of legislation to be assessed properly and as a basis for decisions on any future legislation.
- There is little information at EC level about steps taken to deal with air quality problems, their effectiveness in different circumstances and what further measures might be useful in problem areas.

It was therefore decided that the European Union should undertake a substantial programme to bring air quality limit values up to date and to address the problems revealed by existing legislation. The *goals* for this programme are set out in the Fifth Programme of Action on Sustainable Development and the Environment. They are:

- Provision of effective protection of all citizens against recognised effects of air pollution.
- Establishment of permitted concentration levels of air pollutants which take into account the protection of the environment.

The first legislative result of the programme is Directive 96/62/EC on Ambient Air Quality Assessment and Management, adopted by Council in September 1996 (the Air Quality Framework Directive) (EC 1996a). The second is the Council Decision establishing a reciprocal exchange of information and data from networks and individual stations measuring ambient air pollution within the Member States (EC 1997a). Together these two pieces of legislation build on the key concepts of existing Directives to provide a more comprehensive and transparent framework for action to improve improving ambient air quality in the European Union.

24.1.2 KEY FEATURES OF THE AIR QUALITY FRAMEWORK DIRECTIVE

The four main aims of the Directive are to:

- define and establish objectives for ambient air pollution in the Community designed to avoid, prevent and reduce harmful effects on human health and the environment as a whole;
- assess ambient air quality in Member States on the basis of common methods and criteria;
- obtain adequate information on ambient air quality and ensure that it is made available to the public inter alia by means of alert thresholds;
- maintain ambient air quality where it is good and improve it in other cases.

The Directive is a *framework Directive*. It does not itself set air quality objectives and target dates, nor lay out details of monitoring methods. Instead it provides a basic structure, which must be filled in pollutant by pollutant by means of daughter legislation. Figure 24.1 shows how this basic structure will operate. The emphasis is on information and action.

Once legislation setting limit values comes into force Member States must divide their territory into zones and must assess air quality annually in all of them. The Directive applies throughout the Member States, but special attention is paid to large urban areas or agglomerations.

Daughter Directives setting limit values will include the date by which the limit values must be attained. They may also set temporary margins of tolerance. The margin of tolerance is a new concept in EC legislation on air quality. Despite its name it is not a derogation from a limit value. It is a trigger for action in the period before the limit value must be met.

As Figure 24.1 shows, the *margin of tolerance* is added to the limit value when the legislation setting the limit value comes into force. It is reduced each year to reach zero on the date by which the limit value must be met. The purpose of the margin of tolerance is to identify the zones with the worst air quality. Member States must prepare detailed action plans for these areas (Group 1 in Figure 24.1) showing how the limit value will be met on time. These action plans must be made available to the public and sent to the Commission, which will monitor progress.

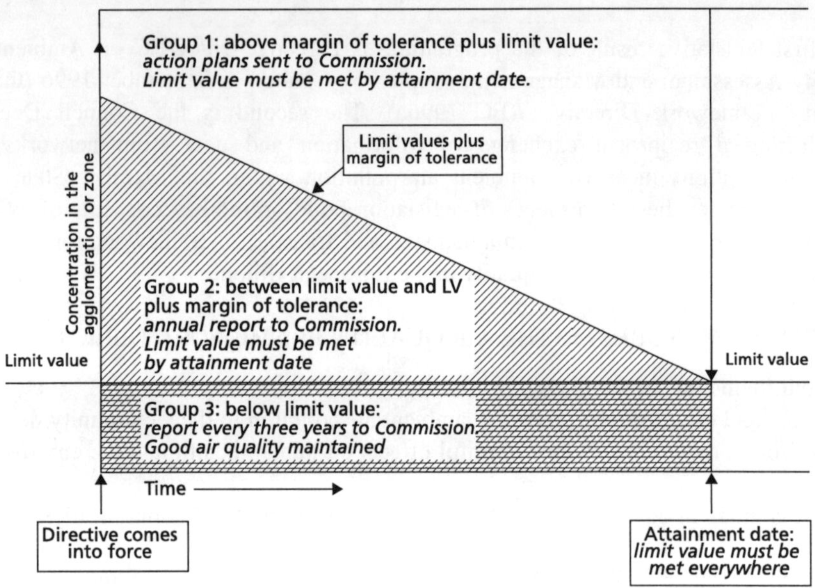

Figure 24.1 Operation of the basic structure of the framework directive for air quality.

Zones where maximum pollution levels are between the limit value and the limit value plus margin of tolerance (Group 2 in Figure 24.1) are not required to forward detailed action plans to the Commission. But they must report concentrations annually to the Commission and must take any necessary steps to ensure that the limit value is met by the attainment date.

The Commission will publish every year a list of the zones in Groups 1 and 2. Zones where maximum pollution levels are below the limit value must maintain or continue to improve their good air quality. Under the Air Quality Framework Directive they will report to the Commission every three years.

All zones will also supply information on concentrations every year under the related Council Decision on exchange of information. The Commission, calling on the expertise of the European Environment Agency will publish a report every three years on air quality in the EU.

The effect of the margin of tolerance will therefore be to ensure that action is taken in problem areas well before the date by which a limit value must be met. The more comprehensive provisions for reporting information under the new framework will ensure firstly that the effectiveness of action plans is monitored and secondly that information is available about air quality in all zones across the European Union, regardless of whether or not they develop action plans.

Although the margin of tolerance is not a level below which pollution must be reduced immediately the level at which it is set is extremely important. If it is set too high, relatively few action plans will be made and problem areas may fail to take action soon enough. If it is set very low then zones which will easily meet limit values on current trends will waste effort and resources making detailed action plans. If no margin of tolerance is set at all then all zones in Group 2 of Figure 24.1 would be obliged to develop and forward detailed action plans to the Commission.

24.2 Air quality objectives

Annex I of the Air Quality Framework Directive lists the pollutants for which air quality objectives are to be set (Table 24.2). It is not an exclusive list. The Directive also includes criteria for judging whether further pollutants should be added once the initial work programme is complete.

Table 24.2 Pollutants listed in Annex I of Directive 96/62/EC.

Pollutants for which there is existing EC legislation	
1.	Sulphur dioxide
2.	Nitrogen dioxide
3.	Lead
4.	Suspended particulate matter
5.	Fine particulate matter such as soot (including PM_{10})
6.	Ozone
Other pollutants	
7.	Benzene
8.	Carbon monoxide
9.	Cadmium
10.	Cadmium
11.	Arsenic
12.	Nickel
13.	Mercury

The Air Quality Framework Directive envisages two types of air quality objective. The normal usual air quality objective will be a *limit value*. It is defined as:

a level fixed on the basis of scientific knowledge with the aim of avoiding, preventing or reducing harmful effects on human health and/or the environment as a whole, to be attained within a given period and not to be exceeded one attained.

The European Commission is required to submit to Council proposals for limit values for 12 of the 13 pollutants listed in the Air Quality Framework Directive. For ozone an alternative is possible. Because ozone is often a transboundary pollutant, resulting from emissions over areas which can be much larger than single Member States, the

Directive allows target values to be set for ozone as well as or instead of limit values. If a target value is set, Member States will be required to see that it is met where possible over a given period. If it is not met they will inform the Commission of the measures they have taken when aiming to reach it. The Commission will then evaluate whether additional measures are necessary at EU level and if so, will submit proposals to the Council. It should be noted that the Commission will in any case submit a strategy for reducing precursors of ozone to the Council at the same time as its proposal for a daughter Directive on air quality objectives for ozone.

The Air Quality Framework Directive also envisages the setting of alert thresholds where appropriate. Alert thresholds are not air quality objectives. The aim is not to reach them. But as with limit values and any target value for ozone they are to be derived on the basis of risk assessment. They are defined as:

a level beyond which there is a risk to human health from brief exposure and at which immediate steps shall be taken by Member States....

The purpose is to limit the effect of high concentrations by providing information to the public and as far as possible taking action to reduce concentrations.

24.2.1 ROLE OF WHO AIR QUALITY GUIDELINES

The recital to the Air Quality Framework Directive states that the numerical values for limit values, alert thresholds and target values are to be based on the findings of work carried out by international scientific groups active in the field.

As in the case of existing EC air quality legislation the work of the World Health Organisation is an important building block. The European Commission has been working co-operatively with the World Health Organisation on air quality since 1993. In particular the Commission has supported WHO in revision of the WHO Air Quality Guidelines for Europe. WHO adopted new guidelines in October 1996 (WHO 1998). They are being used as the starting point for developing new EC air quality limit values. Table 24.3 shows the WHO 1996 Air Quality Guidelines for the pollutants listed in the Air Quality Framework Directive.

It can be seen that WHO Air Quality Guidelines take three forms. Guidelines for the so-called *classical air pollutants* are usually expressed as guideline values. These are concentrations which protect against particular health effects. For *particulate matter* however WHO has expressed its guideline as an effect-response relationship. For *carcinogens* guidelines are expressed as unit risks. A unit risk is the extra risk above baseline of developing a disease if exposed to 1 $\mu g/m^3$ of a pollutant for a lifetime.

All three types of guideline are intended by WHO as background information and guidance to national and international authorities in making risk assessment and risk management decisions. WHO points out that they are based on health or environmental effects only. Other considerations need to be taken into account in developing legislation which sets not only binding standards but also binding timescales for meeting them. This is particularly obvious where guidelines are expressed as unit risks or dose/effect relationships. Even where clear guideline values are available, issues of risk acceptance and the extent of exposure may lead towards standards which are higher or lower than guidelines. Feasibility and costs must also be taken into account, especially in determining timetables, or in the case of particularly difficult pollutants, in deciding whether interim stages are necessary.

Table 24.3 WHO 1996 Guidelines.

Substance	Guideline	Averaging time
Health effects		
Sulphur dioxide	500 $\mu g/m^3$	10 min
	125 $\mu g/m^3$	24 hour
	50 $\mu g/m^3$	annual
Nitrogen dioxide	200 $\mu g/m^3$	1 hour
	40 $\mu g/m^3$	annual
Particulate matter	effect /response (PM_{10} and $PM_{2.5}$)	24 hour
		annual
Lead	0.5 $\mu g/m^3$	annual
Ozone	120 $\mu g/m^3$	8 hour
Benzene	6 x 10^{-6} $\mu g/m^3$	lifetime
Carbon monoxide	100 mg/m^3	15 min
	60 mg/m^3	30 min
	30 mg/m^3	1 hour
	10 mg/m^3	8 hour
Poly-aromatic hydrocarbons (Benzo-a Pyrene)	8.7 x 10^{-5} $\mu g/m^3$	lifetime
Cadmium	5 ng/m^3	annual
Arsenic	1.5 x 10^{-3} $\mu g/m^3$	lifetime
Nickel	3.8 x 10^{-4} $\mu g/m^3$	lifetime
Mercury	1.0 $\mu g/m^3$	annual
Ecotoxic effects		
SO_2 critical level	10 - 30 $\mu g/m^3$ (species dependant)	annual and winter
NO_X ($NO + NO_2$) critical level	30 $\mu g/m^3$	annual

As yet no limit values have been set under the Air Quality Framework Directive. But the European Commission has recently submitted the first proposal for daughter legislation to Council (EC 1997b). It proposes limit values for sulphur dioxide, oxides of nitrogen, particulate matter and lead, and an alert threshold for sulphur dioxide. At the time of writing the proposal is being considered by the European Parliament and Council. It is therefore too early to say what limit values will finally be adopted. Table 24.4 shows however the limit values proposed by the Commission.

It can clearly be seen that the limit values put forward for sulphur dioxide, oxides of nitrogen and lead are very closely linked to WHO Guidelines. Less obviously, the proposed limit values for particulate matter also take account of the effect/response functions provided as guidelines by WHO. A further feature of note is the proposal that limit values should be set for SO_2 and NO_X to protect ecosystems or other vegetation. This is the first time that EC air quality limit values have been proposed with the primary aim of protecting the environment. These environmental limit values would apply differently from health-based limit values. Health based limit values will apply everywhere. But the concentrations needed to protect the environment are lower than those proposed to protect health. The Commission has therefore proposed that the environmental limit values should apply only in the rural background. They will provide a safety net, with Member States free to go further where they feel this is warranted.

Table 24.4 Proposed limit values for SO_2, NO and NO_2, particulate matter and lead (EC 1997b). The proposal is under discussion and same changes have been made[1].

	Averaging period	Limit value	Date by which limit value is to be met
Sulphur dioxide			
1. Hourly limit value for the protection of human health	1 hour	350 $\mu g/m^3$ not to be exceeded more than 24 times per calendar year	1 January 2005
2. Daily limit value for the protection of human health	24 hours	125 $\mu g/m^3$ not to be exceeded more than 3 times per calendar year	1 January 2005
3. Limit value for the protection of ecosystems, to apply away from the immediate vicinity of sources	calendar year and winter (1 October to 31 March)	20 $\mu g/m^3$	two years from entry into force of the Directive
Oxides of nitrogen			
1. Hourly limit value for the protection of human health	1 hour	200 $\mu g/m^3$ NO_2 not to be exceeded more than 8 times per calendar year[1]	1 January 2010
2. Annual limit value for the protection of human health	calendar year	40 $\mu g/m^3$ NO_2	1 January 2010
3. Annual limit value for the protection of vegetation	calendar year	30 $\mu g/m^3$ NO + NO_2	two years from entry into force of the Directive
Particulate matter			
Stage 1			
1. 24-hour limit value for the protection of human health	24 hours	50 $\mu g/m^3$ PM_{10} not to be exceeded more than 25* times per year[1]	1 January 2005
2. Annual limit value for the protection of human health	calendar year	30 $\mu g/m^3$ PM_{10}[1]	1 January 2005
Stage 2			
1. 24-hour limit value for the protection of human health	24 hours	50 $\mu g/m^3$ PM_{10} not to be exceeded more than 7 times per year	1 January 2010
2. Annual limit value for the protection of human health	calendar year	20 $\mu g/m^3$ PM_{10}	1 January 2010
Lead			
Limit value for the protection of human health	calendar year	0.5 $\mu g/m^3$	1 January 2005

[1] The number of accepted exceedance for NO_2 and PM_{10} has been changed to from 8 to 18 and from 25 to 35 respectively. The stage 1 limit value for PM_{10} has been changed from 30 to 40 $\mu g/m^3$.

24.3 Air quality assessment

Air quality assessment is the term used in the Air Quality Framework Directive to cover all methods of obtaining information about air quality. It includes measurement, the compilation of emission inventories and air quality modelling. It is a vital part of the new framework for improving air quality. Previous Directives setting air quality limit values have included harmonised requirements and reporting requirements only for measurement. Measurement will continue to be a key tool for monitoring compliance with limit values and managing air quality under the new framework. Monitoring will be expanded and the quality of results will improve, owing to provisions requiring better quality assurance.

However, even a relatively dense network of monitoring stations cannot represent fully the quality of the air over a large zone, particularly a complex urban area. Firstly, each station may be representative of only a small surrounding area. Furthermore, measurement alone is not sufficient to relate concentrations to sources of emissions nor to allow the likely results of actions to be predicted. These steps are an essential part of developing and implementing air quality action plans. Article 6 of the Framework Directive therefore provides for the use of all appropriate tools for assessing air quality.

Member States are required to divide their territory into zones. The decision as to what shall form a zone is left to Member States, with one exception. The Directive defines an agglomeration as a zone "with a population concentration greater than 250,000" (i.e. a large urban area). The Air Quality Framework Directive relates assessment requirements to the level of pollution in a zone, with special attention paid to agglomerations. Article 6 of the Air Quality Framework Directive identifies two levels of pollution, which are used to relate the intensity of assessment requirements for an agglomeration or other zone to the risk that a limit value might be exceeded. The first Commission proposal for a daughter Directive refers to these two levels as the *upper and lower assessment thresholds*. Table 24.5 summarises its requirements.

Table 24.5 Assessment requirements under Directive 96/62/EC.

Area	Assessment regime, from the most demanding (top) to the lowest (bottom) requirements
1. Where levels are above the upper assessment threshold	Based on high quality measurements - may be supplemented by modelling
2. Where levels are between the upper and lower assessment thresholds	Combination of high quality measurement (but less intensive than in case 1) and modelling allowed
3. Where levels are below the lower assessment threshold	At least one high quality measuring site per agglomeration, combined with modelling, objective estimation, indicative measurements
a. In agglomerations for pollutants for which an alert threshold has been set	At least one high quality measuring site per agglomeration, combined with modelling, objective estimation, indicative measurements
b. In all other cases	Modelling, objective estimation, indicative measurements

The aim is:
- to ensure that the most intensive assessment is carried out in those agglomerations and other zones within which there is the highest risk of a limit value being exceeded;
- to ensure that the least intensive requirements apply only where pollution levels are sufficiently low that there is virtually no risk of an exceedance. It should be noted however that if an alert threshold has been set for a pollutant measurements must be made within agglomerations no matter how low the level of pollution.

Daughter legislation on each pollutant will fill in the framework by including:
- criteria and techniques for measurement, including the location of sampling points, the minimum number of sampling points and reference measurement and sampling techniques;
- criteria for the use of other techniques for assessing ambient air quality, particularly modelling;
- defining the upper and lower assessment thresholds.

24.3.1 PUBLIC INFORMATION

One of the four aims of the Air Quality Framework Directive is to ensure that adequate information is obtained on air quality and that it is made available to the public, inter alia by means of *alert thresholds*. An alert threshold is similar to the public information threshold in the existing Ozone Directive. If an alert threshold is exceeded, Member States are obliged to take steps to inform the public. The information to be given to them will be defined in the daughter Directive which sets the alert threshold. Member States are also required to develop short term action plans to reduce the risk of an alert threshold being exceeded and to limit the duration of such an occurrence. The Directive does not set out the measures which an action plan should include. This will depend on the case and the extent to which measures are likely to prove effective.

Although the Air Quality Framework Directive identifies alert thresholds as only part of the information to be supplied to the public it does not provide further details of what more should be done. The European Commission has included in its proposal for limit values for sulphur dioxide, oxides of nitrogen, particulate matter and lead a requirement that information on these pollutants should be regularly supplied to the public, whether or not a particular threshold has been exceeded.

24.4 Implementing the air quality legislation

The legislation does not spell out who should be the responsible authorities for air quality management in Member States. Clearly this is a matter for their own decision. But in many Member States responsibility will fall in large part to regional and local authorities. Wherever they fall the responsibilities will be considerable. The competent authorities within Member States should already be considering whether they have preliminary information about levels of the pollutants for which new limit values are set. If not, they should take steps to obtain it in time for the implementation of legislation setting limit values.

Once limit values have been set authorities will have to assess air quality in sufficient detail each year, and forecast future trends in order to decide whether the limit values are likely to be met by their attainment dates. If there is a risk that limit values will not be met they will have to take action to meet them. Even where limit values are likely to be met, decisions about new housing, transport and industrial development will have to take into account the impact on future air quality. Authorities will have to make contingency plans for episodes of high pollution and will have to develop good public information systems.

In the more polluted areas this is likely to mean increasing monitoring activity, developing more detailed emission inventories and carrying out more sophisticated air quality modelling than previously. It will also mean gaining public participation and acceptance of action plans for meeting limit values. Authorities in cleaner areas which have not so far undertaken much work on air quality will also have to increase their activity and account for air quality in decision making.

Existing EC legislation setting air quality limit values does not spell out the steps which must be taken to meet them. The Air Quality Framework Directive and its daughter legislation will also prescribe measures. The Framework Directive does however include a list of the information to be included in local, regional or national plans for improving ambient air quality:

- Location of excess pollution;
- General information including an estimate of the polluted area and the population exposed to pollution;
- Responsible authorities;
- Nature and assessment of pollution, including concentrations and techniques used for assessing it;
- Origin of pollution (e.g. transport, industry, cross-border pollution) and the quantity from each source;
- Details of factors responsible for the pollution and measures for improvement;
- Details of measures for improvement which existed prior to the coming into force of this Directive (local, regional, national and international measures);
- Details of further measures which will be taken to meet limit values;
- Details of measures planned or researched for the long term.

It can be seen from this that whilst it is for Member States to decide what are the best means to tackle local problems it is not the case that there will be no further action at EC level to meet limit values. Inevitably meeting limit values will require a combination of measures at EC, national and regional level.

24.4.1 VEHICLE EMISSIONS

EC legislation on vehicle emissions and fuel quality standards has evolved greatly since the first Directive setting emission limits for petrol vehicles in 1970. The early legislation, whilst aiming to reduce pollution was designed to avoid different standards in different Member States which might act as a barrier to trade. Increasingly legislation on vehicle emissions is designed with a view to meeting air quality targets.

The *Auto-oil I programme*, carried out by the European Commission in conjunction with industry set up air quality targets for a number of traffic-related pollutants and assessed the cost-effectiveness of different vehicle technologies and fuel quality standards in helping to meet them in a number of European cities. The target for NO_2 was 100% compliance with the new WHO air quality guideline of 200 $\mu g/m^3$ (EC 1996b). The programme resulted in proposals for new vehicle and fuel standards to take effect in 2000 and 2005. These should enable the target for NO_2 (and the hourly limit value proposed by the Commission) to be met by 2010 in much of the Union without the need for further action. However, there were cities, such as Athens where further action will be needed.

The Commission is now working with Member States, industry and NGOs on a *second Auto-Oil programme*. The results will be used to confirm or modify certain of the standards proposed for 2005. But the programme is also looking at non-technical measures such as road pricing, traffic management and vehicle scrapping schemes which could be used locally to tackle areas of high pollution. It is also looking in more detail than Auto-Oil I at stationary sources. Cost-effective strategies for meeting limit values will often mean apportioning the costs of abatement between different sectors.

24.4.2 INDUSTRIAL EMISSIONS

EC legislation on industrial air pollution has always taken air quality limit values into account. The key piece of legislation dealing with industrial emissions to air was until recently Directive 84/360/EEC on Industrial Plants (EEC 1994). This framework Directive requires that certain types of industrial plants should only operate if they have been authorised. The authorisation must include emission limits for important pollutants, including SO_2, NO_x, CO and heavy metals. An authorisation may only be issued if, amongst other things, the competent authority is satisfied that best available technology not entailing excessive costs has been applied to reduce pollution and that applicable air quality limit values have been taken into account. Directives dealing with large combustion plant and incinerators were subsequently adopted under the framework of Directive 84/360/EEC.

In 1996 Council adopted Directive 96/61/EC on integrated pollution prevention and control (IPPC) (EC 1996c). This Directive is a new departure for EC legislation on pollution. Its purpose is to reduce pollution to the environment as a whole, avoiding transfer from one medium to another. As far as air pollution is concerned, the IPPC Directive will replace Directive 84/360/EEC. It takes effect for new industrial installations in 1999. Existing installations covered by the Directive will be required to upgrade by 2007. Industrial operators will be required to employ Best Available Techniques, taking into account economic considerations. The link with environmental quality standards has been strengthened. Article 10 of the IPPC Directive requires that:

where an environmental quality standard requires stricter conditions that those achievable by the use of the best available techniques, additional measures shall in particular be required in the permit, without prejudice to the measures which might be taken to comply with environmental quality standards.

Developing authorisations for integrated pollution prevention and control is a complex process with judgements to be made about the solution which is best for the environment as a whole. The best answer is likely to vary from installation to installation and may involve development of new techniques.

Those responsible for air quality management will need to predict the likely results for air quality. In some Member States the same authority will be responsible for both air quality management and issuing industrial authorisations. In others two or more authorities may be involved. Where this is the case close liaison will be required, especially on pollutants which are emitted by a mixture of sources. Obvious examples are oxides of nitrogen and particulate matter, both dealt with as priority pollutants under the Air Quality Framework Directive.

24.5 Next steps

The procedure for adopting EC legislation on air quality involves three main institutions: the European Parliament, the Council of Ministers and the European Commission. Under the arrangements now in force under Article 130s of the Treaty the Commission prepares proposals for legislation and submits them to the Council. Council is required to request the opinion of Parliament and the Economic and Social Committee. Council has the primary power to adopted legislation, acting by Qualified Majority Vote. The Air Quality Framework Directive has already been adopted. Member States are required to transpose it into national law by March 1998.

The Commission has presented its proposal for the first daughter Directive setting limit values for sulphur dioxide, nitrogen dioxide, particulate matter and lead. This is now being considered by the European Parliament and Council. Once it is adopted there will be a delay, typically of 18 months or so, for Member States to transpose it into national law. They will then begin assessing air quality in earnest and preparing any necessary action plans.

The Commission is now preparing proposals for daughter Directives on carbon monoxide, benzene and ozone. It is expected to submit these during this year. Proposals for the remaining pollutants (poly-aromatic hydrocarbons, cadmium, arsenic, nickel and mercury) are due by the end of 1999. Once the Amsterdam Treaty has been ratified, proposals under Article 130s will in future be dealt with under the co-decision procedure. This means that legislation must be adopted by both the European Parliament and Council. Some or all of the remaining proposals for air quality limit values will be dealt with under this procedure.

24.6 References

EC (1996b) Auto-oil proposals, *COM (96) 248*.

EC (1997a) Council Decision 97/101/EC establishing a reciprocal exchange of information and data from networks and individual stations measuring ambient air pollution within the Member States (1997), *Official Journal* **L 35**, 14.

EEC (1980) Directive 80/779/EEC on air quality limit values and guide values for sulphur dioxide and suspended particulates (1980), *Official Journal* **L 229**, 38.

EEC (1982) Directive 82/884/EEC on a limit value for lead in air (1982), *Official Journal* **L 372**, 15.

EEC (1994) Directive 84/360/EEC on the combating of air pollution from industrial plants (1994), *Official Journal* **L 188**, 20.

EEC (1995) Directive 85/203/EEC on air quality standards for nitrogen dioxide (1985), *Official Journal* **L 87**, 1.

EEC (1989) Directive 89/427/EEC amending Directive 80/779/EEC air quality limit values and guide values for sulphur dioxide and suspended particulates (1989), *Official Journal* **L 201**, 53.

EC (1996c) Directive 96/61/EC concerning integrated pollution prevention and control (1996b), *Official Journal* **L 257**, 10.

EC (1996a) Directive 96/62/EC on ambient air quality assessment and management (1996a), *Official Journal* **L 296**, 55.

EC (1997b) Proposal for a Council Directive relating to limit values for sulphur dioxide, oxides of nitrogen, particulate matter and lead in ambient air, COM 500 final (1996), *Official Journal* **C 009**, 6.

WHO (1998) air quality guidelines for Europe (to be published).

Chapter 25

AIR POLLUTION IN EUROPEAN CITIES - AN OVERVIEW

LÁSZLÓ BOZÓ
Institute for Atmospheric Physics of the
Hungarian Meteorological Service
P.O.Box 39, Gilice 39, H-1675 Budapest, Hungary

HANS EERENS
RIVM, Laboratory for Air Research (LLO)
P.O.Box 1, NL-3720 BA Bilthoven, The Netherlands

STEINAR LARSSEN
Norwegian Institute for Air Research (NILU)
P.O.Box 100, Institutveien 18, N-2007 Kjeller, Norway

MILLÁN M. MILLÁN
Centro de Estudios Ambientales del Mediterraneo
Parque Tecnologico, Calle 4, Sector Oeste
E-46980 Paterna, Valencia, Spain

NICOLAS MOUSSIOPOULOS and SOPHIA PAPALEXIOU
Department of Mechanical Engineering
Aristotle University, GR-54006 Thessaloniki, Greece

ZISSIS SAMARAS
Laboratory of Applied Thermodynamics
Aristotle University, GR-54006 Thessaloniki, Greece

Compiled and edited by the editors.

25.1 Introduction

Europe is a highly urbanised continent. In 1990, more than 70% of the total population lived in cities. The concentration of human activities on a relatively small area causes an enormous pressure on the natural system leading to numerous environmental problems. Maybe it is the best known, but least understood, problem in terms of impacts to citizens is air pollution.

The actual occurrence and frequency of increased air pollution concentrations depends primarily on magnitude and distribution of emission sources, on local topography (e.g. flat terrain, basin or valley) and local meteorology (e.g. frequency of calm weather conditions, average wind speed, occurrence of inversion layers). The influence of these factors varies between the regions of Europe leading to different levels and types of air pollution in various cities.

In the last decade air quality problems and associated topics like exposure of citizens have been addressed through a number of case studies. Most larger European cities operate monitoring networks for assessment of the air quality. However, the network design and the technical contents of the programmes (the choice of compounds and techniques for measuring as well as the number and location of sites for the monitoring stations) varies widely within Europe (Table 25.1). The more advanced networks have incorporated air pollution dispersion models and use of emission inventories in order to determine the geographical distribution of air pollution in the city area and to quantify the contribution from the different sources. In these cities the authorities may use this information to decide about the measures needed to reduce the air pollution to an acceptable level, and to estimate the costs of the measures. Successful measures in the past for reducing air pollution have been the regulation of the fuel for space heating (low sulphur content in oil/coal, introduction of natural gas instead of coal/oil), reduced lead content in petrol and restriction of the use of cars in inner parts of the cities (e.g. Cologne, Dortmund, Zurich and Copenhagen). During smog episodes, emergency actions are taken by the authorities in some cities (e.g. Rome, Milan and Athens).

Table 25.1 Examples of the technical content of the air pollution monitoring network in some major European cities.

City	Population[2]	Area[3]	Sites	O_3	NO_x	SO_2	Pb	PM(TSP)	CO
Athens	3073	457	11	3	4	3	1	3	3
Barcelona	2398	202	13	3	6	17	2	6	3
Berlin	3475	889	29	7	19	29	≥1	33	≥1
Brussels	1349	485	13	2	5	13	1	13	1
Hamburg	1703	755	24	3	23	23	-	23	8
Madrid	3120	100[1]	>5	1	5	5	-	6	5
Milan	2402	539	10	2	6	6	-	4	8
Munich	1858	310	-	2	6	6	-	3	≥1

[1] Built-up area [2] in thousands [3] in km^2

In the following an overview is given of the air pollution levels in Europe. The main emphasis is on presentation of the general similarities and differences in the air pollution levels in European urban areas. A more complete description can be found, e.g. in (EEA 1998).

25.2 Emissions in different parts of Europe

The main source categories that contribute to the total emission of the components are power generations, space heating, traffic and industry. Traditionally investigated components are SO_2, NO_2, CO, particulate matter, organics and heavy metals. In commercial cities the largest contributions come from local traffic and space heating if oil, coal or wood is used.

The most complete emission inventories for Europe, the CORINAIR inventory (Chapter 6), contains emission data for 11 main categories and for 8 pollutants (SO_2, NO_X, NMVOC, CH_4, CO, CO_2, N_2O, NH_3) with a spatial resolution corresponding to the smallest administrative areas in EU.

25.2.1 REGIONAL DIFFERENCES

For urbanised areas in various EU regions, the contribution from major source categories to emissions of *SO_2 and NO_2* are shown in Figure 25.1 (next page). The figure represents all EU cities with more that 100,000 inhabitants. It shows the dominance of large point sources (LPS: Power plants, large industrial plants) and other industries to SO_2 emissions in urban areas, and the dominance of traffic (mainly road traffic) to the urban NO_X emissions with LPS and industries making up most of the rest. This picture is rather similar in the three EU regions chosen (North, West, South), however, with some significant differences: For SO_2, the relative traffic contribution is much larger in the Southern region, due to relatively higher sulphur contents in motor diesel. Also for NO_X, the relative traffic contribution is somewhat larger in the Southern sector, at the expense of the LPS contribution.

Accurate emission data for many other pollutants are not available. One of the most important is particulate matter, PM_{10} and $PM_{2.5}$. However, strong efforts are - and will be - made to estimate these emissions, since particles are believed to be very important from a human health point of view.

In Figure 25.2 (page 437) three anthropogenic source sectors are shown: Stationary combustion, industrial processes, and transport (incl. road dust re-suspention). Other anthropogenic source sectors are of less importance, but agriculture is probably a source of significance. Natural sources, mainly re-suspension of dust from surfaces is, however, very important in many regions, especially arid ones, e.g. Spain. As natural sources are not included, the PM_{10} emission data in the figure do not necessarily reflect the differences between countries regarding atmospheric PM_{10} levels. For instance, the PM_{10} concentration in Spain tends to be high due to natural sources.

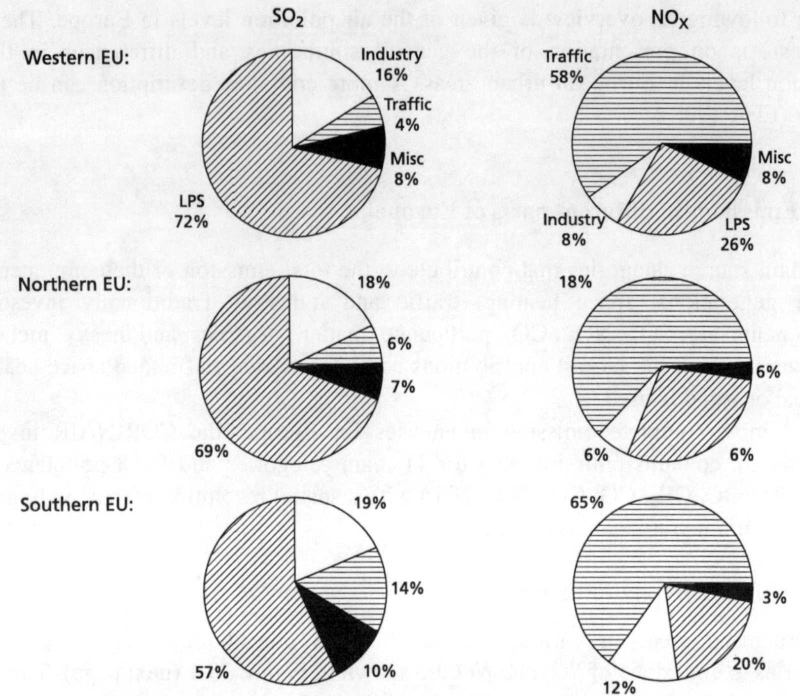

Figure 25.1 Relative contributions from main sectors to SO_2 and NO_x emissions in regions in EU countries containing cities with more than 100,000 inhabitants. LPS: Large point sources (power plants and industrial plants) (EEA 1998).

PM_{10} emissions per inhabitant in 1990 and 1993 show for many countries a fairly even distribution between contributions from the three main sectors, although with some dominance from traffic and stationary combustion. However, especially in Central/Eastern Europe, PM_{10} emissions from stationary combustion are large, and dominate the PM_{10} emissions in these countries. These data can only give an indicative picture of the PM_{10} source contributions in cities in Europe, but indicate the potential for high concentrations in industrialised cities in Central/Eastern Europe. It also indicates a substantial reduction in PM_{10} emissions from 1990 to 1993 in some countries, notably Germany (reductions in former DDR), Bulgaria and Hungary, and a substantial increase in others, e.g. The Czech Republic/Slovakia and Poland. In the EU countries, the changes from 1990 to 1993 are small, except in Finland and Ireland.

Figure 25.3 (page 438) shows emission data for *SO_2 and NO_X* (annual emissions per inhabitant) from a selection of the cities with data from at least two of the years 1985, 1990 and 1995 (or adjacent years), for the main source sectors (traffic, space heating and industry). Data may be missing for a sector in some cities, as is obvious from the figure. The cities are ranged from South to Central, West and North in Europe.

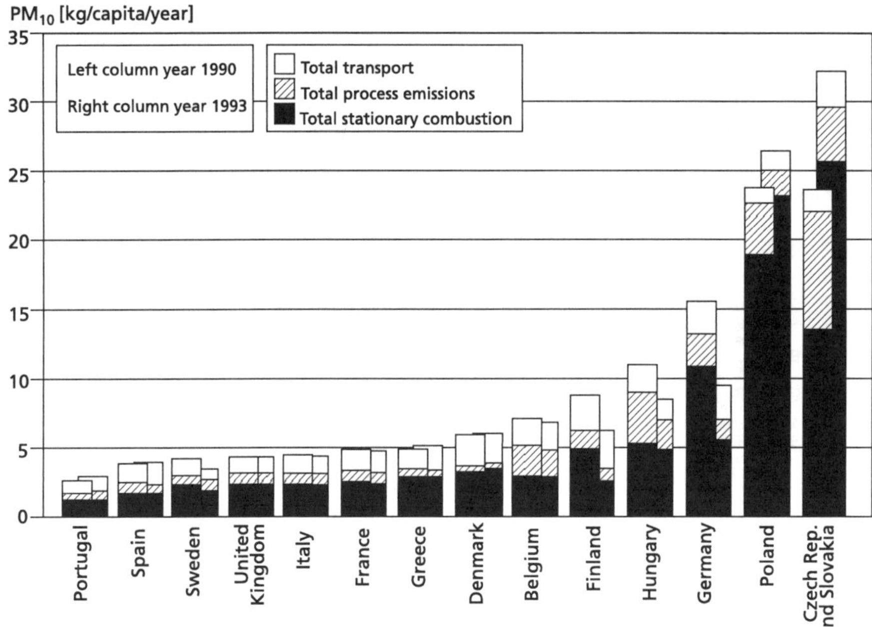

Figure 25.2 National emissions of PM_{10} (kg/capita/year). Main source sector contributions. Unofficial data based upon expert estimates (Berdowski et al. 1996).

Specific *SO_2 emissions* vary more than two orders of magnitude, depending upon the degree of industrialisation, which is the factor dominating the emission situation. In some cities (e.g. Ljubljana and Leipzig) space heating still gives significant SO_2 contributions, while quantification of space heating has substantially reduced such emissions in other cities (e.g. Prague). Industrial use of natural gas and other policies to reduce emissions from industry combustion (e.g. low-sulphur oil) have reduced industrial SO_2 contributions in many cities in Europe over the last decade (e.g. Prague, Sofia, Ljubljana, Leipzig, Berlin, Stockholm, Helsinki and Copenhagen). Reduced industrial activity may also have contributed to this reduction in some of these cities, as well as in Bucharest.

The *NO_X emission* varies less than for SO_2, but also here cities with extensive industrial activities, e.g. steel industry, mining and power production, have high emissions (e.g. Bratislava, Rotterdam, Antwerp and Helsinki). The traffic emissions dominate in most cities, with per capita NO_X emissions typically of the order of 10-20 kg/a. In "harbour" cities such as Rotterdam, ship traffic contributes to the large traffic NO_X emissions.

In most cities, NO_X emissions have been reduced somewhat over the last 5-10 years, and this is mostly due to emission reductions from space heating and/or industry. Traffic emissions have changed very little in general, but significant reductions have been obtained for some cities (Zurich and Stockholm), possibly indicating success in traffic reductions programs. For Athens and Paris, a substantial increase in traffic NO_X emissions is indicated.

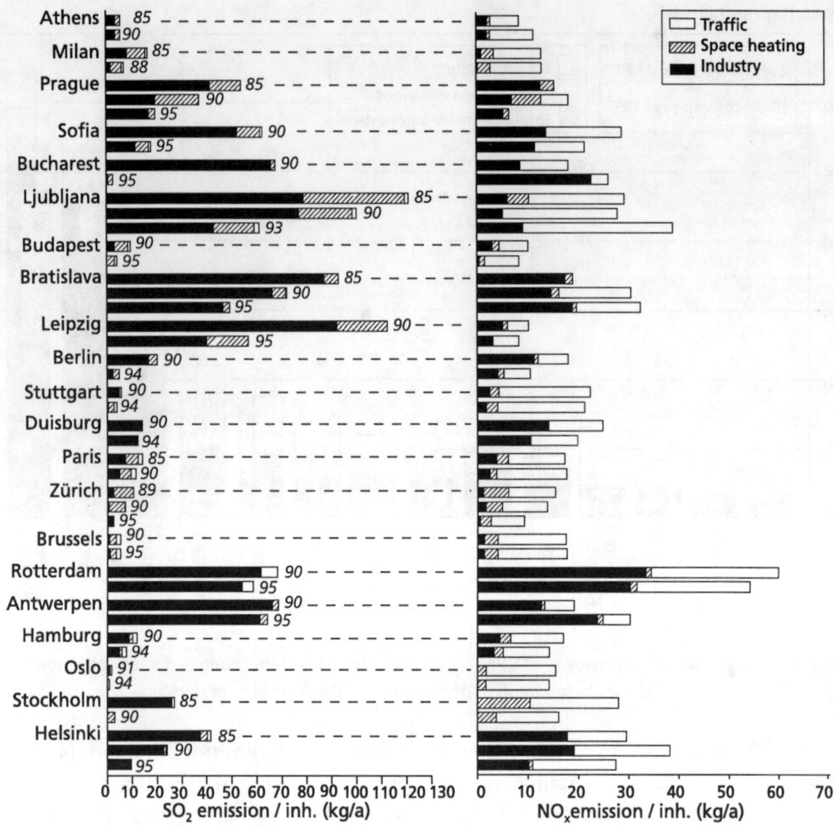

Figure 25.3 Emission estimates (kg/capita/year) of SO_2 and NO_x for selected cities in Europe for 1985, 1990 and 1995 (or some adjacent years). Main source sector contributions. Data reported from the cities as requested as part of the data collection process (EEA 1998).

The emission data (urban emissions to the extent available) show clearly that industry and power plants still play a determining role for the SO_2 pollution levels in European cities. For NO_x it is mainly emissions from passenger cars and diesel trucks/buses, that determines the NO_x levels. This means also that the per capita NO_x emissions vary much less between cities than the per capita SO_2 emissions.

For *PM_{10}*, road traffic, industry and stationary combustion are all significant contributors, with a dominating contribution from stationary combustion in many areas in Central and Eastern European countries. The secondary formation of PM_{10} on the regional scale (e.g. sulphate and nitrate particles) means that the regional PM_{10} background is high, and it may be as high or higher than the city's own contribution to PM_{10} concentrations, especially in the Central parts of Europe. This has a bearing on the abatement strategies for PM_{10}. In these areas, it is equally important to control the regional contribution, as it is to control the city's own emissions.

25.2.2 RELATIVE EMISSIONS

The relative emissions from different types of sources vary from city to city. In the Figures 25.4-25.6 are shown the relative emissions of VOCs, CO and NO_x in some Western European cities. The emissions are shown for passenger cars petrol (PCp), passenger cars diesel (PCd), light duty vehicles (LDV), heavy duty vehicles (HDV), busses (Bus), two wheelers (TW) and stationary sources.

VOC emissions from engines are in general caused by incomplete combustion, which includes phenomena such as flame quenching, misfire, detachment of lubricating oil film, evaporation of fuel etc. Whereas the combustion of fossil fuels is responsible for only one third of total NMVOC emissions, as Figure 25.4, depicts road traffic is responsible for almost 90% of anthropogenic VOC emissions in urban areas. In particular, petrol passenger cars account for almost two thirds of these emissions, while two wheeled vehicles may also have an important contribution in some cities.

At the European level road transport is responsible for over three quarters of the *CO* emitted. At the urban scale, road transport is responsible for more than 90% of the CO emissions, with the petrol passenger cars being responsible for almost 75% (Figure 25.5, next page).

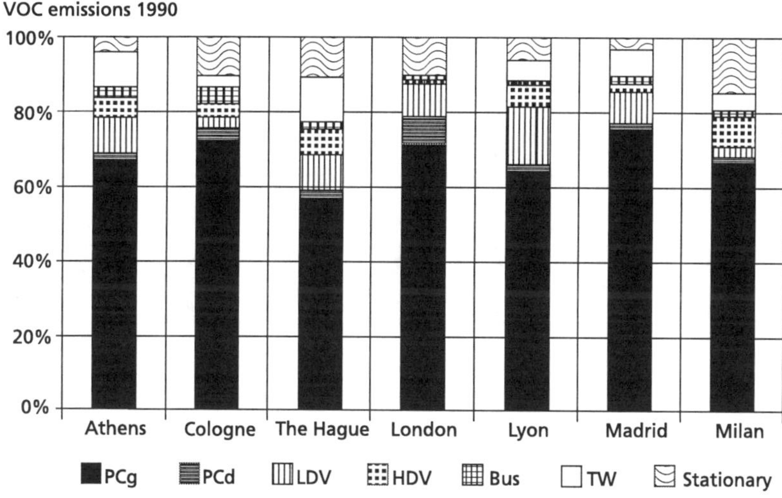

Figure 25.4 Relative emissions of VOC in selected European cities (EC 1997).

Most of the *NO_x emission* from engines is emitted as nitric oxide, especially from petrol powered engines, while in diesel engines NO_2 may represent up to 30% of total NO_x emissions, depending on the operation mode.

Road transport is the main source of urban NO_x emissions. In 1990 at the European scale road vehicles were responsible for about 50% of total NO_x emissions. At the urban scale, road traffic contribution may vary from 50 to 80% depending on the city and its activities (Figure 25.6, next page).

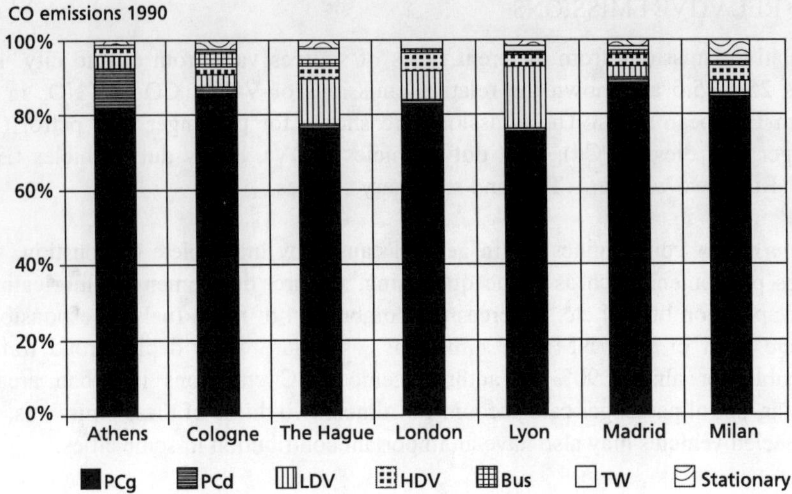

Figure 25.5 Relative emissions of CO in selected European cities (EC 1997).

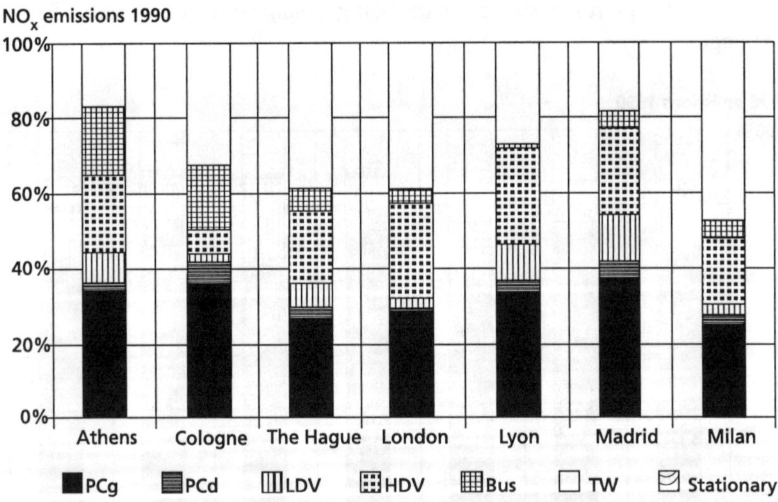

Figure 25.6 Relative emissions of NO_x in selected European cities (EC 1997).

Only very sparse emission data exist for *Eastern Europe* and large changes occurred after the opening between East and West. In Eastern Europe the emissions from road traffic has increased significantly and many outdated industrial installations have closed and in some cases replaced by modern plants. South-eastern cities are a couple of decades behind the Western European cities concerning the road transport organisation. Traffic crossing the city centre by heavy trucks and also private cars increase significantly the number of cars inside the cities. Fortunately, during the last years intensive construction campaigns have been started (e.g. in Budapest and Prague) to build up the ring highways around the cities to enable a lot of cars to avoid entering the cities centre if not needed.

The highest SO_2 emission densities (per area unit) are reported from Sofia, Ljubljana and Bucharest while the lowest from Bratislava, Budapest, Prague and Zagreb. Less significant variation can be detected at the emission densities calculated for nitrogen dioxide. The highest CO and VOC emission densities are reported from Sofia.

25.3 Air quality in large European cities

An overview of the most relevant air quality indices in selected European cities is given in Table 25.2 (next page). The indices are related to urban background levels. However, hot spot air pollution is superposed to the urban background levels and can lead to very significant human exposure. This may vary substantially between different European regions. The table gives three groups of indices:

- Emissions (for winter and summer smog),
- Exceedances (maximum pollution concentrations relative to WHO AQG),
- Exposure (percentage of population exposed to concentrations above WHO AQG).

The indices for the exceedances are based mainly on results from air pollution measurements in each city. The indices for environmental pressure and emissions may be used to explain the differences in air quality from one city to another (EEA 1998).

25.3.1 HOT SPOT AIR POLLUTION

Hot spots of air pollution includes locations like urban streets with busy traffic, and locations influenced by pollution from point sources (industries, power plants and heating plants) in cities.

The *traffic air pollution* in busy streets is typically a factor 5-10 higher than the urban background. This ratio between the levels in streets and urban background varies significantly with local dispersion condition (wind speed), which furthermore is influenced by e.g. the street configuration (building height and street width). It has been estimated that 10-20 million people in Europe spend a considerable time at or near roadside (e.g. commuters, those living or working in roadside buildings or road workers). This exposure is an important contribution to their total exposure to air pollution. Such exposure occurs in all cities, but high concentrations occur more frequently in cities with unfavourable winter smog dispersion conditions.

In all European cities, the road traffic contributes significantly to smog occurrences and long term average concentrations of harmful compounds, e.g. VOCs, benzene, particles and benzo-(a)-pyrene. Road traffic contributes with more than half of the NO_X emissions and in the order of 35% of the VOC emissions. However, at street locations the contribution from road traffic in percentage to the human exposure is often considerable higher.

Table 25.2 Indices describing air pollution conditions in large European cities in 1995.

City	Exceedance[1)] Winter (SO_2+PM)	Exceedance[1)] Summer (O_3)	Exposure[2)] Winter (SO_2+PM)	Exposure[2)] Summer (O_3)	At least one exceedance for assical pollutants
Antwerp	1		1		1
Athens	1	2	1	4	4
Barcelona	3	2	4	4	4
Berlin	2	2	4	4	4
Birmingham	2	2	4	4	4
Bremen	2	2	3	4	4
Brussels[3]	1	2	1	4	4
Budapest	3		4		4
Copenhagen		2		4	4
Dublin	1		2		2
Frankfurt	1	2	1	4	4
Glasgow					1
Hamburg	2	2	4	4	4
Hanover	2	2	3	4	4
Helsinki	2	1	1	1	4
Istanbul	2		4		4
Katowice	3	2	4	4	4
Kharkov	4		4		4
Krakow	2	2	4	4	4
Leeds	2	2	4	4	4
Lille[3]	1	2	1	4	4
Lisboa	1		1	4	4
Liverpool	2	2	4	4	4
Ljubljana	2	2	3	4	4
Lodz	2		4		4
London	2	2	4	4	4
Lyon	1	2	3	4	4
Manchester					1
Milan[3]	1	3	1	4	4
Munich	2	2	1	4	4
Nuremberg[4]	1	2	2	4	4
Oslo	1*	1		1	1
Prague	3	2	4	4	4
Riga	5		4		4
Sarajevo	6		4		4
Sofia	6		4		4
Stockholm	0.5	1	1	1	1
Stuttgart	1*	2	1	4	4
Thessaloniki	2	1	4	1	4
Tirana[4)]	1		1		1
Turin[3)]	3	2	4	4	4
Valencia	2		4		4
Vienna	1	2	1	4	4
Vilnius	6		4		4
Warsaw	3	2	4	2	4
Zurich	1	2	1	4	1

Notes: * = uncertain data

[1)] 0.5: < 0.5 AQG, 1: 0.5-1 AQG, 2: 1-2AQG, 3: 2-3 AQG, 4: 3-4 AQG, 5: 4-5 AQG, 6: >5 AQG.

[2)] 1: 0-5% population, 2: 5-33% population, 3: 33-66% population, 4: >66% population.

[3)] Available data refer to 1996.

[4)] Available data refer to 1992-1993.

A significant local air pollution problem in the Southern countries is *evaporation of fuel* from vehicles. A particular local air pollution problem in the Nordic countries is particles produced by wear of the road surface due to the use of studded tyres in winter.

In some cities, air pollution from road traffic and space heating is supplemented significantly by emissions from local industry. Depending on the height of the stacks and the prevailing wind direction the impacted area is usually a few kilometres from the sources. In some areas, e.g. parts of Eastern Europe, regions in the Mediterranean area and regions in Central Europe, such sources give rise to significant air pollution exposure to the population.

25.3.2 WINTER SMOG; SO_2 AND PARTICULATES

Winter-type air pollution episodes are generally characterised by a high pressure (anti-cyclone) system above Europe which persists for several days. The health effects during winter smog air pollution episodes are characterised by the 24-h. average concentrations of sulphur dioxide (SO_2) and Suspended Particulate Matter (SPM).

During winter smog episodes wind speeds are low and an marked temperature inversion limits the vertical mixing of pollutants to the lowest atmospheric layers. Air pollutants further accumulate due to increased emissions and reduced removal rates. Due to the low temperatures energy demand increases; space heating related emissions up to 70% higher than average in a winter season can be expected during episodes. The deposition of SO_2 and other pollutants is reduced when the soil is frozen and/or covered with snow. Winter type episodes have a regional or meso-scale character. Long range transport on the scale of Europe is indicated by measurements and model calculations (Lübkert 1989; De Leeuw, Van Rheineck Leyssius 1990).

The general circulation patterns in Europe result, on the average, in West-to-East transport. However, under episodic conditions East-West transport is more frequent. For North Western Europe it is estimated that during episodes at least 50% - but probably up to 75% - of the background SO_2-level is of Eastern European origin. For sulphate, and probably also SPM, the Eastern European contribution will be of the order of 80-90%. Note that the yearly average contribution of Eastern Europe to the sulphur deposition in North Western Europe is of the order of only 10%.

Besides SO_2 and SPM, the winter smog mixture comprisis compounds such as carbon monoxide (CO), nitric acid, nitrogen dioxide (NO_2) and various organic components. Particles are largely (ca. 90%) in the respirable fraction and the chemical composition will be highly variable; the major components are soot (black smoke), sulphate, nitrate, and ammonium: minor components are heavy metals (e.g. cadmium, Cd and lead, Pb) and organic matter (e.g. BaP). The particulate matter originates partly from direct emissions, and is partly formed in the atmosphere (so-called secondary aerosol). The concentrations of strong oxidants such as ozone are low during winter type episodes.

The most severe smog episode ever reported in the literature was the London smog episode of December 1952 (Chapter 2). Maximum concentrations of SO_2 and SPM both reached values of 5000 μg/m^3. In a two-week period during and after the smog episode a total of ca. 4000 excess deaths compared to a similar period in previous years was observed. More recently, two episodes (in January 1985 and January 1987) occurred in North Western Europe with maximum 24-h. SO_2 and SPM concentrations of 900 and 700 μg/m^3. Winter type episodes occur more frequently and are more severe in eastern European countries. For the densely populated parts of Czechoslovakia, the former GDR and Southern Poland model calculations, supported by measurements, show yearly average concentrations up to 10 times higher than in Western Europe. Based on these data it is expected that, under winter smog conditions, SO_2 concentrations in rural areas will be well above 400 μg/m^3.

A downward trend in *annual average SO_2 concentrations* was observed in the late eighties and continued in the time period up to 1995 in most cities. In 1995, the long term WHO-AQG (50 μg/m^3) was exceeded only in Katowice and Istanbul (10 cities observed exceedances in 1990). As an example, the long term trend in SO_2 concentrations for a number of cities is presented in Figure 25.7.

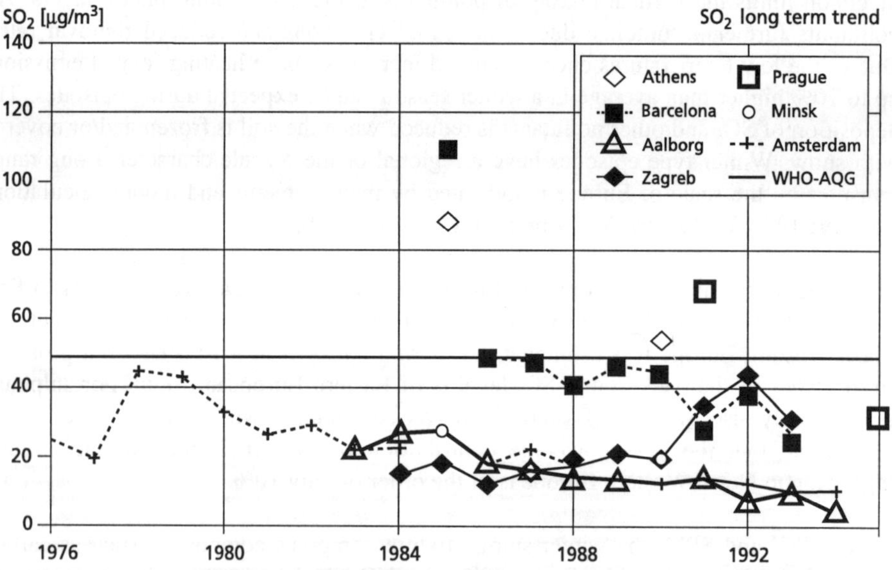

Figure 25.7 Long term trends in urban SO_2 concentrations in selected European cities (Source: APIS, AIRBASE) (EEA 1998).

Annual average SO_2 concentrations are generally lowest in Northern Europe. Highest values are found in Central European cities and some of the Southern European cities. In 37% of the 41 cities for which data was available, the short term WHO-AQG (125 μg/m^3, daily values) was exceeded. In 1990, out of 76 cities 43% reported exceedances. In most urban areas the number of exceedances is confined to only a few days a year with the highest concentrations in Katowice and Sofia (374 and 373 μg/m^3 respectively).

Multiple terms are used in Europe to describe *Particulate Matter (PM)* concentrations. Most common are Black Smoke (BS), Total Suspended Particulates (TSP) or Particulate matter smaller than 10 µm (PM_{10}). During the second half of the eighties annual average concentrations did not show a clear trend, but concentrations have dropped considerably during 1990-1995 in most cities. Highest annual average concentrations were reported from Central European and some Southern European cities. However, both the long term WHO-AQG for BS (50 µg/m^3) and the EU limit value for TSP (150 µg/m^3) are not exceeded in any city. Averaged maximum 24-h. city background concentrations exceed the short term WHO-AQG in 69% of the cities (86% in 1990).

Short-term exceedances of WHO-AQG for SO_2 and/or PM have been taken as indicator for winter-type smog. Highest exceedances are observed in Central European cities. Figure 25.8 presents the potential percentage of citizens exposed to the exceedances of the winter-type smog indicator.

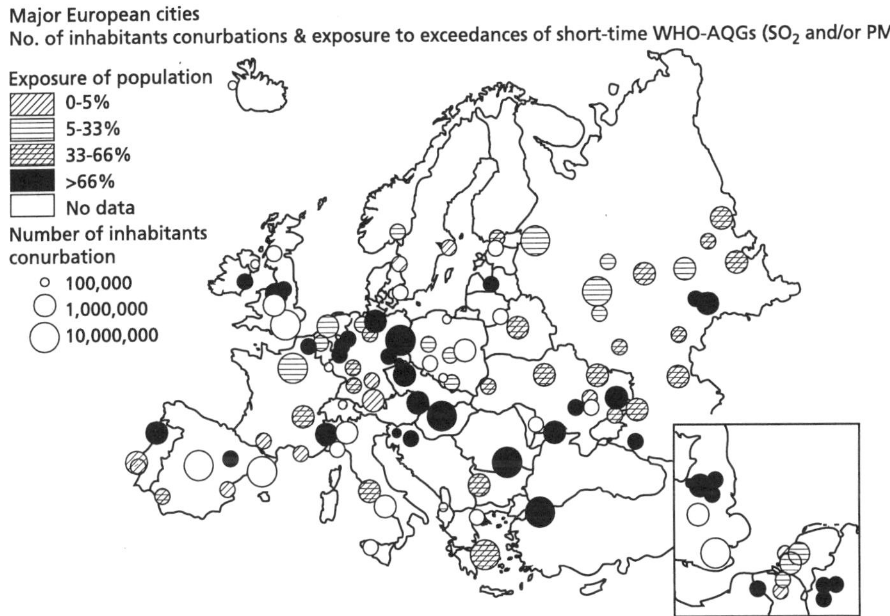

Figure 25.8 Number of inhabitants in conurbations shown as diameter of circles and the percentage of population exposed to exceedances of short-term WHO air quality guidelines for SO_2 and /or particulate matter (Source: APIS, AIRBASE) (EEA 1998).

25.3.3 SUMMER SMOG; OZONE, NO_X, NO_2 AND VOC

The occurrence of *photochemical oxidants*, most frequently studied by the ozone measurements, is a regional phenomenon in Europe (Chapter 8). In most cases regionally elevated ozone concentrations will be partly suppressed in urban areas due to reactions with locally emitted nitrogen oxide (NO), which reacts with ozone to produce nitrogen dioxide (NO_2). Further reactions, involving VOCs and nitrogen oxides (NO_X)

mainly released from road traffic, will result in production and build-up of ozone and oxidants in the city "plume", but these processes are slow and the additionally produced ozone will usually be found downwind of the area where the emissions occurred. An exception is cities located in confined valleys, or coastal cities where the polluted air is trapped in land-sea breeze circulation systems, and the residence time is long enough for significant photo-oxidation to take place. For example, ozone concentrations, up to 400 $\mu g/m^3$ are measured in Athens and Barcelona.

Ozone episodes occur every summer in and around most of European cities. During these periods, many of which have a duration of several consecutive days, ozone concentration rise to several times the ambient average. Episodic concentrations are added to background concentrations and the latter are possibly subject to an upward trend of about 1% per year (Logan 1994). Comparison with historical data suggests that the long term average ozone levels over Europe have doubled since the turn of the century and that most of the increase has occurred since the 1950's (Borrell et al. 1995). The one hour WHO-AQG of 150 $\mu g/m^3$ for ozone has been exceeded in 27 of these cities in 1995. The most affected cities are Athens, Barcelona, Frankfurt, Krakow, Milan, Prague and Stuttgart.

25.3.4 ROAD TRANSPORT POLLUTANTS: NO_2, CO, PB AND BENZENE

Urban street pollution is monitored in most European cities, and available measurements show that short-term maximum concentrations of CO, NO_2 and TSP may occasionally exceed the AQGs by a factor of 2 to 4, depending upon the actual traffic and dispersion condition of the street.

Road transport dominates the *NO_X emissions* in Europe to about the same extent as power production dominates SO_2. Reported annual average nitrogen dioxide concentrations generally do not show a clear trend. Sixteen out of the 38 cities, that have reported on annual NO_x concentrations, exceed the long term AQG of 40 $\mu g/m^3$. Southern European cities seem to experience significantly higher annual average concentrations.

Exceedances of the short-term WHO-AQG (equivalent to 200 $\mu g/m^3$ as a maximum hourly value) at city background locations during the reporting period 1990-1995 were observed at 15 of the 27 cities that gave information on hourly values. However, NO_2 one hour maximum concentrations show a downward trend from 1990 to 1995 with the exception of Helsinki, London, Oslo and Vienna (Figure 25.9). This downward trend is also apparent from Figure 25.10, where the frequency distribution of the cities within each class of NO_2 annual average concentrations is shown.

The main source of *CO pollution in urban areas* is uncontrolled petrol powered passenger cars. Experience has shown that the CO problem is restricted to areas close to the main road network. Available data on annual average carbon monoxide concentrations showed a general Europe-wide downward trend in the period 1990-1995, although changes in CO levels may vary from site to site mainly depending on trends in traffic density and on CO emissions regulations. The short-term WHO-AQG (8-h. average 10 mg/m^3) is exceeded in 13 out of the 27 cities that reported on CO 8-hour values, their majority, however, experiences decreased CO concentration levels in 1995 compared to 1990 with the exception of Ljubljana, Reykjavik, Seville, Stuttgart and Warsaw (Figure 25.11, next page).

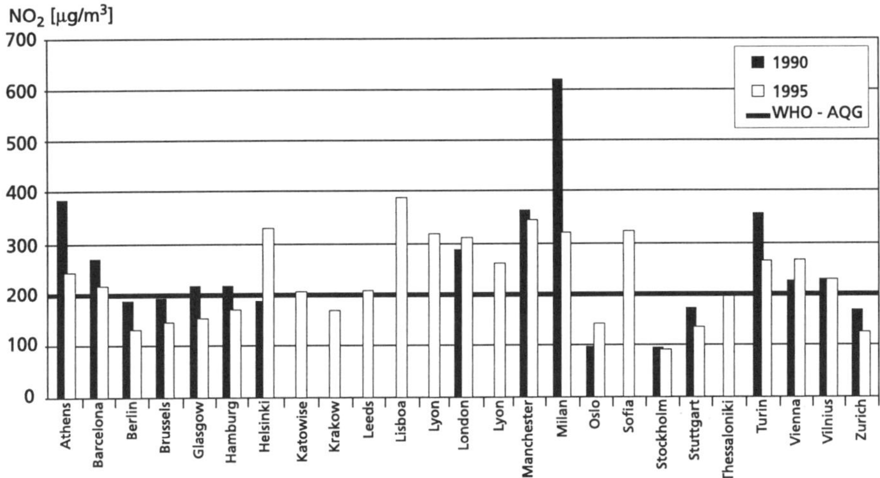

Figure 25.9 Maximum 1 hour NO_2 concentrations for the 25 most affected European cities. Note: Concentration values for Milan and Turin refer to 1996. The data are compared with WHO air quality guideline (WHO-AQG) for maximum 1 hour average at 200 µg/m^3 (EEA 1998).

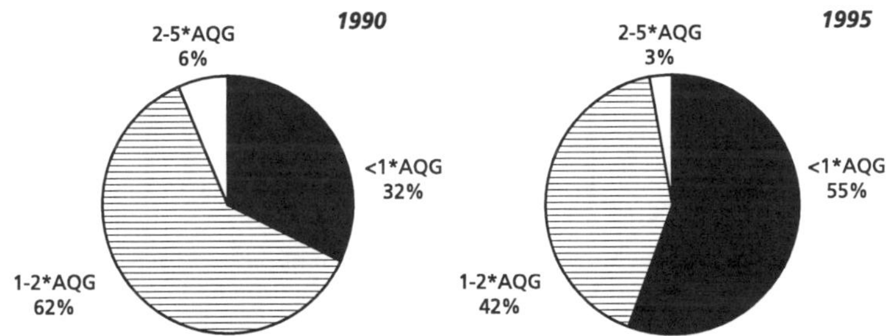

Figure 25.10 Annual average NO_2 concentrations in 1990 and 1995. Percentage of cities within each class of NO_2 concentrations in relation to WHO air quality guideline (AQG) at 40 µg/m^3 (EEA 1998).

Lead in petrol is a dominating source of lead pollution in air in most urban areas. In most European countries the maximum lead content in leaded petrol is now reduced to 0.15 g/l and the market share of unleaded petrol is rapidly increasing. In some Eastern European countries this development may be slower. Annual average lead concentrations in the majority of the European cities for which relevant monitoring data were available have dropped sharply during the period 1990-1995. This clear downward trend is shown in Figure 25.12 (next page). The latter should be attributed to the introduction of lead-free fuels and new vehicle technologies in the European market. However, concentrations in a restricted number of Eastern/Central European cities (e.g. Vilnius) have slightly increased during the last five years, mainly due to the recent increase in road transport emissions and the fact that fuels are still untreated in most of the Eastern/Central European countries. Annual average concentrations at hot spots (mostly busy streets) stand below the lower limit of the WHO-AQG (0.5 μg/m^3); no cities experienced exceedances of the 0.5 μg/m^3 guideline after 1993.

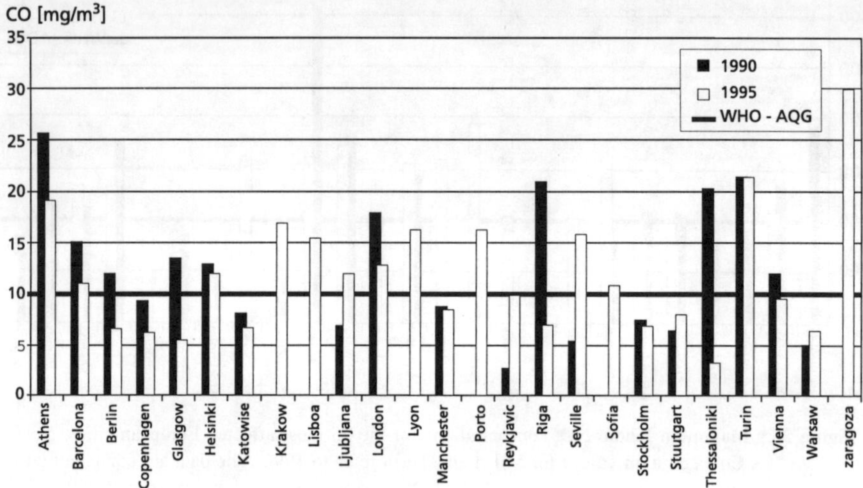

Figure 25.11 Maximum 8-h. CO concentrations for the 25 most affected European cities. Note: Concentration values for Reykjavik and Turin refer to 1996. Concentration value for Berlin refers to 1994. The data are compared with WHO air quality guideline (WHO-AQG) for maximum 8-h. average at 10 mg/m^3 (EEA 1998).

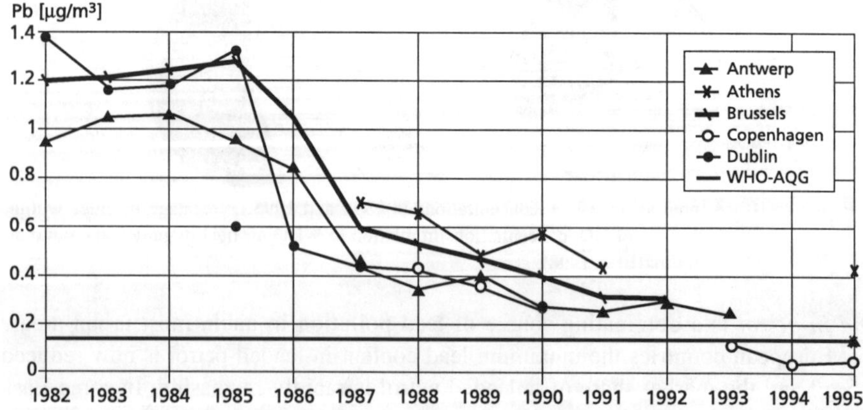

Figure 25.12 Long term trend of annual lead (Pb) concentrations for selected European cities (EEA 1998).

Benzene is a carcinogenic pollutant. Petrol-powered vehicles are the dominating emission source, being responsible for more than 90 per cent of the urban emissions. Most of the benzene is emitted through the exhaust, but 10 per cent is due to evaporation losses. The petrol distribution system and space heating are also responsible for rather significant contributions, while industry emissions may be important locally.

Less benzene measurement data were available within the frame of this report; ten out of the 62 cities are available on benzene. With the exception of Antwerp, exceedances of the long-term WHO-AQG (equivalent to 2.5 $\mu g/m^3$ as an average yearly value) were observed at all cities for which corresponding data were accessible.

25.4 Large scale aspects of urban air pollution

The urban air pollution is not only a local problem. Emissions from urban areas contribute significantly to the emission of green house gases, which is outside the scope of this book. In a smaller scale air pollutants and their reaction products as a result of chemical/physical transformations transported from an urban area can contribute significantly to the air pollution at locations several hundred kilometres away.

25.4.1 LONG RANGE TRANSPORT

Long range transport in Northern Europe is especially important for acidification and eutrophication of lakes, forests and other biotops. These phenomena are not the topics of this book. However, in some cases long range transport of sulphur and nitrogen compounds is a significant contribution to urban air pollution. It is well known that the contribution from neighbour countries to urban air pollution can be dominating during episodes. These episodes are mainly governed by transport processes in regional scales. Figure 25.13 (next page) shows a long range transport episode in Denmark (Kemp et al. 1997).

The ozone episodes in Northern Europe are also determined by transport of ozone and/or ozone precursors in a regional scale. The locally created ozone during episodes is normally only a minor fraction of the total ozone concentration in the Northern European countries. Figure 25.14 (next page) showing an ozone episode in Denmark (Kemp et al. 1997).

25.4.2 REGIONAL PHOTOCHEMICAL AIR POLLUTION IN SOUTHERN EUROPE

In Southern Europe the *photochemical air pollution* is dominating. The Western Mediterranean Basin is surrounded by mountains 1,500 m or higher. On summer days their east and south facing slopes are strongly heated and act like orographic chimneys. These favour the early formation of up-slope winds which reinforce the sea breezes and link the surface winds directly with their return flows aloft, and further, with their compensatory subsidence over the sea. The result is the formation of stacked layers along the coasts with the most recently formed layers at the top and the older ones near the sea (Millán et al. 1992; Millán et al. 1996). The layers act as reservoirs of aged pollutants, and the lower ones can be brought inland by the sea breeze of the next day(s), creating re-circulations, as illustrated in Figures 25.15 and 25.16 (page 452).

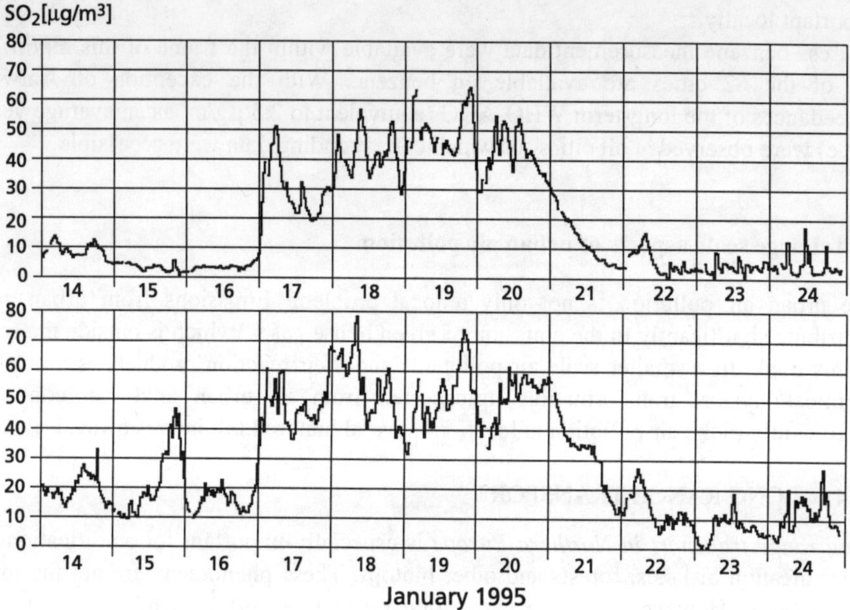

Figure 25.13 An SO_2 episode in Denmark in January 1995 measured at a rural and urban site. During the episode with transport from South East the wind speed was rather constant around 6 m/s. Both particulate sulphate and SO_2 were significantly elevated in the whole country (Kemp et al. 1996).

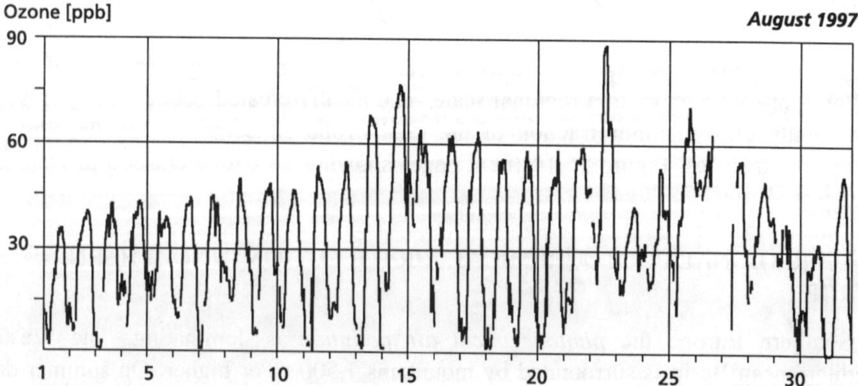

Figure 25.14 An ozone episode observed at Keldsnor in Denmark in August 1997. The episode showed the typical meteorological behaviour with a day to day increase in temperature and more and more stable air transport from South (Kemp et al. 1998).

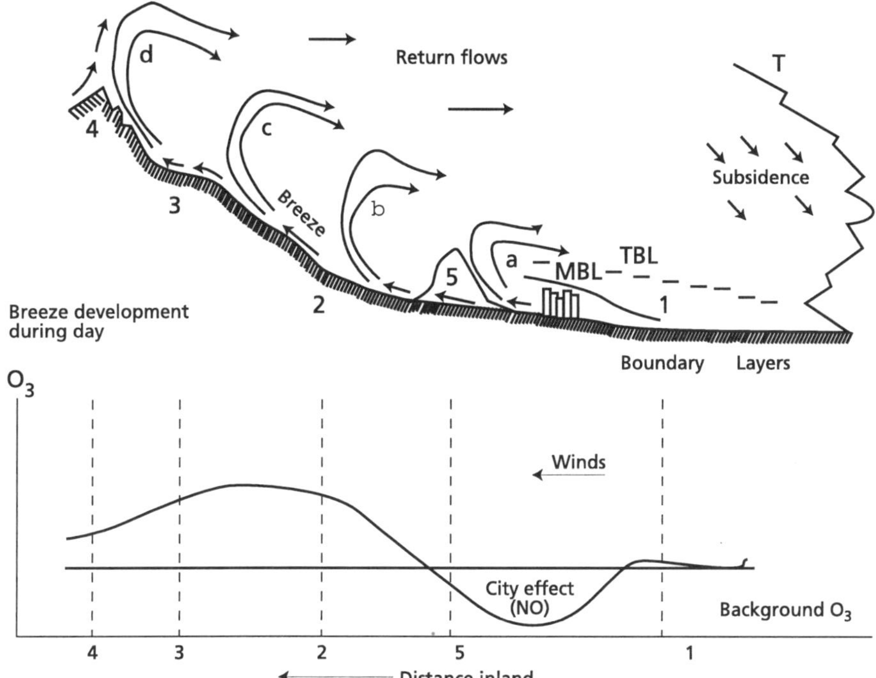

Figure 25.15 The diurnal circulations in coastal regions of the Mediterranean Basin on a summer day. They include the entrance of the sea-breeze and formation of stratified layers aloft. Letters (a) - (d) indicate successive stages of this process. The lower part shows a schematic of the ozone decay, and subsequent production as it interacts with the NO emissions in a coastal city (Derwent, Davies 1994).

These processes have been documented within several EC research projects (Millán et al. 1997), and tracer experiments have shown that turnover times range from 2 to 3 days along the Spanish Mediterranean coast, where the layers reach 2 to 3 km in depth, have variable width over land (up to 100 km), and extend more than 300 km over the sea. The land-based processes die out at night, and the reservoir layers can drift along the coasts and contribute to regional, inter-regional and long range transport of aged pollutants. Similar processes have been documented in other regions within the Mediterranean Basin, and it is also known that there are important differences between the Western and Eastern Mediterranean Basins, which are presently being studied (Kallos et al. 1998).

At the *larger scales*, deep convection in some regions (e.g. Spanish and Turkish central plateaus), or strong up-slope winds in others (e.g. Alps and Atlas mountains), can inject aged air masses directly into the Mid-troposphere (3 to 6 km) and into the upper troposphere, i.e., 10 (+) km, where they can participate in long-range transport processes within Southern and Central Europe, and at the continental-global scales, respectively. Some of these processes have also been documented by EC sponsored research (Millán et al. 1997), and the resulting ozone anomalies over the Mediterranean Basin can be detected by satellites (Fishman et al. 1990).

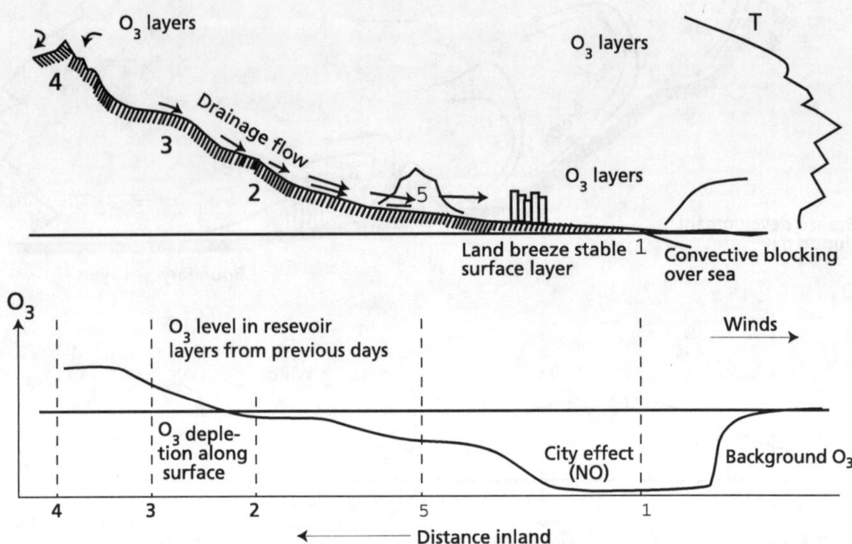

Figure 25.16 The diurnal circulations in coastal regions of the Mediterranean Basin on a summer night. They include the formation of drainage flows, the accumulation of a stable air mass at the bottom of the valleys and over the coastal plains. The draining air mass can become blocked at some distance from the coast, whenever the sea surface temperature is higher than the drainage air temperature. The lower part shows a schematic of the ozone evolution along the path of the draining air. In these processes, stations located high above the coastal plains can remain within the reservoir layers during light (Derwent, Davies 1994).

Under strong insulation, the *coastal re-circulations* become "large natural photo-chemical reactors" where most of the NO_x emissions and other precursors are transformed into oxidants, acidic compounds, aerosols and ozone, which exceed EU limit values for several months, as illustrated in Figures 25.17 and 25.18 (page 454). Relevant aspects of this problem are: (1) that the concept of upwind and downwind of conurbations loses its usual meaning and must be considered in its broadest meaning, (2) that the ozone is generated at the regional scale from emissions in urban centres and other NO_x source areas, and (3) that as much as 60%, or more, of the observed O_3 in any one site may result from advection within the re-circulating air masses.

Figure 25.17 shows the annual evolution of the *24-h. average ozone concentrations* at two rural sites located more than 42 km inland from the Spanish East coast and more than 80 km from the nearest city. The threshold level of EC directive 92/72/CEE for damages to vegetation (65 $\mu g/m^3$, <24-h.>) is exceeded systematically for more than 6 months of the year, the one for human health (110 $\mu g/m^3$, <8-h.>) for at least four months, and the one for information to population (180 $\mu g/m^3$, <1-h.>) can also be exceeded frequently from April to August.

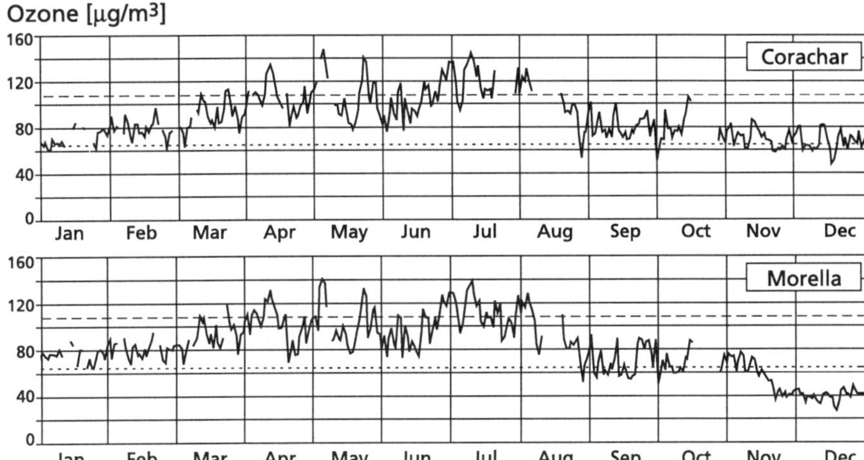

Figure 25.17 Annual evolution of the 24-h. average ozone concentrations for two rural inland stations located at 41 km (Corachar) and 50 km (Morella) from the Spanish East coast in 1995. The EU threshold values for protection of human and vegetation are shown as broken and dotted lines respectively.

Figure 25.18 (next page) shows the *monthly averaged ozone diurnal cycles* for six stations in this general area, which are located at various altitudes and distances from the coast, as indicated by the codes shown in Figure 25.15 (page 451). The existing monitoring network confirmed the existence of the re-circulations described (Millán et al. 1992). The results available to date (1998) further confirm these processes, and illustrate the observed variability within a 100 km square.

The situations shown are the norm, rather than the exception, for the coastal regions surrounding the Western Mediterranean Basin, and illustrate the existence of chronic type ozone episodes in this region, created by atmospheric re-circulations, as compared with peak type episodes in Central and Northern Europe, which are created by atmospheric stagnation. They also reveal: (1) a problem of data interpretation for those responsible for monitoring networks from the local to the EC level, and (2) a source of great annoyance for politicians in Southern Europe, who can do little in response to the new EC directives.

The observed ozone cycles depend strongly on the topographic location of the observing station and its relationship to the reservoir layers, the atmospheric circulations involved, and the chemical processes along each path (Derwent, Davies 1994), as illustrated in the figures. Thus, each ozone monitoring station shows a part of the whole, and could even be considered to represent a specific area, providing the relevant processes are understood for each site, and the site itself has been adequately selected. However, no single station can be considered representative of regional processes, and much less of the whole situation.

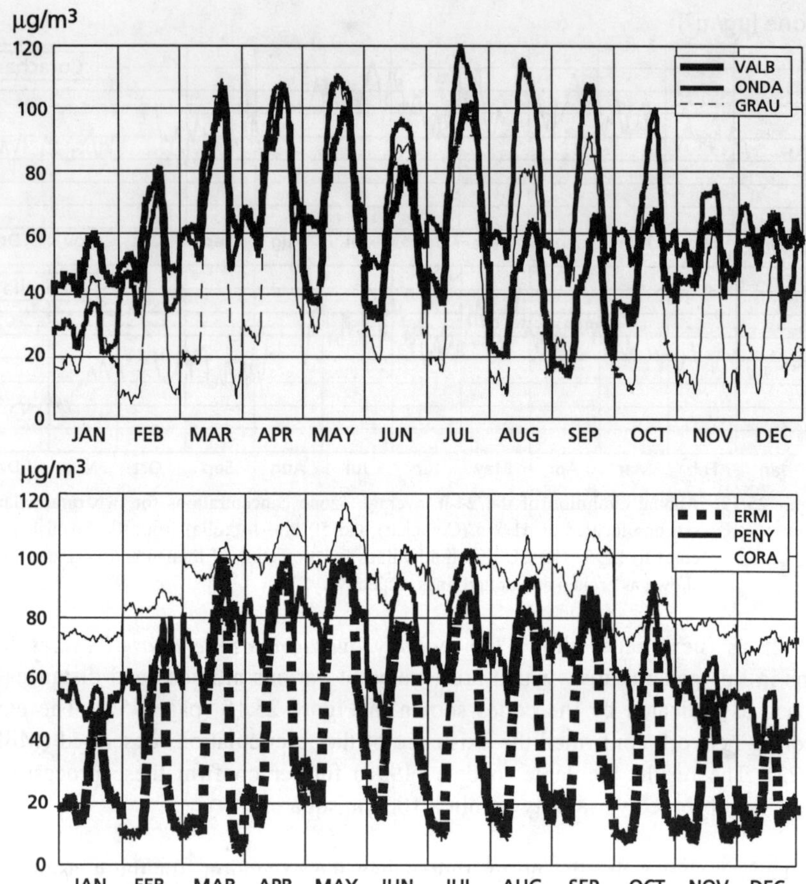

Figure 25.18 Monthly averaged diurnal cycles of ozone recorded at various sites on Spanish East coast in 1997. The sites are located at various distances from the coast and at different altitudes, all within a 100 km. square. The station codes as per numbers in Figure 25.16 (page 452) are VAL (3), ONDA (2), GRAU (1), ERMI (1), OENY (5), CORA (4).

A situation, which is both a cause and a result of present delays, is the layout of the existing *monitoring networks in Southern Europe*, which are mostly urban, and still respond to conventional requirements for public health (SO_2 and particulates). These networks were definitively not instrumental in discovering the ozone problem. Moreover, the awareness of present or potential ozone problems and their regional extent is still weak at the political levels, which still tend to favour the importation of procedures and models from elsewhere, even though they may not be applicable to these regions.

To summarise, in the Western Mediterranean Basin, and in other areas in Southern Europe, the "effects of vehicle emissions in urban areas" extend much further than the urban source areas and, thus, this concept may now represent a dated view inherited from experiences in other climatic regions. This further emphasises the point that to properly address atmospheric pollution problems in Southern Europe, the diffusion-transport scenarios and transformation mechanisms must be well understood at their relevant time and space scales.

25.5 Air quality in different parts of Europe

25.5.1 NORTHERN AND WESTERN EUROPEAN CITIES

The Northern and Western (NW) part of Europe has a long tradition of performing air pollution monitoring. Some examples of long-term trend recordings of particles are given in Figure 25.19 for the UK, Paris, Brussels and Amsterdam. The urban structure of NW-Europe can be characterised as highly populated with the cities scattered all over the area). The NW part of Europe includes Ireland, United Kingdom, Denmark, Netherlands, upper half of Germany (Frankfurt and above), Belgium, Luxembourg, Norway, Sweden, Finland and the northern part of France (Paris and above). A short characterisation of NW-Europe is given below.

Examples will be given here of the air pollution situation as being caused by various pollutants (e.g. SO_2, NO_2, particulates, O_3, lead, CO) in various NW-European cities.

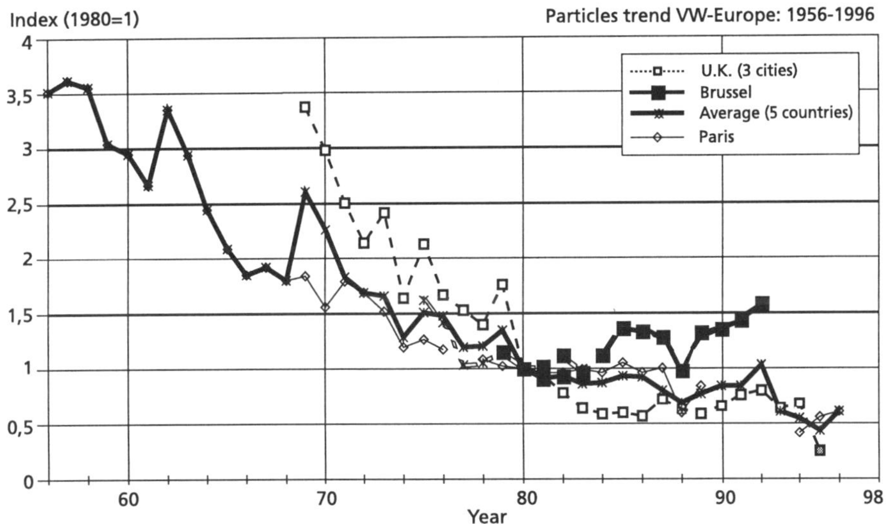

Figure 25.19 Long term trends of particles for UK, Paris, Brussels and Amsterdam. The levels in 1980 was: Amsterdam = 71 $\mu g/m^3$ (SPM), UK = 23.5 $\mu g/m^3$ (BS), Luxembourg = 19.1 $\mu g/m^3$ (BS) and Paris = 41.5 $\mu g/m^3$ (BS).

The *large NW-European cities* (Capitals and cities with more than 500,000 inhabitants) are presented in the table in the chapter introduction "Urban sources of air pollution" (page 33). Totally 38 cities in NW-Europe, approximately one third of Europe's main cities, with a total population of over 50 million inhabitants fulfil the above given criteria. More useful background information about this subject can be found in Sluyter et al. 1995. Large changes in number of inhabitants between 1990 and 1995 is seen for some of the cities. Most of the changes are due to changes in definition of the suburbs/satellite cities belonging to the urban agglomeration. Definitions change when cities expand in time, building new areas, which merge into new, larger agglomerations, leading to a sudden increase in the population of the agglomeration, e.g. Frankfurt am Main, growing from 1.3 million in 1990 to 3.6 million in 1995.

London is a typical example of how air quality standards have reduced traditional air pollution in Western European cities over the last decades. Annual mean sulphur dioxide concentrations have dropped dramatically from the 300 to 400 $\mu g/m^3$ of the 1960's to 20 to 30 $\mu g/m^3$. These levels are well below WHO annual guidelines. Both are the result of the switch in fuel type from fuel oil and coal to smokeless coal and natural gas as well as of the introduction of pollution control technologies in the manufacturing industry. A factor which has contributed to the drastic reduction of these pollutants in the city centre is also the relocation of these activities out of the city. However, high pollution episodes are still a problem during the winter. Exceedances of sulphur dioxide short-term guidelines (500 $\mu g/m^3$ 10-minute means and 350 $\mu g/m^3$ one-hour mean) are still a regular occurrence.

Emission from motor vehicles is the most significant source of carbon monoxide, lead, nitrogen oxides and secondary pollutants such as ozone. These are expected to increase further due to the increase in the total mileage driven. Greater London had 2.7 million motor vehicles in 1988 (one car per three inhabitants) and together with the metropolitan area it accounts for over 6 million cars (35 per cent of all UK cars). According to OECD estimates, road traffic in London has increased by 70 per cent during the last 20 years. A 7 per cent increase has occurred since 1989 alone.

A drastic reduction in *atmospheric lead concentrations* took place in the second half of the 1980's as a results of the phase-out of permissible lead levels in petrol which entered into force in 1986. Annual concentrations are now below WHO guidelines at background sites and roadside concentrations do not exceed the upper limit of WHO guidelines in central London.

Carbon monoxide concentrations show different patterns at different monitoring sites. The number of episodes of the 8-h. WHO guideline exceedances in central London are of major concern. Emissions from road traffic, which account for 95% of total carbon monoxide emissions, are trapped by ground based temperature inversions, building up high local concentrations. High traffic volumes cause also high concentrations of nitrogen oxides. While some sites show stable trends in nitrogen oxides concentrations, other sites indicate marked increases. During 1989, 16 sites in Central London exceeded the EC 98 percentile limit of 200 $\mu g/m^3$ and all 64 sites throughout London but two had 50 or 90 percentiles greater than the EC guide values of 50 and 135 $\mu g/m^3$. Measurements of ambient tropospheric ozone concentrations at a number of sites in central London show no clear trends, but they indicate the frequent occurrence of exceeding the hourly mean value of 214 μg/m^3 (100 ppb), the upper limit of the WHO guideline range.

25.5.2 EASTERN EUROPE

The total number of inhabitants in the eastern region (Albania, Bulgaria, Croatia, Czech Republic, Hungary, Romania, Slovakia, Slovenia and Yugoslavia) is almost 80 million and approximately 15 million of them live in urban areas with more than 500 thousand inhabitants, including the conurbation of the cities. The largest populations are represented mostly by the capitals of the above countries - Tirana, Albania (500,000), Sofia, Bulgaria (1,300,000), Zagreb, Croatia (950,000), Prague, Czech Republic (1,220,000), Budapest, Hungary (3,100,000), Bucharest, Rumania (2,400,000), Bratislava, Slovakia (500,000), Ljubljana, Slovenia (273,000) and Belgrade, Yugoslavia (1,300,000).

Obviously *economic conditions* have strong effect on the atmospheric environment of the cities through the quality of the car fleet and types of environmental protection related installations at the stationary sources (industry, power plants etc.) as well as the type of home heating. These countries belonged to the so called Eastern Block between the mid-40's and late-80's which was characterised by less effective economical production and environmental pollution problems were generally not seriously considered. However, air quality is influenced by several parameters (emission density, orography, meteorology, land-use, organisation of transport etc.) so significant differences can be identified among the South Eastern urban areas investigated.

During *the period of 1960-1970*, the cities in SE region of Europe could be characterised by high sulphur dioxide and soot pollution in fall-winter season (London type smog) as many other cities in the whole European continent. Later, due to the reduction of coal and lignite based pollution sources, sulphur and soot emission decreased significantly. However, the number of vehicles increased rapidly, releasing increasing amount of NO, hydrocarbons and CO, which components lead to photochemical oxidant formation (ozone, PAN etc.). It can generally be stated that at present the dominant pollution source is the traffic in most of the cities of the region. It means that diurnal variation of pollutants is significant (high daytime concentrations and relatively low at night) so the average concentrations presented in this Chapter show a less critical picture than the reality during daytime on the busy streets of the cities. All the data presented below refer to early and mid-90's.

The highest annual average urban background concentrations of *sulphur dioxide* were detected in Sofia, Bucharest and Prague in the range of 38-50 μg/m^3 (Figure 25.20, next page). In central or industrial areas of Sofia and Prague the annual average sulphur dioxide concentration exceeds 60-70 μg/m^3. 98 percentile concentration values are rarely available from the cities considered, but in Prague for example it is close to 300 μg/m^3. During the period of 1990-1995 the annual average concentration of sulphur dioxide is tending to decrease in urban areas of the region. In Budapest for example, an intensive campaign and technological investment were performed so that to reduce the sulphur content of diesel oil consumed by the public buses which are the major users of that type of fuel. In addition, the individual coal-based heaters are being replaced by central natural gas-heating systems.

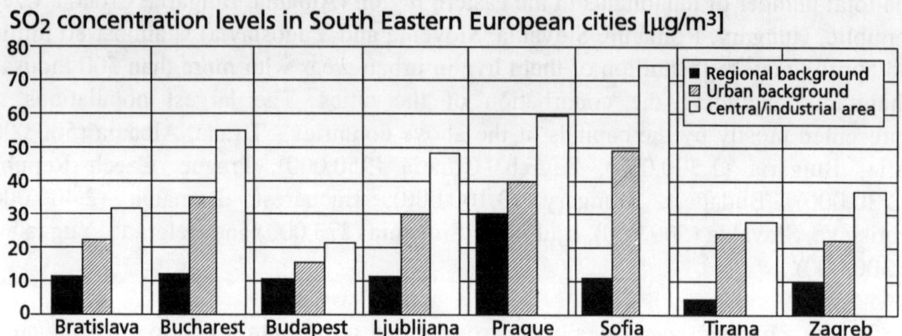

Figure 25.20 The SO_2 concentration levels measured in different Eastern European cities at regional background, urban background and central/industrial monitoring sites.

As regards the nitrogen-dioxide concentrations (Figure 25.21) one can see that both the urban background and central site (downtown) concentrations are close to each other. The latter ones are in the range of 50-70 $\mu g/m^3$. The sources of nitrogen oxides are represented mostly by traffic so the diurnal variation of concentration field follows the temporal variation of traffic emission.

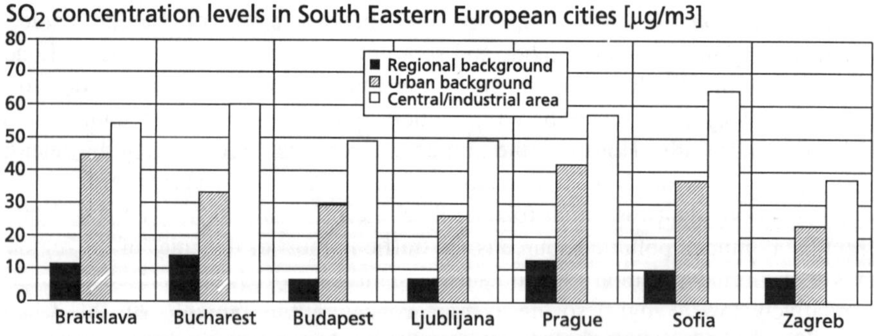

Figure 25.21 The NO_2 concentration levels measured in different Eastern European cities at regional background, urban background and central/industrial monitoring sites.

Suspended particulates can cause several health problems since they can be carriers of toxic organic and inorganic compounds. Their size vary several orders of magnitude, and their physical characteristics (dry deposition velocity, atmospheric residence time) are very different. Most of the pollutants are emitted in vapour form (toxic metals, PAH etc.) which later condense on the surface of aerosol particles mostly of fine size range (r = 0.1-1.0 µm). Wide range size distribution of aerosol particles makes the sampling difficult: some cities report total mass of particulate while others PM_{10} or respirable

$PM_{2.5}$. For this reason the comparison of particulate concentration levels is extremely difficult. Cities in this part of Europe are still troubled by the atmospheric lead (Pb) emitted mostly by vehicles using leaded petrol (Bozó, Horváth 1992). The number of cars equipped with catalytic converters is increasing, and also a larger fraction of petrol consumed is unleaded in the region. At the same time, however, older cars can use only leaded petrol due to their engine construction.

Surface ozone levels are relatively low in the cities of the region. It is simply explained by the fact that large amount of NO and VOCs are emitted from the old and out-dated car fleet and react with ozone. However, in the extended vicinity of urban areas (approx. 70-100 km downwind of the city), the large amount of ozone precursors makes it possible to reach extremely high ozone concentrations under clear and sunny weather conditions in the urban plume (Bozó, Baranka 1996).

With respect to public health the *exceedances of limit values* for different pollutants are determined by data collected from the monitoring network. Detailed information is shown here only from Budapest, but it is assumed that similar characteristics could be found for other large cities in the region. The exceeding 24-h. threshold values in Budapest are shown in Figure 25.22. It can be seen that sulphur-dioxide and ozone do not reach high concentrations in the city, while there is a relatively high percentage of exceeding 24-h. threshold values for NO_2 in non-heating season (April-September) and for suspended particulate in heating season.

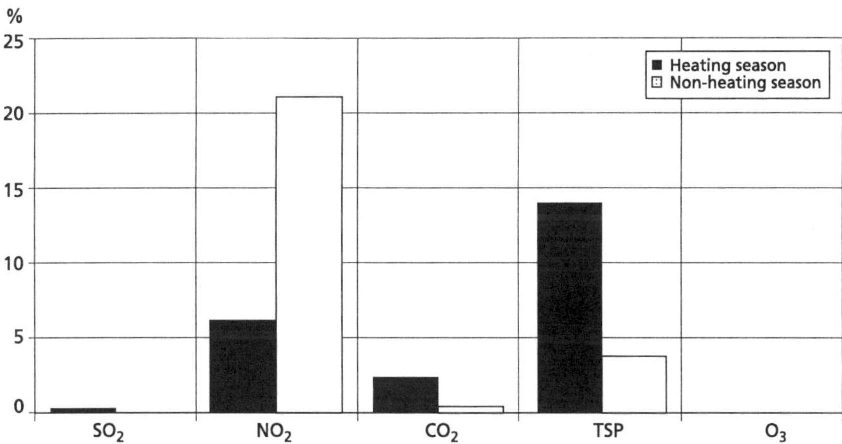

Figure 25.22 Exceedances of Hungarian threshold values for SO_2, NO_2, CO, TSP and O_3 in Budapest during heating and non-heating seasons in 1995.

25.5.3 SOUTHERN EUROPE

The South Europe includes Southern France and Portugal and the Mediterranean countries, e.g. Greece, Spain and Italy.

The high air pollution levels in large cities in Southern Europe, caused by high anthropogenic emissions, are known to be significantly affected by local wind circulation systems which are decisively influenced by the underlying topography. The road traffic is the main air pollution source in e.g. Athens, Milan, Barcelona and Madrid. This is caused by the relatively high degree of motorization and in some countries in conjunction with the large in-city mileage of the cars.

As in most of the large cities of the world, the *Athens area*, suffers under significant air pollution problems. The air pollution phenomenon in this city is known as "nephos" (the cloud), a name which underlines its visible character. Visibility impairment during pollution episodes is due to high burdens of aerosol particles and the yellowish brown colour of the cloud is due to nitrogen dioxide. The air pollution problem becomes worse because of the bad city planning and the overpopulation of the area. Until recently, the problems associated with road traffic emissions were enhanced by the long lifetime of the average Athenian car. In 1990 a new air pollution control strategy was initiated in Athens aiming primarily at a rapid renewal of the motorcar fleet in the city. This strategy was proved very successful: A large fraction of the oldest and most polluting cars disappeared from the streets of Athens, the consequence being drastical emission reductions and noticeable air quality improvements. Specifically, in the last years both downtown CO levels and the exposure to ozone in the northern suburbs of Athens are found to decline significantly. Also, adverse parameters for the accumulation of pollutants are the topography of the region (basin surrounded by mountains) and the locally weather characteristics (intense sunshine, sea breeze circulation). As a result, the city often experiences photochemical air pollution episodes. In conjunction with the Athens bid for the Olympic Games in 2004, the international "Athens 2004 Air Quality Study" was initiated for analysing how air quality in Athens will be affected by ongoing and other planned infrastructure changes in the Greater Athens area (new airport, underground, ring road system) (Moussiopoulos, Papagrigoriou 1998)

The long term trend of air pollutant concentrations in Athens reveals a significant improvement in the 90's, for some pollutants even since the mid-80's. This improvement may, at least partly, be attributed to air pollution abatement measures regarding (i) the mobile sources (gradual substitution of the old vehicles with new technology vehicles equipped with catalytic converters, implementation of a system for controlling car emissions), (ii) the fuel quality (reduction in sulphur content) and (iii) the efficient control of stationary emission sources. As a result, in 1995 air pollution in Athens was 5% of the days at high levels, while the corresponding percentage in 1993 and 1994 was 11%. In addition, in the same year, there has been for the first time after 1984 no violation of the NO_2 98 percentile limit value of 200 $\mu g/m^3$ for the whole of the city area. There is still a problem with the ozone levels. This pollutant remains at high levels during the whole of the summer period due to the high sunshine intensity in the area. Ozone exceedances are recorded mainly at the sites located at the north of the city, where the exceedance of the 8-h. EU guide value can reach 30%.

Figure 25.23 shows one hour annual mean concentrations of NO_2 for the period 1984-1993 and the accumulated exposure to ozone over the threshold value of 60 ppb (in ppb·h) in representative stations summed up for each year of the period 1987-1994. The

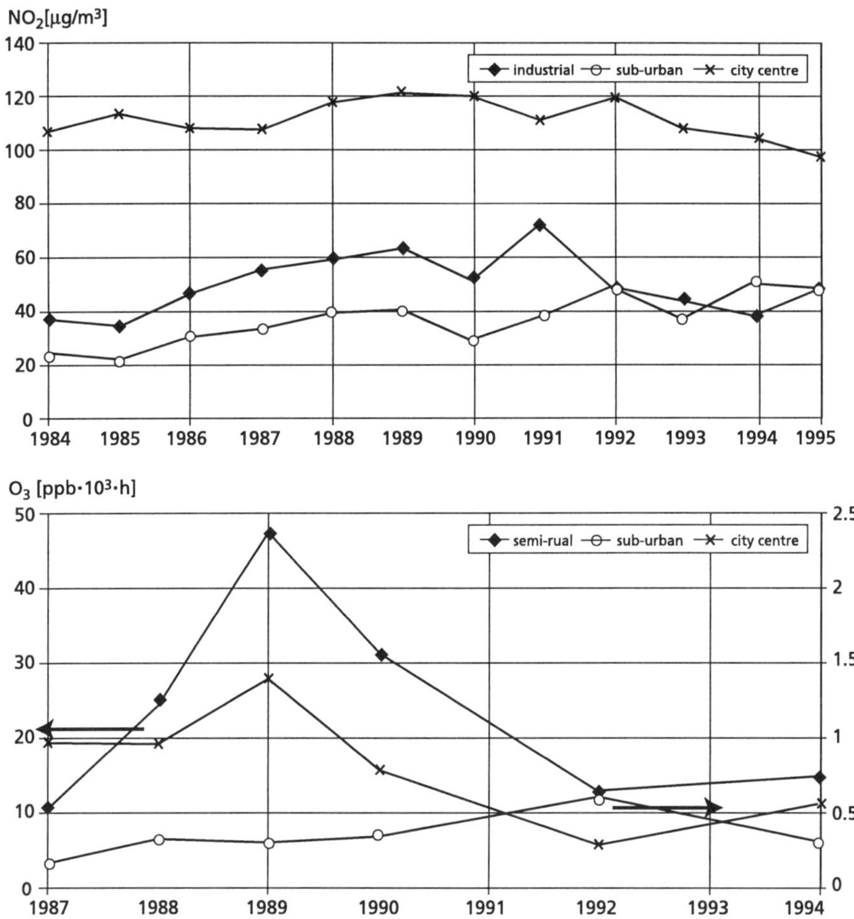

Figure 25.23 Annual mean concentration of NO_2 in Athens during 1984-1995 and accumulated exposure to O_3 over 60 ppb for the period 1987-1994. Note: Curves corresponding to semi-rural and sub-urban stations refer to the left axis, the curve corresponding to city centre refers to the right axis (EEA 1998).

NO_2 graph refers to data collected at stations located in the city centre, i.e. an area characterised by elevated traffic emissions, in an industrial area and in a suburban area, while O_3 was measured in the city centre, in a suburban area and in a semi-rural area. Highest NO_2 values are observed at the station close to the city centre, since traffic constitutes the major source of NO_X. Concentrations fluctuate around 100 μg/m^3 from year to year, far above the WHO-AQG of 40 μg/m^3. The lowest values are recorded at the suburban area. No clear upward or downward trend of the NO_2 levels can be identified before 1990, but a slight reduction appears having been achieved in the last years. In contrast to NO_2, ozone remains at low levels in the city centre, since it is depleted in the presence of high NO traffic emissions. However, the exposure to ozone may be significant at stations located far from the city centre (for instance in the semi-rural area). There appears to be a clear downward trend of the exposure to ozone in such locations since 1989. In the sub-urban area considered ozone levels have remained almost constant in the same period.

25.6 Conclusions

Although *air quality in Europe* and particularly in the large European urban areas has improved in recent decades, air pollution still represents one of the major environmental issues that authorities at all levels have to face. The substitution of coal and heavy fuel oils by cleaner fuels such as gas oil and natural gas and by electricity and heat supplied from large electric power plants and district heating plants had as a result the considerable decrease of SO_2, Pb and TSP emissions which led to a clear downward trend in the ambient concentrations of these pollutants in most European cities. Nevertheless, exceedances of short term SO_2 and TSP air quality standards set by WHO have still been recorded in the vast majority of European cities.

In contrast to the decreasing SO_2, Pb and TSP ambient concentrations, pollutants associated to road transport such as NO_x, CO, VOC and indirectly O_3 have considerably increased during the last decades and only very recently their levels have been stabilised and in some cases (i.e. CO) declined. Exceedances of ozone short term WHO AQG have been recorded in 75% of the cities that provided ozone data whereas the corresponding percentages for NO_2 and CO are 60% and 52% respectively.

The recorded concentrations of the above mentioned pollutants give a good impression of the pollution situation, but the picture is far from being complete. Other pollutants such as PM_{10}, $PM_{2.5}$ or VOCs with characteristic representatives the carcinogenic benzene or the polycyclic aromatic hydrocarbons for which no EU limit or guide values exist may reach very high values. Unfortunately, in the large majority of European cities, monitoring data for such pollutants are not available. In the frame of the present research, only 10 out of the 62 cities that responded to the questionnaire provided information on benzene and with one exception the long-term WHO AQG was exceeded in all these cities.

Although *monitoring of both air quality and emission data* progressed considerably in the majority of the European cities during the last years, more improvements have to be achieved towards harmonisation and evaluation of these data. The important issue of their interconnection to exposure and health effects is still in an unsatisfactory level. Further study is, consequently, necessary in order to quantitatively assess the damages on human health as well as on historic monuments and buildings.

Efforts towards long term abatement have been intensified during the last years. These efforts include the ICLEI initiative, with the purpose to achieve and monitor improvements in global environmental conditions through cumulative local actions, as well as the EEA Urban Monograph for air quality management and capabilities assessment. Meeting the increasingly stringent air pollution targets is, however, a very difficult task as the Auto Oil Programme has proved. According to its striking conclusion, even with the maximum technical package introduced in the EU, for three out of the seven cities taken as examples, the emission targets would not be achieved. In summary, the efforts for the air pollution abatement have to be continued and reinforced. There is still a long way ahead and unless very drastic actions are taken, a clean urban atmosphere is not to be expected for the near future in many European cities.

25.7 References

Berdowski, J.J.M., Mulder, W., Veldt, C., Visschedijk, A.J.H., Zandveld, P.Y.J. (1996) *Particulate emissions (PM10 2,5 PM0.1) in Europe in 1990 and 1993*. First Draft, August 1996.

Borrell, P., Builtjes, P., Grennfelt, P., Hov, O., Van Aalst, R., Fowler, D., Mégie, G., Moussiopoulos, N., Warneck, P., Volz-Thomas, A., Wayne, R. (1995) Photo-oxidants, Acidification and Tools: Policy Applications of EUROTRAC Results, in: Power, H., Moussiopoulos, N., Brebbia, C.A. (editors), *Air Pollution III*, Computational Mechanics Publications, Southampton **1**, 19-26.

Bozó, L., Baranka, G. (1996) Evaluation of the Air Quality in Budapest Based on the Data of the Monitoring Network and Measurement Campaigns, in: Allegrini, I., Santis, F.D. (editors), *Urban Air Pollution*, Monitoring and Control Strategies NATO ASI Series, Springer Verlag Berlin.

Bozó, L., Horváth, Zs. (1992) Atmospheric Concentration and Budget of Lead and Cadmium over Hungary, *Ambio*, **21**, 324-326.

Coote, A.T., Yates, J.T.S., Chakrabarti, S., Bigland, D.J., Ridal, J.P., Butlin, R.N. (1991) *Evaluation of decay to stone tables: Part 1. After exposure for 1 year and 2 years*, Building Research Centre, Garston, United Kingdom. (Convention on Long-Range transboundary Air Pollution. UN/ECE international co-operative programme on effects of materials, including historic and cultural monuments.).

De Leeuw, F.A.A.M., Van Rheineck Leyssius, H.J. (1990) Modeling study of SO_x and NO_x transport during the January 1985 smog episosde. *Water, Air, Soil Poll.*, **51**, 357-371.

Derwent, R.G., Davies, T.J. (1994) Modelling the impact of NO_x or hydrocarbon control on photochemical ozone in Europe, *Atmospheric Environment*, **28**, 2039-2052.

ECOTEC Research and Consulting (1986) *Identification and Assessment of Materials Damage to Buildings and Historic Monuments by Air Pollution*, Birmingham, United Kingdom.

EEA (1997) *Air Pollution in Europe 1997*, report prepared by the European Topic Centre on Air Quality and the European Topic Centre on Air Emissions.

EEA (1998) *Europe's Environment - The second assessment*. European Environment Agency, Copenhagen, 293 pp.

EC (1997) *Methodologies for Estimating Air Pollutant Emissions from Transport "Emission Factors and Traffic Characteristics Data Set Deliverable 21"*, Draft Final Report to EC.

Fishman, J., Watson, C.E., Larsen, J.C., Logan, J.A. (1990) The distribution of tropospheric ozone determined from satellite data. *J. Geophys. Res.*, **95**, 3599-3617.

Kallos, G., Kotroni, V., Lagouvardos, K., Papadopoulos, A. (1998) On the long-range transport of air pollutants from Europe to Africa. *Geophysical Research Letters*, (In press).

Kemp, K., Palmgren, F., Manscher, O. (1996) The Danish Air Quality Monitoring Programme. Annual Report for 1995. National Environmental Research Institute, Roskilde, Denmark. *NERI Technical Report 180*.

Kemp, K., Palmgren, F., Manscher, O. (1998) The Danish Air Quality Monitoring Programme. Annual Report for 1997. National Environmental Research Institute, Roskilde, Denmark. *NERI Technical Report 245*.

Kucera, V., Henriksen, J., Knotkova, D., Sjostrom, Ch. (1992) Model for Calculations of Corrosion Cost Caused by Air Pollution and its Application in Three Cities, in: Costa, J.M., Mercer, M.D. (editors), *Progress in the Understanding and Prevention of Corrosion*, pp. 24-32, London, The Institute of Materials.

Lahmann, E. (1990) Winter Smog, with special regard to the situation in Germany, WHO Collaborating Centre for Air Quality Management and Air Pollution Control, Berlin, Germany, *Air Hygiene Report 3, ISSN 0938-9822*.

Lübkert, B. (1989) Characteristics of the mid-January 1985 SO_2 smog episode in Central Europe: report from an international workshop, *Atmospheric Environment*, **23**, 611-613.

Logan, J. (1994) "Trends in the vertical distribution of ozone: an analysis of ozone sonde data", *Journal of Geophysical Research*, **99**, 25.553-25.585.

Millán, M.M., Artíñano, B., Alonso, L., Castro, M., Fernandez-Patier, R., Goberna, J. (1992) Meso-meteorological Cycles of Air Pollution in the Iberian Peninsula, (MECAPIP). *Air Pollution Research Report 44, (EUR Nº 14834)* CEC-DG XII/E-1, Rue de la Loi, 200, B-1040, Brussels.

Millán, M.M., Salvador, R., Mantilla, E., Artíñano, B. (1996) Meteorology and photochemical air pollution in southern Europe: experimental results from EC research projects, *Atmospheric Environment*, **30**, 1909-1924.

Millán, M.M., Salvador, R., Mantilla, E., Kallos, G. (1997) Photo-oxidant Dynamics in the Western Mediterranean in Summer: Results from European Research Projects, *J. Geophys. Res.*, **102**, D7, 8811-8823.

Moussiopoulos, N., Papagrigoriou, S. (editors) (1998) *Athens 2004. Air Quality*, CD-ROM edition, FiatLux Publications.

Moussiopoulos, N., Sahm, P., Kessler, Ch. (1995) Numerical simulations of photochemical smog formation in Athens, Greece - A case study, *Atmospheric Environment*, **29**, 3619-3632.

OECD (1995) *OECD Environmental Data: compendium 1995*, OECD, Paris.

RIVM (1992) The Environment in Europe: a Global Perspective, *RIVM Bilthoven, report no. 481505001*.

RIVM (1995, Air Quality in Major European Cities: Part I, *RIVM Bilthoven, report no. 722401004*.

Sluyter, R.J.C.F., Larssen, S., Eerens, H.C., Van Velze, K., Burn, J., Gronskei, K., Bezuglaya, E., Smeets, W., Henriksen, J.F., Lubkert-Alcamo, B. (1995) Air Quality in major European cities, *RIVM/NILU, Bilthoven, reportnr 722401004*.

UNEP/WHO (1997) *Air Quality Management and Assessment Capabilities in 20 major European Cities*. WHO/UNEP-Montoring and Assessment Research Centre, London.

WHO (1987) *Air quality guidelines for Europe*, WHO Regional Publications, European Series no. 23, Copenhagen.

WHO (1996) *Revised WHO Air Quality Guidelines*, Unedited Version, 5 November 1996, WHO European Centre for Environment and Health, Copenhagen.

Chapter 26

SUMMARY AND CONCLUSIONS

JES FENGER, OLE HERTEL and FINN PALMGREN
National Environmental Research Institute
Department of Atmospheric Environment
P.O.Box 358, DK-4000 Roskilde, Denmark

Cities are by nature concentrations of humans, materials and activities. They therefore exhibit both the highest levels of pollution and the largest targets of impact. In 1990 more than 70% of the European population lived in urban areas, which contain a corresponding amount of buildings and materials of economic and cultural value. As a result of national legislation as well as international agreements, the general air quality in European cities has improved in recent decades - often in spite of an increase in population density and standard of living. Nevertheless, with the present demands on human health and the continuously strengthening of the environmental protection, urban air pollution still represents one of the major issues that authorities at all levels have to face. There are, however, marked differences between the conditions and the pollution levels in various European regions.

26.1 Urban infrastructure

The major part of *Northern, Central and Western Europe* is highly industrialised. The road traffic is fairly dense in most of the urban areas. However, generally the car fleet is modern and the road traffic is often well regulated, especially with restricted traffic in the city centres. The most dense traffic is located at ring roads and highways and radial approach roads and highways.

In *Southern Europe, the Mediterranean urban areas* are highly industrialised and with dense traffic. The emission density from road traffic is thus high. The cities are in many cases located in valleys, resulting in serious air pollution problems due to low wind speed and formation of photochemical smog.

South-eastern urban areas have a partly outdated car fleet and are decades behind in the organisation of road traffic, and only recently have efforts been put into the construction of ring highways to reduce unnecessary crossing of the city centres.

The information on *Eastern Europe* is modest, but until the collapse of the Soviet Union most of the countries had a rather outdated and ineffective industry; it is now partly closed down or substituted.

26.2 Dispersion and transformation

Basically the pollution levels depend not only upon the distribution and magnitude of emissions, but also upon the conditions for dispersion and transformation of the emitted compounds. Dispersion of pollution in the atmosphere is governed by the wind pattern. In Europe the large scale phenomena are influenced by three main factors: the temperature difference between the polar air in the North and the subtropical air in the South; the distribution of land and sea with the Atlantic Ocean to the West, Asia to the East and the Mediterranean Sea and Africa to the South; and finally the barriers constituted by the mountain systems like the Alps, the Pyrenees and the Scandinavian chain.

North of about 40°N (i.e. in most of Europe) the wind regime consists mainly of eastward moving weather systems. Therefore pollution is predominantly transported from the West to the East; consequently the United Kingdom is in general cleaner than its emissions would suggest; Denmark "imports" some pollution, but "exports" more, whereas the opposite holds for Sweden.

In cities, the air pollution levels are influenced by a regional contribution, but mainly determined by direct sources, and a general urban background.

In urban areas, the presence of buildings strongly effects the advection and dispersion of air pollutants from local sources. The complex flow patterns around single buildings and in streets govern to a large extend the pollution levels. However, the urban air pollution phenomena cannot be considered only as governed by local scale processes. Heat island effects may influence the flows on a scale as large as the total city area. Mesoscale phenomena like land-sea breezes and mountain-valley flows strongly affect local urban pollution conditions, e.g. in many Mediterranean cities.

In Northern Europe *long range transport of air pollution* leads not only to eutrophication and acidification, but also during episodes to significantly elevated concentrations of air pollutants e.g. NO_2 and SO_2, in cities. Typically this is the case during high wind speed conditions with high pressure over Central/Eastern Europe. In summer, the transport of air pollution can lead to ozone episodes.

Photochemical episodes occur frequently in *Southern Europe*. The episodes are regional scale phenomena, characterised by anti cyclonic situations with low wind speeds and circulation of air masses governed by land-sea breezes and mountain-valley air flows.

26.3 Emissions and levels

The emissions of *sulphur dioxide* per capita varies two orders of magnitude between different regions depending upon the degree of industrialisation. In all regions the dominant emissions are due to large point sources i.e. power plants and large industrial plants, in urban areas accounting for about 70% in Western and Northern Europe and nearly 60% in Southern Europe.

During the last decade the emissions have been reduced in most areas due to increasing use of low-sulphur oil and gas as fuel for space heating and industry. In the eastern part of Europe, a reduction in industrial activity after the collapse of the Soviet Union also play a significant role. In some (Eastern European) cities, however, space heating still gives a significant contribution.

The general reduction in emissions has led to a corresponding reduction in urban sulphur pollution levels, which in most cities are now down at acceptable levels - both with respect to human health and material damage. Thus in 1995 the long term WHO-AQG of 50 $\mu g/m^3$ was only exceeded in two cities compared with exceedances in 10 European cities in 1990.

The differences in emissions per capita of *nitrogen oxides* are smaller, although high values are seen in industrialised areas. Traffic is the major source in urban areas accounting for nearly two thirds of the emissions. Some reductions in total emissions within the last 5-10 years have mainly been within other sectors; therefore the urban pollution concentrations have not shown a clear tendency. In many cities, especially in the South of Europe the WHO limit value of 40 $\mu g/m^3$ is regularly violated.

Traffic is to an even larger extent the major source of *carbon monoxide,* in cities accounting for approx. 90% - gasoline passenger cars alone accounting for approx. 75%. Introduction of three-way catalytic converters have lead to a general reduction in emissions and a corresponding reduction in concentrations. Still, however, in 13 out of 27 investigated major cities the 8 hour limit value of 10 mg/m^3 is violated.

Lead was formerly an important additive in petrol and thus the dominating source of lead pollution in urban areas. It is now being phased out. Most EU-countries have a maximum lead content of 0.15 g/l, and the market share of lead-free petrol is increasing. In some Eastern European cities the development is slower due to a temporary increase in the number of older cars without catalytic converters. In no cities, however, the long term limit value of 0.5 $\mu g/m^3$ as yearly average appears to be violated.

In all European cities, the road traffic contributes significantly to smog occurrences and long term average concentrations of harmful compounds, e.g. *Volatile organic compound* (VOCs), benzene, particles and PAH, e.g. benzo-(a)-pyrene. Road traffic contributes with more than half of the NO_X emissions and in the order of 35% of the VOC emissions. However, at street locations the contribution from road traffic in percentage to the human exposure is often considerable higher.

The natural emissions of VOCs from vegetation are especially in Southern Europe - important for the photochemical air pollution.

A significant local air pollution problem in the southern countries is *evaporation of fuel* from vehicles.

Photochemical oxidants - including ozone - are secondary pollutants and thus particularly dependent upon meteorological and topographical conditions. In Southern and Central Europe, which have a hot and sunny weather and where some regions are surrounded by mountains keeping the reagents together urban ozone pollution episodes is a frequent phenomenon, sometimes lasting several days. In 1995 the 1 hour WHO-AQG of 150 µg/m^3 was exceeded in 27 European cities. The production of ozone in - or downstream - the cities adds to the general background, which have been doubled since the turn of the century with the largest increase after 1950.

In the northern part of Europe the situation is different. The local production of ozone is modest and episodes are normally related to long range transport from the South. Urban levels are generally lower than in rural areas due to reactions between ozone and nitrous oxide. Therefore isolated national reductions of the primary pollutants may have no local effect or may even result in higher levels of ozone - as it is sometimes observed during weekends with lower urban traffic.

The classical winter smog of combined sulphur dioxide and particle pollution is still observed - mainly in Central and Eastern Europe - but it is a diminishing problem. *Particles* as such however, are receiving increasing attention as a health risk in urban areas both as pollutants per se and as carriers of heavy metals or organic compounds. Earlier measurements concerned total amounts of particles expressed as "black smoke" (BS) or "total suspended particles" (TSP). The concentrations have dropped considerably in the period 1990-95 in most European cities. The highest values are registered in Central and Southern Europe, but the long term WHO-AQG for black smoke (50 µg/m^3) and EU limit for TSP (150 µg/m^3) are not exceeded. In Central and Eastern Europe PM_{10} from stationary sources dominate. The pro capita emission is several times higher than in Western Europe.

The different size fractions of particles are deposited in different parts of the human air ways. In addition, the chemical composition may be important for assessment of the impact. From a health point of view *fine and ultra fine particles* are believed to be the most important. In order to assess the health impact the size distributions of the must therefore be known. However, so far only few detailed data are available and only for shorter campaigns - and the effects are not well known.

Exhaust from diesel and petrol engines appears to be the main source for fine and ultra fine particles and considerable effort has been put on description on these particle. Future improvement of engine technology may reduce the particle mass which may be seen in TSP, PM_{10} and $PM_{2.5}$, but may also lead to larger number of ultra fine particles (which have only little impact on particle mass).

26.4 Present regulations

The principal legal basis for the regulation of *air quality* in EU is a set of EU-directives, which so far comprise limit values for sulphur dioxide, particulate matter, nitrogen dioxide and lead. Threshold values for ozone for information and warning to the public are also regulated by EU-directives. A new framework directive for management and assessment of air quality has been approved and the corresponding so-called daughter directives are under preparation for SO_2, NO_2/NO_X, particles, ozone, benzene and CO with new, generally more strict limit values. A series of WHO-guidelines serves as a supplement and the scientific basis for setting of new standards, but has itself no legal standing. Corresponding limit values are set in many other European countries.

Emissions are regulated by numerous EU-directives, they include directives for vehicles, industrial sources, fuels etc. The Auto-Oil programmes include traffic related air pollution and assessment of cost effectiveness of measures.

Traffic restrictions and city planning are tools for local reduction of air pollution from traffic, including traffic restriction, parking restriction, road pricing etc. In some large cities the traffic can be reduced by law during serious air pollution episodes.

26.5 Future efforts

Long term abatement has been intensified during recent years. The efforts include the ICLEI initiative, with the purpose to achieve and monitor improvements in global environmental conditions through cumulative local actions , as well as the publication of the EEA Urban Monograph for air quality management and capabilities assessment.

Meeting the increasing stringent air pollution targets is, however, a very difficult task as the "Auto Oil Programme" has proved. According to its striking conclusion, even with the maximum technical package introduced in the EU, for three out of seven cities taken as examples, the emission targets would not be achieved.

Traffic pollution in urban streets represent in general the most important pollution for European air pollution exposure of the population. The pollution exposure of the population has often been assessed through crude assumptions e.g. that levels observed at a single or few street monitoring stations in the urban monitoring programmes are representative for the exposures of the entire population in the urban area. A more promising approach seems to be a combined use of monitoring data, personal exposure measurements on smaller selected groups and application of air pollution (exposure) models. Given that time- and activity pattern of the population may be obtained the exposure can be assessed. This type of approach is believed to find increasing use in health impact assessment studies in the future.

26.6 Literature

The following books and reports offer presentations of the general aspects of urban air pollution and the special conditions in Europe. Information on specific countries can be found in national reports published by environmental authorities.

Butterwick, L., Harrison, R., Merritt, Q. (1991) *Handbook for Urban Air Improvement*, Commission of the European Communities, Brussels.

EEA (1998) *Europe's Environment - The second assessment*, European Environment Agency, Copenhagen.

EC (1997) Proposal for a Council Directive relating to limit values for sulphur dioxide, oxides of nitrogen, particulate matter and lead in ambient air, Com 500 final (1996), *Official Journal* **C 009**, 6.

UNEP, WHO (1992) *Urban Air Pollution in Megacities of the World.* Blackwell Publishers, Oxford UK.

INDEX

Common for all chapters

- Compounds are indexed after their name with formulas or short names in parentheses
- Abbreviations and acronyms are written in full in parentheses
- Page references are in most cases given under the full name
- Composite names are often listed under the main name